EXPERIMENTAL METHODS IN THE PHYSICAL SCIENCES

Robert Celotta and Thomas Lucatorto, *Editors in Chief*

Founding Editors

L. MARTON
C. MARTON

Experimental Methods in the Physical Sciences

VOLUME 31

VACUUM ULTRAVIOLET SPECTROSCOPY I

Volume 31

Vacuum Ultraviolet Spectroscopy I

Edited by

J. A. R. Samson
Department of Physics and Astronomy
University of Nebraska
Lincoln, Nebraska

and

D. L. Ederer
Physics Department
Tulane University
New Orleans, Louisiana

ACADEMIC PRESS

626776

San Diego London Boston New York Sydney Tokyo Toronto

This book is printed on acid-free paper. ∞

Copyright © 1998 by Academic Press

All Rights Reserved.
No part of this publication may be reproduced or transmitted in any form or by any means, electronic or mechanical, including photocopy, recording, or any information storage and retrieval system, without permission in writing from the Publisher.
The appearance of the code at the bottom of the first page of a chapter in this book indicates the Publisher's consent that copies of the chapter may be made for personal or internal use of specific clients. This consent is given on the condition, however, that the copier pay the stated per-copy fee through the Copyright Clearance Center, Inc. (222 Rosewood Drive, Danvers, Massachussetts 01923) for copying beyond that permitted by Sections 107 or 108 of the U.S. Copyright Law. This consent does not extend to other kinds of copying, such as copying for general distribution, for advertising or promotional purposes, for creating new collective works, or for resale. Copy fees for pre-1998 chapters are as shown on the title pages. If no fee code appears on the title page, the copy fee is the same as for current chapters.
1079-4042/98 $25.00

Academic Press

525 B Street, Suite 1900, San Diego, CA 92101-4495, USA
1300 Boylston Street, Chestnut Hill, MA 02167, USA
http://www.apnet.com

United Kingdom Edition published by
Academic Press Limited
24-28 Oval Road, London NW1 7DX

International Standard Serial Number: 1079-4042/98

International Standard Book Number: 0-12-475978-5

PRINTED IN THE UNITED STATES OF AMERICA
98 99 00 01 02 EB 9 8 7 6 5 4 3 2 1

CONTENTS

CONTRIBUTORS   xiii

VOLUMES IN SERIES   xv

1. Synchrotron Radiation Sources
STEVEN L. HULBERT and GWYN P. WILLIAMS

1.1. General Description of Synchrotron Radiation	1
1.2. Theory of Synchrotron Radiation Emission	2
1.3. Insertion Devices (Undulators and Wigglers)	12
1.4. Transverse Spatial Coherence	21
1.5. Fourth-Generation Sources	24
References	25

2. Configuration of a Typical Beamline
JOHN B. WEST

2.1. Introduction	27
2.2. Design Procedure	27
2.3. Beamline Example	30
2.4. Use of Ray Tracing	32
2.5. Ancillary Components	35
References	36

3. Glow Discharges and Wall Stabilized Arcs
JAMES R. ROBERTS

3.1. Introduction	37
3.2. Glow Discharges	37

3.3. Dielectric Barrier or Silent Discharges	41
3.4. Wall Stabilized Arc Source	45
3.5. Electron Cyclotron Resonance Source	52
3.6. Electron-Beam Ion Trap Sources	54
3.7. Inductively Coupled Plasma Sources	57
3.8. Capillary Discharges	59
References	61

4. Hollow Cathode, Penning, and Electron-Beam Excitation Sources
MICHAEL KÜHNE

4.1. Hollow Cathode Sources	65
4.2. Penning Discharge Sources	72
4.3. Electron-Beam Excitation Sources	76
References	80

5. Laser Produced Plasmas
MARTIN RICHARDSON

5.1. Laser Plasma Sources of VUV Radiation	83
References	90

6. Transition Radiation
ARTHUR J. BRAUNDMEIER, JR and EDWARD T. ARAKAWA

6.1. Transition Radiation	93
References	99

7. Vacuum Ultraviolet Lasers
PIERRE JAEGLÉ

7.1. Introduction	101
7.2. XUV Lasers	101

7.3. High-Order Harmonic Generation	109
References	115

8. Radiometric Characterization of VUV Sources
MICHAEL KÜHNE

8.1. Quantitative VUV Radiometry by Use of Source Standards	119
8.2. Primary Standards	120
8.3. Radiometric Characterization of Secondary Source Standards	125
References	143

9. Imaging Properties and Aberrations of Spherical Optics and Nonspherical Optics
JAMES H. UNDERWOOD

9.1. Need for Mirror Optics in Vacuum and Extreme Ultraviolet Spectroscopy	145
9.2. The Law of Reflection	147
9.3. Paraxial Optics	148
9.4. Geometrical Optics with Finite Apertures and Objects	150
9.5. Nonaxisymmetrical Systems	158
9.6. Toroidal and Wolter Optics	173
9.7. Fabrication of Mirrors	175
References	180

10. Reflectometers
W. R. HUNTER

10.1. Introduction	183
10.2. Specular Reflectance Measurements	184
10.3. Diffuse Reflectance	200
10.4. Reflectometry Facilities	203
References	203

11. Reflectance Spectra of Single Materials
W. R. HUNTER

11.1.	Introduction	205
11.2.	Aluminum	206
11.3.	Silicon	207
11.4.	Beryllium	208
11.5.	Gold	209
11.6.	Silver	210
11.7.	Copper	211
11.8.	Platinum	212
11.9.	Iridium	213
11.10.	Osmium	214
11.11.	Rhodium	215
11.12.	Tungsten	216
11.13.	Nickel	217
11.14.	Chromium	218
11.15.	Molybdenum	219
11.16.	Tantalum	219
11.17.	Zinc Sulfide	221
11.18.	SiO_2	222
11.19.	Al_2O_3	222
11.20.	Carbon	223
11.21.	Silicon Carbide	224
	References	225

12. Polarization
W. R. HUNTER

12.1.	Introduction	227
12.2.	Linear Polarization	228

12.3. Circular Polarization	246
References	253

13. Optical Properties of Materials
E. M. GULLIKSON

13.1. Introduction	257
13.2. Optical Constants	257
13.3. Wave Propagation in a Solid	260
13.4. Reflection and Transmission	264
13.5. Determination of Optical Constants	267
References	269

14. Reflecting Optics: Multilayers
EBERHARD SPILLER

14.1. Introduction	271
14.2. Multilayer Theory	273
14.3. Multilayer Design	279
14.4. Multilayer Fabrication and Performance	285
References	287

15. Zone Plates
YULI VLADIMRSKY

15.1. Coherent Imaging	289
15.2. Diffractive Focusing with a Zone Plate	293
15.3. Multilayer Zone Plates	296
15.4. Zone Plate Aberrations	296
15.5. Zone Plate Fabrication	298
References	302

16. Windows and Filters
W. R. HUNTER

16.1. Introduction	305
16.2. Dielectric Materials as Transmitting Filters	305
16.3. Thin Solid Films as Transmitting Filters	320
16.4. Gases as Transmitting Filters	326
16.5. Thin Films on Substrates as Reflecting Filters	329
16.6. Multiplate Resonant Reflectors	340
References	343

17. Diffraction Gratings
TAKESHI NAMIOKA

17.1. Introduction	347
17.2. VLS Ruled Grating with Curved (or Straight) Grooves	348
17.3. Holographic Grating Recorded with Spherical Wavefronts	350
17.4. Holographic Grating Recorded with Aspheric Wavefront	352
17.5. Transmission Grating	353
17.6. Theory and Basic Properties of Diffraction Gratings	355
References	375

18. Multilayer Gratings
W. R. HUNTER

18.1. Introduction	379
18.2. Designing Multilayers for Gratings	380
18.3. Efficiency Measurements of Multilayer-Coated Gratings	386
References	398

19. Crystal Optics
ECKHART FÖRSTER

19.1. Introduction	401
19.2. Basic Equations	403
19.3. Flat Crystal Spectrometers	405
19.4. Bent Crystal Spectrometers	408
References	411

INDEX 415

CONTRIBUTORS

Numbers in parentheses indicate the pages on which the authors' contributions begin.

E. T. ARKAWA (93), *111 Amherst Lane, Oak Ridge, Tennessee 37830*
ART BRAUNDMEIER (93), *Department of Physics, Southern Illinois Univeristy at Edwardsville, Edwardsville, Illinois 62026*
ECKHARD FÖRSTER (401), *University of Jena, Institute of Optics and Quantumelectronics, Max-Wien-Platz1, 07743, Jena, Germany*
ERIC M. GULLIKSON (257), *Center for X-Ray Optics, Materials Sciences Division, Lawrence Berkeley National Laboratory, Berkeley, California*
S. L. HULBERT (1), *National Synchrotron Light Source, Brookhaven National Laboratory, Upton, New York, 11973*
W. R. HUNTER (183, 205, 227, 305, 379), *SFA Inc. 1401 McCormick Drive, Largo, Maryland 20774*
PIERRE JAEGLÉ (101), *Laboratoire de Spectroscopie Atomique et Ionique, Bat 350, Universite Paris-Sud, 91405 Orsay Cedex, France*
MICHAEL KÜHNE (65, 119), *Physikalisch-Technische Bundesanstalt, Abbestrasse 2-12, D10587 Berlin, Germany*
TAKESHI NAMOIKA (347), *4-2-221, Takamori, Izumi-Ku, Sendai, 981-3203, Japan*
MARTIN RICHARDSON (83), *Laser Plasma Laboratory, CREOL/University of Central Florida, 4000 Central Florida Boulevard, Orlando, Florida 32861*
JAMES R. ROBERTS (37), *National Institute of Standards and Technology, Gaithersburg, Maryland 20899*
EBERHARD SPILLER (271), *IBM T. J. Watson Research Center, Yorktown Heights, New York 10598*
JAMES H. UNDERWOOD (145), *Center for X-Ray Optics, Lawrence Berkeley Laboratory, One Cyclotron Road, Berkeley, California 94720*
YULI VLADIMIRSKY (289), *University of Wisconsin-Medicine, CXRL, 3731 Schneider Drive, Stoughton, Wisconsin 53589*
JOHN B. WEST (27), *Synchrotron Radiation Department, Daresbury Laboratory, Daresbury, Warrington, Cheshire, WA4 4AD, United Kingdom*
G. P. WILLIAMS (1), *National Synchrotron Light Source, Brookhaven National Laboratory, Upton, New York 11973*

VOLUMES IN SERIES
EXPERIMENTAL METHODS IN THE PHYSICAL SCIENCES

(formerly Methods of Experimental Physics)

Editors-in-Chief
Robert Celotta and Thomas Lucatorto

Volume 1. Classical Methods
Edited by Immanuel Estermann

Volume 2. Electronic Methods, Second Edition (in two parts)
Edited by E. Bleuler and R. O. Haxby

Volume 3. Molecular Physics, Second Edition (in two parts)
Edited by Dudley Williams

Volume 4. Atomic and Electron Physics—Part A: Atomic Sources and Detectors; Part B: Free Atoms
Edited by Vernon W. Hughes and Howard L. Schultz

Volume 5. Nuclear Physics (in two parts)
Edited by Luke C. L. Yuan and Chien-Shiung Wu

Volume 6. Solid State Physics—Part A: Preparation, Structure, Mechanical and Thermal Properties; Part B: Electrical, Magnetic and Optical Properties
Edited by K. Lark-Horovitz and Vivian A. Johnson

Volume 7. Atomic and Electron Physics—Atomic Interactions (in two parts)
Edited by Benjamin Bederson and Wade L. Fite

Volume 8. Problems and Solutions for Students
Edited by L. Marton and W. F. Hornyak

Volume 9. Plasma Physics (in two parts)
Edited by Hans R. Griem and Ralph H. Lovberg

Volume 10. Physical Principles of Far-Infrared Radiation
By L. C. Robinson

Volume 11. Solid State Physics
Edited by R. V. Coleman

Volume 12. Astrophysics—Part A: Optical and Infrared Astronomy
Edited by N. Carleton

Part B: Radio Telescopes; Part C: Radio Observations
Edited by M. L. Meeks

Volume 13. Spectroscopy (in two parts)
Edited by Dudley Williams

Volume 14. Vacuum Physics and Technology
Edited by G. L. Weissler and R. W. Carlson

Volume 15. Quantum Electronics (in two parts)
Edited by C. L. Tang

Volume 16. Polymers—Part A: Molecular Structure and Dynamics; Part B: Crystal Structure and Morphology; Part C: Physical Properties
Edited by R. A. Fava

Volume 17. Accelerators in Atomic Physics
Edited by P. Richard

Volume 18. Fluid Dynamics (in two parts)
Edited by R. J. Emrich

Volume 19. Ultrasonics
Edited by Peter D. Edmonds

Volume 20. Biophysics
Edited by Gerald Ehrenstein and Harold Lecar

Volume 21. Solid State: Nuclear Methods
Edited by J. N. Mundy, S. J. Rothman, M. J. Fluss, and L. C. Smedskjaer

Volume 22. Solid State Physics: Surfaces
Edited by Robert L. Park and Max G. Lagally

Volume 23. Neutron Scattering (in three parts)
Edited by K. Sköld and D. L. Price

Volume 24. Geophysics—Part A: Laboratory Measurements; Part B: Field Measurements
Edited by C. G. Sammis and T. L. Henyey

Volume 25. Geometrical and Instrumental Optics
Edited by Daniel Malacara

Volume 26. Physical Optics and Light Measurements
Edited by Daniel Malacara

Volume 27. Scanning Tunneling Microscopy
Edited by Joseph Stroscio and William Kaiser

Volume 28. Statistical Methods for Physical Science
Edited by John L. Stanford and Stephen B. Vardeman

Volume 29. Atomic, Molecular, and Optical Physics—Part A: Charged Particles; Part B: Atoms and Molecules; Part C: Electromagnetic Radiation
Edited by F. B. Dunning and Randall G. Hulet

Volume 30. Laser Ablation and Desorption
Edited by John C. Miller and Richard F. Haglund, Jr.

Volume 31. Vacuum Ultraviolet Spectroscopy I
Edited by J. A. R. Samson and D. L. Ederer

1. SYNCHROTRON RADIATION SOURCES

S. L. Hulbert and G. P. Williams

Brookhaven National Laboratory
Upton, New York

1.1 General Description of Synchrotron Radiation

Synchrotron radiation is a very bright, broadband, polarized, pulsed source of light extending from the infrared to the x-ray region. It is an extremely important source of vacuum ultraviolet radiation. Brightness is defined as flux per unit area per unit solid angle and is normally a more important quantity than flux alone particularly in throughput-limited applications, which include those in which monochromators are used.

It is well known from classical theory of electricity and magnetism that accelerating charges emit electromagnetic radiation. In the case of synchrotron radiation, relativistic electrons are accelerated in a circular orbit and emit electromagnetic radiation in a broad spectral range. The visible portion of this spectrum was first observed on April 24, 1947, at General Electric's Schenectady, New York facility by Floyd Haber, a machinist working with the synchrotron team, although the first theoretical predictions were by Liénard [1] in the latter part of the 1800s. An excellent early history with references was presented by Blewett [2] and a history covering the development of the utilization of synchrotron radiation was presented by Hartman [3].

Synchrotron radiation covers the entire electromagnetic spectrum from the infrared region through the visible, ultraviolet, and into the x-ray region up to energies of many tens of kilovolts. If the charged particles are of low mass, such as electrons, and if they are traveling relativistically, the emitted radiation is very intense and highly collimated, with opening angles of the order of 1 mrad. In electron storage rings there are three possible sources of synchrotron radiation: dipole (bending) magnets; wigglers, which act like a sequence of bending magnets with alternating polarities; and undulators, which are also multiperiod alternating magnet systems but in which the beam deflections are small, resulting in coherent interference of the emitted light.

In typical storage rings used as synchrotron radiation sources, several bunches of up to $\sim 10^{12}$ electrons circulate in vacuum, guided by magnetic fields. The bunches are typically several tens of centimeters long, so that the light is pulsed, being on for a few tens to a few hundreds of picoseconds, and off for several tens to a few hundreds of nanoseconds depending on the particular machine and the radio-frequency cavity, which restores the energy lost to synchrotron radiation. However, for a ring with a 30-m circumference, the revolution time is 100 ns, so that each bunch of 10^{12} electrons is seen 10^7 times per second, giving a current of ~1 A.

The most important characteristic of accelerators built specifically as synchrotron radiation sources is that they have a magnetic focusing system which is designed to concentrate the electrons into bunches of very small cross section and to keep the electron transverse velocities small. The combination of high intensity with small opening angles and small source dimensions results in very high brightness.

The first synchrotron radiation sources to be used were operated parasitically on existing high-energy physics or accelerator development programs. These were not optimized for brightness and were usually accelerators rather than storage rings, meaning that the electron beams were constantly being injected, accelerated, and extracted. Owing to the successful use of these sources for scientific programs, a second generation of dedicated storage rings was built starting in the early 1980s. In the mid-1990s, a third generation of sources was built, this time based largely on special magnetic insertions called *undulators* and *wigglers*. A fourth generation is also under development based on what is called *multiparticle coherent emission*, in which coherence along the path of the electrons, or longitudinal coherence, plays the major role. This is achieved by microbunching the electrons on a length scale comparable to or smaller than the scale of the wavelengths emitted. The emission is then proportional to the square of the number of electrons, N, which, if N is 10^{12}, can be a very large enhancement. These sources can reach the theoretical diffraction limit of source emittance (the product of solid angle and area).

1.2 Theory of Synchrotron Radiation Emission

1.2.1 General

The theory describing synchrotron radiation emission is based on classical electrodynamics and can be found in the works of Tomboulian and Hartman [4] (1956), Schwinger [5] (1949), Jackson [6] (1975), Winick [7] (1980), Hofmann [8] (1980), Krinsky, Perlman, and Watson [9] (1983), and Kim [10] (1989). A quantum description is presented by Sokolov and Ternov [11] (1968).

Here we present a phenomenological description in order to highlight the general concepts involved. Electrons in circular motion radiate in a dipole pattern as shown schematically in Fig. 1a. As the electron energies increase and the particles start traveling at relativistic velocities, this dipole pattern appears different to an observer in the rest frame of the laboratory. To find out how this relativistic dipole pattern appears to the observer at rest, we need only appeal to

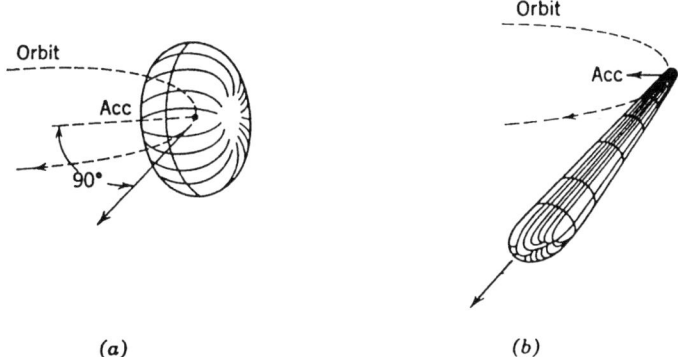

FIG. 1. Conceptual representation of the radiation pattern from a charged particle undergoing circular acceleration at (a) subrelativistic and (b) relativistic velocities.

standard relativity theory. This tells us that angles θ_t in a transmitting object are related to those in the receiving frame, θ_r, by:

$$\tan \theta_r = \frac{\sin \theta_t}{\gamma(\cos \theta_t - \beta)}, \quad (1)$$

with γ, the ratio of the mass of the electron to its rest mass, being given by E/m_0c^2, E being the electron energy, m_0 the electron rest mass, and c the velocity of light; β is the ratio of electron velocity, v, to the velocity of light, c. Thus for electrons at relativistic energies, $\beta \approx 1$ so the peak of the dipole emission pattern in the particle frame, $\theta_t = 90°$, transforms to $\theta_r \approx \tan \theta_r \approx \gamma^{-1}$ in the laboratory frame as shown in Fig. 1b. Thus γ^{-1} is a typical opening angle of the radiation in the laboratory frame.

Now for an electron viewed in passing by an observer, as shown in Fig. 2, the duration of the pulse produced by a particle under circular motion of radius ρ will be $\rho/\gamma c$ in the particle frame, or $\rho/\gamma c \times 1/\gamma^2$ in the laboratory frame owing to the time dilation. The Fourier transform of this function will contain frequency components up to the reciprocal of this time interval. For a storage ring with a radius of 2 m and $\gamma = 1000$, corresponding to a stored electron beam energy of ~500 MeV, the time interval is 10^{-17} s, which corresponds to light of wavelength 30 Å.

1.2.2 Bending Magnet Radiation

It is useful to define a few quantities in practical units because these will be used in the calculations that follow. For an electron storage ring, the relationship between the electron beam energy E in GeV, bending radius ρ in meters, and field B in T is

$$\rho \,[\text{m}] = \frac{E\,[\text{GeV}]}{0.300 B\,[\text{T}]}. \tag{2}$$

The ratio γ of the mass of the electron to its rest mass is given by

$$\gamma = E/m_0 c^2 = E/0.511\,\text{MeV} = 1957 E\,[\text{GeV}], \tag{3}$$

and λ_c, which is defined as the wavelength for which half the power is emitted above and half below, is

$$\lambda_c = 4\pi\rho/(3\gamma^3) \quad \text{or} \quad \lambda_c\,[\text{Å}] = 5.59\rho\,[\text{m}]/E^3\,[\text{GeV}^3] = 18.6/(B\,[\text{T}]E^2\,[\text{GeV}^2]). \tag{4}$$

The critical frequency and photon energy are

$$\omega_c = 2\pi c/\lambda_c = 3c\gamma^3/(2\rho) \quad \text{or} \quad \varepsilon_c\,[\text{eV}] = \hbar\omega_c\,[\text{eV}] = 665.5 E^2\,[\text{GeV}^2]\,B\,[\text{T}]. \tag{5}$$

The angular distribution of synchrotron radiation emitted by electrons moving through a bending magnet with a circular trajectory in the horizontal plane is

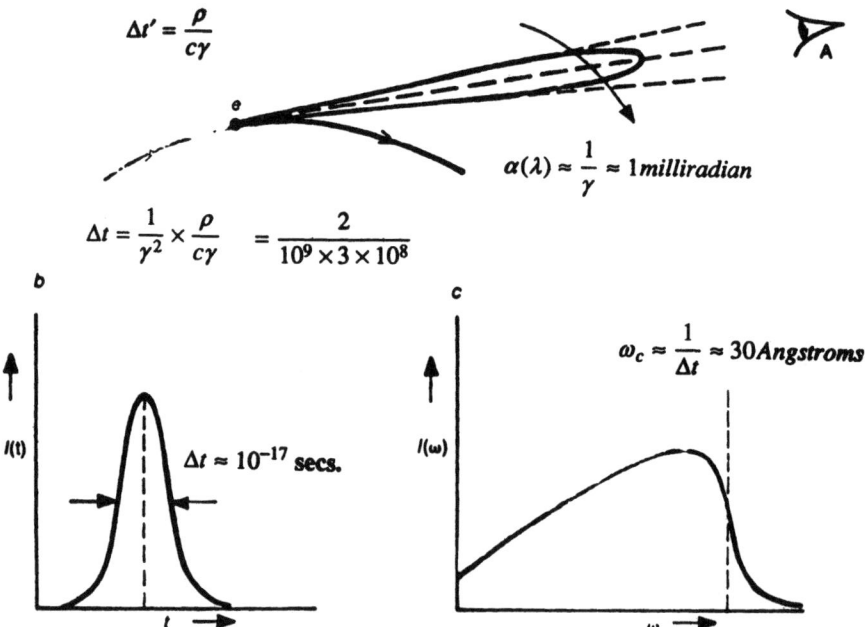

FIG. 2. Illustration of the derivation of the spectrum emitted by a charged particle in a storage ring.

given [9] by

$$\frac{d^2 F_{bm}(\omega)}{d\theta\, d\psi} = \frac{3\alpha}{4\pi^2} \gamma^2 \frac{\Delta\omega}{\omega} \frac{I}{e} \left(\frac{\omega}{\omega_c}\right)^2 (1 + \gamma^2\psi^2)^2 \left[K_{2/3}^2(\xi) + \frac{\gamma^2\psi^2}{1 + \gamma^2\psi^2} K_{1/3}^2(\xi) \right], \quad (6)$$

where F is the number of photons per second, θ the observation angle in the horizontal plane, ψ the observation angle in the vertical plane, α the fine structure constant (1/137), ω the light frequency, I the beam current, and $\xi = (\omega/2\omega_c)(1 + \gamma^2\psi^2)^{3/2}$. The subscripted K's are modified Bessel functions of the second kind. The $K_{2/3}$ term represents light linearly polarized parallel to the electron orbit plane, while the $K_{1/3}$ term represents light linearly polarized perpendicular to the orbit plane.

If one integrates over all vertical angles, then the total intensity per radian is

$$\frac{dF_{bm}(\omega)}{d\theta} = \frac{\sqrt{3}}{2\pi} \alpha\gamma \frac{\Delta\omega}{\omega} \frac{I}{e} \frac{\omega}{\omega_c} \int_{\omega/\omega_c}^{\infty} K_{5/3}(y)\, dy. \quad (7)$$

In practical units these formulas become:

$$\frac{d^2 F_{bm}(\omega)}{d\theta\, d\psi} = 1.326 \times 10^{13} E^2\, [\text{GeV}^2]\, I\, [\text{A}](1 + \gamma^2\psi^2)^2 \left(\frac{\omega}{\omega_c}\right)^2$$
$$\times \left[K_{2/3}^2(\xi) + \frac{\gamma^2\psi^2}{1 + \gamma^2\psi^2} K_{1/3}^2(\xi) \right] \quad (8)$$

in units of photons/s/mrad^2/0.1% bandwidth, and

$$\frac{dF_{bm}(\omega)}{d\theta} = 2.457 \times 10^{13} E\, [\text{GeV}]\, I\, [\text{A}] \frac{\omega}{\omega_c} \int_{\omega/\omega_c}^{\infty} K_{5/3}(y)\, dy \quad (9)$$

photons per second per milliradian per 0.1% bandwidth.

The Bessel functions can be computed easily using the algorithms of Kostroun [12]:

$$K_\nu(x) = h \left\{ \frac{e^{-x}}{2} + \sum_{r=1}^{\infty} e^{-x \cosh(rh)} \cosh(\nu rh) \right\} \quad (10)$$

and

$$\int_x^\infty K_\nu(\eta)\, d\eta = h \left\{ \frac{e^{-x}}{2} + \sum_{r=1}^{\infty} e^{-x \cosh(rh)} \frac{\cosh(\nu rh)}{\cosh(rh)} \right\} \quad (11)$$

for all x and for any fractional order ν, where h is some suitable interval such as 0.5. In evaluating the series, the sum is terminated when the rth term is small, $<10^{-5}$ for example.

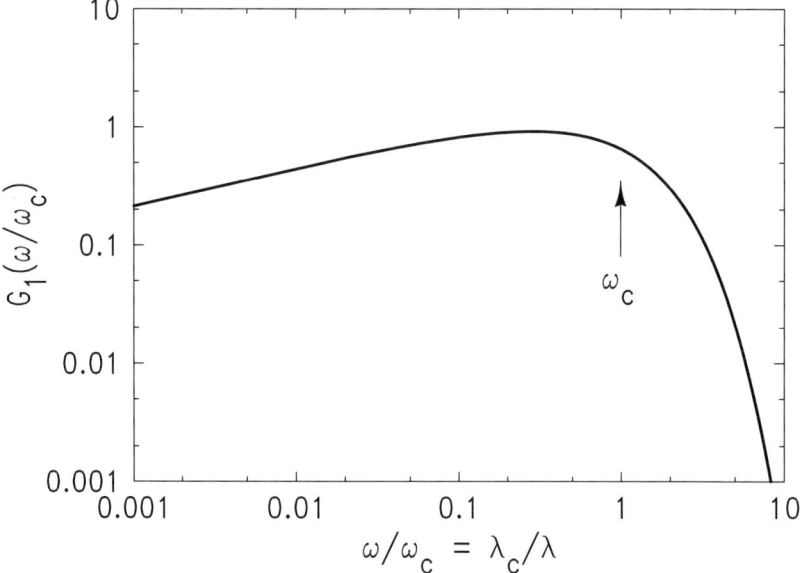

FIG. 3. Universal synchrotron radiation output curve.

In Fig. 3 we plot the universal function

$$G_1\left(\frac{\omega}{\omega_c}\right) = \frac{\omega}{\omega_c} \int_{\omega/\omega_c}^{\infty} K_{5/3}(y)\, dy$$

from Eq. (7) or (9), so that the photon energy dependence of the flux from a given ring can be calculated readily. It is found that the emission falls off exponentially as $e^{-\lambda_c/\lambda}$ for wavelengths shorter than λ_c, but only as $\lambda^{-1/3}$ at longer wavelengths.

The vertical angular distribution is more complicated. For a given ring and wavelength, there is a characteristic natural opening angle for the emitted light. The opening angle increases with increasing wavelength. If we define ψ as the vertical angle relative to the orbital plane, and if the vertical angular distribution of the emitted flux is assumed to be Gaussian in shape, then the rms divergence σ_ψ is calculated by taking the ratio of Eqs. (7)/(6) evaluated at $\psi = 0$:

$$\sigma_\psi = \sqrt{\frac{2\pi}{3}} \frac{1}{\gamma} \left(\frac{\omega}{\omega_c}\right)^{-1} \frac{\int_{\omega/\omega_c}^{\infty} K_{5/3}(y)\, dy}{K_{2/3}^2(\omega/2\omega_c)}. \qquad (12)$$

In reality, the distribution is not Gaussian, especially in view of the fact that the distribution for the vertically polarized component vanishes in the horizontal plane ($\psi = 0$). However, σ_ψ defined by Eq. (12) is still a simple and useful measure of the angular divergence. Equation (12) is of the form:

$$\sigma_\psi = \frac{1}{\gamma} C(\omega/\omega_c), \tag{13}$$

and the function $C(\omega/\omega_c)$ [10] is plotted in Fig. 4. At $\omega = \omega_c$, $\sigma_\psi \approx 0.64/\gamma$. The asymptotic values of σ_ψ can be obtained from the asymptotic values of the Bessel functions and are

$$\sigma_\psi \approx \frac{1.07}{\gamma} \left(\frac{\omega}{\omega_c}\right)^{-1/3}; \quad \omega \ll \omega_c \tag{14}$$

and

$$\sigma_\psi \approx \frac{0.58}{\gamma} \left(\frac{\omega}{\omega_c}\right)^{-1/2}; \quad \omega \gg \omega_c. \tag{15}$$

In Fig. 5 we show examples of the normalized vertical angular distributions of both parallel and perpendicularly polarized synchrotron radiation for a selection of wavelengths.

1.2.3 Circular Polarization and Aperturing for Magnetic Circular Dichroism

Circularly polarized radiation is a valuable tool for the study of the electronic, magnetic, and geometric structures of a wide variety of materials. The dichroic response in the soft x-ray spectral region (100 to 1500 eV) is especially

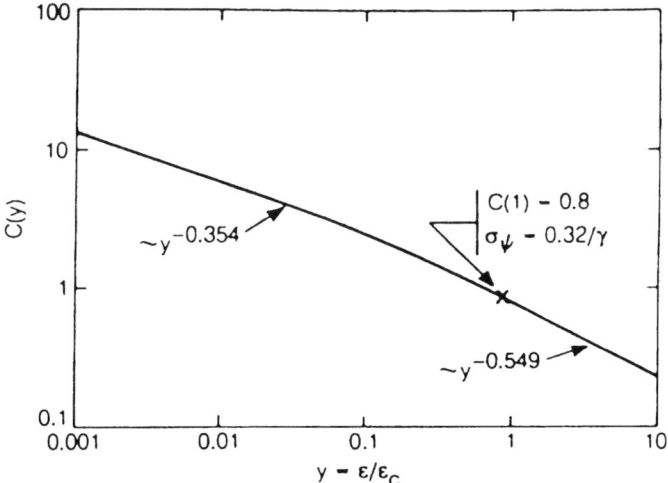

FIG. 4. Plot of the function $C(y)$ defined in Eq. (13).

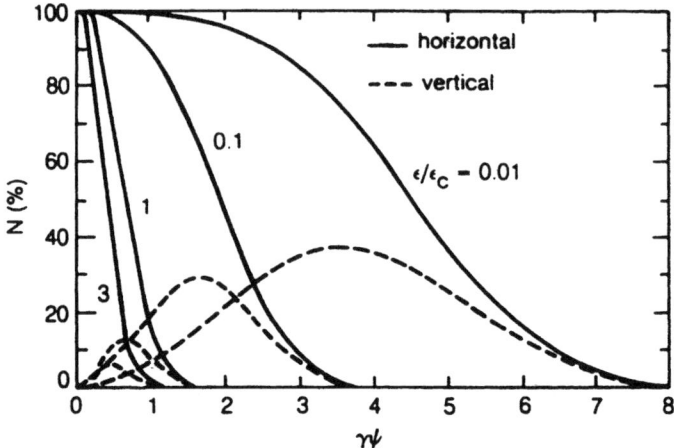

FIG. 5. Normalized intensities of horizontal and vertical polarization components, as functions of the vertical observation angle for different photon energies ε in units of the critical energy ε_c.

important because in this energy range almost every element has a strong dipole transition from a sharp core level to its lowest unoccupied state [13].

The production of bright sources of circularly polarized *soft x-rays* is therefore a topic of keen interest, and is a problem which has seen a multitude of solutions, from special insertion devices (crossed undulators, helical undulators, elliptically polarized undulators/wigglers) to optical devices (multiple-bounce reflectors/multilayers, and quarter-wave plates). However, standard bending magnet synchrotron radiation sources are good sources of elliptically polarized soft x-rays when viewed from either above or below the orbital plane.

As discussed by Chen [13], a practical solution involves acceptance of a finite vertical angular range, $\psi_{\text{off}} - \Delta\psi/2 < \psi < \psi_{\text{off}} + \Delta\psi/2$ centered about any vertical offset angle $\psi = \psi_{\text{off}}$ or, equivalently, about $\psi = -\psi_{\text{off}}$. This slice of bending magnet radiation exhibits a circular polarization [14]:

$$P_c = -\frac{2A_h A_v}{(A_h^2 + A_v^2)}, \tag{16}$$

where $A_h = K_{2/3}(\xi)$ and $A_v = \gamma\psi/(1 + \gamma^2\psi^2)^{1/2} K_{1/3}(\xi)$ are proportional to the square roots of the horizontally and vertically polarized components of bending magnet flux [Eq. (8)], that is, A_h and A_v are proportional to the horizontal and vertical components of the electric field, respectively. The value of P_c depends on the vertical angle ψ, electron energy γ, and, through ξ, the emitted photon energy ω/ω_c. In Fig. 6 we plot values of P_c vs. $\gamma\psi$ and ω/ω_c for $\gamma = 1565$ ($E = 0.8$ GeV) and $\rho = 1.91$ m ($h\nu_{\text{crit}} = 594$ eV).

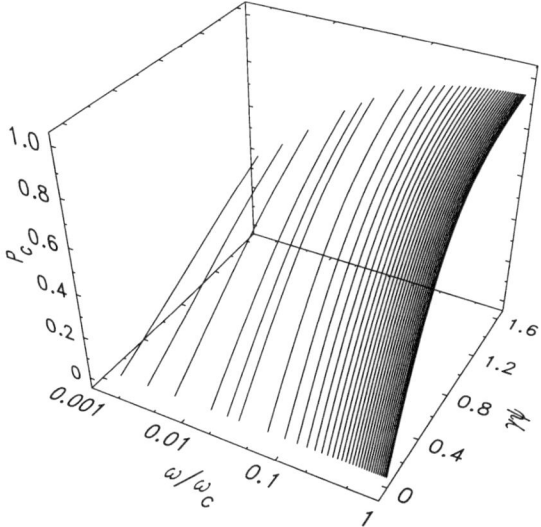

FIG. 6. P_c vs $\gamma\psi$ vs ω/ω_c for $E = 0.8$ GeV and $\rho = 1.91$ m.

Magnetic circular dichroism (MCD) measures the normalized difference of the absorption of right circular and left circular light. Assuming no systematic error, the signal to noise ratio in such a measurement defines a figure of merit

$$\text{MCD figure of merit} = (\text{average circular polarization})$$
$$\times (\text{flux fraction})^{1/2}, \quad (17)$$

where

$$\text{average circular polarization} = \frac{\int_{\psi_{\text{off}}-\Delta\psi/2}^{\psi_{\text{off}}+\Delta\psi/2} P_c(\psi)(dF/d\psi)\,d\psi}{\int_{\psi_{\text{off}}-\Delta\psi/2}^{\psi_{\text{off}}+\Delta\psi/2} (dF/d\psi)\,d\psi} \quad (18)$$

and the fraction of the total (vertically integrated) flux emitted into the vertical slice $\psi = \psi_{\text{off}} \pm \Delta\psi/2$ is

$$\text{flux fraction} = \frac{1}{dF_{bm}(\omega)/d\theta} \int_{\psi_{\text{off}}-\Delta\psi/2}^{\psi_{\text{off}}+\Delta\psi/2} \frac{d^2 F_{bm}(\omega)}{d\theta\,d\psi}\,d\psi. \quad (19)$$

Here $d^2 F_{bm}(\psi)/d\theta\,d\psi$ is the angular dependence of the bending magnetic flux from Eq. (8) and $dF_{bm}(\omega)/d\theta$ is the vertically integrated flux from Eq. (9). For an 0.8-GeV storage ring (e.g., NSLS VUV), the best choices of ψ and $\Delta\psi$ are 0.5

and 0.66 mrad, respectively. This yields a flux fraction ~0.3, a circular polarization of ~0.65, and a figure of merit of ~0.35.

1.2.4 Bending Magnet Power

Integration of $I/e\hbar\omega d^2F_{bm}(\omega)/d\theta\,d\psi$ from Eq. (8) over all frequencies ω yields the angular distribution of power radiated by a bending magnet:

$$\frac{d^2P_{bm}}{d\theta\,d\psi} = \frac{I}{e}\int_0^\infty \hbar\omega\,\frac{d^2F_{bm}(\omega)}{d\theta\,d\psi}\,d\omega = \frac{I}{e}\frac{\alpha hc\gamma^5}{2\pi\rho}\frac{7}{16}F(\gamma\psi), \qquad (20)$$

which is independent of the horizontal angle θ as required by symmetry, and the vertical angular dependence is contained in the factor

$$F(\gamma\psi) = \frac{1}{(1+\gamma^2\psi^2)^{5/2}}\left[1 + \frac{5}{7}\frac{\gamma^2\psi^2}{(1+\gamma^2\psi^2)}\right]. \qquad (21)$$

The first term in $F(\gamma\psi)$ represents the component of the bending magnet radiation parallel to the orbital plane; the second represents the perpendicular polarization component. $F(\gamma\psi)$ and its polarization components are plotted vs $\gamma\psi$ in Fig. 7. Note that the area under the F_{parallel} curve is approximately seven times greater than that for $F_{\text{perpendicular}}$.

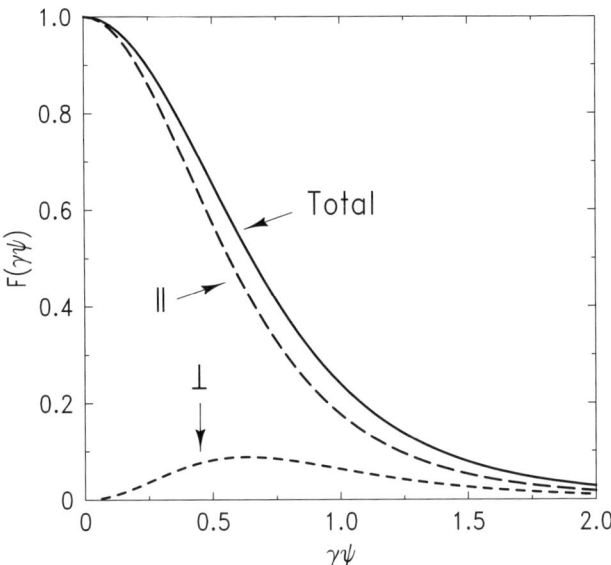

FIG. 7. Vertical angle dependence of bending magnet power, $F(\gamma\psi)$ vs. $\gamma\psi$.

In practical units,

$$\frac{d^2 P_{bm}}{d\theta\, d\psi} \text{ (W/mrad}^2\text{)} = \frac{18.082 E^5 \,[\text{GeV}^5]\, I\,[\text{A}]}{\rho\,[\text{m}]} F(\gamma\psi). \tag{22}$$

Integrating Eq. (20) over the out-of-orbital-plane (vertical) angle ψ yields the total power radiated per unit in-orbital-plane (horizontal) angle θ:

$$\frac{dP_{bm}}{d\theta} = \frac{1}{3\pi} \frac{I}{e} \frac{hc\alpha\gamma^4}{\rho} \tag{23}$$

or, in practical units,

$$\frac{dP_{bm}}{d\theta} \text{ (W/mrad)} = \frac{14.080 E^4 \,[\text{GeV}^4]\, I\,[\text{A}]}{\rho\,[\text{m}]}. \tag{24}$$

For example, a 1.0-GeV storage ring with 2-m-radius bends generates 7.04 W/mrad per ampere of stored current. By contrast, a 2.5-GeV machine with 7-m-radius bends generates 78.6 W/mrad/A and a 7-GeV machine with 39-m-radius bends generates 867 W/mrad/A.

1.2.5 Bending Magnet Brightness

Thus far we have calculated the emitted flux in photons per second per square milliradian of solid angle. To calculate the brightness, we need to include the source size. In these calculations we calculate the central (or maximum) brightness, for which we use the natural opening angle to define both the horizontal and vertical angles. Using vertical angles larger than this will not increase the flux as there is no emission. Using larger horizontal angles will increase the flux proportionately as all horizontal angles are filled with light, but owing to the curvature of the electron trajectory, the *average* brightness will actually be less. The brightness expression [15, 16] is

$$B_{bm} = \frac{d^2 F_{bm}/(d\theta\, d\psi)|_{\psi=0}}{2\pi \Sigma_x \Sigma_y}, \tag{25}$$

where:

$$\Sigma_x = [\varepsilon_x \beta_x + \eta_x^2 \sigma_E^2 + \sigma_r^2]^{1/2} \tag{26}$$

and

$$\Sigma_y = \left[\varepsilon_y \beta_y + \sigma_r^2 + \frac{\varepsilon_y^2 + \varepsilon_y \gamma_y \sigma_r^2}{\sigma_\psi^2}\right]^{1/2}, \tag{27}$$

where ε_x and ε_y are the electron-beam emittances in the horizontal and vertical directions, respectively; β_x and β_y are the electron-beam betatron functions in the horizontal and vertical planes; η_x is the dispersion function in the horizontal plane; and σ_E is the rms value of the relative energy spread. All the electron beam parameters are properties of a particular storage ring. The diffraction-limited source size is $\sigma_r = \lambda/4\pi\sigma_\psi$. The effective source sizes (Σ_x and Σ_y) are photon energy dependent via the natural opening angle σ_ψ and the diffraction-limited source size σ_r.

1.3 Insertion Devices (Undulators and Wigglers)

1.3.1 General

Insertion devices are periodic magnetic structures installed in straight sections of storage rings, as illustrated in Fig. 8, in which the vertical magnetic field varies approximately sinusoidally along the axis of the undulator. The resulting motion of the electrons is also approximately sinusoidal, but in the horizontal plane. We can understand the nature of the spectra emitted from these devices by again studying the electric field as a function of time, and this is shown in Fig. 9. This shows that the electric field and hence its Fourier transform, the spectrum, depend critically on the magnitude of the beam deflection in the device. At one extreme, when the magnetic fields are high, as in Fig. 9a, the deflection is large and the electric field is a series of pulses similar to those obtained from a dipole. Such a device is termed a *wiggler*. The Fourier transform for the wiggler is N times that from a single dipole. At the other extreme, as in Fig. 9b, the deflection of the electron beam is such that the electric field as a function of time is sinusoidal, and the Fourier transform is then a single peak with a width proportional to the inverse of the length of the wavetrain, L^*, according to $\lambda^2/\Delta\lambda = L^*$, where L^* is obtained by dividing the real length of the device, L, by γ^2 because of relativistic effects. Thus for a meter-long device emitting at a wavelength $\lambda = 10$ Å (or 1 nm) in a machine of energy 0.5 GeV ($\gamma \sim 1000$), we get $\lambda^2/\Delta\lambda = 10^{-6}$ m, and $\lambda/\Delta\lambda = 1000$.

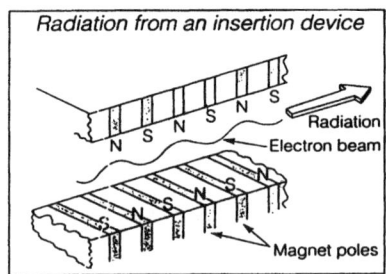

FIG. 8. Schematic of an insertion device.

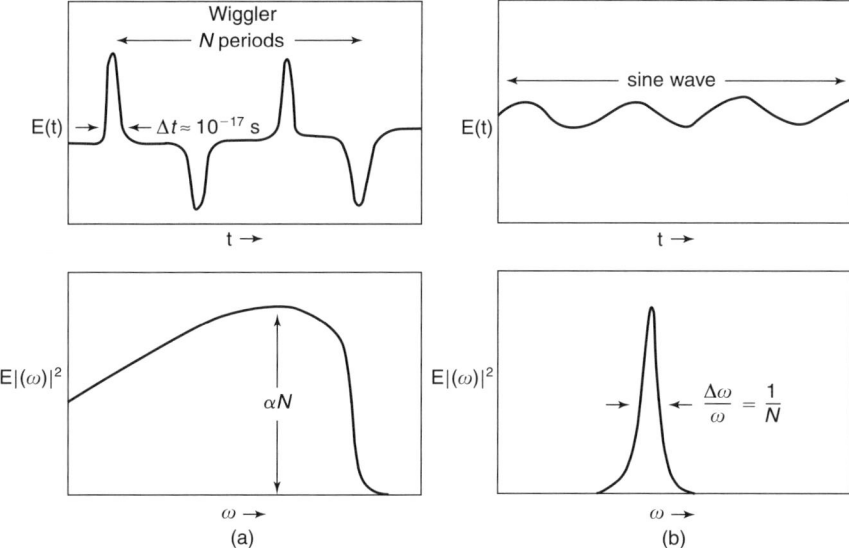

FIG. 9. Conceptual representation of the electric fields emitted as a function of time by an electron in (a) a wiggler and (b) an undulator, with the corresponding intensity spectra.

Interference occurs in an undulator since the electric field from one part of the electron path is added coherently to that from adjacent parts.

1.3.2 Formal Treatment

We assume that the motion of an electron in an insertion device is sinusoidal, and that we have a field in the vertical (y) direction varying periodically along the z direction, with:

$$B_y = B_0 \sin(2\pi z/\lambda_u), \qquad 0 \leq z \leq N\lambda_u, \tag{28}$$

where B_0 is the peak magnetic field, λ_u is the period length, and N the number of periods. By integrating the equation of motion, the electron transverse velocity $c\beta_x$ is found:

$$\beta_x = \frac{K}{\gamma} \cos(2\pi z/\lambda_u), \tag{29}$$

where

$$K = eB_0 \lambda_u/2\pi mc = 0.934 \lambda_u \text{ [cm]} B_0 \text{ [T]} \tag{30}$$

is a dimensionless parameter which is proportional to the deflection of the electron beam. The maximum slope of the electron trajectory is

$$\delta = \frac{K}{\gamma}. \tag{31}$$

In terms of δ, we define an undulator as a device in which $\delta \leq \gamma^{-1}$, which corresponds to $K \leq 1$. When K is large, the device is called a *wiggler*. In most insertion devices the field can be changed either electromagnetically or mechanically, and in some cases K can vary between the two extremes of undulator and wiggler operation.

1.3.3 Wigglers

For the wiggler, the flux distribution is given by $2N$ (where N is the number of magnetic periods) times the appropriate bending magnet formulas in Eqs. (8) and (9). However, ρ or B must be taken at the point in the path of the electron which is tangent to the direction of observation. For a horizontal angle θ,

$$\varepsilon_c(\theta) = \varepsilon_{c\,\mathrm{max}} \sqrt{1 - (\theta/\delta)^2}, \tag{32}$$

where

$$\varepsilon_{c\,\mathrm{max}} \, [\mathrm{keV}] = 0.665 E^2 \, [\mathrm{GeV}] \, B_0 \, [\mathrm{T}]. \tag{33}$$

Integration over θ, which is usually performed numerically, gives the wiggler flux.

The calculation of the brightness of wigglers needs to take into account the depth-of-field effects, that is, the contribution to the apparent source size from different poles. The expression for the brightness of wigglers is

$$B_W = \frac{d^2 F_W}{d\theta\,d\psi} \sum_{\pm} \sum_{n=-(1/2)N}^{(1/2)N} \frac{1}{2\pi} \times \frac{\exp[-\frac{1}{2}(x_0^2/(\sigma_x^2 + z_{n\pm}^2 \sigma_x'^2))]}{[(\sigma_x^2 + z_{n\pm}^2 \sigma_x'^2)(\varepsilon_y^2/\sigma_\psi^2 + \sigma_y^2 + z_{n\pm}^2 \sigma_y'^2)]}, \tag{34}$$

where $z_{n\pm} = \lambda_W(n \pm \frac{1}{4})$, λ_W is the wiggler period, and σ_ψ is identical to Eq. (11), but evaluated, in the wiggler case, as the instantaneous radius at the tangent to the straight-ahead ($\theta = \psi = 0$) direction (i.e., minimum ρ, maximum ε_c); $\sigma_x = \sqrt{\varepsilon_x \beta_x}$ and $\sigma_y = \sqrt{\varepsilon_y \beta_y}$ are the rms transverse beam sizes, while $\sigma_x' = \sqrt{\varepsilon_x/\beta_x}$ and $\sigma_y' = \sqrt{\varepsilon_y/\beta_y}$ are the angular divergences of the electron beam in the horizontal and vertical directions, respectively. The exponential factor in Eq. (34) arises because wigglers have two source points separated by $2x_0$, where

$$x_0 = \frac{K \lambda_w}{\gamma \, 2\pi}. \tag{35}$$

The summations in Eq. (34) must be performed for each photon energy because σ_ψ is photon energy dependent.

1.3.4 Undulators

The interference which occurs in an undulator, that is when K is moderate ($K \leq 1$), produces sharp peaks in the forward direction at a fundamental ($n = 1$) and all odd harmonics ($n = 3, 5, 7, \ldots$) as shown for a zero emittance ($\varepsilon = 0$) electron beam in Fig. 10a (dotted line). In the $\varepsilon = 0$ case, the even harmonics ($n = 2, 4, 6, \ldots$) peak off-axis. For real ($\varepsilon \neq 0$) electron beams, the sharpness of the peaks and indeed the peak brightness are strongly dependent on the emittance and energy spread of the electron beam as well as the period and magnitude of the insertion device field.

In general, the effect of electron beam emittance is to cause all harmonics to appear in the forward direction (solid line in Fig. 10a). The effect of angle integration on the spectrum in Fig. 10a is shown in Fig. 10b, a spectrum which is independent of electron-beam emittance except for the presence of "noise" in the zero emittance case. The effect of electron-beam emittance on the angular distribution of the fundamental, second, and third harmonics of this device is shown in Fig. 10c, which also nicely demonstrates the dependence on harmonic number.

The peak wavelengths of the emitted radiation, λ_n, are given by

$$\lambda_n = \frac{\lambda_u}{2n\gamma^2}\left(1 + \frac{K^2}{2} + \gamma^2\theta^2\right), \tag{36}$$

where λ_u is the undulator period length. They soften as the square of the deviation angle θ relative to the forward direction.

Of particular interest is the intense central cone of radiation. An approximate formula for flux integrated over the central cone is (for the odd harmonics)

$$F_u(K, \omega) = \pi\alpha N \frac{\Delta\omega}{\omega} \frac{I}{e} Q_n(K), \quad n = 1, 3, 5, \tag{37}$$

where

$$Q_n(K) = \left(1 + \frac{K^2}{2}\right)\frac{F_n(K)}{n}, \quad n = 1, 3, 5 \tag{38}$$

and

$$F_n(K) = \frac{K^2 n^2}{(1 + K^2/2)^2}\left\{J_{(n-1)/2}\left[\frac{nK^2}{4(1 + \frac{1}{2}K^2)}\right] - J_{(n+1)/2}\left[\frac{nK^2}{4(1 + \frac{1}{2}K^2)}\right]\right\}^2. \tag{39}$$

In practical units, the flux in photons/s/0.1% bandwidth is given by

$$F_u(K, w) = 1.431 \times 10^{14} N Q_n(K) I \text{ [A]}, \tag{40}$$

To calculate the undulator flux angular distribution and spectral output into arbitrary solid angle, one can use freely available codes such as Urgent [17]

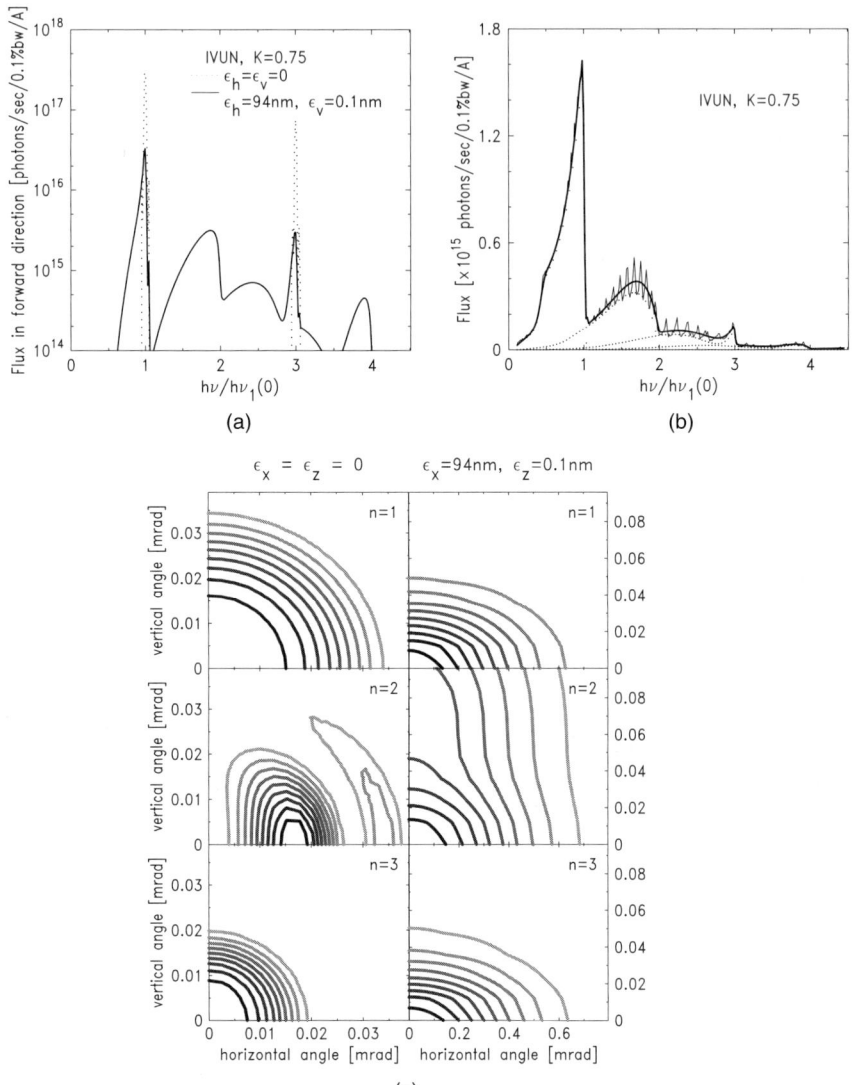

FIG. 10. Spectral output and angular distribution of the emission from the NSLS in-vacuum undulator (IVUN) for $K = 0.75$. (a) Spectral output in the forward direction, with (solid line) and without (dotted line) the effect of electron beam emittance. (b) Angle-integrated spectral output with (solid line) and without (faint solid line) the effect of electron beam emittance, and the decomposition into harmonics ($n = 1, 2, 3, 4$) (dotted lines). (c) Angular distribution of the first three harmonics ($n = 1, 2, 3$) with and without the effect of electron-beam emittance. The emittance of the NSLS x-ray ring is 94 nm horizontal and 0.1 nm vertical.

(R. P. Walker and B. Diviacco). To include magnetic field errors (e.g. measured values), use Ur [18] (R. J. Dejus and A. Luccio).

The brightness of an undulator, B_u, is approximated by dividing the central cone flux by the effective angular divergence, $\Sigma'_x(\Sigma'_y)$, and by the effective source size, $\Sigma_x(\Sigma_y)$, in the horizontal (vertical) directions. These are given by convolution of the Gaussian distributions of the electron beam and the diffraction-limited photon beam, in both space and angle:

$$\Sigma_{x'} = \sqrt{\sigma_{x'}^2 + \sigma_{r'}^2}, \qquad \Sigma_{y'} = \sqrt{\sigma_{y'}^2 + \sigma_{r'}^2}, \tag{41}$$

$$\Sigma_x = \sqrt{\sigma_x^2 + \sigma_r^2}, \qquad \Sigma_y = \sqrt{\sigma_y^2 + \sigma_r^2}. \tag{42}$$

Thus, B_u is given by:

$$B_u = \frac{F_u}{(2\pi)^2 \Sigma_x \Sigma_y \Sigma'_x \Sigma'_y}. \tag{43}$$

The diffraction-limited emittance of a photon beam is the minimum value in the inequality

$$\varepsilon_r = \sigma \sigma_{r'} \geq \frac{\lambda}{2} = \frac{\lambda}{4\pi}, \tag{44}$$

where ε is the photon emittance and λ is the wavelength, in direct analogy to the Heisenberg uncertainty principle in nonrelativistic quantum mechanics. The space versus angle separation of this minimum emittance is energy and harmonic dependent [19]. For the exact harmonic frequency in the forward direction, given by Eq. (36) with $\theta = 0$, there appears to be consensus that σ_r and $\sigma_{r'}$ are given by

$$\sigma_r = \frac{\sqrt{2\lambda L}}{4\pi}, \qquad \sigma_{r'} = \sqrt{\frac{\lambda}{2L}}. \tag{45}$$

On the other hand, at the peak of the angle-integrated undulator spectrum, which lies a factor of $(1 - (1/nN))$ below the exact harmonic energy, σ_r and $\sigma_{r'}$ are given by

$$\sigma_r = \frac{\sqrt{\lambda L}}{4\pi}, \qquad \sigma_{r'} = \sqrt{\frac{\lambda}{L}}. \tag{46}$$

It is clear from Eqs. (41) and (42) that the choice of expression for σ_r and $\sigma_{r'}$ can have a nonnegligible effect on the undulator brightness value, especially for small beam size and opening angle. Lacking a functional form for σ_r and $\sigma_{r'}$ as a function of photon energy, we use Eq. (46) in evaluating the expression for undulator peak spectral brightness from Eq. (43).

In Figs. 11 and 12 we show calculations based on the preceding equations for the output flux and brightness for three synchrotron radiation facilities; the

NSLS, Brookhaven National Laboratory, Upton, NY, the Advanced Light Source, Lawrence Berkeley National Laboratory, Berkeley, CA, and the Advanced Photon Source, Argonne National Laboratory, Argonne, IL, all in the United States. These sources are representative of most of the facilities around the world. Their characteristics are summarized in Table I. For undulators the brightness in the first harmonic is plotted.

1.3.5 Insertion Device Power

The Schwinger [5] formula for the distribution of radiated power from an electron in a sinusoidal trajectory, which applies with reasonable approximation to undulators and, to a lesser extent, wigglers, reduces [20] to

$$\frac{d^2 P}{d\theta\, d\psi} = P_{\text{total}} \frac{21\gamma^2}{16\pi K} G(K) f_K(\gamma\theta, \gamma\psi), \qquad (47)$$

where the total (angle-integrated) radiated power is

$$P_{\text{total}} = \frac{N}{6} \frac{Z_0 I 2\pi e c}{\lambda_u} \gamma^2 K^2 \qquad (48)$$

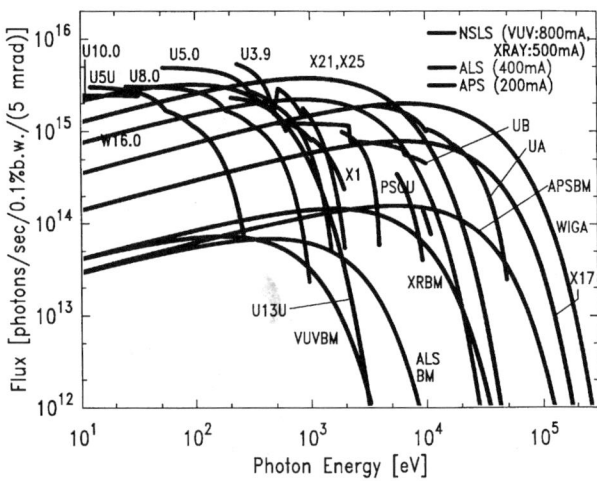

FIG. 11. Spectral flux as a function of photon energy for selected undulators, wigglers, and bending magnets at the NSLS, ALS, and APS. The wiggler and bending magnet flux curves represent the vertically integrated flux within a 5-mrad horizontal collection angle. Each undulator curve is the locus of narrow peaks of radiation, tuned by altering the undulator magnetic strength parameter K between 0.5 and 3, and represents the envelope of the first, third, and fifth harmonics. All of the curves correspond to standard injection current values for each storage ring (shown in the key).

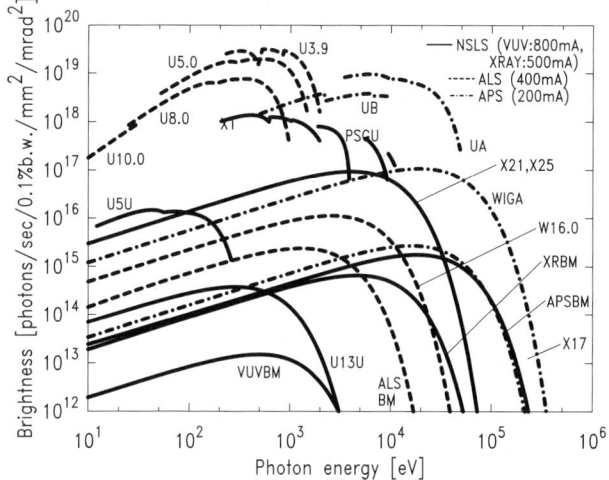

FIG. 12. On-axis spectral brightness as a function of photon energy for selected undulators, wigglers, and bending magnets at the NSLS, ALS, and APS. Each undulator curve is the locus of narrow peaks of radiation, tuned by altering the undulator magnetic strength parameter K between 0.5 and 3, and represents the envelope of the first, third, and fifth harmonics. All of the curves correspond to standard injection current values for each storage ring (shown in the key).

or, in practical units,

$$P_{\text{total}} [\text{W}] = 633.0 E^2 [\text{GeV}^2] B_0^2 [\text{T}^2] L [\text{m}] I [\text{A}], \quad (49)$$

where N is the number of undulator or wiggler periods, Z_0 is the vacuum impedance (377 Ω), I is the storage ring current, e is the electronic charge, c is

TABLE I. Characteristics of a selection of insertion devices at various synchrotron radiation facilities in the US.

ID	E (GeV)	B_0 (T)	I (A)	λ_u (cm)	N	K	P_{total} (W)	$\dfrac{d^2P}{d\theta\,d\psi}(\theta=0, \psi=0)$ (W/mrad^2)
NSLS U5	0.8	0.25	0.8	7.5	27	1.75	41	23.4
ALS U8.0	1.9	0.23	0.4	8.0	27	1.75	104	343
ALS U5.0	1.9	0.37	0.4	5.0	44	1.75	275	898
NSLS X1	2.584	0.23	0.5	8.0	35	1.75	313	1900
APS UA	7.0	0.57	0.2	3.3	36	1.75	2394	104,315

the speed of light, $L = N\lambda_u$ is the length of the insertion device,

$$G(K) = \frac{K}{(1+K^2)^{7/2}}\left(K^6 + \frac{24}{7}K^4 + 4K^2 + \frac{16}{7}\right) \tag{50}$$

and

$$f_K(\gamma\theta, \gamma\psi) = \frac{16K}{7\pi G(K)}\int_{-\pi}^{\pi}d\alpha\left(\frac{1}{D^3} - \frac{4(\gamma\theta - K\cos\alpha)^2}{D^5}\right)\sin^2\alpha, \tag{51}$$

where

$$D = 1 + \gamma^2\psi^2 + (\gamma\theta - K\cos\alpha)^2. \tag{52}$$

The integral in the expression for f_K is best evaluated numerically.

For $K > 1$, which includes all wigglers and much of the useful range of undulators, an approximate formula for the angle dependence of the radiated power is

$$f_K(\gamma\theta, \gamma\psi) = \sqrt{1 - \left(\frac{\gamma\theta}{K}\right)^2}\, F(\gamma\psi), \tag{53}$$

where $F(\gamma\psi)$ is the bending magnet formula from Eq. (21). This form clearly indicates the strong weakening of insertion device power as θ increases, vanishing at $\theta = \pm K/\gamma$.

Since $f_K(0,0)$ is normalized to unity, the radiated power density in the forward direction (i.e., along the undulator axis) is

$$\frac{d^2P}{d\theta\, d\psi}(\theta = 0, \psi = 0) = P_{\text{total}}\frac{21\gamma^2}{16\pi K}G(K) \tag{54}$$

or, in practical units,

$$\frac{d^2P}{d\theta\, d\psi}(\theta = 0, \psi = 0)\ [\text{W/mrad}^2] = 10.84 B_0\ [\text{T}]\ E^4\ [\text{GeV}^4]\ I\ [\text{A}]\ NG(K). \tag{55}$$

The total radiated power and forward power density values for selected undulators at NSLS, ALS, and APS are shown in Table I. As is clearly evident from this table, the forward power density from undulators can be quite high, owing to their extremely narrow angular spread, while their total radiated power is relatively small.

1.3.6 Polarization of Undulators and Wigglers

The polarization properties of the light emitted by wigglers is similar to that of dipoles. For both sources the radiation is elliptically polarized when observed at some angle away from the orbital plane as given by Eq. (6). For radiation from planar undulators, however, the polarization is always linear. The polarization direction, which is in the horizontal plane when observed from that plane, rotates in a complicated way at other directions of observation.

A comprehensive analysis of the polarization from undulators has been carried out by Kitamura [21]. The linear polarization of the undulator radiation is due to the symmetry of the electron trajectory within each period. The polarization can, in fact, be controlled by a deliberate breaking of this symmetry. Circularly polarized radiation can be produced by a helical undulator, in which the series of dipole magnets is arranged each rotated by a fixed angle with respect to the previous one. For a variable polarization capability, one can use a pair of planar undulators oriented at right angles to each other. The amplitude of the radiation from these so-called crossed undulators is a linear superposition of two parts, one linearly polarized along the x direction and another linearly polarized along the y direction, x and y being orthogonal to the electron-beam direction. By varying the relative phase of the two amplitudes by means of a variable-field magnet between the undulators, it is possible to modulate the polarization in an arbitrary way. The polarization can be linear and switched between two mutually perpendicular directions, or it can be switched between left and right circularly polarized. For this device to work, it is necessary to use a monochromator with a sufficiently small bandpass, so that the wave trains from the two undulators are stretched and overlap. Also the angular divergence of the electron beam should be sufficiently small or the fluctuation in relative phase will limit the achievable degree of polarization. A planar undulator whose pole boundaries are tilted away from a right angle with respect to the axial direction can be used as a helical undulator if the electron trajectory lies a certain distance above or below the midplane of the device.

1.4 Transverse Spatial Coherence

As shown by Kim [22] and utilized in the brightness formulas given earlier, in wave optics the phase-space area of a radiation beam is given by the ratio of flux (F_0) to brightness (B_0). A diffraction-limited photon beam (no electron size or angular divergence contribution) occupies the minimum possible phase-space area. From Eqs. (41)–(45) this area is

$$(2\pi\sigma_r\sigma_{r'})^2 = (2\pi\varepsilon)^2 = \left(\frac{\lambda}{2}\right)^2. \tag{56}$$

Thus, the phase space occupied by a single Gaussian mode radiation beam is $(\lambda/2)^2$, and such a beam is referred to as completely transversely coherent. It then follows that the transversely coherent flux of a radiation beam is

$$F_{\text{coherent}} = \left(\frac{\lambda}{2}\right)^2 B_0 \tag{57}$$

and the degree of transverse spatial coherence is

$$\frac{F_{\text{coherent}}}{F_0} = \left(\frac{\lambda}{2}\right)^2 \frac{B_0}{F_0}. \tag{58}$$

Conversely, the number of Gaussian modes occupied by a beam is

$$\frac{F_0}{F_{\text{coherent}}} = \frac{F_0}{B_0(\lambda/2)^2}. \tag{59}$$

Transverse spatial coherence is the quantity which determines the throughput of phase-sensitive devices such as Fresnel zone plates used for x-ray microscopy.

The degree of transverse spatial coherence of the various sources representative of synchrotron radiation shown in Fig. 11 is plotted in Fig. 13. It is clear from this figure that undulators on the lowest emittance storage rings provide the highest degree of transverse coherence and are therefore the source of choice for x-ray microscopy.

In this chapter, we have attempted to compile the formulas needed to calculate the flux, brightness, polarization (linear and circular), and power produced by the three standard storage ring synchrotron radiation sources: bending magnets, wigglers, and undulators. Where necessary, these formulas

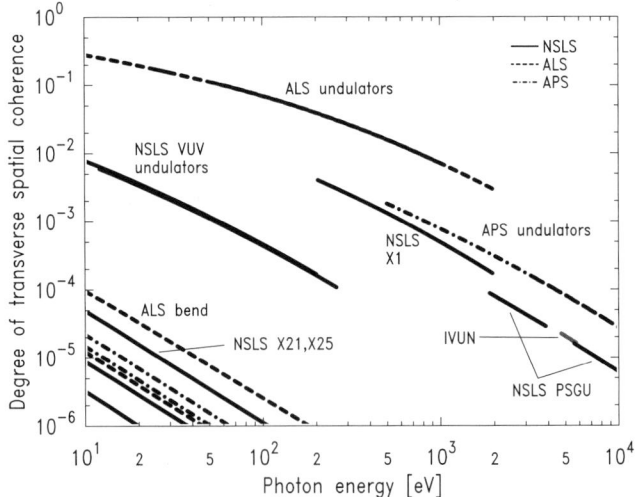

FIG. 13. Degree of transverse spatial coherence.

have contained reference to the emittance (ε) of the electron beam, as well as to the electron-beam size (σ) and its divergence (σ'). For all three types of sources, the source phase-space area, that is, the spatial and angular extent of the effective (real) source, is a convolution of its electron and photon components. Modification of this effective beam size and divergence by drift along the direction of propagation and via interaction with optical elements (slits, pinholes, mirrors, gratings, crystals, zone plates, etc.) is most conveniently displayed by means of phase-space diagrams. We discuss here the storage ring machine parameters which determine the source phase-space area at any point around the ring, and leave the treatment of the other components, that is, the beamlines, to be discussed in Chapter 2.

The electron beam in a storage ring at a given point z around the ring may be described in the vertical (y-y') plane by the phase ellipse (see Fig. 14):

$$\gamma y^2 + 2\alpha y y' + \beta y'^2 = \text{beam emittance } \varepsilon, \tag{60}$$

where $\alpha = -\beta/2 = -d\beta/2dz$ [23] and $\gamma = (\alpha^2 + 1)/\beta$ [22] with α, β, and γ being Twiss parameters [24] characterizing the beam.

As discussed earlier, electrons in circular orbits emit radiation with an energy spread depending on their energy and the radius of their orbit. There is also a characteristic vertical angular spread given by Eq. (12) which defines an rms divergence, σ_ψ, that depends on the photon energy and is smaller for higher photon energies. For a particular wavelength, and using a one-σ_ψ contour, a photon phase ellipse such as that shown in Fig. 15 may be constructed from the electron phase ellipse depicted in Fig. 14, through a convolution of the angular distributions of the electrons in the beam pipe and the photons emitted by each electron [23]. The equation for this new ellipse is

$$\left(\gamma + \frac{\sigma^2}{\varepsilon}\right) y^2 + 2\alpha y y' + \beta y'^2 = \varepsilon + \beta\sigma^2. \tag{61}$$

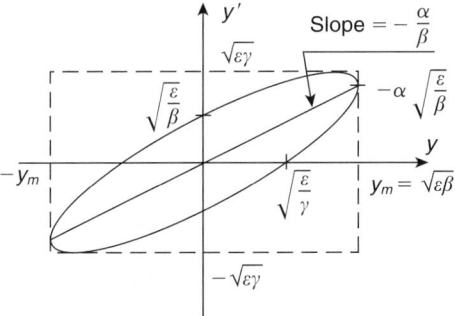

FIG. 14. Storage ring electron-beam phase-space ellipse.

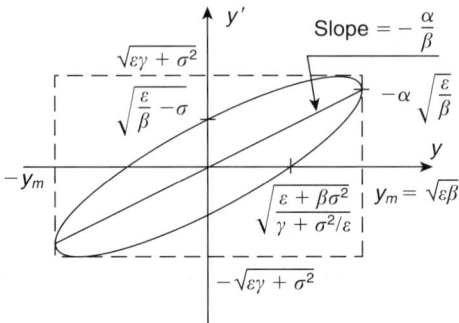

FIG. 15. Photon phase-space ellipse.

Notice that the electron and photon ellipses have common spatial widths $2(\beta E)^{1/2}$, and common diameters, but that the photon beam has a larger angular divergence than the electron beam. This is the result of the addition in quadrature of the synchrotron radiation natural opening angle σ_ψ; the common diameter indicates that each electron emits photons symmetrically in angle with respect to the direction of its motion [25].

A photon beam that has propagated a distance z from the storage ring can be described by a photon phase ellipse with area $S_{sy} = \pi(\varepsilon[\varepsilon + \beta\sigma^2])^{1/2}$, maximum angular divergence $\Delta y'_m = 2(\varepsilon\gamma + \sigma^2)^{1/2}$, maximum spatial width $\Delta y_m = 2([\varepsilon\gamma + \sigma^2]z^2 - 2\alpha\varepsilon z + \beta\varepsilon)^{1/2} = 2([y'_m z]^2 - 2\alpha\varepsilon z + \beta\varepsilon)^{1/2}$, and constant source size (as seen by an observer looking along the z axis) $\Delta y_0 = 2S_{sy}/\pi y'_m$. These photon ellipses are described for one particular wavelength. Ellipses will differ for different wavelengths, due to a variation in the natural opening angle σ_ψ. Proper treatment of the electron beam phase space involves integrating the convolved electron-beam/photon beam sizes/angular divergences over the source depth accepted by the beamline. See Green [23] and West and Padmore [26] for rather complete descriptions of this integration.

1.5 Fourth-Generation Sources

For completion we discuss fourth-generation sources at least conceptually. These sources are of even higher brightness than the devices discussed in the preceding text and are based on multiparticle coherence which can be understood as follows. In Fig. 1 the electric field induced by one electron, and hence the intensity, is proportional to the charge on an electron. If N electrons are circulating together in a storage ring, the emission is simply proportional to N times the emission of a single electron. However, when the electrons circulating in the storage ring, or passing through an insertion device, are close together compared to the wavelength of the light being emitted, the electric fields add coherently, so that the intensity scales like N^2.

The electrons can be forced to microbunch when they are in the presence of the electric field of a superimposed light beam and a magnetic field. The degree of multiparticle enhancement depends on the degree to which the microbunching occurs. Under certain circumstances, the electron beam will self bunch, the so-called self-amplified spontaneous emission (SASE) mode. New, even brighter sources of VUV radiation are being planned based on these principles.

References

1. Liénard, A. *L'Eclairage Electrique* **16**, 5 (1898).
2. Blewett, J. P. *Nucl. Instrum. Methods* **A266**, 1 (1988).
3. Hartman, P. C. *Nucl. Instrum. Methods* **195**, 1 (1982).
4. Tomboulian, D. H., and Hartman, P. L. *Phys. Rev.* **102**, 1423 (1956).
5. Schwinger, J. *Phys. Rev.* **75**, 1912 (1949).
6. Jackson, J. D. *Classical Electrodynamics*, Wiley, New York, 1975.
7. Winick, H. *Synchrotron Radiation Research*. Plenum Press, New York, Chap. 2, 1980.
8. Hofmann, A. *Phys. Reports* **64**, 253 (1980).
9. Krinsky, S., Perlman, M. L., and Watson, R. E. Chap 2 in *Handbook of Synchrotron Radiation* (E. E. Koch, ed.). North-Holland, Amsterdam, 1983.
10. Kim, K. J. *AIP Proc.* **184**, 565 (1989).
11. Sokolov, A. A., and Ternov, I. M. *Synchrotron Radiation*. Cambridge University Press, U.K., 1968.
12. Kostroun, V. O. *Nucl. Instrum. Methods* **172**, 371 (1980).
13. Chen, C. T. *Rev. Sci. Instrum.* **63**, 1229 (1992).
14. Born, M., and Wolf, E. *Principles of Optics*. Pergamon Press, New York, 1964.
15. Lawrence Berkeley Laboratory Publication 643, Rev. 2 (1989).
16. Hulbert, S. L., and Weber, J. M. *Nucl. Instrum. Methods* **A319**, 25 (1992).
17. Walker, R. P., and Diviacco, B. *Rev. Sci. Instrum.* **63**, 392 (1992).
18. Dejus, R. P., and Luccio, A. *Nucl. Instrum. Methods* **A347**, 61 (1994).
19. Howells, M. R., and Kincaid, B. M. Lawrence Berkeley Laboratory Report 34751 (1993).
20. Kim, K. J. *Nucl. Instrum. Methods* **A246**, 67 (1986).
21. Kitamura, H. *Japan. J. Appl. Phys.* **19**, L185 (1980).
22. Kim, K. J. *Nucl. Instrum. Methods* **A246**, 71 (1986).
23. Green, G. K. Brookhaven National Laboratory 50522 (1976); G. K. Green, Brookhaven National Laboratory 50595 (1977).
24. Courant, E.D., and Snyder, H. S. *Ann. Phys.* **3**, 1 (1958).
25. Matsushita, T., and Kaminaga, U. *J. Appl. Cryst.* **13**, 464 (1980).
26. West, J. B., and Padmore, H. A. "Optical Engineering," in *Handbook on Synchrotron Radiation*, G. V. Marr (ed.), North-Holland, Vol. 2, p. 21, 1987.

2. CONFIGURATION OF A TYPICAL BEAMLINE

J. B. West
Daresbury Laboratory
United Kingdom

2.1 Introduction

With the rapidly growing use of synchrotron radiation sources during the last 30 years, the development of the beamline, the primary means of bringing the radiation to the experiment, has advanced to a high degree. Having started out as a simple pipe connected to the tangent vessel in the accelerator in the early days of parasitic use of electron synchrotrons, it is now a sophisticated construction where the source, beamline, and monochromator with associated optics are combined to create a fully integrated system. The advent of purposely-built synchrotron radiation sources, with parameters that have been optimized to provide a stable light source, has required corresponding care in beamline design to make the best use of the radiation available.

Although no beamline can be described as typical—beamlines vary considerably depending on which research program they are intended to serve—there are nevertheless certain general aspects that are common to all beamlines. These can be summarized as follows:

Optical matching to the source
Interface to the accelerator vacuum system
Radiation protection
Stability of components
Alignment and mechanical adjustment

2.2 Design Procedure

These topics were covered in some detail in a review of a few years ago [1], but the technology of the synchrotron radiation source, and of the dispersing focusing elements, has continued to advance. The basic design principle remains the same, however: Decide on the wavelength range to be covered and on the characteristics required for the experiment envisaged. For example, are there requirements for high photon flux, high resolution, and high spectral purity? Is a scanning instrument or a static (spectrograph type) instrument required, and is constant deviation essential? Rarely can all of these requirements be met simultaneously, and compromises have to be made. The design process begins with the wavelength range required, the central element being the diffraction grating. The focal equations for this and their solution using Fermat's principle have long been known [2]:

$$\frac{\cos^2 \alpha}{r} - \frac{\cos \alpha}{R} + \frac{\cos^2 \beta}{r'} - \frac{\cos \beta}{R} = 0 \quad \text{dispersion plane,} \quad (1)$$

$$\frac{1}{r} - \frac{\cos \alpha}{R} + \frac{1}{r'} - \frac{\cos \beta}{R} = 0 \quad \text{sagittal plane.} \quad (2)$$

where the various parameters are shown in Fig 1. In general, three different types of diffraction gratings are used:

1. *Spherical concave grating*: One solution to Eq. (1) is $r = R \cos \alpha$, $r' = R \cos \beta$, which is used in the Rowland circle mount, widely used in concave grating spectrometers. For the far VUV, that is, for photon energies greater than 30 eV where normal incidence mounts are inefficient, so-called spherical grating monochromators (SGMs) have been developed. In these instruments scanning of the instrument is achieved just by rotating the grating about an axis through its pole parallel to its rulings (axis z in Fig. 1). In effect this rotates the Rowland circle and thus the spectrometer immediately goes out of focus, but with the high ruling densities and the large size which can be accommodated on modern synchrotron radiation sources, the large dispersion available still provides adequate resolution.

2. *Toroidal concave grating*: Such gratings correct for the astigmatism inherent in a grazing incidence mount by having different radii of curvature R for the dispersion (or meridional) plane and the plane perpendicular to it, the sagittal plane. The first synchrotron radiation (SR) source on which they were used was

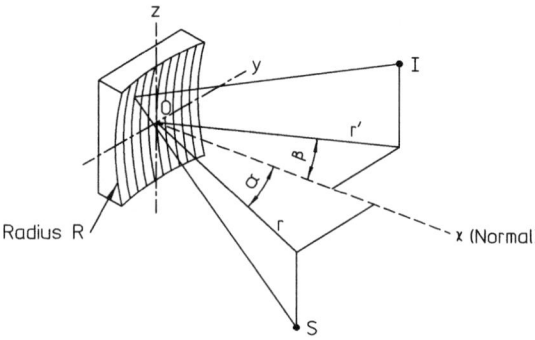

FIG. 1 Concave diffraction grating, imaging a monochromatic source S to an image I.

the NBS storage ring [3], resulting in a very simple toroidal grating monochromator (TGM). They are highly efficient, but are limited in general to medium-resolution requirements, because the aberrations, primarily astigmatic coma resulting from "coupling" between the two radii of curvature, dominate their performance particularly if large apertures are required. There are well-established optimization procedures for the design of such instruments [4], in which the various aberrations are minimized over the chosen wavelength range. However, TGMs have been largely supplanted now by SGMs, for two main reasons.

The first reason is that the requirement for focusing in the two perpendicular planes is met by using a "separated" or "crossed" optical system, in which the spherical grating provides the focusing in the plane of dispersion and a pre-mirror, usually also spherical, provides focusing in the sagittal plane. This has the advantage of removing the astigmatic coma aberration by decoupling the two radii of the equivalent toroid, and also allows for different demagnifications in the two planes. This can be an advantage, particularly for high-resolution applications where the source may need more demagnification in the dispersion plane, in order to match it to a small entrance slit.

The second reason is that, with the increasing demand for high resolution, the surface figure of the optical components is becoming a limiting factor, particularly in view of the high brilliance (small source size and divergence) and stability available from the modern generation of SR sources. Such tolerances, of less than 1 arcsec, can be achieved at reasonable cost only for spherical surfaces. Furthermore, with increasing interest in higher photon energies (100 eV and higher), grazing incidence (large values of α and β) is required. Inspection of Eqs. (1) and (2) will reveal that there is very little sagittal focusing in such cases, so it is straightforward to separate the focusing for the two different planes.

Offsetting the use of separated optical components in this way is the fact that more reflecting surfaces are required, placing higher demands on the reflectivity of those surfaces and the accuracy of their positioning relative to each other. Substantial improvements on the smoothness of optical surfaces (better than 5 Å rms can now routinely be achieved) and on the alignment and mounting stability of optical components have largely overcome this disadvantage.

3. *Plane grating*: These are an alternative to SGMs for use at the high-energy end of the VUV, where particular properties such as suppression of higher orders or the ability to remain "on blaze" are required.

The wavelength range to be covered by the beamline will generally determine the value of the included angle ($\alpha + \beta$), simply from reflectivity considerations, and this value will be the limiting factor for all reflections in the beamline. The resolution required effectively determines the scale of the instrument by setting the value of R and also of d, the grating spacing contained in the well-known grating formula:

$$n\lambda = d(\sin\alpha + \sin\beta), \tag{3}$$

where n is the spectral order and λ the wavelength. Both R and d determine the wavelength dispersion of the system. The scale and overall layout of the system having been decided, further steps can be taken to minimize the aberrations, primarily coma and spherical aberration. The former can to some extent be corrected by careful choice of mirror combinations; because the latter is very much a function of the aperture accepted by the optical system, the highly collimated beams from undulators on present SR sources are of great benefit in reducing it.

From this point on the layout of the monochromator and its associated prefocusing and postfocusing optics will be determined by physical constraints, such as the space available, and by experimental requirements. Constant deviation is usually essential, because most experiments cannot easily be moved along a focal curve or follow a moving exit beam. This is one of the primary reasons for the choice of a simple rotation mechanism for the diffraction grating. Experimental requirements vary considerably in terms of physical parameters such as spot size required on a sample, beam divergence, and beam height, so in these respects there is no typical beamline but the design considerations outlined earlier will be typical.

2.3 Beamline Example

The following example, for an undulator beamline on the ASTRID storage ring at the Institute for Physics and Astronomy at Aarhus in Denmark, shows how the preceding design process was carried out, leading to the layout which was eventually built and installed.

The experiment for which this beamline was designed was the measurement of absolute photoionization cross sections of atomic ions. The method to be used was the merged ion and photon beam technique [5], in view of the low density of the ion beam. A merged pathlength of up to 1 m was required, and the photon energy range to be covered was 40 to 200 eV. High resolution was not a priority, a resolving power over this range between 500 and 1000 being specified. However, in view of the fact that absolute measurements were to be made, order sorting was a priority. This biased the design toward a plane grating instrument based on the Miyake principle[6], which has been established to provide an adequate degree of higher order rejection [7], is of relatively simple design, and has constant deviation; in fact, this instrument produces an output beam parallel to the (horizontal) input beam. The main compromises in using such a design were resolving power, vertical acceptance, and dependence of the resolution on the source size: This design has no entrance slit. For the experiments envisaged, the resolving power was adequate; the limitation on vertical acceptance could, however, have been a serious disadvantage. On a bending magnet beamline, this would have resulted in very poor use of the available radiation because the instrument cannot be placed close to the source: Its focusing conditions demand that it be placed at least 10 m from the source. On an undulator beamline this situation is far less serious due to the much smaller opening angle of the radiation. This will be seen readily during the analysis of our example beamline, taking a photon energy of 125 eV in the middle of the intended range.

The beamline layout shown in Fig. 2, together with the physical parameters for the various components, was chosen to meet the requirements outlined. There are two focal positions for the mirror M2, for the geometry shown, giving two ranges of order sorting as pointed out by Miyake *et al.* [6]. The position for the high-energy range, 100 to 180 eV, corresponds to the dimensions given. Mirror M1 is designed to focus the source in the middle of the 1-m-long interaction region, after the monochromator as shown. Mirror M2 focuses the

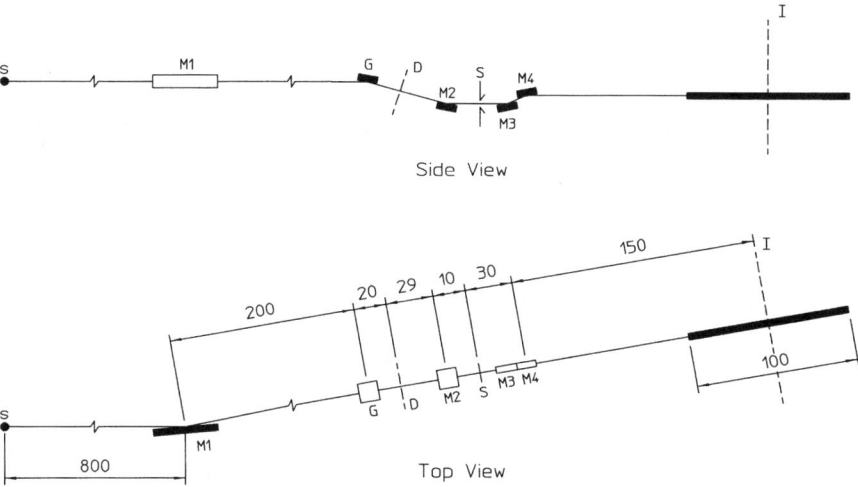

FIG. 2 Layout for the undulator beamline at the ASTRID storage ring at the Institute of Physics and Astronomy, Aarhus University, Denmark. G: diffraction grating, 1200 lines/mm, dimensions 2.5 × 2.5; combined angle of incidence plus diffraction: variable over the range of 130° to 170°, 162° for the position shown. M1: sperical mirror, radius 15,000, dimensions 2 × 25; angle of incidence 87.5°. M2: cylindrical mirror, tangential radius 240, dimensions 2.5 × 2.5; angle of incidence: variable over the range 65° to 86°, 81° for the position shown. M3: plane mirror, dimensions 8 × 12; angle of incidence 85°. M4: spherical mirror, radius of curvature 690, dimensions 8 × 12; angle of incidence 85°. See text for relevance of planes D and I; all dimensions in centimeters unless otherwise stated.

source, following dispersion by the diffraction grating G, onto the exit slit. Mirror M4, in effect focusing in the vertical direction only because of the large value of the angle of incidence, is chosen to generate a parallel beam with the exit slit as its source. M3 is a plane mirror used in combination with M4 to provide a horizontal beam.

In this way a beam of low divergence over the 1-m pathlength of the interaction region should be achieved, but the presence of both spherical aberration and coma will affect this. The resolution can be derived by differentiating the grating equation, and in the first diffraction order is given by

$$\left(\frac{d\lambda}{ds}\right)_{\alpha=\text{const.}} = \frac{d}{F}\cos\beta, \qquad (4)$$

where F is the distance from the diffraction grating to the exit slit, and s refers to the slit width. This equation assumes that the diffraction grating and focusing element are in the same place, for example, a concave diffraction grating. When they are not, as in this case, the equation has a (usually small) correction term [1], and becomes

$$\left(\frac{d\lambda}{ds}\right)_{\alpha=\text{const.}} = \frac{d}{F}\left(1 + \frac{P}{r'}\right)\cos\beta, \qquad (5)$$

where P is the distance between the plane grating and focusing mirror and F is now the distance from the mirror to the exit slit; r' is the virtual source to grating distance calculated from Eq. (1), remembering that for a plane grating $R = \infty$. The source size (FWHM) at ASTRID is 4 mm horizontally × 0.57 mm vertically; Eqs. (1) and (5) give a dispersion of 5.7 Å/mm. Using Eqs. (1) and (3) to calculate the size of the focused image of the source on the exit slit yields a value of 0.026 mm, giving a resolution of 0.15 Å or a resolving power of ~660 at 125 eV.

2.4 Use of Ray Tracing

Finally, because in this region of the spectrum the optics are not diffraction limited, the whole system can be checked for its efficiency by using a geometric ray-tracing program such as SHADOW [8], and further optimization carried out. Using these procedures, and the considerable amount of data available on the optical constants for the reflecting materials used in the VUV spectral region [9, 10], it is possible to predict with some certainty the performance of a particular beamline design. In Fig. 3a the spot diagram shows rays incident on a plane perpendicular to the beam between the grating and M2, the screen D in Fig. 2, for the undulator source on ASTRID. It is worth noting that the low divergence of the undulator radiation matches the vertical aperture of this monochromator quite well, as can be seen from the number of rays lost. For a bending magnet source on ASTRID 82% of the light would be lost, almost all of it in the vertical direction. Figure 3b shows the monochromatic image on the exit

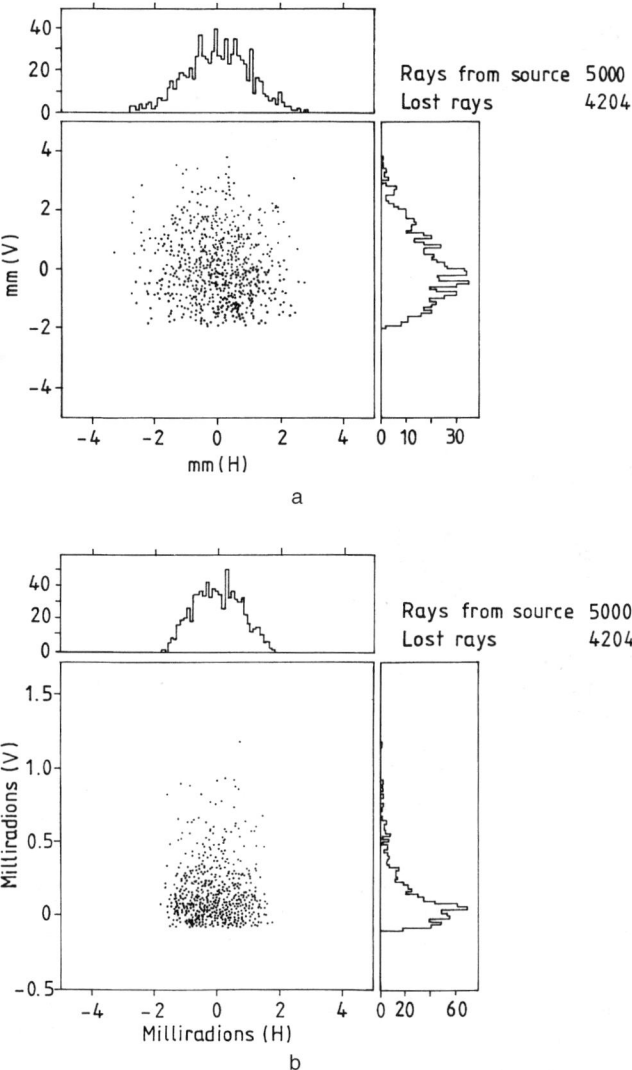

FIG. 3 (a) Spot diagram showing the spatial distribution of the image at plane D in Fig. 2. V, vertical dimensions; H, horizontal dimensions. (b) As in part (a) but at the exit slit plane S in Fig. 2.

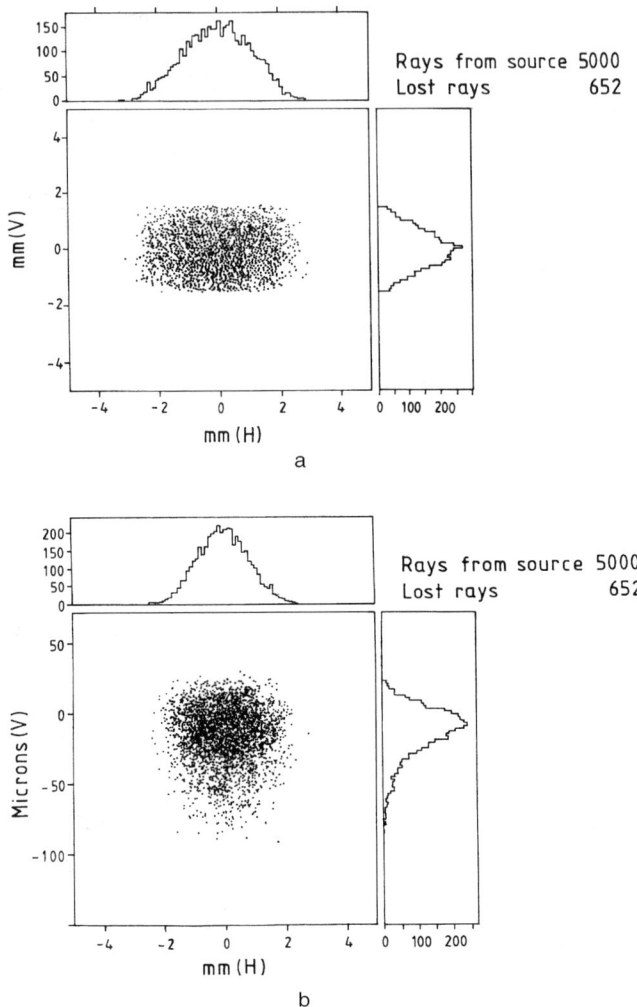

FIG. 4 (a) Spot diagram showing the spatial distribution of the image at plane I in Fig. 2. V, vertical dimensions; H, horizontal dimensions. (b) As in part (a) but showing the angular divergences of rays passing through the plane I.

plane of the monochromator, where it can be seen that the vertical image size is close to the one calculated, defining the source size limited resolution.

Because dispersion across the exit plane is continuous, an exit slit has to be placed there to select the wavelength required. There is, of course, no advantage to be gained in making this slit smaller than the defocused image of the source, 0.03 mm (FWHM) in this case. Fig. 4a shows the image size in the middle of the ion-beam interaction region, I in Fig. 2, and Fig. 4b the angular divergences of the rays passing through the plane I, for a slit width of 0.03 mm. For these ray traces a polychromatic source was used, and the large number of rays lost is due to the fact that the slit selects only a small fraction of these corresponding to the bandwidth for which it has been set. Although the coma aberration is clearly evident in Fig. 4a, it can be seen that a beam cross section of 2×2 mm, parallel within 1 mrad over the 1-m pathlength required, has been achieved.

The purpose of the ray-tracing procedure is to make fine adjustments to the dimensions of the optical system, optimizing it to meet the required specification. It is also useful in highlighting the effects of aberrations, since inspection of the spot diagrams can reveal the nature of these. Steps can then be taken to minimize them: In the case of spherical aberration, this can be reduced by reducing the acceptance aperture. In general, coma can be reduced by choosing subsequent refocusing optics appropriately [11]. The requirement for a horizontal output beam prevented such a solution from being applied in the case of the example given here, and is typical of the compromises that must be made when designing a beamline system.

2.5 Ancillary Components

With the basic layout decided—although much remains to be done—the remainder of the design is straightforward. Careful thought in the design and siting of the various ancillary components will have a pronounced effect on the ease of use of the beamline. The following is a summary of some of the most important aspects.

It is necessary to calculate the heat loading on the optical components, particularly those first in the beam, to determine whether cooling is required and which materials should be used. For the ASTRID source no cooling is required, but on other sources where a large x-ray flux must be absorbed it is essential, particularly if a VUV beamline is being installed on a high energy storage ring. Also, where high-resolution is a priority, implying high precision and stability in the optical components, temperature uniformity and control will be essential, and in the engineering design the use of finite analysis techniques has proved its worth. The beamline should also include cooled apertures and slits to collimate the radiation, to reduce the scattered light component in the output. As a general rule, vacuum valves should be placed on either side of major optical components such as the first mirror and the monochromator, so that access to these is possible without letting the entire beamline up to atmospheric pressure. A fast flap valve, closing in a few milliseconds, is generally incorporated at the storage ring end of the beamline in the event of a vacuum accident in the beamline itself. An acoustic delay line may also be added for further protection. A beam height monitoring system, which detects the central part of the beam, is also essential to assist in the steering of the electron beam in the storage ring; nowadays these are included in a closed-loop alignment system to ensure that the beamline is correctly illuminated. A means of surveying the beamline into the correct position relative to the source is also needed, and can be provided using external marks or reference plates on the chambers containing the optical components. These can then be viewed using accurate telescopes and leveling devices to locate components accurately using standard surveying techniques. In some instances, it may also be helpful to install a laser alignment permanently in the beamline.

Radiation protection is a major concern, with two major hazards in particular being present: exposure to scattered x-rays and to high-energy brehmsstrahlung. Fortunately, in the VUV all the equipment can be contained in a stainless steel vessel under vacuum, which also contains the soft x-ray scatter. However, windows and bellows usually require additional shielding, and vacuum interlocks are used to prevent exposure to x-rays while working on any internal part of the beamline at atmospheric pressure. High-energy brehmsstrahlung are generally contained along the central axis of the beamline, so there must be substantial shielding at the end of the beamline and also an interlock system preventing access to central components when the beam is on. Again, the VUV region has an advantage in this respect because the beam is generally deflected away from the central axis by the first mirror, making work on the following monchromator or experiment less hazardous. Even so, radiation monitoring of a beamline after construction is essential to ensure safe working conditions. Other hazards, such as exposure to VUV radiation directly, and injury from remotely controlled mechanisms, are also present and require protection measures.

References

1. J. B. West and H. A. Padmore, "Handbook on Synchrotron Radiation," Vol 2 (G. V. Marr, ed.), Elsevier, Amsterdam, p. 21 (1987).
2. T. Namioka, *J. Opt. Soc. Am.* **49**, 440 (1959); **49**, 951 (1959).
3. R. P. Madden and D. L. Ederer, *J. Opt. Soc. Am.* **62**, 722A (1972).
4. W. R. McKinney and M. R. Howells, *Nucl. Instrum. Methods* **172**, 149 (1980).
5. I. C. Lyon, B. Peart, J. B. West, and K. Dolder, *J. Phys. B: Atom Mol. Phys.* **17**, L345 (1984).
6. K. P. Miyake, R Kato, and H. Yamashita, *Sci. Light* **18**, 39 (1969).
7. J. B. West, K Codling, and G. V. Marr, *J. Phys. E: Sci. Instrum.* **7**, 137 (1974).
8. B Lai, K Chapman, and F Cerrina, *Nucl. Instrum. Methods A* **266**, 544 (1988).
9. H. J. Hagemann, W. Gudat, and C. Kunz, Deutsches Elektronen Synchrotron Int. Rep. DESY SR74-7 (1974).
10. B. L. Henke, P. Lee, T. J. Tanaka, R. L. Shimabukuro, and B. J. Fujikawa, *Atom. Data Nucl. Data Tables* **27**, 1 (1982).
11. T. Namioka, H. Noda, K. Goto and T. Katayama, *Nucl. Instrum. Methods A* **266**, 544 (1988); see also Ref. 1, p. 54.

3. GLOW DISCHARGES AND WALL STABILIZED ARCS

James R. Roberts
Contribution of the National Institute of Standards and Technology
Gaithersburg, U.S.A.

3.1 Introduction

This chapter of the part on Laboratory Sources deals with glow discharges, dielectric barrier discharges, wall stabilized arcs, electron cyclotron resonance sources, electron-beam ion trap sources, inductively coupled plasma sources, and capillary discharges. Descriptions of hollow cathode, Penning, and electron-beam excitation discharges are presented in other chapters. All of these discharges may have different ways to excite bound atomic or molecular states and to ionize constituents of the discharge, but they will all radiate both spectral line radiation and continuum radiation. In some cases, the source conditions can be controlled so one or the other of these types of radiation may dominate. The choice of a source often depends on which type of radiation is desired. For example, if a source exhibits a stable well-characterized continuum spectrum in the vacuum ultraviolet VUV) region, then this source could be used as a secondary radiometric calibration standard throughout this spectral region [1]. If the discharge emits primarily narrow, well-known spectral lines, then it might be used as a wavelength calibration source for spectrometers. If the source emits spectral lines of certain shapes (spectral profiles), then definite characteristics of the plasma discharge may be determined. For example, if the plasma exhibits line shapes with a Gaussian profile, this may be due to the random thermal motion of the emitting atom (or ion), which is proportional to its thermal temperature [2]. Likewise, other spectral line shapes may indicate broadening by plasma electrons, atoms, and ions, and if an appropriate theoretical description is available, a plasma electron density may bedetermined [3].

Some of the sources discussed in this chapter were designed and used as sources of spectra, that is, both line and continuum spectra as in the case of the wall stabilized arc. However, some sources that are described were designed for uses other than sources of spectra; for example, the electron cyclotron resonance tool has been primarily used as an ion source.

3.2 Glow Discharges

A source is described in Refs. [4], [5], and [6] that produces enhanced neutral and ion line emission in the VUV spectral region by exploiting the principle of the charged particle oscillator [7]. The original use of the charged particle oscillator was the production of ion beams. Comparison of the ion production efficiency of this source with a similar ion source is reviewed in Ref. [8] and its use as an electrostatic ion gun with different modes of operation is detailed in Ref. [9].

The nature of the source is that free electrons inside a cylindrical electrode (cathode) follow oscillatory trajectories, passing between two wires (anodes) which are coaxial to the cylinder. Because of this oscillating mode, the electron collisional mean free path can be much longer than the dimensions of the cylinder. This permits the discharge to be maintained at low pressures in the space between and orthogonal to the plane of the wires.

As described by Ref. [5] (see Fig. 1), this device was utilized as a source in photoelectron spectroscopy and consisted of 0.5-mm-diameter tungsten wires as anodes, separated by 3 mm, held under tension by a spring, concentric to a cylindrical, 16-mm-i.d., stainless steel cathode. The wires are insulated from the cylinder by boron nitride spacers.

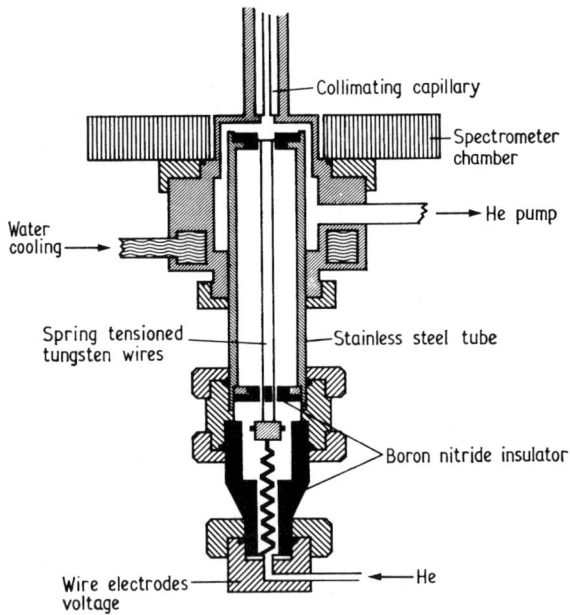

FIG. 1. Schematic diagram of VUV, cold cathode, glow discharge source for photoelectron spectroscopy. (Reprinted with permission from the *Journal of Physics E (Scientific Instruments)*, Vol. 8, no. 5, pp. 420–422 (1975), F. Burger and J. P. Maier, Figure 1 of "Vacuum ultraviolet source of line radiation of the rare gas ions suitable for photoelectron spectroscopy," Institute of Physics Publishing Ltd.)

The discharge operates in two distinct modes: (1) as a glow discharge at pressures >0.1 Pa and (2) in an oscillating mode at pressures <10^{-2} Pa with a transition mode of operation between these pressures depending on the type of gas. Typical oscillating mode discharge conditions, with currents of 10 to 25 mA, a 0.6- to 2.5-kV anode potential, and a pressure of 0.003 to 0.01 Pa, produced spectra of He I, He II, Ar II, Ar III, Ne I, Ne II, and Ne III. An improved, water-cooled variant, based on prototypes [4, 5], was also built to improve stability and long-term behavior [10].

The photoelectron spectrum of H_2O using Ne in the discharge is shown in Fig. 2 [5]. Although these are photoelectron spectra, it demonstrates the source as a generator of the line spectra of several resonance transitions in rare gases and their ions. This source, operating in the oscillating mode with Ne, shows a dominant Ne II spectrum as seen in the portion of Fig. 2 labeled C. This spectrum was obtained with a discharge current of 10 mA and a 2.5-kV anode potential at 0.01 Pa. This source also functioned in the ion trapping mode, with

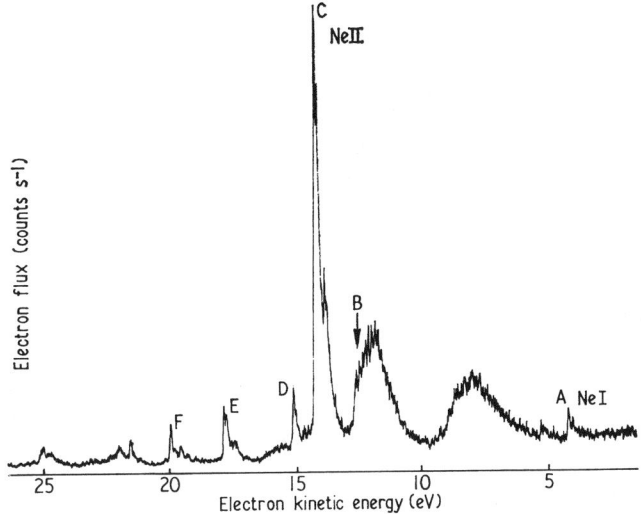

FIG. 2. The photoelectron spectra of H_2O from Ne emission produced by a cold cathode, glow discharge. The bands correspond to the $^2B_1$ state of H_2O^+ and refer to the following photoionizing lines: A, Ne I (73.6 nm); B, Ne III (49.0 nm C, Ne II (40.1 and 46.2 nm); D, Ne II (44.6 nm, average value); E, Ne II (40.6 and 40.7 nm); F, Ne III (37.9 nm). The spectrum was obtained with a discharge current of 10 mA, a 2.5-kV anode potential, 0.01-Pa pressure, and 500 counts/sec for the maximum of band C. (Reprinted with permission from the *Journal of Physics E (Scientific Instruments)*, Vol. 8, no. 5, pp. 420–422 (1975), F. Burger and J. P. Maier, Figure 3 of "Vacuum ultraviolet source of line radiation of the rare gas ions suitable for photoelectron spectroscopy," Copyright 1975, Institute of Physics Publishing Ltd.)

the central wires operated as cathodes and the cylinder as the anode. At a pressure of 0.27 Pa and 0.5 kV in He or at 0.04 Pa and 1.8 kV in Ne, atomic line radiation dominated.

In the case of Ref. [6], the source uses tungsten rods and water cooling, which facilitate power dissipation for higher He I and He II resonance line intensities. This source was operated in transition mode [9] also for photoelectron spectroscopy. By means of a differential pumping system, the pressure of the spectrometer chamber was maintained at a pressure of 2.7×10^{-8} Pa while the source operated in this mode. This source produced very high intensities of He I and He II lines with discharge currents of 120 mA and an anode potential of 3.0 kV. Experiments in He with constant discharge currents show the abrupt change between operating modes for this source as a function of pressure. Figure 3 [6] shows a sharp change in operation of the source at approximately 13 Pa, demonstrating the transition between the two modes for He discharges.

Other types of devices utilized as ion sources, which emit VUV radiation by electron impact ionization and excitation of gases at low pressures, have also been described. As an example, a source used to study H^- ions [11] exhibited

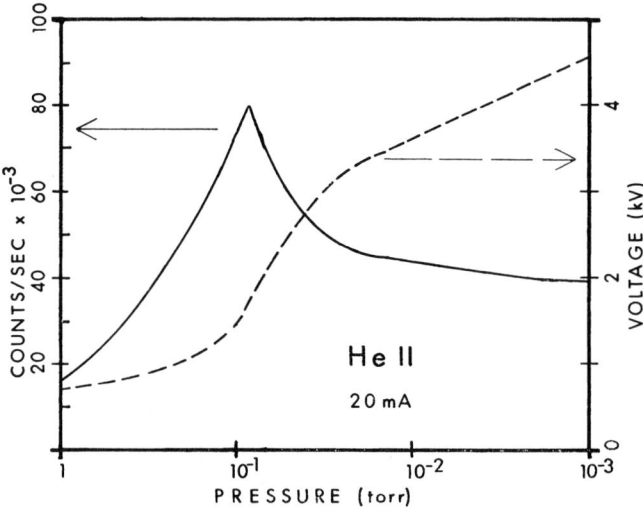

FIG. 3. The He II photoelectron count rate and voltage across a cold cathode, glow discharge source as a function of discharge pressure at a constant current of 20 mA. (Reprinted from the *J. Electron Spectrosc. Phenom.*, Vol. 14, pp. 143–153 (1978), G. Lancaster, J. Taylor, A. Ignatiev, and J. Rabalais, Figure 5 of "Vacuum ultraviolet resonance line radiation source from rare gas atoms and ions for uhv photoelectron spectroscopy," with kind permission of Elsevier Science—NL, Sara Burgerharstraat 25, 1055 KV Amsterdam, The Netherlands.)

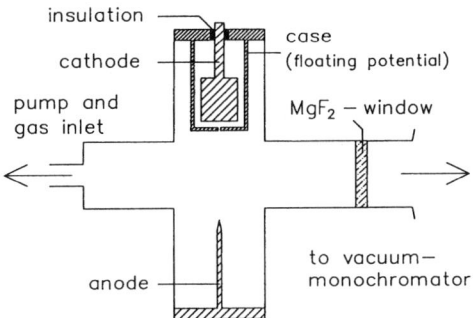

FIG. 4. Schematic diagram of VUV, constricted glow discharge source. (Reprinted with permission from the *Rev. Sci. Instrum.*, Vol. 59, pp. 565–568 (1988), D. Lindau and H. Döbele, Figure 1 of "The constricted glow discharge: A stationary source of vacuum-uv rare-gas excimer continua." Copyright 1988 American Institute of Physics.)

the typical VUV spectrum of molecular hydrogen in the wavelength region from 80 to 170 nm [12].

A VUV source used to produce rare gas excimer continua is reported in Ref. [13]. This source is described as a constricted glow discharge. A diagram of this source is shown in Fig. 4. It operates in the pressure range from 0.1 to 1.0 Pa. The unique characteristic of this discharge is the development of a sharp plasma boundary as a result of the constriction. There is a critical current for the constriction, below which the discharge is diffuse. This current is dependent on the type of gas and its filling pressure. If the source electrodes are mounted vertically, the constricted column forms a filament between the two electrodes. Typical electrode separation is varied up to 100 mm. The radiation from this discharge consists of the line spectra characteristic of the filling gas as well of excimer continuum radiation when rare gases are used. The power requirements are typically less than 1 kV and up to 50 mA. In the transition between a diffuse glow discharge to the constricted mode, the voltage–current characteristics change from a negative dV/dI to a positive one. Therefore, a voltage or current regulating power supply is desirable.

3.3 Dielectric Barrier or Silent Discharges

The origin of the dielectric barrier discharge, or silent discharge as it is also called, is attributed to Siemans in 1857 [14]. Its use as a source of VUV radiation ranges from the generation of excimer radiation (see, e.g., Refs. [15], [16], and [17]) to the production of ozone for industry, (see, e.g., Refs. [18] and [19]). It has also been used as a source for metal deposition [20]. A summary of the concepts, modeling, and applications of this unique discharge are presented in, for example, Refs. [21] and [22].

The dielectric barrier discharge is a nonequilibrium discharge created between two electrodes, at least one of which is covered by an insulator. The spacing between the electrode is typically a few millimeters and the pressure of operation ranges from 0.01 to 10 MPa. As voltage is applied to the gap, breakdown is attributed to the buildup of high local electric fields in the space charge along the surface of the insulator. This can produce an avalanche of electrons resulting in a thin, nearly cylindrical conductive filament from the cathode to the anode (see, e.g., Refs. [22] and [23]). These filaments are each approximately 100 μm in diameter. The local breakdown, and resultant current, usually lasts from 1 to 100 ns. These filamentary discharges occur over the entire surface of the insulator, and when a time-varying voltage is applied to the electrodes, breakdown can be induced several times per cycle depending on the voltage amplitude and the waveform. Thus, when the discharge is viewed between the electrode–insulator–electrode space, it is the integrated output of an enormous number of individual filamentary discharges. Each discharge will transport tens to thousands of picoCoulombs of charge with a current density of several hundred amperes per square centimeter. This results in a plasma with characteristics of electron density from approximately 10^{14} cm^{-3} to greater than 10^{15} cm^{-3} and with mean electron energies ranging from 1 eV to greater than 10 eV. Plasmas characterized by these types of conditions produce a prominent spectral output in the VUV region. Typically, a sinusoidal voltage, ranging from a few hundred volts to several kilovolts in amplitude, is applied with a frequency of a few hertz up to several megahertz. More typically the frequency used is from a few kilohertz to a few hundred kilohertz. Use in this range is due primarily to the convenience and cost of available power amplifiers and transformers. Also it has been shown at higher frequencies that heating of the dielectric material will become important [24].

A typical electrical circuit arrangement is depicted in Fig. 5 [25]. The generator is a typical function generator and the audio amplifier is a unit capable of delivering several watts of output power, for example, 100 W. The high-voltage (HV) transformer in this type of experiment is the output transformer from a HV switching power supply capable of delivering several kilovolts. The dielectric discharge for this experiment was an open-ended, 25-cm-long,

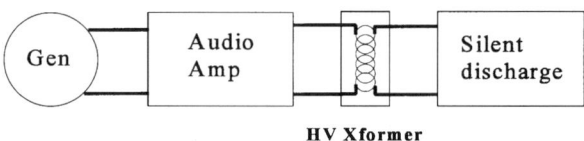

FIG. 5. Schematic of the electrical circuit for a dielectric barrier discharge.

1-mm-diameter glass capillary tube with one end for gas input. The tube was coated with a conducting paint on one-quarter of its surface on two opposite sides for nearly the entire length of the tube. Wires were mechanically attached for the electrical connections. The discharge occurred within the capillary by flowing gas through the tube, thus generating a 25-cm-long, 1-mm-diameter plasma. The preliminary experiments were done with rare gases.

In the case of rare gas dielectric barrier discharges, the spectrum consists of resonance spectral line and excimer continua. The generation of the various spectral components of a Xe dielectric barrier discharge is depicted in Fig. 6 [17]. The spectral output of this type of glow discharge is shown in Fig. 7 [20], which shows the pressure dependence of the emission spectral distribution of a dielectric barrier discharge in Xe. In general for the rare gases, when the pressure is lowered substantially, for example, from 0.1 MPa to approximately 0.01 MPa, the maximum of the spectral distribution shifts toward shorter wavelengths and the 1st excimer continuum dominates over the 2nd. To obtain resonance spectral lines without self-absorption, the pressure of the discharge must be reduced to pressures below 100 Pa. At pressures between this lower range and 0.1-MPa pressure, the spectral content consists of resonance radiation and the 1st excimer continuum, as well as contributions from the 2nd excimer continuum as demonstrated in Fig. 6. Also, it has been demonstrated that the spectral distribution of the excimer continua of a single rare gas can be modified by the introduction of two or more rare gases [26]. In this case the spectral distribution was considerably broadened to approximately 30 nm FWHM.

The geometry of the discharge can take many forms. The dielectric barrier can be placed in contact with the HV electrode and the plasma will fill the space

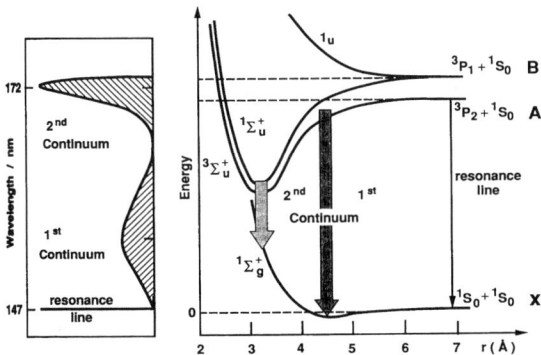

FIG. 6. Partial potential energy diagram of Xe with corresponding 1st and 2nd excimer continua. (Reprinted with permission from *Appl. Phys. B*, Vol. 52, pp. 14–21 (1991), B. Gellert and U. Kogelschatz, Figure 5 of "Generation of excimer emission in dielectric barrier discharges." Copyright 1991 Springer-Verlag, New York, Inc.)

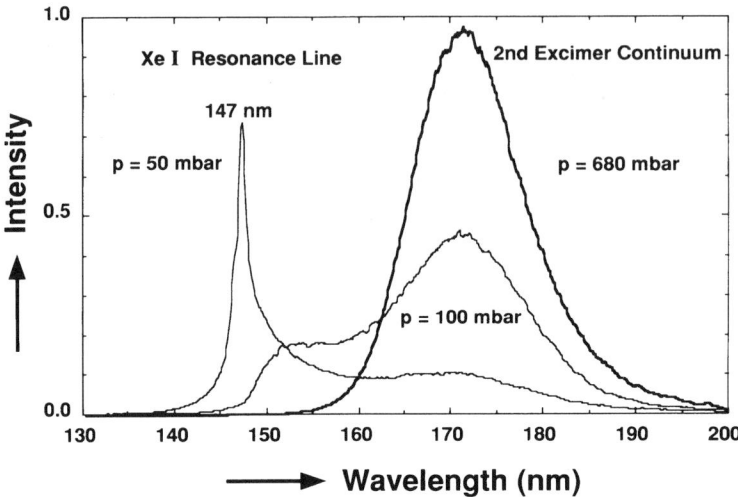

FIG. 7. Pressure-dependent emission spectra from a dielectric barrier discharge. (Reprinted from *Appl. Surf. Sci.*, Vol. 54, pp. 440–444 (1992), H. Esrom and U. Kogelschatz, Figure 3 of "Metal deposition with a windowless vuv excimeter source," with kind permission of Elsevier Science—NL, Sara Burgerharstraat 25, 1055 KV Amsterdam, The Netherlands.)

between the dielectric surface and the ground electrode. The dimension of this space varies, usually a few millimeters, and depends on the type of gas and the gas pressure. Other configurations are possible and several different geometrical concepts are outlined in Ref. [27]. A configuration for large-area illumination is depicted in Fig. 8 [20]. One of the attractive characteristics of a dielectric barrier discharge is the versatility of geometric design. One can imagine slit-like configurations for optimizing a spectrometer throughput to multiparallel and large-area arrays for illumination and all of it possible in the VUV spectral region.

An experiment using a flat-panel configuration [28] depicts a power source consisting of a sinusoidal signal generator (50 Hz to 100 kHz), several 100-W power amplifiers, a tunable matching network, a HV setup transformer capable of delivering 20 kV_{p-p}. A buffer capacitor is utilized in this circuit in parallel to the lamp to prevent a significant drop after the lamp is ignited. These experiments with this large-area dielectric discharge in Xe of 40-mm-diameter × 5 mm gap, pressure of 2×10^4 Pa, $V_{p-p} = 3.8$ kV, demonstrated an average surface density of 2 filaments/cm^2. The 172-nm Xe excimer radiation was observed at two different frequencies, 5 and 22 kHz. The results indicate that at 5 kHz the filament current pulses were <100 ns in width and preceded the light output by approximately 500 ns. For 22 kHz the current pulses were approximately 1 μs in duration and overlapped the excimer light emission, which was essentially the same

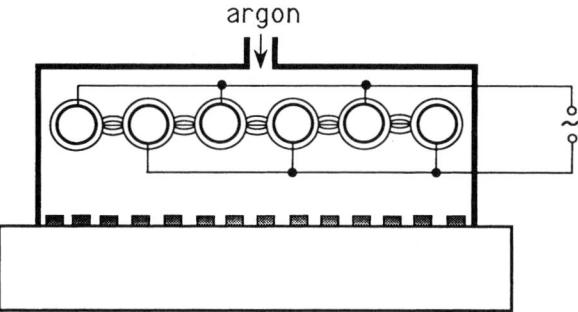

FIG. 8. Schematic diagram of a multielectrode, windowless dielectric barrier discharge configuration for generation of large-area VUV excimer radiation applications. (Reprinted from *Appl. Surf. Sci.*, Vol. 54, pp. 440–444 (1992), H. Esrom and U. Kogelschatz, Figure 8 of "Metal deposition with a windowless vuv excimeter source," with kind permission of Elsevier Science—NL, Sara Burgerharstraat 25, 1055 KV Amsterdam, The Netherlands.)

temporal shape at each frequency. The results also indicate for this excimer continuum centered at 170 nm, the spectral distribution was not sensitive to pressure changes from 5×10^3 Pa to 8×10^4 Pa.

Because of the versatility in design configurations, as well as the variable spectral distributions, the dielectric barrier discharge also offers a wide variety of potential applications. For example, its use as a high-intensity VUV light source for selective photolytic dissociation can be applied in the deposition of amorphous semiconductor material [29]. Two very different dielectric barrier plasma sources were designed and experiments were performed to determine their effectiveness for this type of application. The first was a planar configuration with the powered electrode covered with quartz and the light is transmitted through a fine wire mesh ground electrode. The other configuration was a parallel plate device with 15 quartz-encapsulated electrodes. The power is applied to alternating plates and the plates between the powered ones were grounded. The light was emitted from one edge of the parallel plate configuration. The excimer spectra of Ar, Kr, and Xe, as well as Kr/Ar and Xe/Ar mixtures, were observed at various total pressures ranging from 5×10^4 Pa to 2×10^5 Pa and mixture ratios ranging from 0.003 to 0.25 at 2×10^5 Pa. As in another experiment [26], the excimer continua of each rare gas were obtained and widened spectral features were obtained with the rare gas mixtures.

3.4 Wall Stabilized Arc Source

The wall stabilized arc is capable of generating stable plasmas in the temperature range from 6000 to 25,000 K and electron densities up to 1×10^{18} cm^{-3}.

It has been used to investigate the properties of plasmas, including their stability. Their high radiative outputs have made wall stabilized arcs an important tool in the studies of thermal equilibrium, atomic physics, and quantitative spectroscopy. The details of arc properties have been summarized in Refs. [30] and [31]. The development of the modern wall stabilized arc and its modular, insulated disk form is attributed to Maecker [32]. The extension of the wall stabilized arc into the VUV spectral region was pioneered by Boldt [33] and the introduction of a differential pumping system was introduced by Morris and Garrison [34]. A detailed cross section of a wall stabilized arc design is shown in Fig. 9 [35].

The column in which the arc is formed consists of several, liquid-cooled disks (usually copper), with a coaxial hole of a few millimeters in diameter, each disk separated by insulating ring gaskets. These gaskets can be made of any insulating material that can tolerate contact with the hot gas of the nearby arc and function as gastight seals. Gases (single components or admixtures) are introduced into the arc through these insulating disks as well as the electrode regions (see Fig. 9), usually at atmospheric pressure. The gas flow is usually controlled by the use of flow meters with typical flow rates ranging from 1 to 5 μmol/s. Care must be taken not to have the flow rate too small or the external atmospheric constituents may diffuse into the arc channel. Also if the flow rate is too large, the arc may be pressured to greater than the external atmospheric pressure. These can be critical factors in the emitted output since the radiation is a direct function of the gas constituents and pressure.

The arc is often started with a tungsten rod, which is either attached to or touching the grounded electrode. By pushing the rod into the arc channel until it contacts the powered electrode, the electrodes are momentarily short-circuited, and by withdrawing the rod slowly the arc is formed from one electrode to the other. Because there can be multiple sections for this style of wall stabilized arc, the region within the arc channel where the gases are introduced can be controlled. For example, if gases that may adversely affect the electrodes are to be studied, they can be introduced only into the midsection of the arc while other gases, such as argon, can flow near the electrodes. Care must be taken to adjust the respective flow rates and pressures so diffusion of the gases is not a factor in the length of the plasma to be investigated. Also in this way, gases that are difficult to ignite for the full length of the arc (e.g., helium) can be investigated. This condition can arise when the ionization potential of the gas to be investigated is large and the power requirements to maintain a full-length discharge are not appropriate for the power supply being used. If this were to occur, admixtures of gases can be used, but care should be taken not to produce toxic or explosive chemical by-products. Another consideration when admixing input gases to the arc is to be aware that there might be spectral lines produced from the plasma that would interfere with observations, and thus the interpretation of the results. A method of introducing material into the arc is the use

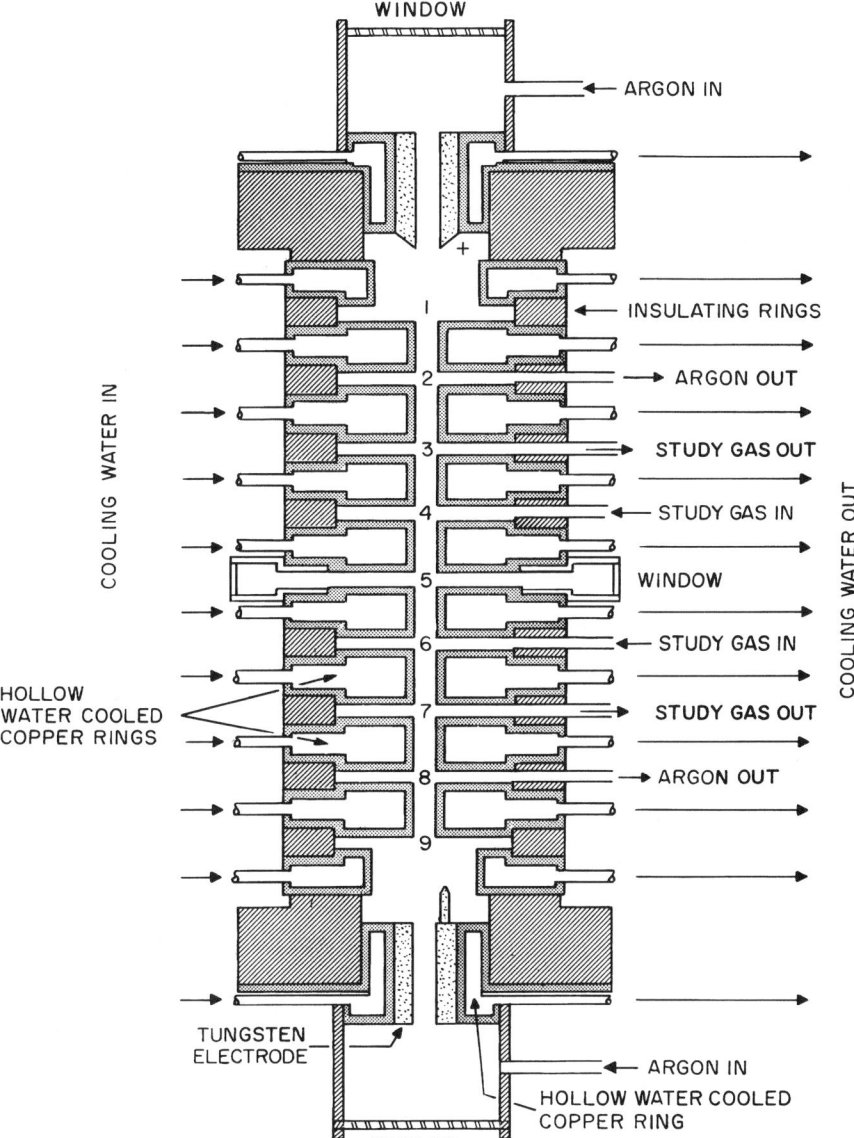

FIG. 9. Schematic diagram of a wall stabilized arc. (Reprinted with permission from the *Phys. Rev.*, Vol. 129, pp. 1225–1232 (1963), W. Wiese, D. Paquette, and J. Solarski, Jr., Figure 1 of "Profiles of Stark-broadened Balmer lines in a hydrogen plasma." Copyright 1963 American Physical Society.)

of the insulating rings. These rings can be designed to extend sufficiently near the arc that in practice they can contribute minute components to the arc constituents by evaporation. Thus, by adjusting the cooling flow to the conducting arc disks, a small controlled amount of impurity species can be introduced into the arc column by evaporation from the insulating disks. This can be a method of introducing known trace elements that are normally not gases or do not have a high vapor pressure at room temperature. An example of this would be boron nitride; even though gases containing boron exist (e.g., BF_3), they may be too toxic, produce interfering spectral lines, or destroy the electrodes. Another method is to pass the incoming gas through an airtight oven containing a volatile compound of the trace element of interest (e.g., Fe_2Cl_3) to obtain Fe. By adjusting the temperature of the oven, the vapor pressure of the compound will rise so the vapor will be carried along with the input gas into the arc chamber. Care has to be taken in the design of the oven so the vapor does not condense and clog the orifice in the side of the arc.

The radiation from the arc column can be observed either side-on through windows in the insulator rings or by observing end-on along the length of the plasma (see Fig. 9). Because the plasma will normally have radial distributions of the species densities and temperature, the side-on measurements must be made with sufficient spatial resolution to permit an Abel inversion procedure [35, 36] so the radial distributions can be determined. For the same reason, the apparatus making end-on measurements should have a small angular and physical extent so the gradients in the radial distributions do not affect the analysis of the observations.

The reproducibility and stability in the radiometric output of the wall stabilized arc are very dependent on the properties of the applied voltage and current. In general, arcs often require some form of electrical stabilization because of their voltage–current (V-I) characteristics. The V-I characteristics of a DC low-pressure electrical discharge are traditionally divided into three categories; the dark discharge, the glow discharge, and the arc (see Ref. [37], p. 253). The arc region is characterized by currents typically greater than 1 A, and by a V-I characteristic that may exhibit a negative derivative, dV/dI, for currents from this lower transition current to currents approaching 100 A. Because of this, unstable operation may occur and thus it may be necessary to include a ballast resistor in series with the arc to modify the electrical circuit. This introduces a positive derivative for stable operation in the region of interest. Even then dV/dI will usually be near zero, so only small changes in the power supply voltage are needed to make significant changes in the current. Therefore, well-regulated power is desirable for stable operation. In the case of a wall stabilized arc, if it were not for the wall stabilization of the arc column, the arc would expand its diameter to compensate for any increase in current. Instead, any increase in current changes the thermodynamic properties of the plasma. Thus, the stabilizing wall allows changes in the plasma properties as the current changes.

High voltages, high-current power supplies utilizing modern methods of regulating the current electronically, are the present power sources of choice for arcs demanding the most stabilized radiometric output. It is necessary for this type of power supply to respond sufficiently fast to fluctuations in the arc impedance, so the thermodynamic plasma properties are not significantly perturbed. This is entirely dependent on the use of the arc. A typical arrangement of the electrical circuit for a wall stabilized arc used as a transfer standard in radiometric calibrations is shown in Fig. 10. Because the arc current is one of the most sensitive parameters of the thermodynamic condition of the plasma, it is important to measure it continuously and with an accuracy commensurate with the arc's use. The current, I, is usually monitored by inserting a stable, high-accuracy resistor of 0.001 to 0.1 Ω and measuring the voltage drop across this resistor with an accurate digital voltmeter. The current through the arc is usually measured with sufficient accuracy (typically $1:10^6$) to keep it monitored within the power supply's regulating capabilities.

Arc plasmas usually exhibit a state of equilibrium of the plasma where the radiation is characterized by spectral lines and continuum emission. This is called local *thermodynamic equilibrium*, in contrast to total thermal equilibrium where the plasma radiation is dominated by black body-like radiation. Black bodies emit according to the Plank radiation law given by

$$dM_e(T)/d\lambda = 2\pi hc^2 \lambda^{-5} \times 1/[\exp)hc/(kT\lambda)) - 1] \quad (\text{W/m}^3), \qquad (1)$$

where M_e is the spectral exitance at the blackbody temperature, T, in watts per unit area per unit wavelength.

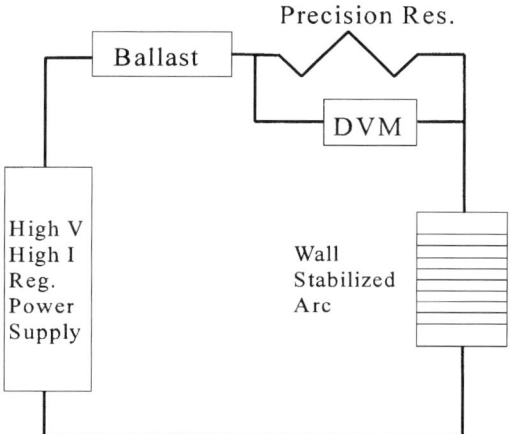

FIG. 10. Diagram of a wall stabilized arc electrical circuit.

An example of the output from a wall stabilized arc, which exhibits both continuum and line spectra in the VUV region depending on the gas constituents, arc current, and pressure is shown in Fig. 11 [38]. This type of arc has been used primarily in the VUV spectral region with a transmitting window as the high-pressure–vacuum interface. The extension of the wall stabilized arc into the VUV spectral region with a differential pumping interface has also been done with the rare gases and hydrogen from 30 to 100 nm [39] and with hydrogen from 50 to 100 nm [40]. The spectrum typically consists of a recombination continuum, possibly excimer continua with rare gases, as well as line spectra. The case of Ref. [40] is typical, where the radiation consists of resonance lines of hydrogen, Ly$_{\alpha-\varepsilon}$, the dominant free-bound Lyman recombination continuum to the $n = 1$ state of hydrogen, He I lines, as well as lines from the atoms and ions of ambient impurities such as N_2, O_2, and Ar.

Because this type of arc source has demonstrated a very reproducible and stable output, its continuum emission can be calibrated with respect to a primary radiometric standard and used as a transfer radiometric source standard. Their use as radiometric standard sources in the VUV spectral region has been summarized [38, 41]. These types of arcs have also been used for the emission of stable line spectra to investigate atomic quantities such as oscillator strengths (e.g., see Refs. [42] through [46]), line broadening parameters (e.g., see Ref. [47]), and plasma equilibrium conditions due to their stable properties. As seen in the spectrum of Fig. 11, an argon arc exhibits a relatively smooth continuum with isolated spectral lines, in this case due to impurities of oxygen, nitrogen and carbon, over a broad range of the VUV spectral region. The use of the wall stabilized arc as a more intense source of continuum spectra in the VUV region has also been summarized [48].

Because the wall stabilized arc discussed here operates at or near conditions where the plasma exhibits local thermal (or partial local) thermodynamic equilibrium (e.g., see Ref. [49], p. 108), this implies that the populations between the bound states within one species can be expressed by the Boltzmann equation;

$$n(j)/n(i) = g(j)/u(i) \times \exp[-E(j)/kT_e], \qquad (2)$$

where i and j represent the lower and upper bound levels, $n(i, j)$ their population densities, $g(j)$ the upper level's statistical weight ($= 2J + 1$), $u(i)$ the partition function (see Ref. [49], p. 113), $E(j)$ the level's upper state energy, and kT_e the plasma electron temperature. Also, the relationship between like-constituent ion stages is given by the Saha equation (see Ref. [49], p. 118);

$$n(z+1) \times n_e/n(z) = u(z+1)/u(z) \times 2(2\pi m k T_e/h^2)^{3/2}$$
$$\times \exp[\chi(z,g)/kT_e] \qquad (\text{cm}^{-3}), \qquad (3)$$

where z is the ion charge ($=$ zero for neutrals), n_e is the plasma electron density,

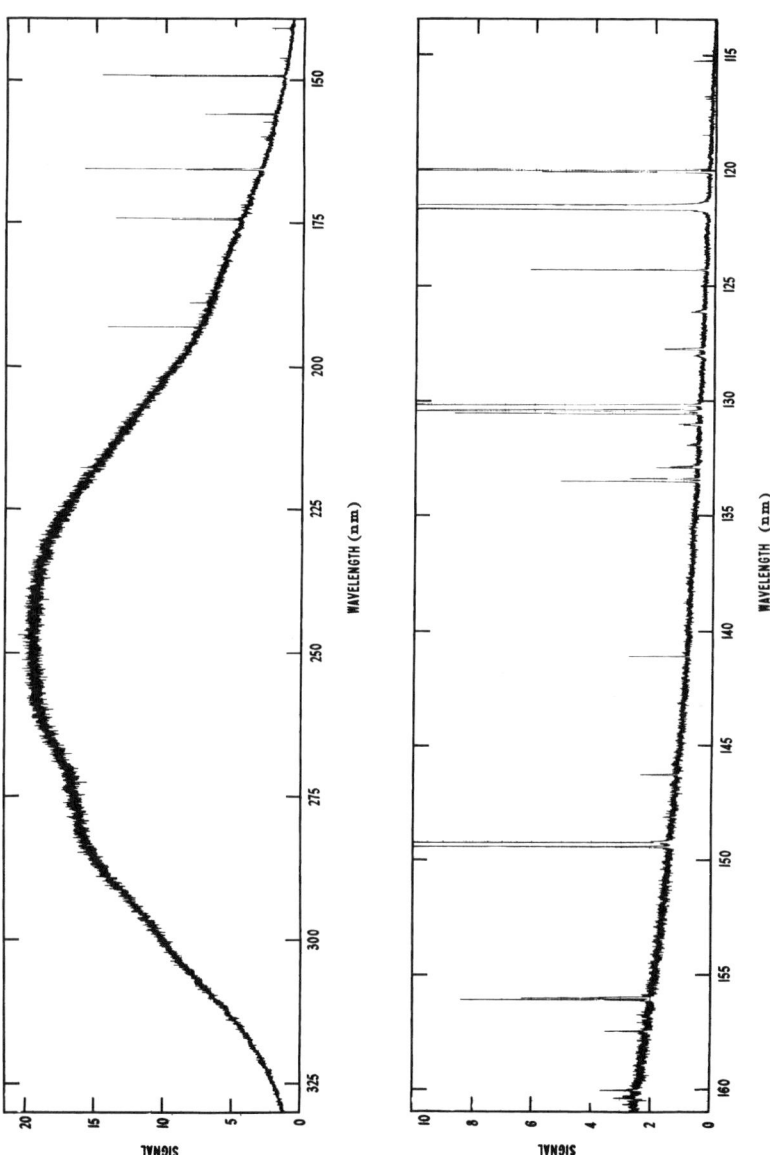

FIG. 11. A photoelectric (solar-blind photomultiplier) scan of the spectrum of an argon wall stabilized arc between 115 and 320 nm, with 0.01-nm spectral resolution. Reprinted with permission from the *J. Res. Natl. Bur. Stand.*, Vol. 93, pp. 21–39 (1988), J. Bridges, J. Klose, and W. Ott, Figure 10 of "Radiometric calibrations of portable sources in the vacuum ultraviolet," National Institute of Standards and Technology.)

and $\chi(z, g)$ is the ionization potential of the ion of charge z. In the case of an optically thin medium, the total spectral line intensity is given by

$$I(j, i) = 1/4\pi \int n(j)A(j, i)h\nu(j, i)\, dl \quad \text{(W/cm}^2/\Omega\text{)}, \qquad (4)$$

where $A(j, i)$ is the atomic transition probability [50] of a photon of energy $h\nu(j, i)$. The integral is over the length of the emitting medium viewed by the detector and is measured in units of watts per unit area per unit solid angle, Ω. In fact by using the property of optical thin emission for spectral lines, where the transition probabilities are accurately known, the temperature associated with the excited state population within the arc (usually equal to the free electron temperature as well) can be determined. If two (or more) spectral line intensities are measured, by plotting the $\log_e\{I(j, i)/[g(j)A(j, i)]\}$ vs $E(j)$, the value of the slope of a straight line though the points is equal to $-1/[kT_e]$. This is often referred to as a *Boltzmann plot*. In practice, several lines are measured and a least-squares straightline fit is made to the data, since the values of the $A(j, i)$'s may not be sufficiently accurate. The larger the upper state energy difference, the more accurate the temperature can be determined, for a given set of $A(j, i)$ values. This is because the uncertainty in the temperature, $\Delta T_e/T_e$, is proportional to the reciprocal of the upper state energy differences of the lines measured.

The use of a wall stabilized arc to investigate fundamental atomic parameters has demonstrated its use as a source of a variety of VUV line spectra. The quality of the spectral emission can be seen in Fig. 12 [51] and from other publications where oscillator strengths of neutral atoms are measured in emission [44, 45, 52, 53]. In these experiments, emission lines were observed using arcs at atmospheric pressures utilizing windows for wavelengths above the transmission cutoff of the window material and differential pumping systems for wavelengths below window transmission cutoffs.

Another example of the use of the wall stabilized arc in the VUV spectral region was the determination of the photoionization cross section of C I [54]. In this application, the arc plasma's local thermal equilibrium properties are utilized to determine the absolute emission coefficient of the carbon continuum from which the absorption coefficient is determined and thus the photoionization cross section can be deduced.

3.5 Electron Cyclotron Resonance Source

When a magnetic field is applied in conjunction with a microwave frequency electric field to a gas, special conditions can occur that enhance the plasma breakdown. The absorbed power supplied to such magnetized plasmas reaches a maximum near the plasma electron cyclotron resonance (ECR) frequency [37, p. 395]. The ECR frequency, v_{ec}, is given in terms of the applied magnetic induction by

$$v_{ec} = eB/2\pi m_e = 27.9922 \times B \quad \text{(GHz)}. \quad (5)$$

At the common rf frequency of 2.45 GHz, the value of B is 87.5 mT for this condition. Therefore, in principle, the power input to a rf generated plasma can be enhanced by applying a relatively small magnetic field. This concept is known as electron cyclotron resonance plasma generation and has been applied to make ion sources of highly charged ions, to provide heating in fusion plasmas, and to produce low-temperature, high-density (electron and ion) plasmas for semiconductor processing. In these capacities, it has also produced VUV radiation as reported by different authors [55, 56]. The significant characteristic in these types of plasmas is the enhanced ionization of plasma constituents that can be accomplished. As a result of higher ion stages, the spectra are more varied and emit at shorter and shorter wavelengths. A microwave discharge [57] used in photoelectron spectroscopy and enhanced by the application of an external magnetic field [58] has demonstrated the effectiveness of this

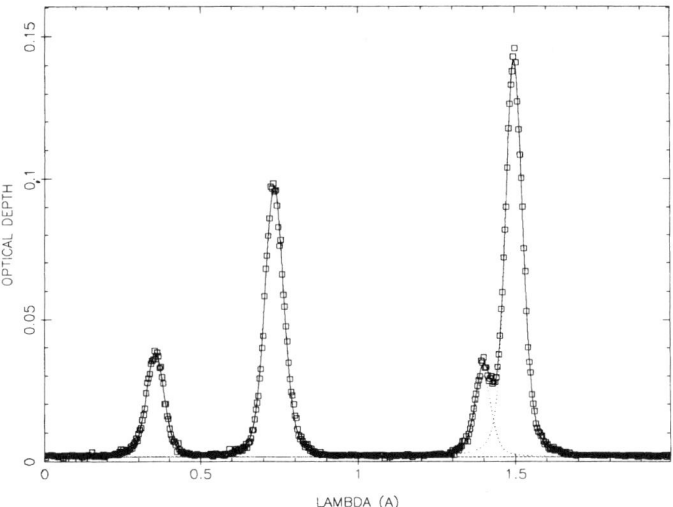

FIG. 12. Spectral scan of C I multiplet 3; 156.031-nm, 156.068–156.071-nm, and 156.134–156.137-nm blended pairs and 156.144-nm lines. Th solid curve is the analytical least-squares fit through the data points (open squares), obtained by summing individual Voigt profiles and a constant continuum background (dashed curves). (Reprinted with permission from *Astron. Astrophys.*, Vol. 181, pp. 203–209 (1987), C. Golbach and G. Nollez, Figure 1e of "Oscillator strength measurements in the vacuum-ultraviolet." Copyright 1987 Springer-Verlag, New York, Inc.)

method to generate higher degrees of ionization and thus shorter wavelengths in the VUV region.

In the application of Ref. [55], the authors describe the use of an ECR ion source with a VUV spectrometer to investigate the Be-like isoelectronic sequence, specifically, the intercombination line ratio of $I(2s2p\ ^3P-2p^2\ ^3P)/I(2s2p\ ^1P-2p^2\ ^1P)$ for C III, O V, and Ne VII. This ratio is predominantly electron density dependent and could be used as a plasma diagnostic. In Ref. [56], another plasma phenomenon was investigated, the enhanced radiation from Ar I spectral lines at 80.6 and 81.6 nm. This study indicated that the enhancement was attributed to a double electron transfer process between the Ar^{+2} ion and neutral Cu from the copper ECR source resonance cavity.

An ECR device, appropriate for plasma film deposition and etching, has also been used as a source to produce VUV spectra (see Fig. 13 [59]). With pressures ranging from 0.07 to 0.7 Pa and absorbed microwave powers ranging from 200 to 600 W, spectra were obtained using mixtures of Cl_2 and Ar, CF_4 and O_2, CF_4 and H_2 and Ar, as well as with single component gases of Ar, Cl_2, C_2F_6, and CH_4. Because of the spectrometer's short- wavelength cutoff, only spectra down to 110 nm were observed, which included lines of C II at 133.6 nm and excited state transitions from Cl II and Ar II. Several molecular spectra were also observed in this spectral region depending on the gas constituents. The measured electron density was approximately 5×10^{11} cm^{-3}. A strong correlation was observed between the radiated intensity of Cl II lines and the ECR condition occurring at the position of the maximum microwave power input. The appearance of radiation from Ar II spectral lines suggests a component of the electron temperature of approximately 10 eV, whereas the Boltzmann population for neutral Ar excited states indicates the gas temperature to be less than 1 eV, implying nonequilibrium conditions for these plasmas.

In Ref. [60], two different ECR ion sources provided spectral emission in the wavelength range from 10 to 100 nm. A 3-m grazing incidence spectrometer was used for these observations. These sources are capable of producing ion beams of Ar ions up to Ar^{+9} and oxygen ions up to O^{+6}. The plasma electron density was estimated using the intensity ratio of the two O V lines, $I(2s2p\ ^3P-2p^2\ ^3P)/I(2s^2\ ^1S-2p^2\ ^1P)$, at 76 and 63 nm, respectively.

3.6 Electron-Beam Ion Trap Sources

Another type of source utilized to generate highly charged ions and their spectra is the electron-beam ion trap (EBIT) [61]. This device uses a very high current density electron beam, accelerated up to 200 kV [62], directed through a drift tube region of high magnetic field (>3 T). The radial potential, due to the combination of the high electron current density and the high magnetic field,

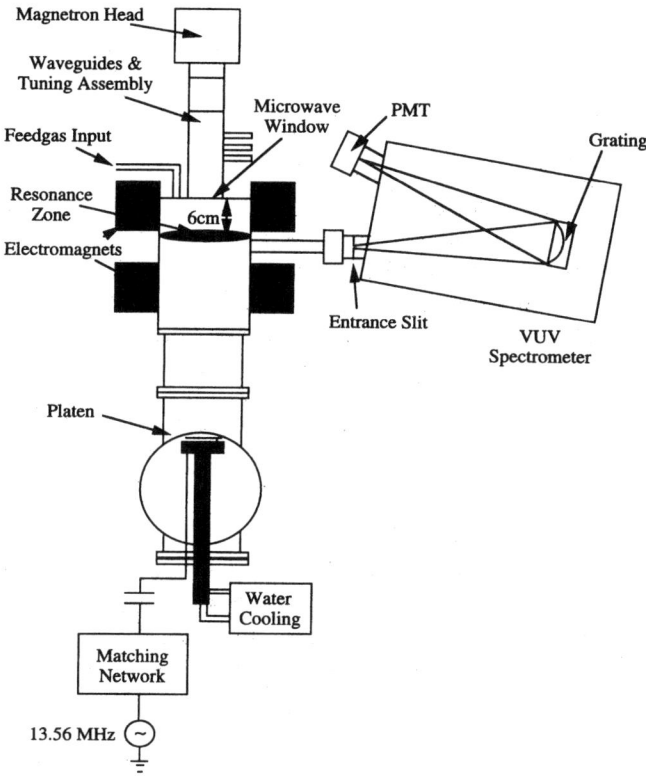

FIG. 13. Schematic diagram of an ECR microwave plasma source showing the location of the resonance zone and the VUV observation port. (Reprinted with permission from the *J. Appl. Phys.*, Vol. 78, pp. 6421–6426 (1995), G. Mehlman, C. Eddy, and S. Douglas, Figure 1 of "Characterization of electron cyclotron resonance plasmas by vacuum ultraviolet spectroscopy." Copyright 1995 American Institute of Physics.)

radially traps the ions into a small filament, approximately 100 μm in diameter. A longitudinal trap, approximately 3 cm long, is also incorporated by dividing the drift tube into three insulated hollow tube assemblies, with the two end pieces at a greater positive potential than the center piece (see Fig. 14 [61]). As a result, ions that are initially trapped undergo successive high-energy electron collisions, and ionization continues until (1) the electron-beam energy is smaller than the energy necessary to ionize to the next stage of ionization, or (2) the ion is fully stripped, or (3) it receives sufficient kinetic energy to escape the trap. In this manner, fully stripped uranium, U^{+92}, has been produced and trapped for up to several hours [63]. Typically for highly ionized species, the spectra that are usually investigated are in the x-ray spectral region [64], but other spectral regions have also been predicted [65].

The ultraviolet spectra of magnetic dipole (M1) transitions within the ground configuration of Ti-like and V-like highly charged ions have been investigated in the EBIT [66, 67]. These investigations have included those transitions in the spectral region of interest to plasma diagnostics where the wavelengths are transmitted in air (320 to 420 nm). As Fig. 15 and Table I show, for Ti-like Xe^{+32}, there is also a VUV line that radiates from this source. Since the A (transition probability) value is usually larger for the VUV lines than the visible or ultraviolet lines, in general this will make them observable as well. This source is capable of generating nearly any highly charged ion and populating the levels of the ground terms, thus generating spectral lines throughout the VUV region. Because of the relative ease of introducing species into the EBIT either by gas injection or by vapor injection from a pulsed arc source, the EBIT is envisioned to be a source of a variety of spectral lines in many portions of the spectrum.

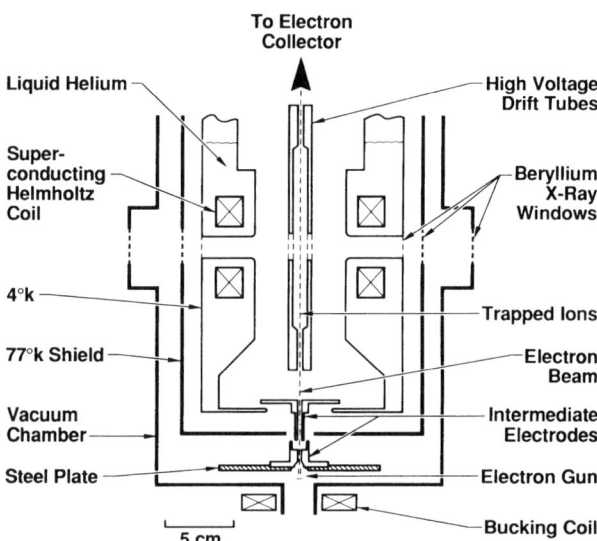

FIG. 14. Schematic diagram of EBIT, showing essential components. (Reprinted with permission from *Phys. Rev. Lett.*, Vol. 60, pp. 1715–1718 (1988), R. Marrs, M. Levine, J. Henderson, D. Knapp, and J. Henderson, Figure 1 of "Measurement of electron-impact-excitation cross sections for very highly charged ions." Copyright 1988 American Physical Society.)

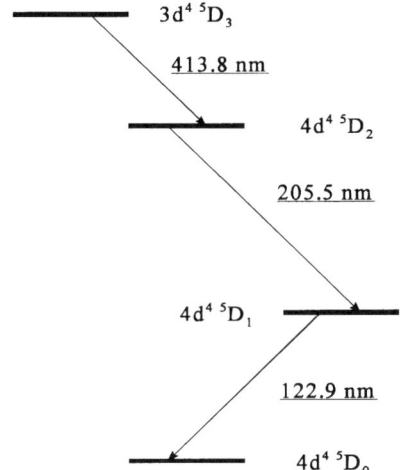

FIG. 15. Partial energy level diagram of Ti-like ions, showing the $4d^4\,^5D$ ground configuration of Xe^{+32}.

TABLE I. Transitions with the Ground Configuration of Ti-Like Xe^{+32}

Transition	Type	Wavelength (nm)	A value (sec^{-1})
$3d^4\,^5D_3-^5D_2$	M1	413.8	416
$3d^4\,^5D_2-^5D_1$	M1	205.6	4,204
$3d^4\,^5D_1-^5D_0$	M1	122.9	16,980

3.7 Inductively Coupled Plasma Sources

Inductively coupled plasma (ICP) device development has matured to the point that many different commercial devices are now available. They are primarily used in the generation of spectra for species identification and concentration determination (e.g., see Refs. [68], [69], [70], and [71]). The concept for producing spectra from an ICP is generally that the material to be studied is introduced into the flow of a carrier gas, usually argon, by gas injection, aerosol or evaporation. The plasma is established in an enclosed region surrounded by an inductive coil powered by a rf power supply (see Fig. 16 [68]). A flowing plasma is generated and the spectroscopy is usually done in the hottest part of the plasma near the enclosed region (see, e.g., Ref. [72]).

Several investigations have produced VUV spectra with an ICP. In Ref. [73] an ICP powered at 27.5 MHz was used to investigate the spectrum of arsenic at 189 nm by depositing As_2O_3 on a tantulum foil or the spectrum of iodine at 183 nm by introducing an I_2/methanol solution within the discharge tube.

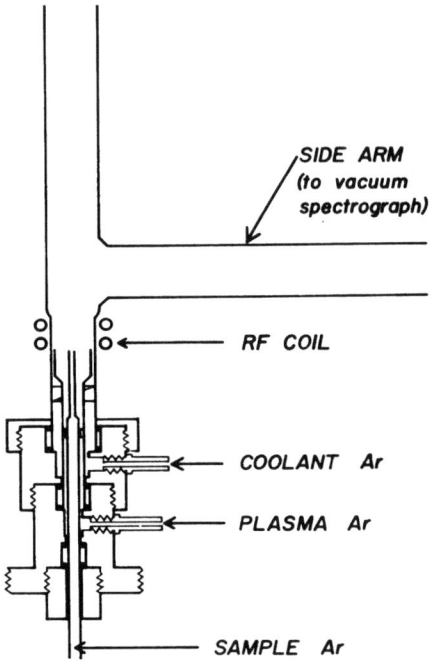

FIG. 16. Schematic diagram of ICP source with sidearm coolant tube. Vacuum ultraviolet emission is viewed along the side arm to the spectrograph. (Reprinted with permission from *Appl. Spectrosc.*, Vol. 34, pp. 595–598 (1980), D. Heine, J. Babis, and M. Denton, Figure 1 of "Quantitative aspects of an inductively coupled plasma in the spectral region between 120 and 185 nm." Copyright 1980 Society for Applied Spectroscopy.)

Aluminum was studied [74] using an ICP by introducing lake water as an aerosol to determine its quantitative abundance levels. Several VUV emission lines were observed, ranging from 167 to 186 nm. The emission spectra from an ICP source have also been investigated from 85 to 200 nm [75]. The interest here was to use the source as a quantitative analytical tool for nonmetallic species identification and as a source of emission of argon spectral lines emanating from metastable $^3P_2$ and $^3P_1$ states in argon. The source used is a commercial device and it is coupled to a recording spectrometer that is purged with helium for observations of wavelengths down to 85 nm. An example of the spectra generated in this manner is shown in Fig. 17. Another commercial ICP source was used to generate VUV spectra in ions [76] that do not have strong ultraviolet or visible spectral lines. Elements were introduced into the plasma by the aerosol method, and the ionic spectra of B, Al, Ga, In, Tl, Si, Ge, Sn, Pb, and Bi were observed and compiled in the spectral region from 150 to 270 nm. In a similar manner, a large number of atomic and ionic spectra were generated by a

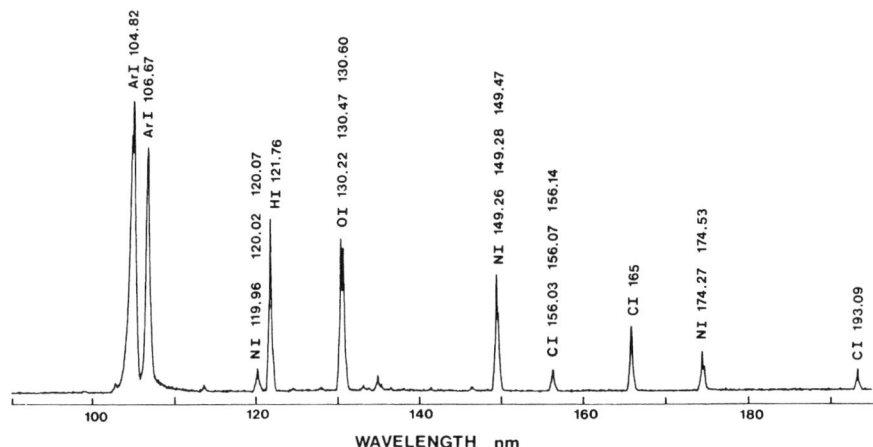

FIG. 17. Spectral scan of the VUV background emission from an ICP source in the region from 90 to 195 nm. The argon and other impurity lines are from the carrier gas. Reprinted from *Spectrochem. Acta*, Vol. 39B, pp. 567–574 (1994), J. Carr and M. Blades, Figure 3 of "Emission spectra of an argon inductively coupled plasma in the vacuum ultraviolet: background spectra from 85 to 200 nm," with kind permission of Elsevier Science—NL, Sara Burgerharstraat 25, 1055 KV Amsterdam, The Netherlands.)

commercial ICP source [77] for purposes of species identification and concentration determination. The spectra of Be, B, C, N, Mg, Al, Si, P, S, Ca, Ti, V, Cr, Mn, Fe, Co, Ni, Cu, Ga, Ge, As, Se, Br, Sr, Zr, Nb, Mo, Ru, Pd, Ag, In, Sn, Sb, Te, I, Ba, Hf, W, Re, Pt, Au, Hg, Tl, Pb, and Bi were generated, and wavelengths from 160 to 200 nm, the stage of ionization, and an intensity estimate were listed. The intensity estimate was specific to ICPs and was done to aid other ICP source observations in making more rapid and definitive identifications by comparing intensities as well as wavelengths. This could aid in the identification of, for example, blends of different spectra. In the experiments of Ref. [78], the viewing spectroscopic apparatus was entirely in a vacuum and observed the ICP through a small orifice, so wavelength observations were made down to 80 nm. In this experiment the Ar resonance transitions at 106.7 and 104.8 nm were observed, as well as those at 88.0, 87.6, 87.0, and 86.7 nm. Features down to 78.8 nm, corresponding to transitions from higher lying levels approaching the ionization limit, were also observed but not resolved.

3.8 Capillary Discharges

A high-flux capillary discharge was utilized for photoelectron spectroscopy of gaseous targets and was differentially pumped for use at wavelengths below 104 nm [79]. This source was compared with others for absolute flux because the specific experiment being performed depended very sensitively on the VUV emission flux. The total VUV flux was measured to be approximately 10^{13} photons/s. The discharge was operated at a pressure of 10^2 Pa with a vacuum in the experimental chamber of 10^{-2} Pa. The discharge capillary is made of Al_2O_3 with a 2 mm diameter hole, 60 mm long bored into the ceramic (see Fig. 18). The source is constructed in such a manner that it can be baked. The capillary is removed and cleaned by acid etching approximately every 200 h due to the deposit of sputtered cathode material on its inner walls. Other mechanical details are discussed in Ref. [79]. The power supply utilized to sustain a discharge in the capillary was a variable voltage and current-stabilized supply up to 6 kV and 200 mA. To maintain stable operation, a 2.2-kΩ series ballast resistor was required. The discharge was primarily operated in He or Ne and produced both the He I and He II or Ne I and Ne II line spectra.

Another capillary discharge, capable of producing a total VUV flux in excess of 10^{11} photons/s, has been discussed in the context of photoelectron spectroscopy [80]. This discharge is differentially pumped and operates in a pressure range from 250 to 550 Pa. The lamp's operating parameters are a discharge current of 100 mA and an anode voltage of +365 V. The power is supplied through a 1320-Ω ballast resistor. Operating in helium, the discharge emits the principal resonance line ($1s$–$2p$) of helium at 58.4 nm and emits smaller fluxes of the 53.7-nm line ($1s$–$3p$) and the 52.2 nm line ($1s$–$4p$).

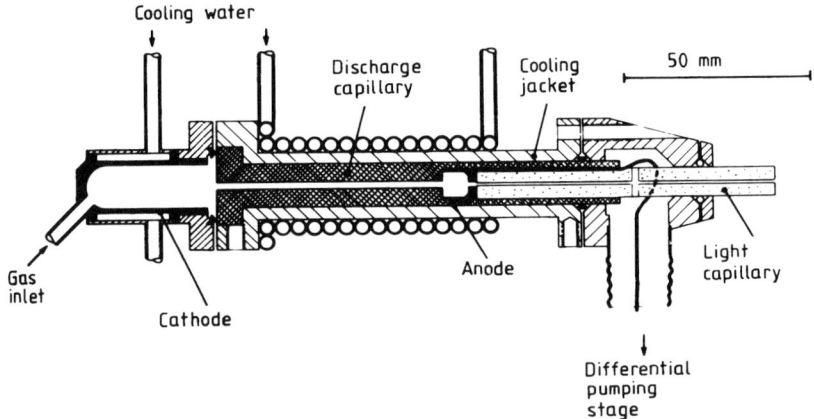

FIG. 18. Schematic diagram of a capillary VUV emission source, with a differential pumping stage. (Reprinted with permission from the *Journal of Physics E (Scientific Instruments)*, Vol. 16, pp. 74–82 (1983), G. Schönhense and U. Heinzmann, Figure 1 of "A capillary discharge tube for the production of intense vuv resonance radiation," Institute of Physics Publishing Ltd.)

References

1. W. Ott, J. Bridges, and J. Klose, *Opt. Lett.* **5**, 225–227 (1980).
2. H. Greim, in "Plasma Spectroscopy," McGraw-Hill, New York, pp. 293–394 (1964).
3. H. Greim, in "Plasma Line Broadening," Academic Press, New York (1974)
4. F. Burger and J. Maier, *J. Electron Spectrosc. Relat. Phenom.* **5**, 783–789 (1974).
5. F. Burger and J. Maier, *J. Phys. E: Sci. Instrum.* **8**, 420–422 (1975).
6. G. Lancaster, J. Taylor, A. Ignatiev, and J. Rabalais, *J. Electron Spectrosc. Relat. Phenom.* **14**, 143–153 (1978).
7. A. McIlraith, *Nature, London* **212**, 1422–1424 (1966).
8. R. Fitch and S.Venkatesh, *Vacuum* **29**, 19–21 (1979).
9. G. Rushton, K. O'Shea, and R. Fitch, *J. Phys. D: Appl. Phys.* **6**, 1167–1172 (1973).
10. F. Burger and J. Maier, *J. Electron Spectrosc. Relat. Phenom.* **16**, 471–474 (1979).
11. W. Graham, *J. Phys. D: Appl. Phys.* 16, 1907–1915 (1983).
12. W. Graham, *J. Phys. D: Appl. Phys.* **17**, 2225–2231 (1984).
13. D. Lindau and H. Döbele, *Rev. Sci. Instrum.* **59**, 565–568 (1988).
14. W. Siemans, *Ann. Phys. Chem.* **102**, 66–122 (1857).
15. D. Neeman and M. Brennan, *Aust. J. Phys.* **48**, 543–556 (1995).
16. U. Kogelschatz, *Pure Appl. Chem.* **62**, 1667–1674 (1990).
17. B. Gellert and U. Kogelschatz, *Appl. Phys. B* **52**, 14–21 (1991).
18. L. Rosocha and W. McCulla, in "Proc. 44th Ann. Gaseous Electronics Conf.," Albuquerque, NM, 1991, **QB-2**, p. 191 (1991).
19. U. Kogelschatz,. in "Process Technologies for Water Treatment—Advanced Ozone Generation" (S. Stucki, ed.)., Plenum Press, New York and London (1988).
20. H. Esrom and U. Kogelschatz, *Appl. Surf. Sci.* **54**, 440–444 (1992).
21. U. Kogelschatz, in "Proc. 10th Int. Conf. on Gas Discharges and Their Applications" (W. Terry Williams, ed.), Swansea, 1992, **2**, pp. 972–980 (1992).
22. B. Eliasson and U. Kogelschatz, *IEEE Trans. Plasma. Sci.* **19**, 309–323 (1991).
23. V. Narory, P. Drallos, and W. Williamson, Jr., *J. Appl. Phys.* **77**, 3645–3656 (1995).
24. H. Janus, W. Jele−ski, J. Musielok, and J. Kusz, *Appl. Phys. B* **52**, 376–379 (1991).
25. M. Walhout, private communication (1996).
26. S. Kubodera, M. Honda, M. Kitahara, J. Kawanaka, W. Sasaki, and K. Kurosawa, *Jpn. J. Appl. Phys.* **34**, L618–L620 (1995).
27. U. Kogelschatz, *Appl. Surf. Sci.* **54**, 410–423 (1992).
28. K. Stockwald and M Neiger, *Contrib. Plasma. Phys.* **35**, 15–22 (1995).
29. F. Kessler and G. Bauer, *Appl. Surf. Sci.* **54**, 430–434 (1992).
30. W. Finkelnburg and H. Maecker, in "Hanbuch der Physik" (S. Flügge, ed.), Vol. 22, Springer, Berlin (1956).
31. W. Wiese, in "Methods of Experimental Physics" (B. Bederson and W. Fite eds.), Academic Press, New York, pp. 307–353 (1968).
32. H. Maecker, *Z. Naturforsch.* **11a**, 457–458 (1956).
33. G. Boldt, in "Proc. Int. Conf. on Ionization Phenomena in Gases, 5th Conf.," Munich, 1961, **1**, 925–928 (1961).
34. J. Morris and R. Garrison, *J. Quant. Spectrosc. Radiat. Transfer* **6**, 899–902 (1966).
35. W. Wiese, D. Paquette, and J. Solarski, Jr., *Phys. Rev.* **129**, 1225–1232 (1963).

36. G. Pretzler, H. Jäger, T. Neger, H. Phillip, and J. Woisetschläger, *Z. Naturforsch.* **47a**, 955–970 (1992).
37. J. Roth, in "Industrial Plasma Engineering, Vol 1: Principles," Institute of Physics Publishing, Bristol and Philadelphia (1995).
38. J. Klose, J. Bridges, and W. Ott, *J. Res. Natl. Bur. Stand.* **93**, 21–39 (1988).
39. M. Leavy and R. Huffman, *Appl. Opt.* **9**, 41–47 (1970).
40. K. Behringer and P. Thoma, *Appl. Opt.* **18**, 2586–2594 (1979).
41. J. Bridges, in "Ultraviolet Technology IV," Society of Photo-Optical Instrumentation Engineers, Bellingham, WA, **1764**, 262–270 (1992).
42. W. Ott, *Phys. Rev. A* **4**, 245–251 (1971).
43. D. Stuck and B. Wende, *Phys. Rev. A* **9**, 1–8 (1974).
44. C. Goldbach, T. Lüdtke, M. Martin, and G. Nollez, *Astron. Astrophys.* **266**, 605–612 (1992).
45. J. Molino Garcia, W. Bötticher, and M. Kock, *J. Quant. Spectrosc. Radiat. Transfer* **55**, 169–179 (1996).
46. J. Musielok, J. Bridges, S, Djurovic, and W. Wiese, *Phys. Rev. A* **53**, 3122–3128 (1996).
47. H. Nubbemeyer, *Phys. Rev. A* **22**, 1034–1040 (1980).
48. A. Wilbers and D. Schram, *J. Quant. Spectrosc. Radiat. Transfer* **46**, 299–308 (1991).
49. Lawrence H. Aller, "Astrophysics—The Atmosphere of the Sum and Stars," The Ronald Press Company, New York, p. 111 (1963).
50. W. Wiese, in "'Atomic Processes in Plasmas, 10th Topical Conf." (A. Osterheld and W. Goldstein, eds.), San Francisco, 1996, **381**, 177–186 (1996).
51. C. Goldbach and G. Nollez, *Astron. Astrophys.* **181**, 203–209 (1987).
52. C. Goldbach, M. Martin, and G. Nollez, *Astron. Astrophys.* **221**, 155–160 (1989).
53. C. Goldbach and G. Nollez, *J. Phys. Colloque CI* **1**, C1-83–C1-87 (1991).
54. W. Hoffman and G. Weissler, *J. Opt. Soc. Am.* **61**, 223–230 (1971).
55. W. Pöffel and K.-H. Schartner, *Rev. Sci. Instrum.* **61**, 613–615 (1990).
56. M. Jogwich, B. Huber, and K Wiessenamm, *Z. Phys. D* **17**, 171–179 (1990).
57. F. Fehsenfeld, K. Evenson, and H. Broida, *Rev. Sci. Instrum.* **36**, 294–298 (1965).
58. T. Vorburger, B. Waclawski, and D. Sandstrom, *Rev. Sci. Instrum.* **47**, 501–504 (1976).
59. G. Mehlman, C. Eddy, and S. Douglas, *J. Appl. Phys.* **78**, 6421–6426 (1995).
60. B. Jettkant, R. Berreby, D. Hitz, and M. Druetta, *Rev. Sci. Instrum.* **67**, 1258–1260 (1996).
61. R. Marrs, M. Levine, J. Henderson, D. Knapp, and J. Henderson, *Phys. Rev. Lett.* **60**, 1715–1718 (1988).
62. R. Marrs, P. Beiersdofer, and D. Schneider, *Phys. Today* **47**, 27–34 (1994).
63. R. Marrs, in "Experimental Methods in the Physical Sciences," Vol. 29A (F. B. Dunning and Randall G. Hulet, eds.), Academic Press, New York, pp. 391–420, (1995).
64. J. Gillaspy, *Physica Scripta* **T65**, 168–174 (1996).
65. U. Feldman, P. Indelicato, and J. Sugar, *J. Opt. Soc. Am.* **8**, 3–5 (1991).
66. C. Morgan, F. Serpa, E. Takács, E. Meyer, J. Gillaspy, J. Roberts, C Brown, and U. Feldman, *Phys. Rev Lett.* **74**, 1716–1719 (1995).
67. F. Serpa, E. Meyer, C. Morgan, J. Gillaspy, J. Sugar, J. Roberts, C. Brown, and U. Feldman, *Phys. Rev. A* **53**, 2220–2224 (1996).
68. D. Heine, J. Babis, and M. Denton, *Appl. Spectrosc.* **34**, 595–598 (1980).
69. T. Nakahara and T. Tamotsu, *Appl. Spectrosc.* **41**, 1238–1242, (1987).

70. M. Denton, M. Pilon, and J. Babis, *Appl. Spectrosc.* **44**, 975–978 (1990).
71. A. Montaser, in "Inductively Coupled Plasmas in Analytical Atomic Spectroscopy" (A. Montaser and D. W. Golightly, eds.), VCH Publishers, New York (1992).
72. J. Babis, M. Pilon, and M. Denton, *Appl. Spectrosc.* **44**, 1281–1289 (1990).
73. G. Dreher and C. Clark, *Appl. Spectrosc, Notes* **28**, 191–192 (1974).
74. T. Uehiro, M. Morita, and K. Fuwa, *Anal. Chem.* **56**, 2020–2024 (1984).
75. J. Carr ans M. Blades, *Spectrochem. Acta* **39B**, 567–574 (1994).
76. T. Uehiro, M. Morita, and K. Fuwa, *Anal. Chem.* **57**, 1709–1713 (1985).
77. D. Nygaard and D. Leighty, *Appl. Spectrosc.* **39**, 968–976 (1985).
78. R. Houk, V. Fassel, and B. LaFreniere, *Appl. Spectrosc.* **40**, 94–100 (1986).
79. G. Schönhense and U. Heinzmann, *J. Phys. E: Sci. Instrum.* **16**, 74–82 (1983).
80. D. Mason, D. Mintz, and A. Kuppermann, *Rev. Sci Instrum.* **48**, 926–933 (1977).

4. HOLLOW CATHODE, PENNING, AND ELECTRON-BEAM EXCITATION SOURCES

Michael Kühne

Physikalisch-Technische Bundesanstalt
Berlin, Germany

4.1 Hollow Cathode Sources

Because no solid window material is available for wavelengths shorter than about 105 nm (cutoff of LiF), plasma vacuum ultraviolet (VUV) radiation sources below 105 nm have to be differentially pumped. Typical examples of such plasma sources are wall stabilized arcs (see preceding chapter) which operate at about atmospheric pressures and need powerful and, therefore, expensive differential pumping systems. At such high pressures continuum emission is available from the dense plasma. To reduce the size and the cost of the differential pumping system, plasma sources operating at lower pressures are very desirable. While high-pressure sources emit mainly continuum radiation, the radiation emission of low-pressure discharges is basically limited to atomic and ionic emission lines from the components of the plasma.

> **Warning:** This is a short review and cannot provide complete information. The information contained in this chapter and in the references is neither intended nor sufficient to design, construct, or operate the types of sources described. It is also not within the scope of this chapter to provide information concerning safety aspects. It is the responsibility of the reader to consult and establish appropriate safety practice and determine the applicability of regulatory limitations prior to using the information provided in this chapter or in its references.

For additional information on the design and operation procedures and on the commercial availability of the described sources, please contact the authors of the referenced literature, in particular, S. Bowyer, J. Hollandt, or M. Kock.

One type of discharge that has been found particularly suitable as a VUV radiation source is the windowless high-current hollow cathode source. Hollow cathode discharges, known since 1916 [1], are a special type of glow discharge. The characteristic feature of these sources is its dominant negative glow. The cathode is shaped like a hollow cylinder with a diameter of approximately 10 mm. The buffer gas pressure is chosen so that the negative glow fills the volume of the hollow cathode to create a thin cathode fall covering the interior surface of the cathode with a thickness of typically 1 mm or less. Most of the electric potential of the discharge is dropped within the cathode fall, which is small compared to the diameter of the hollow cathode, leaving the inner volume of the cathode nearly field free. Electrons emitted from the surface of the cathode, due to ion or VUV radiation bombardment, acquire nearly the full potential energy by passing through the cathode fall. Because the pressure is chosen so that the mean free path of the electrons is larger than the diameter of the hollow cathode these electrons, the so-called *beam electrons*, oscillate between the two opposing cathode falls until their kinetic energy is transferred to the gas atoms by collision.

Typical electron energies of these beam electrons are in the range of a few hundred electron volts. This energy is sufficient to ionize the atoms of the buffer gas (mostly noble gases) that fills the interior of the hollow cathode. The ions are accelerated toward the cathode surface and obtain sufficient kinetic energy to sputter cathode material from the surface into the discharge. By direct collision with the electrons and by charge transfer collisions with the buffer gas ions the sputtered atoms are also ionized. A detailed theory that relates the particle densities to the discharge current density can be found in [2, 3].

Two different discharge regimes can be distinguished: the low-current regime with current densities below 0.1 A cm^{-2} where the sputter rate is low and the discharge is dominated by buffer gas ions. If the current density on the cathodes exceeds approximately 0.1 A cm^{-2} the sputter rate increases significantly and the discharge becomes dominated by the cathode material ions. In this case the erosion of the hollow cathode due to ion bombardment will limit the useful lifetime of the hollow cathode.

Hollow cathode discharges are typically non-LTE plasmas. The buffer gas temperature and the ion temperature as determined from spectroscopic observations [4] is of the order of only 1000 to 1500 K. Aside from the beam electrons with energies close to the cathode–anode potential drop the majority of the electrons have a temperature of typically between 1 and 4 eV. With electron densities of the order of 10^{12} to 10^{14} cm^{-3} the emission lines are mainly Doppler broadened with typical half widths of the order of 1 to 5×10^{-3} nm.

Although the aim of this chapter is to familiarize the reader with windowless high-current hollow cathode sources that emit well below 100 nm, we should mention that a low-current hollow cathode source with magnesium fluoride window and a Pt/Cr cathode has been calibrated by Klose and Bridges [5] at the NIST as a radiance standard. The lines cover the spectral range from about 125 to 725 nm with radiances ranging from 3.5×10^{-6} to $5.8 \times 10^{-3} \text{ W cm}^{-2} \text{ sr}^{-1}$ generally increasing with wavelength. The calibration uncertainties range from 13% below 140 nm and 9% between 140 and 250 nm to 5% above 250 nm.

An early version of a windowless hollow cathode type radiation source was described by Paresce *et al.* [6] in 1971. The source consists of a hollow cathode made out of elkonite (copper–tungsten alloy) with an inner diameter of 10 mm and a high-purity aluminum anode of cylindrical shape which surrounds the cathode (see Fig. 1). Both electrodes are water cooled.

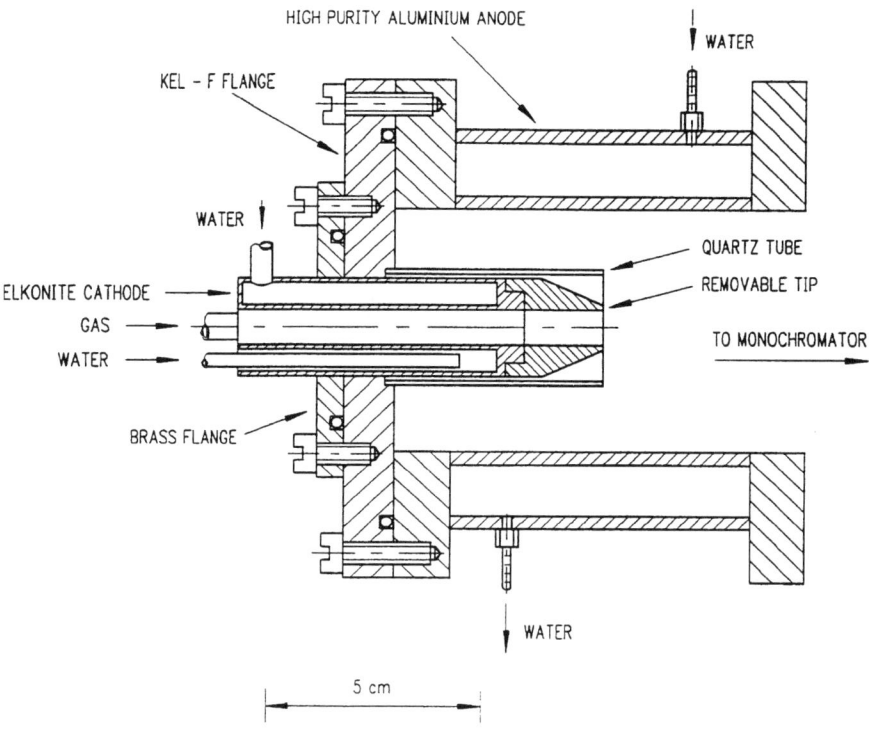

FIG. 1. Schematic diagram of a hollow cathode source. [Reprinted with permission from F. Paresce, S. Kumar, and C.S. Bowyer, *Appl. Opt.* **10**, 1904–1908 (1971). Copyright © 1971 Optical Society of America.]

The source is operated with a ballast resistor of 3 kΩ in series with discharge currents between 50 and 800 mA at voltages between 500 and 2000 V. The source contains no pumping system of its own but is mounted to a monochromator using the entrance slit as a differential pumping aperture. It is operated with noble gas pressures between 0.08 and 13 mbar within the source and 7×10^{-5} to 7×10^{-3} mbar in the monochromator. The radiation emission consists of emission lines from the neutral atoms as well as the singly and doubly ionized atoms. Using a 1-m Seya-Namioka monochromator with a 1200 lines/mm grating and an entrance slit of 400 μm the authors report photon fluxes behind the exit slit of the order of 10^8 photons/s for the most intense noble gas lines.

During the mid-1980s another type of high-current hollow cathode source was developed jointly by the Institut für Plasmaphysik (IPP) at the University of Hannover and the Physikalisch-Technische Bundesanstalt in Berlin (PTB) with the aim of providing a stable source of VUV line radiation that can be calibrated

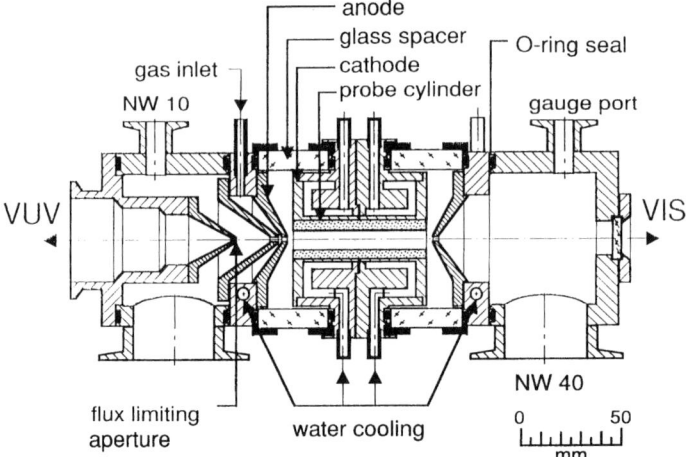

FIG. 2. Longitudinal section through a hollow cathode source with an integrated two-stage differential pumping system. [Reprinted with permission from K. Danzmann, M. Günther, J. Fischer, M. Kock, and M. Kühne, *Appl. Opt.* **33**, 68–74 (1994). Copyright © 1994 Optical Society of America.]

against a primary source standard, for example, an electron storage ring, and can serve as a radiometric secondary source standard in the laboratory [7–10].

A cross section through the source in shown in Fig. 2. Two conical shaped anodes are symmetrically located opposite both ends of a cylindrical hollow cathode with a length of 60 mm and a bore of 8 mm. The electrodes are water cooled and separated from each other by cylindrical glass spacers. The anodes contain central bores with a diameter of 4 mm to allow on-axis observation of the discharge. The distance from the cathode ends to the anodes is chosen to suppress the development of the positive column. Behind one of the anodes a two-stage differential pumping system is located. It is equipped with exchangeable flux-limiting aperture stops with a diameter ranging from 0.6 to 1.2 mm serving as a second differential pumping stage. To reduce significantly the clogging of the flux-limiting aperture by sputtered cathode material, the gas inlet is positioned between the differential pumping aperture of 1.5 mm of the first pumping stage and the anode bore. The source is sealed with Viton O-rings and the first differential pumping stage is pumped by a 50 liters/s turbomolecular pump. The vacuum vessel to which the hollow cathode is connected is providing the pumping for the second differential pumping stage. Because of its modular design the source can be easily disassembled to allow cleaning of the anodes and the glass spacers.

The source is operated with a ballast resistor of at least 50 Ω in series with a typical voltage drop across the electrodes of 400 to 500 V. Buffer gases like

TABLE I. Purity and Typical Pressure of the Employed Buffer Gases When Operating the Hollow Cathode at 1 A and 500 V[a]

Gas	Purity (%)	Pressure (Pa)
Helium	99.996	120
Neon	99.990	80
Argon	99.998	50
Krypton	99.998	60

[a] Pressure was measured at the gauge port of the back anode.
Reprinted with permission from J. Hollandt, M. Kühne, and B. Wende, *Appl. Opt.* **33**, 68–74 (1994). Copyright © 1994 Optical Society of America.

helium, argon, or krypton are used at pressures of typically 100 Pa. For details see Table I.

An operating current of typically 1 or 2 A is set with a regulated power supply. To set the desired operating voltage, the discharge pressure is regulated by means of an electrically operated needle valve in the gas inlet. At the back anode port the source is moderately pumped to establish a slow but constant gas flow through the hollow cathode and through the opening of the other anode cone. Because the voltage drop is sensitively dependent on the discharge pressure, the voltage can be kept constant to within ± 1 V by using the electrically operated needle valve to regulate the gas input.

Aluminum is used as material for the cathode insert. A spectrum containing the emission lines of neutral, singly, and doubly ionized buffer gas atoms as well as those of the cathode material is observed. Because the respective ionization potentials of aluminum are lower than those of the buffer gases, lines of triply ionized aluminum can be observed, particularly at higher currents. The dependence of the radiant intensity on the operating current for several emission lines from different levels of ionization is shown in Fig. 3. Although the radiant intensity of the two neutral lines is only weakly dependent on the discharge current, the buffer gas ion line emission increases approximately linearly and quadratically with the discharge current.

Due to the cathode surface bombardment with ions, erosion of the cathode by sputtering takes place, which slowly changes the shape of the cathode. Material is sputterd away mainly at the outer sections of the cathode, while at the center material is deposited reducing the diameter of the cathode bore. The cathode will eventually be completely blocked in the center. To avoid the influence of the changing shape of the cathode on the spectral emission, the cathode insert has to be replaced after about 30 to 40 h of operation depending mainly on the operating current. Figure 4 shows the reproducibility of the spectral line emission over a period of 70 operating hours including several changes of the buffer

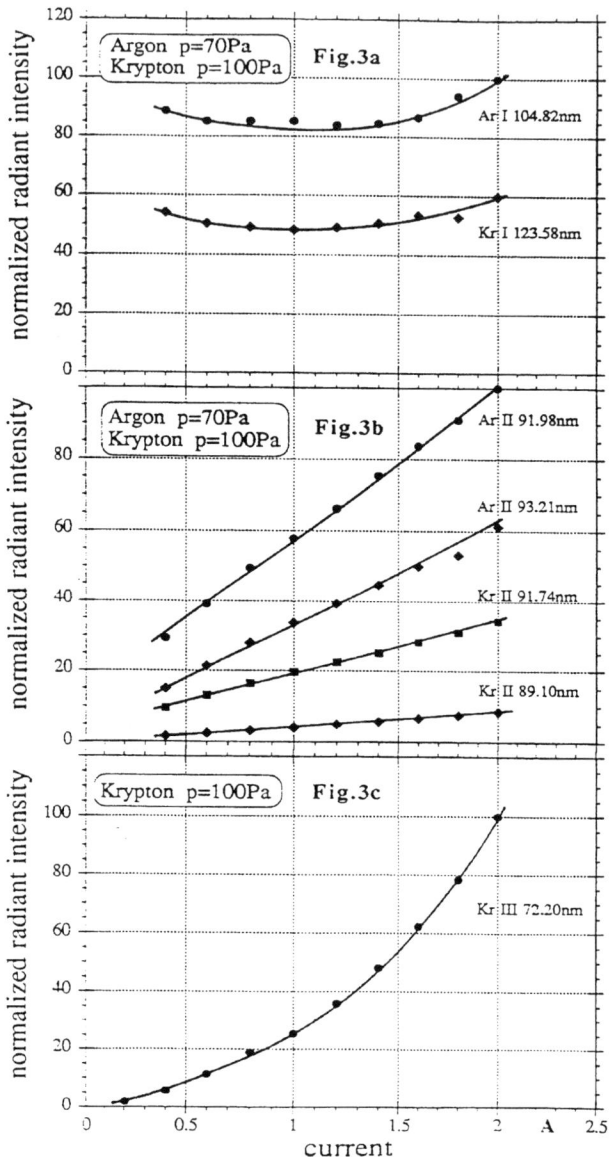

FIG. 3. Dependence of the radiant intensity on the discharge current of several hollow cathode emission lines. Although the radiant intensity of the two neutral gas lines varies only weakly with the discharge current, the radiant intensity of the buffer gas ions increases approximately linearly and quadratically with the discharge current. [Reprinted with permission from J. Hollandt, M. Kühne, and B. Wende, *Appl. Opt.* **33**, 68–74 (1994). Copyright © 1994 Optical Society of America.]

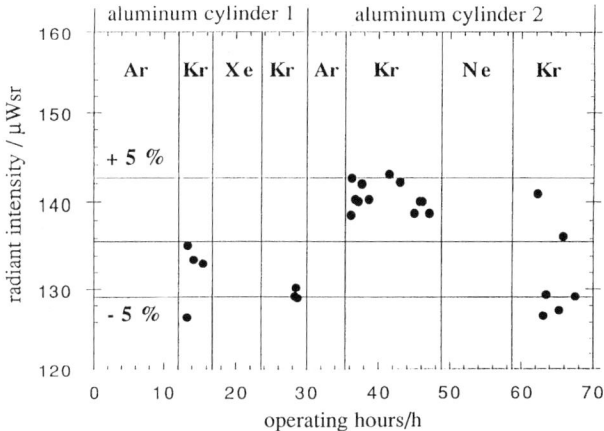

FIG. 4. Radiometric long-term reproducebility of the Kr I 123.58 nm emission line. Starting with argon, the buffer gas was changed several times during operation. After 30 h of operation the hollow cathode was cleaned to remove sputtered aluminum from the anodes and the glass spacers, and the cathode insert was replaced. [Reprinted with permission from J. Hollandt, M. Kühne, and B. Wende, *Appl. Opt.* **33**, 68–74 (1994). Copyright © 1994 Optical Society of America.]

gas and a cathode replacement after 30 h of operation. The observed reproducibility of the spectral line emission is of the order of 5%.

At PTB's radiometric laboratory at the Berlin electron storage ring, BESSY, the high-current hollow cathode has been characterized as a radiometric standard source by comparison with the calculable radiation emission of the electron storage ring [8–9]. In the grazing incidence region, the radiance of several emission lines was determined for operating conditions of 2 A and 400 V (see Table II). These operating parameters were chosen to maximize the sputter rate and therefore the output in the aluminum emission lines. For longer wavelengths mainly in the normal incidence range (see Table III) the operating conditions were set to 1 A and 500 V. For details about the radiometric comparison and the determination of the uncertainty budget see Chapter 8.3 of this book. The radiant intensities have been obtained with a flux-limiting aperture of 1.2 mm. For the angular emission characteristic of the source, see Fig. 5.

One of the major applications of secondary source standards is the spectral sensitivity calibration of spectrometers. In some cases (e.g. solar and stellar VUV astronomy), we need to calibrate complete telescope-spectrometer systems before they can be used in space. Because these systems collect radiation from a practically infinite distance, the required source standard has to be combined with imaging optics to produce the necessary parallel beam. For the spectral

TABLE II. Radiant Intensity and Radiance of Selected Spectral Lines of the Hollow Cathode for Operating Parameters $I = 2$ A and $U = 400$ V Anode

λ (nm)	Ion	Buffer gas	I (W/sr)	L (W/m^2 sr)	Uncertainty (%)
13.1	Al IV	Ar	1.5×10^{-7}	0.13	25
16.0/16.2	Al IV	Ar	1.91×10^{-6}	1.69	13
16.0/16.2	Al IV	He	1.53×10^{-6}	1.35	13
24.3	He II	He	4.65×10^{-5}	41.1	13
25.6	He II	He	1.47×10^{-4}	130	13
30.4	He II	He	6.76×10^{-4}	598	13
58.4	He I	He	1.53×10^{-3}	1350	13

Reprinted with permission from K. Danzmann, M. Günther, J. Fischer, M. Kock, and M. Kühne, *Appl. Opt.* **27**, 4947–4951 (1988). Copyright © 1988 Optical Society of America.

sensitivity calibration of two telescope-spectrometer systems named Solar Ultraviolet Measurements of Emitted Radiation (SUMER) and Coronal Diagnostics Spectrometer (CDS) flying onboard the Solar and Heliospheric Observatory (SOHO), hollow cathodes of the type described earlier have been mated to normal incidence and grazing incidence optics (Figs. 6a and b).

For the calibration of SUMER with a spectral range of 50 to 160 nm, a spherical mirror has been used and for CDS with a spectral range 16 to 80 nm a backside illuminated Wolter type II telescope (supplied by NASA) has been used. The details of the design and values for photon fluxes in the collimated beam of 57 spectral lines are given by Hollandt *et al.* [10]. The spectral lines cover the wavelength range between 15 and 150 nm with photon fluxes ranging from 10^4 to 10^9 photons/s. Using BESSY as a primary standard for the calibration uncertainties for the photon flux of 6 to 8% (typical values) are reported.

4.2 Penning Discharge Sources

Penning discharges were originally investigated for use as ionization gauges for low-pressure measurements [11]. The characteristic feature of a Penning discharge is the application of a strong magnetic field parallel to the electric field of the cathode fall. The electrons in the discharge are forced by the magnetic field to gyrate around the magnetic field lines when they move along the electric field. In this way the effective pathlength of the electron in the discharge is greatly enhanced and thus the collision probability with a neutral atom is increased.

Penning discharges for applications in VUV spectroscopy are generally high-current discharges with a high sputter rate from the cathode producing a cathode material dominated discharge. Compared with the high-current hollow cathode source, the Penning discharge source has several advantages. Due to the greatly enhanced path of the electrons in the discharge the buffer gas pressure can be reduced to about 10^{-2} Pa. This is typically 4 orders of magnitude lower than the pressure in a hollow cathode source, thus easing significantly the requirements for a differential pumping system or eliminating the need for such a system altogether. Discharge voltages of the order of up to 2 kV can be applied, creating a high sputter rate of the cathodes caused by the high kinetic energies of the impacting ions. The produced electrons gain nearly the full energy from the discharge potential so that they can produce multiply charged ions when colliding with the buffer gas or the cathode material atoms. The line emission

TABLE III. Radiant Intensity of Spectral Emission Lines of the Hollow Cathode for Operating Conditions of 1 A and 500 V Anode

λ (nm)	Atom/ion	I (μW/sr)
40.59/40.71	Ne II	141
46.07/46.24	Ne II	1025
48.81-49.11	Ne III	99.0
53.70	He I	30.1
54.29/54.32	Ar II	28.3
58.34	Ar II	6.9
58.43	He I	1410
66.19	Ar II	13.5
71.81	Ar II	13.2
72.20	Kr III	21.1
72.34	Ar II	24.2
72.55	Ar II	18.2
73.09	Ar II	16.1
73.59	Ne I	577
74.03	Ar II	34.8
74.37	Ne I	381
74.49/74.53	Ar II	25.6
76.92	Ar III	15.6
89.10	Kr II	36.8
91.74	Ar II	182
91.98	Ar II	388
93.21	Ar II	238
96.50	Kr II	201
104.82	Ar I	211
106.67	Ar I	175
116.49	Kr I	49.5
123.58	Kr I	136

Reprinted with permission from J. Hollandt, M. Kühne, and B. Wende, *Appl. Opt.* **33**, 68–74 (1994). Copyright © 1994 Optical Society of America.

spectrum of a Penning discharge therefore contains more and stronger short-wavelength lines than a hollow cathode discharge.

A drawback is the fast erosion of the cathodes, which reduces the usable lifetime to a few hours or less. Because of the fast erosion the geometry of the discharge changes with time and causes slow changes of the discharge parameters, for example, the discharge voltage, when using a constant current power supply.

The use of the Penning discharge as a source of VUV radiation dates back to Deslattes *et al.* [12] in 1963. They describe the basic design features (see Fig. 7) that today most Penning discharge sources follow [13–18]. A central cylindrical anode is opposed by two cathodes separated from the anode by isolating spacers. The discharge is viewed and pumped side-on through ports at the anode, which is at ground potential. All electrodes are water cooled. Behind each cathode a permanent magnet is located, producing a magnetic field in the discharge region in the order of 0.1 T. The sources are operated with a ballast resistor in series. Typical discharge voltages are in the 0.7- to 1.5-kV range with currents of the order of 0.3 to 2.5 A. Figure 8 shows a spectrum obtained by Bowyer [17] operating a Penning discharge source with neon as buffer gas and magnesium as cathode material.

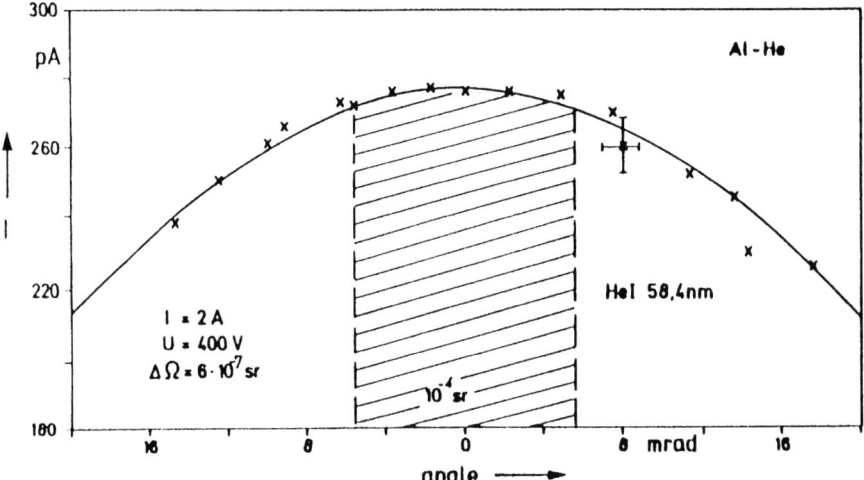

FIG. 5. Angular characteristic of radiation emission. To obtain the characteristic the hollow cathode was pivoted around an axis passing through the flux-limiting aperture perpendicular to the direction of observation. The angular resolution was 6×10^{-7} sr. [Reprinted with permission from K. Danzmann, M. Günther, J. Fischer, M. Kock, and M. Kühne, *Appl. Opt.* **27**, 4947–4951 (1988). Copyright © 1988 Optical Society of America.]

Absolute fluxes have been determined by Heise *et al.* [18] for a discharge with aluminum as cathode material and neon as buffer gas (Table IV). Comparing the spectral line emission of this Penning source with the emission of the hollow cathode source characterized by Danzmann *et al.* [8] for the aluminum doublet at 16 nm and taking the different flux-limiting apertures into account (2 mm for the Penning, 1.2 mm for the hollow cathode), the Penning emits (at $I = 2$ A and $U = 600$ V) a factor of 170 times more radiation than the hollow cathode source ($I = 2$ A and $U = 400$ V). If the current of the Penning dis-

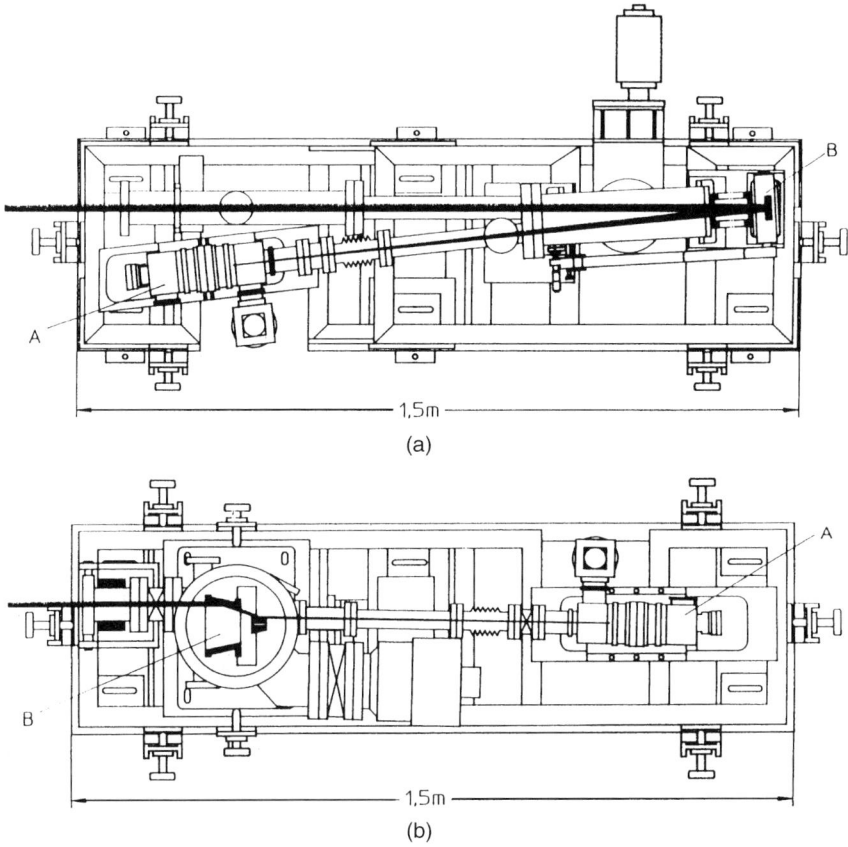

FIG. 6. (a) Schematic drawing of a collimated beam normal-incidence source standard. The diverging radiation from the hollow cathode source (A) is collimated by a concave gold-coated mirror (B). (b) Schematic drawing of a collimated beam grazing-incidence source standard. The diverging radiation from the hollow cathode source (A) is collimated by a Wolter type II telescope (B). [Reprinted with permission from J. Hollandt, M. Kühne, M.C.E. Huber, and B. Wende, *Astron. Astrophys. Suppl. Ser.* **115**, 561–572 (1996). Copyright © 1996 Astronomy & Astrophysics.]

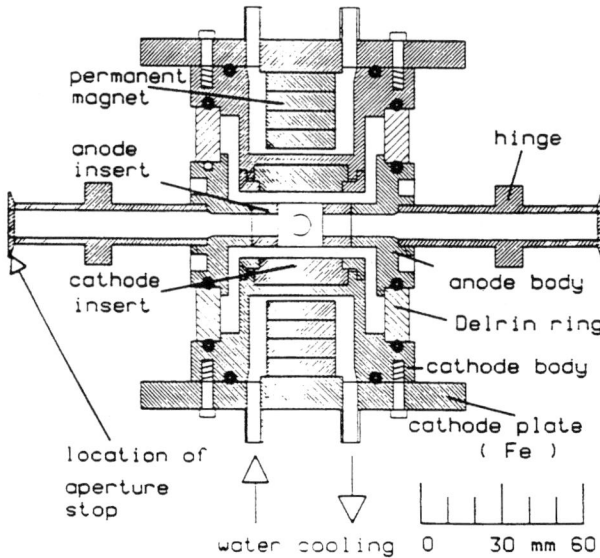

FIG. 7. Sectional drawing of a double-cathode Penning discharge source. Reprinted with permission from C. Heise, J. Hollandt, R. Kling, M. Kock, and M. Kühne, *Appl. Opt.* **33**, 5111–5117 (1994). Copyright © 1994 Optical Society of America.]

charge is increased to 2.5 A, this factor increases to 360. At short wavelengths the Penning discharge is brighter then the hollow cathode source by at least 2 orders of magnitude while operating at pressures about 4 orders of magnitude lower. However, this advantage is bought at the expense of a much higher erosion rate of the cathode. For example, the source of Heise *et al.* requires the cathodes to be changed after an operating time of about 5 h.

4.3 Electron-Beam Excitation Sources

In the case of the hollow cathode or the Penning discharge sources, the electrons in the discharge responsible for the excitation and ionization of the buffer gases obtain the required energy by passing through the cathode fall. The cathode fall is also responsible for the cathode erosion due to ion sputtering. Although the produced metal ions in the discharge are helpful to extend the line emission spectrum to shorter wavelengths, the erosion of the cathode limits the operating time of the sources and causes a slow change in the operating parameters and thus in the radiation output.

When the high-energy beam electrons in the hollow cathode or Penning discharge are replaced by electrons produced from an electron gun no erosion of the electrode takes place and a highly stable source of VUV line radiation can be created. With known excitation cross sections the absolute flux emitted by such a source can be determined from the operating parameters and the geometry of the source. First attempts to build such sources and determine the absolute excitation cross sections with spectrometers calibrated with synchrotron radiation were performed in the early 1980s by Flaig et al. [19] and McPherson et al. [20]. The use of an electron-beam excitation source for a relative VUV intensity calibration using calculated line intensities from the hydrogen molecule is described by Ajello et al. [21]. Detailed descriptions of an electron-beam excitation source as radiometric standards are given by Risley and Westerveld [22] and Jans et al. [23, 24].

The electron-beam excitation source of Jans et al. [23] is shown in Fig. 9. The source consists of an electron-gun chamber and a target gas chamber each

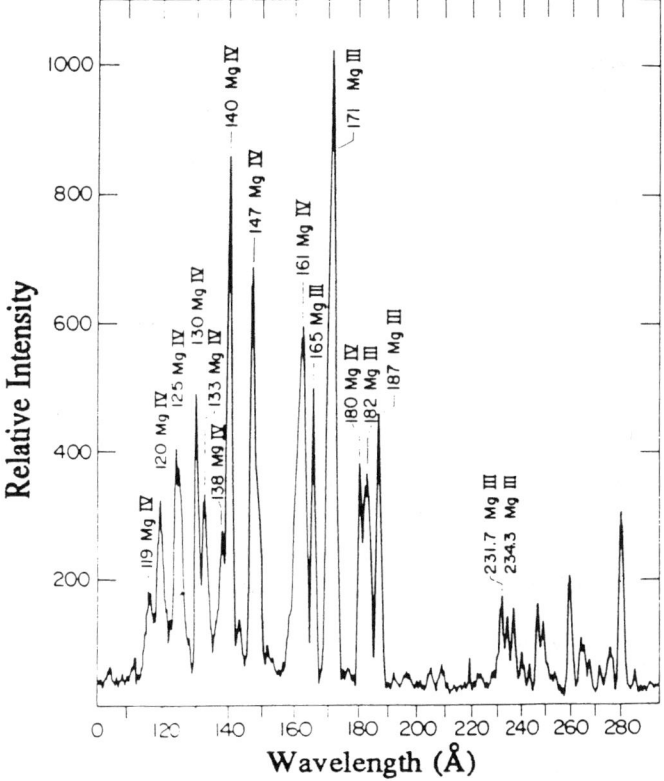

FIG. 8. Line emission spectrum from a Penning source with magnesium cathodes operated with neon as buffer gas. [Reprinted with permission from C. Bowyer, *Appl. Opt.* **32**, 6930–6933 (1993). Copyright © 1993 Optical Society of America.]

TABLE IV. Radiant Intensity and Radiance of Selected Al Emission Lines of a Penning Discharge Source

Spectral line	Discharge current (A)	Radiant intensity (W/sr $\times 10^{-6}$)	Radiance [W/(m^2 sr)]
Buffer gas Ne at 2×10^{-2} Pa			
10.38–10.45 nm. Al V	1.5	1.12	0.36
	2.0	2.91	0.93
12.40–12.61 nm. Al IV/Al V	1.5	7.08	2.25
12.97–13.32 nm. Al IV/Al V	1.5	11.2	3.57
	2.0	41.4	13.2
16.01–16.17 nm. Al IV	0.5	9.91	3.16
	1.0	105	33.5
	1.5	356	114
	2.0	892	284
	2.5	1922	612

Reprinted with permission from C. Heise, J. Hollandt, R. Kling, M. Kock, and M. Kühne, *Appl. Opt.* **33**, 5111–5117 (1994). Copyright © 1994 Optical Society of America.

pumped by a turbomolecular pump. The target gas chamber could be pumped down to a residual gas pressure of 3×10^{-6} Pa. The target gas density was typically of the order of 7×10^{-2} Pa corresponding to a gas density of 1.7×10^{13} cm^{-3}. A system of apertures kept the pressure in the electron-gun chamber below 1×10^{-4} Pa all the time. Details of the interaction region of the target chamber are shown in Fig. 10. The electron-beam is inserted into the target chamber from the top and is intercepted by a Faraday cup at the bottom. A commercial TV electron gun was used for the measurements producing 70 μA at a voltage of 2 kV. The observation of the gas target electron-beam interaction region was performed side-on through a differential pumping system. The first aperture of that system with a size of 2.7×11.5 mm^2 was located a distance of 10 mm from the interaction region. A length of about 4.8 mm was observed for the electron beam. A spinning rotor gauge was used to measure the gas density close to the interaction region.

By comparison with the calculable synchrotron radiation emission of the electron storage ring BESSY, excitation cross sections for 18 lines or line pairs of Ar, Kr, and Ne in the wavelength range from 46 to 96 nm have been determined with uncertainties mostly between 4 and 6% [24]. For details of the radiometric comparison and of the uncertainty budget, see Chapter 8.3 of this book. To use an electron-beam excitation source as a radiometric source standard, the following parameters need to be known with sufficient accuracy: target gas density from pressure and temperature measurements, energy of the exciting electrons, current of the exciting electrons, observed length of the electron beam, and solid angle of observation. For resonance lines the target gas pressure has to be kept small enough so that multiple collisional excitations are avoided and reabsorbtion can be neglected. The electron current must be sufficiently low to prevent nonlinear plasma effects from taking place. These boundary conditions cause the electron-beam excitation source to be a rather weak source compared with the two previously discussed plasma sources. Even when using an $f/6$ spectrometer with a grating efficiency of 10% and a detector counting efficiency of 10%, a typical count rate of only a few hundred counts/s can be expected [22]. If such a count rate, however, is sufficient for an application, then the electron-beam excitation source can be the radiometric source standard with the lowest available uncertainty for laboratory use.

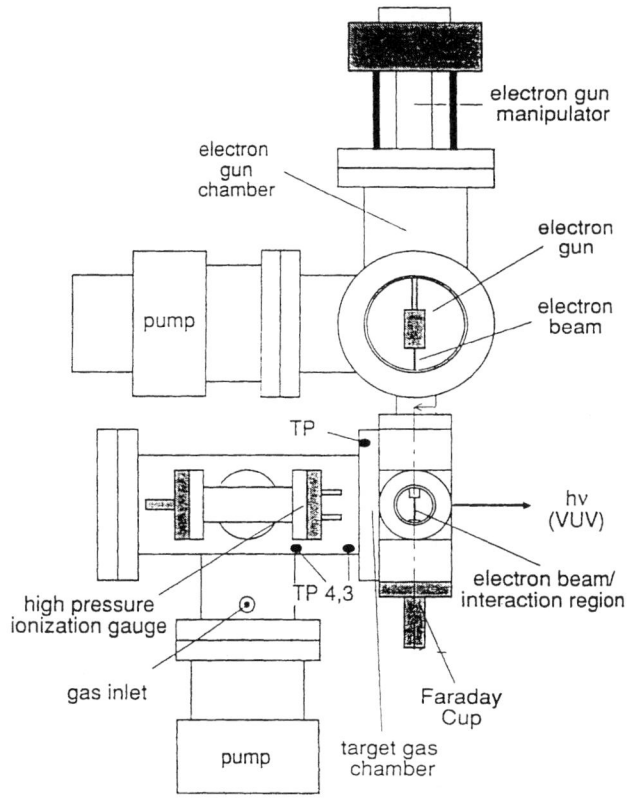

FIG. 9. Schematic diagram of an electron beam excitation source. [Reprinted with permission from W. Jans, B. Möbus, M. Kühne, G. Ulm, A. Werner, and K.-H. Schartner, *Appl. Opt.* **34**, 3671–3680 (1995). Copyright © 1995 Optical Society of America.]

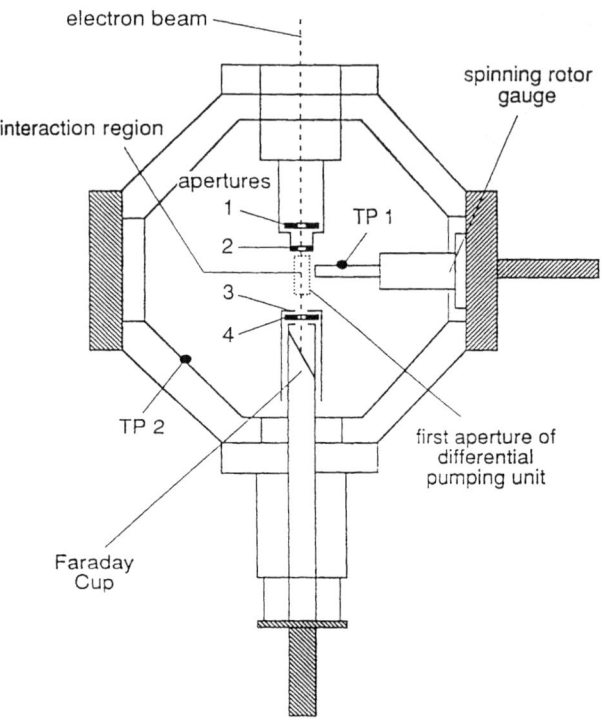

FIG. 10. Cross section of the interaction region of an electron beam excitation source. [Reprinted with permission from W. Jans, B. Möbus, M. Kühne, G. Ulm, A. Werner, and K.-H. Schartner, *Appl. Opt.* **34**, 3671–3680 (1995). Copyright © 1995 Optical Society of America.]

References

1. F. Paschen, *Ann. Phys.* **50**, 901–940 (1916).
2. B. E. Warner, K. B. Persson, and G. J. Collins, *J. Appl. Phys.* **50**, 5694–5703 (1979).
3. H. Koch and H. J. Eichler, *J. Appl. Phys.* **54**, 4939–4946 (1983).
4. D. Gerstenberger, R. Solanki, and G. J. Collins, *IEEE J. Quantum Electron.* **QE-16**, 820–834 (1980).
5. J. Z. Klose and J. M. Bridges, *Appl. Opt.* **26**, 5202–5203 (1987).
6. F. Paresce, S. Kumar, and C. S. Bowyer, *Appl. Opt.* **10**, 1904–1908 (1971).
7. K. Danzmann, J. Fischer, and M. Kühne, *J. Phys. D: Appl. Phys.* **18**, 1299–1305 (1985).
8. K. Danzmann, M. Günther, J. Fischer, M. Kock, and M. Kühne, *Appl. Opt.* **27**, 4947–4951 (1988).
9. J. Hollandt, M. Kühne, and B. Wende, *Appl. Opt.* **33**, 68–74 (1994).

10. J. Hollandt, M. Kühne, M.C.E. Huber, and B. Wende, *Astron. Astrophys. Suppl. Ser.* **115**, 561–572 (1996).
11. E. M. Penning, *Physica* **4**, 71–75 (1937).
12. R. D. Deslatte, T. J. Petersen, Jr., and D. Tomboulian, *J. Opt. Soc. Am.* **53**, 302–304 (1963).
13. E. S. Warden and H. W. Moos, *Appl. Opt.* **16**, 1902–1904 (1977).
14. D. S. Finley, S. Bowyer, F. Paresce, and R. F. Malina, *Appl. Opt.* **18**, 649–654 (1979).
15. D. S. Finley, P. Jelinsky, S. Bowyer, and R. F. Malina, *Proc. SPIE* **689**, 6–9 (1986).
16. M. Finkenthat, A. Litman, P. Mandelbaum, D. Stutman, and J. L. Schwob, *J. Opt. Soc. Am. B* **5**, 1640–1644 (1988).
17. S. Bowyer, Appl. Opt. **32**, 6930-6933 (1993).
18. C. Heise, J. Hollandt, R. Kling, M. Kock, and M. Kühne, *Appl. Opt.* **33**, 5111–5117 (1994).
19. H. J. Flaig, K.-H. Schartner, and H. Kaase, *Nucl. Instrum. Methods* **208**, 405–407(1983).
20. A. McPherson, N. Rouze, W. B. Westerveld, and J. S. Risley, *Ann. Isr. Phys. Soc.* **6**, 20–22 (1983).
21. J. M. Ajello, D. E. Shemansky, B. Franklin, J. Watkins, S. Srivastava, G. K. James, W. T. Simms, C. W. Hord, W. Pryor, W. McClintock, V. Argabright, and D. Hall, *Appl. Opt.* **27**, 890–914 (1988).
22. J. S. Risley and W. B. Westerveld, *Appl. Opt.* **28**, 389–400 (1989).
23. W. Jans, B. Möbus, M. Kühne, G. Ulm, A. Werner, and K.-H. Schartner, *Appl. Opt.* **34**, 3671–3680 (1995).
24. W. Jans, B. Möbus, M. Kühne, G. Ulm, A. Werner, and K.-H. Schartner, *Phys. Rev. A.* **55**, 1890–1898 (1997).

5. LASER PRODUCED PLASMAS

Martin Richardson

Laser Plasma Laboratory
CREOL/University of Central Florida
Orlando, Florida

5.1 Laser Plasma Sources of VUV Radiation

The development of high-power pulsed lasers, with their ability to produce dense hot plasmas, has provided the field of vacuum ultraviolet (VUV) spectroscopy with a new source of short-wavelength radiation. The characteristics of this source are radically different from conventional sources used for UV spectroscopy. As opposed to the electron driven plasma sources discussed in previous chapters, laser plasmas are pulsed, usually produced by the output of lasers having a duration of several nanoseconds or less. They are point sources, having spatial extents of usually less than a millimeter, that are precisely located in a vacuum chamber, allowing easy access and precision alignment to optical instrumentation. Like discharge devices, their spectrum is rich in line emission, which can be easily varied by manipulation of the laser plasma irradiation conditions. Until recently, one of their principal drawbacks has been the generation of target debris, high-velocity projectiles of minute solid particles or aerosol of target material emanating from the plasma that cause the damage or destruction of delicate optical elements. However, with the development of mass limited target geometries this limitation, common to many extreme ultraviolet (XUV) plasma sources, has been removed. Moreover, recent advances in several aspects of laser technology will provide new capabilities to laser-plasma-based XUV spectroscopy. Improvements in laser pumping technology, in particular, the use of high-power diode pumping, will permit the fabrication of compact high-repetition-rate (>1 kHz) laser plasma sources. In addition the recent development of high-power lasers with pulse duration of <100 fs will lead to bright laser plasma XUV sources of extremely short duration. In this review of laser plasma XUV sources we summarize their characteristics and some of their applications.

The recognition that the Q-switched laser could produce focused laser intensities ($>10^{12}$ W/cm^2) sufficient to ionize most materials provided atomic physicists and spectroscopists with a new plasma source. From the late 1960s, many investigators used high-resolution normal and grazing incidence spectrometers to search the spectra of plasmas created from every available element in the 50 to 600-Å range for a new emission lines from highly ionized species [1]. As the development of high-power laser systems progressed, driven in part as a consequence of the laser-driven inertial fusion program, higher focused lasers were achieved, allowing XUV spectroscopists to generate plasmas with higher plasma temperatures and thereby observe line emission from higher states of ionization. Some studies systematically identified specific sequences [2]. Some of these measurements have led to important corrections to models of the energy levels of highly stripped ions [3].

Thus, from its beginning, the laser plasma was used as a source for creating line emission from highly ionized states of various elements. This had advantages for several fields. It provided calibrated line transitions for highly excited ions of specific elements, allowing the identification of these elements and their approximate plasma conditions in astronomical sources, particularly the sun, and in thermonuclear plasmas, where impurity spectra can be a signature of plasma loss to the containment vessel [4]. As the characterization of the XUV emission from laser plasmas became better understood and more broadly known, its use as a source in other investigations was suggested: as an illuminating source for absorption spectroscopy [5], as a backlighting x-ray source for investigating shock waves and instabilities in dense plasmas [6], and as bright XUV light sources for x-ray microscopy [7] and x-ray lithography [8]. As improvements are made in the development of high-power pulsed lasers, and particularly their repetition rate, then the usefulness of laser plasmas as sources for XUV radiation will expand [9]. Higher repetition rate XUV sources will evolve with the development of compact, efficient, diode-pumped solid-state lasers [10]. Developmental versions of 1-kHz Nd:YAG 1-kW average power lasers, capable of focused intensities in excess of 10^{12} W/cm^2, are now available [11]. Although coherent laser light sources are now available in the XUV region, either from short-wavelength lasers or through high harmonic generation, laser plasma XUV light sources will still play an increasing role in science and technology because of the broad availability of wavelengths they offer, because of their cost benefit, and because many applications require the use of incoherent light.

The capability of high-power pulsed lasers to produce small localized high-density plasmas can readily be understood by considering the basic energetics of the interaction of a focused high-power laser pulse with dense matter. A typical laser plasma formed by a 10-ns-duration laser pulse, focused to a spot six of ~100 μm will ablate <1 μm of material during the interaction. This implies energies of $>10^4$ eV per atom for each joule absorbed. These energy densities are sufficient to create plasmas of highly stripped ions at temperatures in excess of 1,000,000 K that radiate XUV emission in all directions.

Physically, the laser light is absorbed on steep plasma density front a few tens of microns thick extending from the target. Several physical mechanisms can couple laser light to the plasma, depending on the characteristics of the laser pulse and the plasma [12]. However, the primary absorption mechanism, and

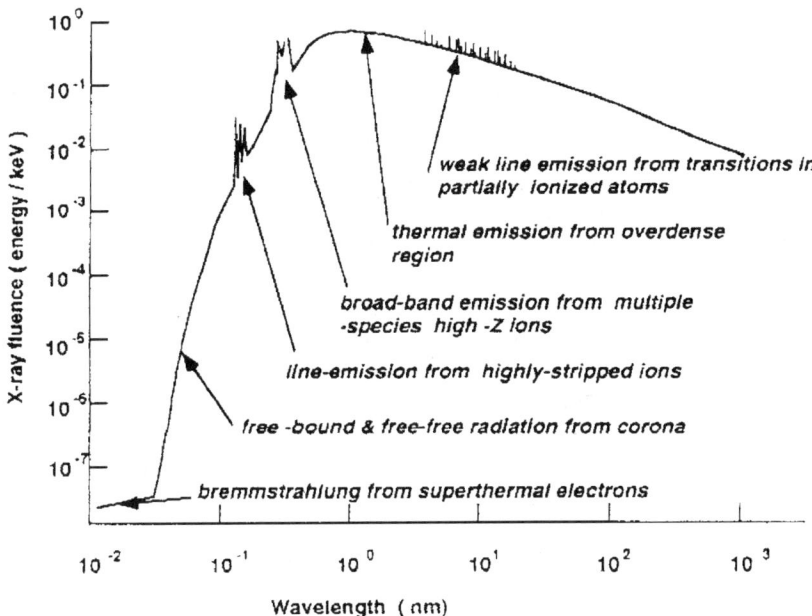

FIG. 1. Short-wavelength emission from laser plasmas.

that preferred for XUV generation, in inverse bremmstrahlung absorption in which light is absorbed at plasma densities just below the critical electron density [$n_e \sim 10^{21}\lambda$ (μm)$^{-2}$ electrons/cm^3), where λ (μm) is the laser wavelength. Collisional absorption of this energy is then dissipated in ablation, plasma expansion, and in the emission short-wavelength radiation (Fig. 1). The direction, intensity, spectrum, and temporal duration of the XUV emission depends on the target and irradiation conditions. Whereas other sounds of XUV emission rely primarily on one physical process, such as synchrotron emission or electron bombardment, the spectrum of the XUV emission from a laser plasma is the result of several processes. As a consequence, manipulation of the target and irradiation conditions can change substantially the relative preponderance of one process to another. Soft x-ray and XUV emission is produced primarily from the overdense region of the plasma ($n_c > n_e$), a region characterized by a steep gradient in the plasma temperature from the absorption region ($T_e \sim 1000$ eV) to the ablation region (few tens of eV). Thus, the resulting XUV source usually has a physical size of only several tens of microns or less in thickness, and a width that depends on the focused laser beam size and can range from a few microns to several hundreds of microns in extent. The XUV emission from laser plasmas has two principal components: continuum radiation resulting from free–free and free–bound transitions, and line radiation emitted by radiative decay of excited ions [4]. The relative level of these two contributions can vary depending on the irradiation and target conditions. More specifically it depends on the plasma temperature and density, which of course vary in space and time. As a consequence of this rich variability in the spectrum of the plasma's emission, its dependence on irradiation conditions, and the fact that this point source is pulsed and, depending on target geometry, can vary in isotropy of emission, the absolute flux calibration of laser plasma sources is not so well characterized and documented as other sources.

Early quantitative measurements of this emission used a combination of broadband filters and either x-ray film or x-ray diodes as a detector [13]. Accurate measurements through this technique are difficult, because of both the poor spectral resolution of the technique ($E/\Delta E < 10$) and the calibration reliability of the x-ray film or x-ray diode [14]. Spectrally dispersive instruments were used in many of the analyses of the spectral features of laser plasmas. These studies were motivated more by their rich production of new emission lines from excited ionic states generated for the first time rather than by an interest in characterizing the plasma as a radiation source. Reliable fluence measurements of the spectral emission from laser plasmas were made later, only after careful calibration of the dispersive element (either a reflective or transmissive grating) and recording materials (usually film) used. Detailed fluence measurements of laser plasmas have been performed by a number of investigators. Eidmann and Kishimoto used a calibrated transmission grating spectrometer to produce a survey of the spectral emission of a large number of target materials [15]. Figure 2 shows a selection of these measurements. The spectrum of the emitted radiation depends strongly on the target material. Observation of the same emission with a spectrometer having a higher spectral resolution [16] would show that each of the prominent spectral features shown actually consist of a tight set of narrow lines originating from individual transitions. Even for a given target material, a change in the temperature and density conditions within the plasma created by a change in the irradiation conditions will result in a modification of the spectral emission, as the relative population of different ion species is changed. Even today, there are insufficient data on the measured XUV fluence in specific lines from laser plasmas. However, as the use of XUV sources becomes more common, and with calibrated active array detectors in this region becoming increasingly available, this inadequacy will disappear.

A major drawback to the use of laser plasmas as XUV and soft x-ray sources has been the collateral damage caused by microscopic target material debris. This effect is of little consequence for low-repetition-rate or single-shot exposures where the risk of damage from debris is minimal. However, as high-repetition-rate sources become both possible and desirable, the need to combat the effects of target debris becomes more crucial. From a normal laser plasma the debris is in several forms, high-velocity ions from the plasma core itself

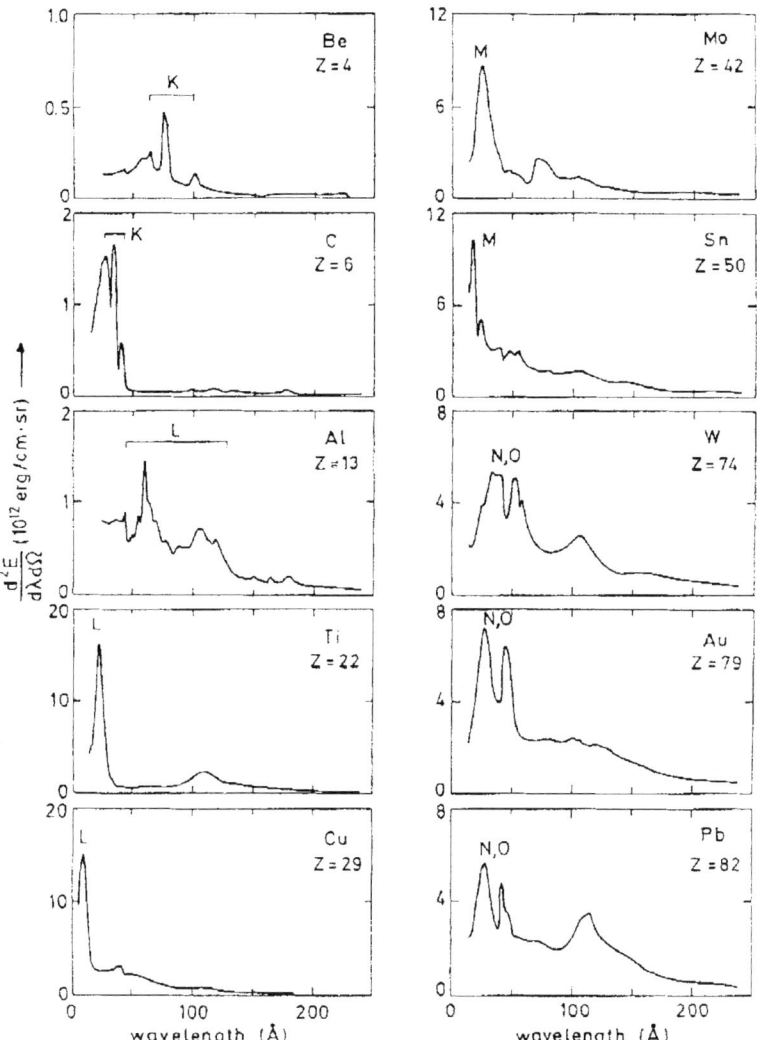

FIG. 2. Typical emission spectra from laser-produced plasmas from various targets [15]

having ion sound velocities in excess of 10^5 cm/s, neutral atoms and low-velocity ions generated by low-intensity light at the periphery of the laser spot, and large solid chunks or liquid aerosols that are spalled or boiled off the target by the inward propagating shock and thermal waves long after the laser light interaction [17]. This particulate matter produces different effects on optical elements in its path. The low-velocity atoms agglomerate as a quasi-uniform coating of target material on the surface, while the high-velocity ions are implanted below the surface.

The large particulate debris, which can have velocities up to several tens of meters per second and dimensions approaching 100 μm can damage or stick to the surface of optical elements and puncture thin-film filters. A number of techniques have been devised to blind sensitive XUV optics from the effects of this debris. These include the use of deflecting magnetic fields or a buffer gas [18], the use of mechanical shutters [19], and the use of thin protective filters [20]. The debris can also be mitigated by using a thin target, perhaps in the form of tape [21], or by attempting to direct the debris to benign regions of the vacuum system.

The use of cryogenic targets as a solution to the debris problem was suggested as early as 1967 [22]. At that time, millimeter-size cryogenic hydrogen targets were being used for laser fusion experiments and for plasma-filling approaches for magnetic fusion systems. Later cryogenic halogen gas targets were proposed as a laser plasma x-ray target for lithography [23]. Recently, a frozen water laser plasma target has been developed for generating line emission in the EUV spectral region [24]. Although one would expect this approach to eliminate particulate debris by evaporation, shocks generated in the crystalline material are effective in producing shards with high velocities that have the range to damage sensitive components in their line of sight. Solid Xe has been shown to produce bright XUV emission in the 13 nm region [25]. The elimination of particulate emission from laser plasmas produced from solid targets can only be guaranteed with the use of a continuous train of mass-limited targets. Under this definition [26], the mass of the target should be no larger than the mass of the total number of EUV radiators required. For sources having laser energies of a few joules, this implied targets of <100 μm in size with masses of <1 μg. Many laser plasma studies have utilized mass-limited targets. They have been exclusively used for laser fusion experiments; however, these are precision targets individually aligned to the laser beam.

A particular simple mass-limited target configuration uses liquid droplets. First proposed as a laser plasma x-ray target in 1989 [27], it has recently received considerable attention as a high-repetition-rate source for microscopy [28]. The general configuration of this target source is shown in Fig. 3. A liquid jet system, either of the ink-jet design or a droplet-on-demand system, injects a train of liquid droplets into the vacuum. The droplets freeze by evaporation and the train passes through the laser focal spot. Unused targets can be collected on a cold finger and, hence, the vacuum integrity is maintained. Long-time exposure studies have shown immeasurable levels of debris generation of over many millions of laser shots [29]. The spectral emission of liquid droplet target sources may be varied by changing the composition of the droplet liquid. High Z materials can be added either in molecular form or as minute particles in

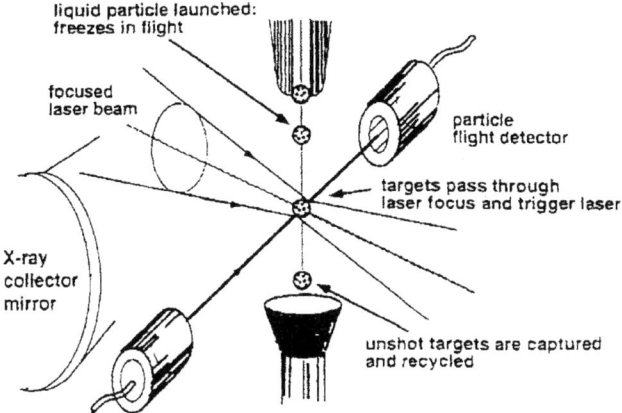

FIG. 3. Continuous droplet target for debrisless operation.

suspension. An alternative debrisless target approach utilize a gas jet [30]. This target system also ensures the creation of minimal target debris.

Up to the present time applications of laser plasma XUV sources have largely been limited to the laboratory. At this level they have been extremely useful, providing combinations of brightness, emission duration, and source size not easily attained by other sources. Many applications have been suggested for XUV radiation from laser plasmas. Much of their use has been in spectroscopy. Eugene Kennedy and his colleagues have pioneered their use as sources for absorption spectroscopy [31]. One of the most successful techniques is resonant laser driven ionization (RLDI) [32]. This technique and the use of dual laser plasmas [33] have allowed high-resolution measurements of photoionization cross sections [34], giant resonances [35], and detailed analysis of sub-shell absorption spectra [36]. Laser plasma sources are also used as transition calibration and transfer sources for such applications as detector calibration and the characterization of XUV optical elements [37]. They are also being used as sources for high-resolution x-ray microscopy, particularly for the analysis of biological organisms [38], where the pulsed nature of the emission has an inherent advantage over synchrotron sources, in that with sufficient flux, a single shot image can be acquired in a time so short that no radiation damage can structurally alter the specimen. Among other applications, recently there has been considerable interest in the use of laser plasma sources for EUV and soft x-ray lithography. A laser plasma source operating at 13.0 nm (or 11.6 nm) is at present the source of choice for projection EUV lithography [39]. Although synchrotron sources are still favoured for proximity soft x-ray lithography, which requires emission at the 1.0-nm wavelength, a laser plasma will be a strong contender if operated at high enough repetition rates. These applications of laser plasmas would signal the first major transition of the laser plasma source from a laboratory tool to an application in manufacturing.

As further improvements are made in the efficiency and average power of pulsed laser systems, and as reductions occur in their cost, complexity, and size, particularly with the introduction of high-power diode pumping, laser plasma short-wavelength sources will find more applications outside the laboratory. The recent development of debrisless target sources eliminates one of the most serious drawbacks to the use of high-repetition-rate laser plasma sources. Thus it is highly likely that their use will, be more widespread in the future.

References

1. C. Colin, Y. Durand, F. Floux, D. Guyot, P. Langer, and P. Veyrie, *J. Appl. Phys.* **39**, 2991 (1968); B. C. Boland, F. E. Irons, and R. W. P. McWhirter, *J. Phys. B* **1**, 1180 (1968); G. Tondelo, E. Jannitti, and A. M. Malvezzi, *Phys. Rev. A* **16**, 1705 (1977); B. C. Fawcett, A. H. Gabriel, F. E. Irons, N. J. Peacock, and P. A. H. Saunders, *Proc. Phys. Soc. London* **88**, 1051 (1966); A. W. Ehlers and G. L. Weissler, *Appl. Phys. Lett.* **8**, 89 (1966); C. Breton and R. Papoular, *J. Opt. Soc. Am.* **63**, 1275 (1973); P. J. Malozzi, H. M. Epstein, R. G. Jung, D. C. Applebaum, B. P. Firand, and W. J. Gallagher, in "Fundamental and Applied Laser Physics" (M. S. Fred, A. Javan, and N. A. Kurnet, eds.), Wiley-Interscience, New York, p. 165 (1973); P. G. Burkhalter, D. J. Nagel, and R. R. Whitlock, *Phys. Rev. A* **9**, 2331 (1974).
2. V. A. Boiko, A. Ya. Faenov, and S. A. Pikus, *J. Quant. Spectrosc. Radiat. Transfer* **19**, 11 (1978); V. A. Bioko, S. A. Pikus, U. I. Safronova, and A. Ya. Faenov, *J. Phys. B* **10**, 1253 (1977); P. K. Carroll and G. O'Sullivan, *Phys. Rev. A* **25**, 275 (1982); G. O'Sullivan and P. K. Carroll, *J. Opt. Soc. Am.* **71**, 227 (1981); J. F. Seely, U. Feldman, C. M. Brown, W. E. Behring, and M. C. Richardson, *Proc. SPIE* **831**, 25 (1987) and references there included.
3. J. F. Seely, J. O. Ekberg, C. M. Brown, U. Feldman, W. E. Behring, J. Reader, and M. C. Richardson, *Phys. Rev. Lett.* **23**, 2924 (1986).
4. C. DeMichelis and M. Mattioli, *Nucl. Fusion* **21**, 677 (1981).
5. P. K. Carroll, E. T. Kennedy, and G. O'Sullivan, *Appl. Opt.* **19**, 1454 (1980); T. B. Lucatorto, T. J. McIlrath, J. Sugar, and S. M. Younger, *Phys. Rev. Lett.* **47**, 1124 (1981).
6. J. Grun, M. H. Emery, S. Kacenjar, C. B. Opal, E. A. McClean, S. P. Obenschain, B. H. Ripin, and A. Schmitt, *Phys. Rev. Lett.* **53**, 1352 (1984).
7. R. J. Rosser, K. G. Baldwin, D. Bassett, A. Coles, and R. W. Eason, *J. Microscopy* **138**, 311–320 (1985).
8. J. Nagel, R. R. Whitlock, J. R. Greig, R. E. Pechacek, and M. C. Peckarar, *Proc. SPIE* **135**, 46–53 (1978).
9. D. J. Nagel, C. M. Brown, M. C. Pekarar, M. L. Ginter, A. Robinson, T. J. McIlrath, and P. K. Carroll, *Appl. Opt.* **23**, 1428 (1984).
10. W. Koechner, "Solid State Laser Engineering," 4th ed., Springer-Verlag, Berlin (1977).
11. M. Hermann, J. Honig, and L. Hackel, *Proc. OSA Top. Mtg. EUV Lithography* **23**, 248 (1994).

12. M. Kruer, "The Physics of Laser Plasma Interactions," Frontiers in Physics, Addison Wesley, Reading, MA (1987).
13. J. L. Bobin, F. Floux, P. Langer, and H. Pignerol, *Phys. Lett.* **28A**, 398–399 (1968); M. D. Rosen, D. W. Phillion, V. C. Rupert, W. C. Mead, W. L. Kruer, J. J. Thomson, H. N. Kornblum, V. W. Slivinsky, G. J. Caparaso, M. J. Boyle, and K. G. Tirsell, *Phys. Fluids* **22**, 2020 (1979).
14. R. H. Day, P. Lee, E. B. Saloman, and D. J. Nagel, *J. Appl. Phys.* **52**, 6965 (1981).
15. K. Eidmann and T. Kishimoto, *Appl. Phys. Lett.* **49**, 377 (1986).
16. W. Schwanda, K. Eidmann, and M. Richardson, *J. X-Ray Sci. Technol.* **4**, 8–17 (1993).
17. W. T. Silfvast, M. C. Richardons, H. Bender, A. Hanzo, V. Yanovsky, F. Jin, and J. Thorpe, *J. Vac. Sci. Technol.* **B10**, 3126 (1992).
18. F. Bjkerk, E. Lois, L. Shmaenok, H. J. Voorma, M. J. Van der Weil, I. C. E. Turcu, and G. J. Talents, *Proc. SPIE* **1848**, 516 (1986).
19. W. T. Silfvast, H. Bender, A. M. Eligon, D. O'Connell, A. Hanzo, and M. Richardson, *Proc. OSA Top. Mtg. on EUV Lithography* **18**, 117 (1993).
20. R. C. Spitzer and D. P. Gaines, *Proc. OSA Top. Mtg. on EUV Lithography* **23**, 243 (1995).
21. S. J. Haney, K. W. Berger, G. D. Kubiak, P. D. Rockett, and J. Hunter, *Appl. Opt.* **32**, 6934 (1993).
22. T. P. Hughes, "Plasmas and Laser Light," Wiley, New York (1975).
23. T. Mochizuki, R. Kodama, and C. Yamanaka, *Proc. SPIE* **733**, 1234 (1986).
24. M. Richardson, K. Gabel, F. Jin, and W. T. Silfvast, *Proc. OSA Top. Mtg. on EUV Lithography* **18**, 156 (1993).
25. S. J. Haney, K. W. Berger, G. D. Kubiak, P. D. Rocket, and J. Hunter, *Appl. Opt.* **32**, 6934 (1993).
26. M. Richardson, W. T. Silfvast, H. Bender, A. Hanzo, V. P. Yanovsky, F. Jin, and J. Thorpe, *Appl. Opt.* **32**, 6901 (1993).
27. J. Traill, "A Compact Scanning Soft X-Ray Microscope," Ph.D. thesis, Stanford University (1989).
28. L. Rymell and H. M. Hertz, *Opt. Comm.* **103**, 105 (1993); M. C. Richardson and F. Jin, *Proc. OSA Top. Mtg. on EUV Lithography* **23**, 265 (1995).
29. D. Torres, F. Jin, M. Richardson, and C. DePriest, *Trends Optics Photon.* **IV**, 75–79 (1996).
30. G. D. Kubiak, L. J. Bernardez, K. D. Krenz, D. J. O'Connell, R. Gutowski, and A. M. Todd, *Trends Optics Photon.* **IV**, 66 (1996).
31. E. T. Kennedy, J. T. Costello, J.-P. Mosnier, A. A. Cafolla, M. Collins, L. Kiernan, U. Koble, M. H. Sayyad, M. Shaw, B. F. Sonntag, and R. Barchewitz, *Opt. Eng.* **33**, 3984 (1994); J. T. Costello, J.-P. Mosnier, E. T. Kennedy, P. K. Carroll, and G. O'Sullivan, *Physica Scripta* **T34**, 77 (1991).
32. T. B. Lucatorto, T. J. McIlrath, J. Sugar, and S. M. Younger, *Phys. Rev. Lett.* **47**, 1124 (1981).
33. P. K. Carroll and E. T. Kennedy, *Phys. Rev. Lett.* **38**, 1068 (1977).
34. D. Cubaynes, S. Diehl, L. Journel, B. Rouvellou, J-M. Bizau, S. A. Moussalami, F. J. Wuilleumier, N. Berrah, L. VoKy, P. Faucher, A. Hibbert, C. Blanchard, E. Kennedy, T. J. Morgan, J. Bozek, and A. S. Schlachter, *Phys. Rev. Lett.* **77**, 2194 (1996).
35. U. Koble, L. Kiernan, J. T. Costello, J.-P. Mosnier, E. T. Kennedy, V. K. Ivanov, V. Kupchenko, and M. S. Shendrik, *Phys. Rev. Lett.* **74**, 2188 (1995).

36. G. O'Sullivan, C. McGuinness, J. T. Costello, E. T. Kennedy, and B. Weinmann, *Phys. Rev. A* **53**, 3211 (1996).
37. E. M. Gullicksen, J. H. Underwood, P. C. Batson, and V. Nikiyin, *J. X-Ray Sci. Technol.* **3**, 283 (1992); Y. Orikawa, K. Nagai, and Y. Yketaki, *Opt. Eng.* **33**, 1721 (1994).
38. M. Richardson, K. Shinohara, K. A. Tanaka, Y. Kinjo, N. Ikeda, and M. Kado, *Proc. SPIE* **1741**, 133 (1992); J. M. Rajyaguru, M. Kado, K. Nekula, M. Richardson, and M. J. Muzynski, *Microbiology* **143**, 733 (1997).
39. N. Ceglio, A. M. Hawryluk, and G. Sommargren, *Appl. Opt.* **32**, 7050 (1993).

6. TRANSITION RADIATION

Arthur J. Braundmeier, Jr
Department of Physics
Southern Illinois University, Edwardsville
Edwardsville, Illinois

Edward T. Arakawa
Oak Ridge National Laboratory
Oak Ridge, Tennessee

When an energetic charged particle makes a transition from vacuum into a metal there is an emission of electromagnetic radiation known as *transition radiation*. This radiation appears whenever a charged particle crosses an interface separating media of different dielectric constants. The radiation is produced at the interface separating the two media and is strongly polarized in the plane containing the normal to the interface and the direction of the emitted photon. This process was first studied theoretically by Frank and Ginsburg [1] in 1945, later by Ritchie and Eldridge [2], and more recently by Ashley [3] and Kroger [4]. A discussion of the development of transition radiation theory can be found in a review by Frank [5].

In the case of a charged particle crossing the interface between vacuum and a metal, there will be a strong radiation peak in the spectral region where the metal becomes transparent, the plasma wavelength. This radiation emanates from the excitation and subsequent radiative decay of a plasma oscillation involving the free electrons of the metal. However, a continuum of transition radiation will also be present at wavelengths other than the plasma wavelength. The properties of the transition radiation are strongly governed by the optical constants of the metal, the thickness of the metal, the incidence angle of the impinging charged particle, and the particle's velocity.

In the case of a thin metal foil, where the charged particle traverses the foil boundaries, transition radiation will be produced at both interfaces and the radiation intensity will show an oscillatory dependence on the foil thickness for wavelengths at or near the plasma wavelength. Because the foil is strongly absorbing elsewhere, the periodic dependence on foil thickness disappears at wavelengths away from the plasma wavelength.

Figure 1 shows the theoretical and experimental spectra from a self-supported, 66-nm-thick vacuum-deposited Ag foil when bombarded by 40-keV normally incident electrons [6]. The radiation was observed at an angle of 30°

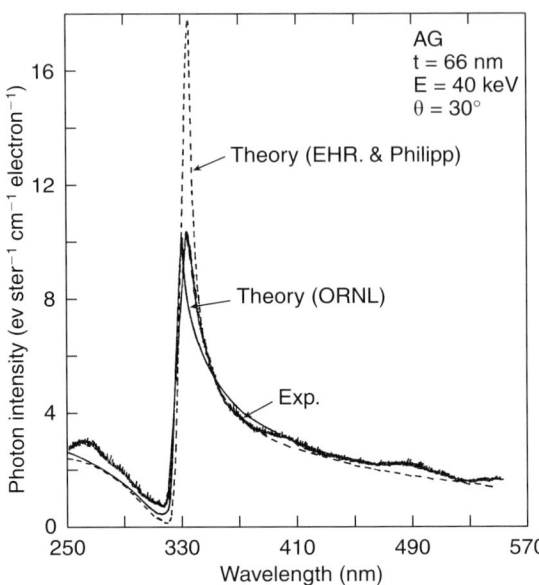

FIG. 1. Spectral distribution of transition radiation from an electron-bombarded silver foil.

from the film normal in the forward direction. The strong peak near 330 nm is the radiation from the excitation of the plasma oscillation in the metal foil. The theoretical curves were obtained from the Ritchie–Eldridge theory using the optical constants measured from bulk Ag [7] and from vacuum-evaporated Ag [8]. Good agreement with the experimental spectrum is obtained if the optical constants of vacuum-evaporated Ag are used in the theory.

Figure 2 shows the measured oscillatory behavior of the 330-nm wavelength transition radiation emitted from Ag foils of various thicknesses and for several different electron energies [6]. In each case the electrons were normally incident on the foils. Also included is the predicted behavior according to the Ritchie–Eldridge theory for normally incident particles.

Braundmeier and Arakawa have measured the transition radiation from a vacuum-evaporated Be foil deposited on a stainless steel substrate [9]. Figure 3 shows the measured radiation and a theoretical fit to the experimental data using the theory of Ashley and the optical constants of Toots *et al.* [10]. The electrons were normally incident on the foil and the transition radiation was observed at 30° from the electron beam in the forward direction. A strong peak in the transition radiation was observed at a wavelength of 66 nm, which was attributed to the excitation and subsequent radiation of the plasma oscillation.

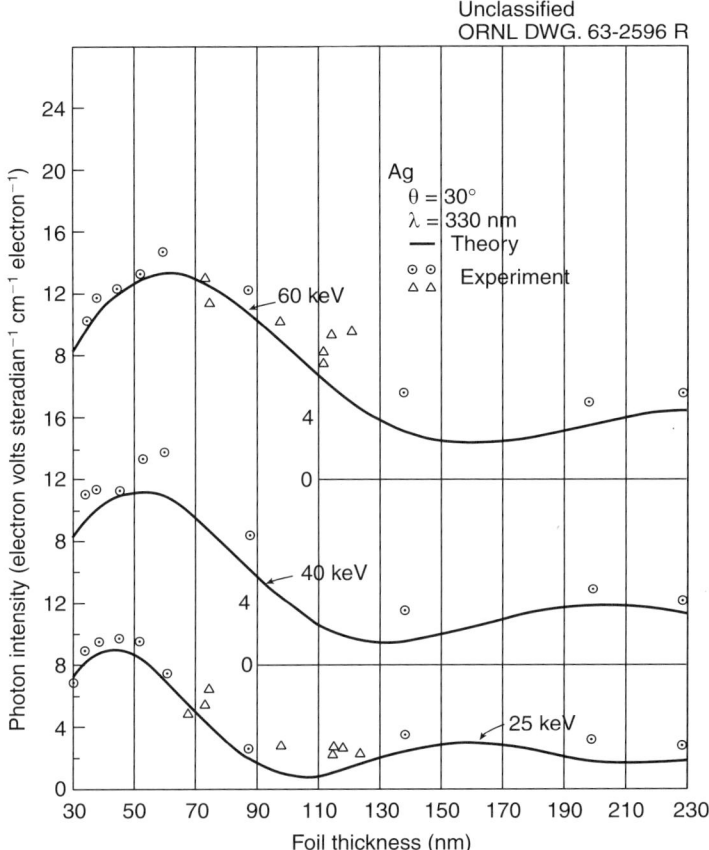

FIG. 2. Measured photon intensity at $\lambda = 330$ nm emitted from silver foils bombarded by normally incident electrons of various energies. The photons were observed at 30° from the foil normal as a function of foil thickness.

Ashley [3] extended the theory of transition radiation from a self-supported thin foil to nonnormally incident particles and thus allowed the determination of the optimum experimental parameters for producing intense transition radiation. As expected, the predicted photon intensities are dependent on the optical constants used in the calculation. Because the theory of Ashley is concerned only with self-supported foils the theoretical plots shown in Figs. 4, 5, and 6 treat this case. However, transition radiation can be observed from foils deposited on substrates as seen in Fig. 3.

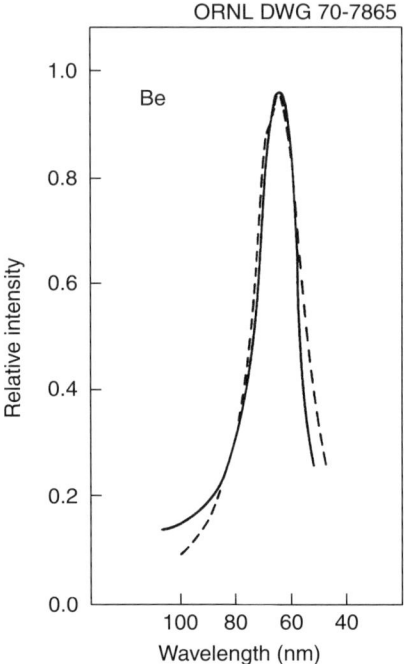

FIG. 3. Transition radiation from a Be film bombarded by 80-keV electrons. The solid curve is the experimentally measured emission; the dashed curve is the theoretical spectrum for a 10-nm-thick foil.

Figure 4 shows the spectrum from 30-nm-thick Ag and Al foils bombarded by normally incident 80-keV electrons. The angle of photon observation is at 30° from the foil normal in the forward direction as in Fig. 1.

A strong source of transition radiation in the vacuum ultraviolet can be obtained if the metal foil is Al. Figure 5 shows a contour plot of the intensities at the 80-nm peak wavelength from a self-supported, 30-nm Al foil as a function of both the incident-electron angle and the photon-observation angle. The thickness of 30 nm was chosen because it represents an easily produced free-standing Al foil.

From curves such as this, one can determine the optimum angles for the incident electrons and the emitted photons in order to obtain the maximum transition radiation for a chosen electron energy and foil thickness. These optimum angles are also dependent on the foil thickness and the energy of the incident electrons. As the electron energy increases the maximum transition radiation intensity increases and moves to slightly higher values of the incident-electron angle. In general, as the foil thickness increases the transition radiation

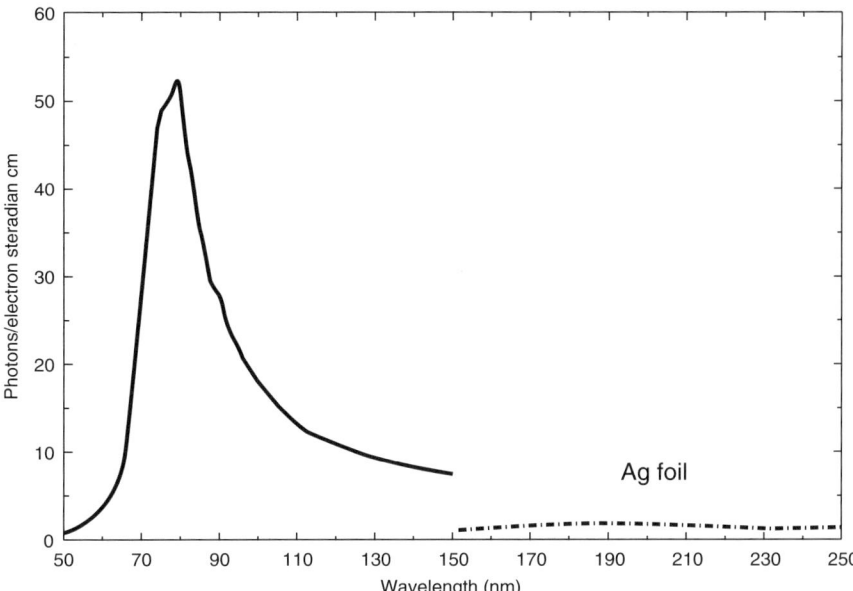

FIG. 4. Theoretical spectral distribution of p-polarized transition radiation from 30-nm-thick Al and Ag foils.

decreases in intensity and the maximum intensity peak moves to higher values of the electron incident angle. The optimum observation angle for the transition radiation is weakly dependent on the foil thickness and electron incident angle.

Figure 6 shows the theoretical spectral distribution of transition radiation from a self-supported, 30-nm Al foil bombarded by 80-keV electrons at an incidence angle of 20°. The photons were observed at an angle of 20°, in accordance with the optimum conditions shown in Fig. 5. Although much reduced in intensity from the maximum, there is significant transition radiation extending from 50 to 150 nm.

It is clear from the foregoing that transition radiation can be used as a source of ultraviolet and extreme ultraviolet radiation. Although this source has not been widely used for experimental purposes, there are situations where its use may be advantageous. In comparison with other laboratory sources such as gas discharge sources and laser-produced plasmas, transition radiation is a "clean" source with no residues or particles emitted which may contaminate the sample being studied. It can be operated and performs best under the ultrahigh vacuum conditions compatible with most experimental chambers. In addition, transition radiation is nearly 100% p-polarized. This feature can be quite useful in a laboratory source since polarizers in this wavelength region are very inefficient, which can result in a 90 to 95% decrease in intensity if polarized light is required.

Transition radiation sources probably will not be able to compete with the intensities available from synchrotrons but several advantages are clearly seen. For example, most of the radiation from transition radiation is emitted in a narrow wavelength band unique to the metals being irradiated. This is in contrast to the wideband nature of synchrotron radiation and its attendant higher order interference effects in the spectrometers. These higher orders must be dealt with either by optical filters or by the use of premonochromators.

Transition radiation sources can be built in a volume as small as 1 cm^3 with a simple electron emitting filament and a target material. Electron energies as low as 20 eV have been used to excite transition radiation [11]. Other metals which have been used in studies of transition radiation include Cd [12], Zn [12], and Mg [13] with peaks at 136, 144, and 140 nm, respectively. A convenient, miniature source that covers a wide range of wavelengths can be constructed quite easily with the use of a multitarget source.

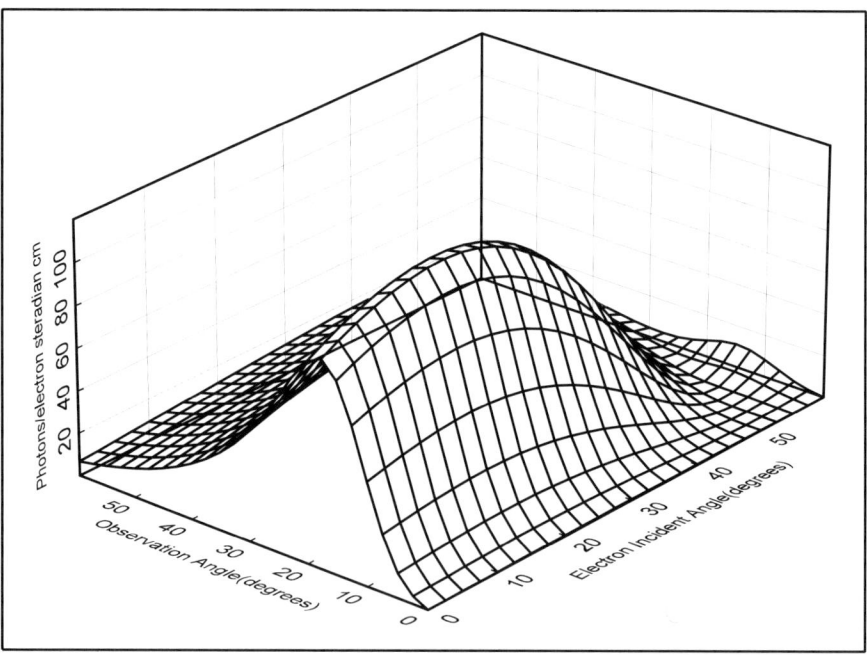

FIG. 5. Theoretical p-polarized transition radiation at $\lambda = 80$ nm emitted from a 30-nm-thick Al foil bombarded by 80 keV electrons.

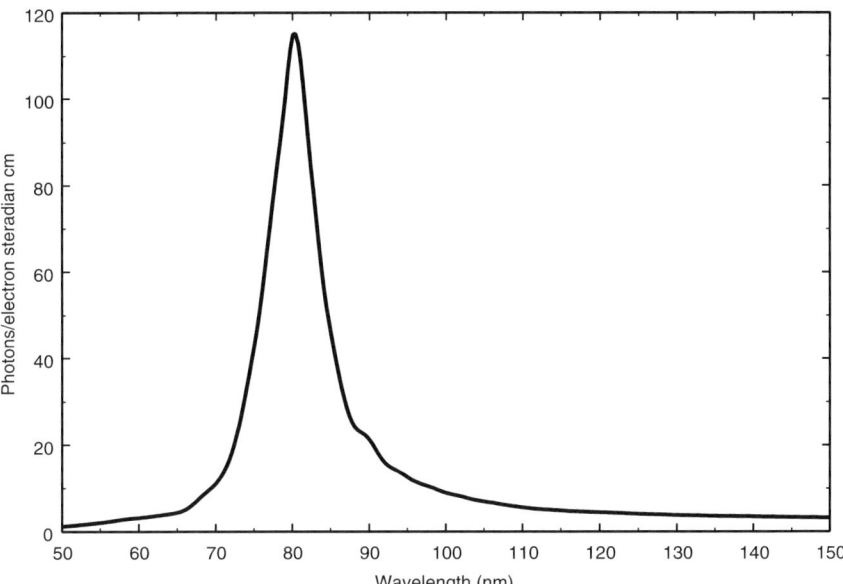

FIG. 6. *P*-polarized transition radiation from a self-supported, 30-nm-thick Al foil bombarded by 80-keV electrons at 20° from the foil normal. The photons are observed at 20° from the foil normal on the side opposite the incident electrons.

Hughes *et al.* [14] present spectrograms from 220 to 2 nm of optical emission produced by a simple source using 3- to 10-keV electrons on W and Ta targets. Their investigation was directed toward the use of their source as a portable standard of radiation in the vacuum ultraviolet and soft x-ray region of the spectrum.

References

1. I. Frank and V. Ginsburg, *J. Phys. USSR* **9**, 353–362 (1945).
2. R. H. Ritchie and H. B. Eldridge, *Phys. Rev.* **126**, 1935–1947 (1962).
3. J. C. Ashley, *Phys. Rev.* **155**, 208–210 (1967).
4. E. Kroger, *Z. Physik* **235**, 403–421 (1970).
5. I. M. Frank, *Usp. Fiz. Nauk* **87**, 189–210 (1965) [English transl.: *Sov. Phys. Usp.* **8**, 729–742 (1966)].
6. E. T. Arakawa, N. O. Davis, and R. D. Birkhoff, *Phys. Rev.* **135**, A224–A226 (1964).
7. H. Ehrenreich and H. R. Philipp, *Phys. Rev.* **128**, 1622–1629 (1962)
8. R. H. Huebner, E. T. Arakawa, R. A. MacRae, and R. N. Hamm, *J. Opt. Soc. Am.* **54**, 1434–1437 (1964).
9. A. J. Braundmeier, Jr., and E. T. Arakawa, *Opt. Comm.* **2**, 257–258 (1970).
10. J. Toots, H. A. Fowler, and L. Marion, *Phys. Rev.* **172**, 670–676 (1970).

11. M. S. Chung, T. A. Calcott, E. T. Kretschmann, and E. T. Arakawa, *Surf. Science* **91**, 245–263 (1980).
12. E. T. Arakawa, R. N. Hamm, W. F. Hanson, and T. M. Jelinek, in "Optical Properties and Electronic Structure of Metals and Alloys" (F. Abeles, ed.), North-Holland, Amsterdam, pp. 374–385 (1965).
13. E. T. Arakawa, R. J. Herikhoff, and R. D. Birkhoff, *Phys. Rev. Lett.* **12** 319–320 (1964).
14. R. H. Hughes, T. A. Heumier, and P. M. Griffin, *Appl. Opt.* **20**, 1350–1354 (1981).

7. VACUUM ULTRAVIOLET LASERS

Pierre Jaeglé
Laboratoire de Spectroscopie Atomique et Ionique
Université Paris-Sud
Orsay, France

7.1 Introduction

XUV lasers (XRL) and high-order harmonic generation (HHG) are new sources whose applicability to scientific and technological works appeared practical only about the middle of the 1990s [1]. For both sources the short-wavelength limit of the physical process is about a few nanometers, but the intensity and other beam qualities required for most practical uses are achievable only at longer wavelengths. X-ray lasers of large power have been demonstrated up to the 15-nm wavelength. However, size and cost of experimental facilities leave the giant scale only for wavelengths above 20 nm. As for HHG, the conversion efficiency in the vacuum ultraviolet (VUV) range is approximately constant with decreasing wavelength but this *plateau* falls to smaller values below 30 to 40 nm.

Visible or UV pulsed lasers of large power are an unavoidable part of the technical environment of x-ray lasers and frequency conversion systems as well. For XRLs the pump pulse duration is 0.1 to 1 ns, whereas it ranges around 25 fs to 50 ps for HHG. This difference in pulse duration makes the driving lasers smaller in the second case. The development of laser technology already allows us to consider new laser systems which will largely reduce XRL driver size in the future. At present, development of a 46.9-nm wavelength x-ray laser, pumped by an electrical discharge, is progressing quickly [2].

The principle of the two sources is described in the next sections with examples of contemporary realizations. We show values of beam intensity, duration, brightness and coherence parameters. In addition, x-ray laser applications, which are already in progress, will be briefly described.

7.2 XUV Lasers

7.2.1 Population Inversions in Hot Plasmas

The multiple charged ions of hot plasmas are intense sources of XUV radiation, which is emitted by spontaneous transitions between ion optical levels. Moreover, spectral studies of laser-produced plasmas have already shown population anomalies suggesting possible contributions of stimulated emission in experimental spectra [3,4]. It is a fact that the laser-produced plasmas present characteristics which are exceptionally propitious for VUV and XUV stimulated processes to appear. Even in local thermodynamic equilibrium plasmas, at temperatures of a few hundreds of electron volts, Boltzmann population ratios of levels separated by a few tens of electron volts only are rather close to unity. Because photon mean free paths, for many spectral lines, are much shorter than the plasma size, the emerging line intensities include not only spontaneous emission and photon absorption but also a significant contribution of stimulated emission. Now these plasmas are submitted to very fast density and temperature variations which push them out of equilibrium and may raise population ratios to inversion. When this occurs the plasma yields amplified emissions. Note that the achievement of a large amplification requires the plasma to be in an elongated cylindrical form of a few centimeter length. This is obtained by cylindrical focusing of the driving laser beam, as shown in Fig. 1. Many alternative focusing configurations can be used according to the structure of the target and the number of pumping beams.

As a matter of fact, two sets of processes involving plasma hydrodynamics and atomic physics have proved to be efficient in producing population inversions. The first to be recognized is the so-called recombination scheme of hydrogenic or lithium-like ions [5–8]. In this scheme the driving laser energy is used to heat the plasma to temperatures where atoms are completely stripped, in the case of hydrogenic ion pumping, or brought to the helium-like configuration, in the case of lithium-like ion pumping. After the end of the driving laser pulse, the plasma expands freely in the vacuum and cools down. As long as the electronic density remains large, the ion upper levels are rapidly populated by two-electron recombination while the lower levels are much more slowly populated by radiative recombination. This leads to population inversions between, for instance, 3-2 levels of hydrogenic ions and between 5-3, 4-3, and 5-4 levels of lithium-like ions. The amplified radiation wavelength is 18.2 nm

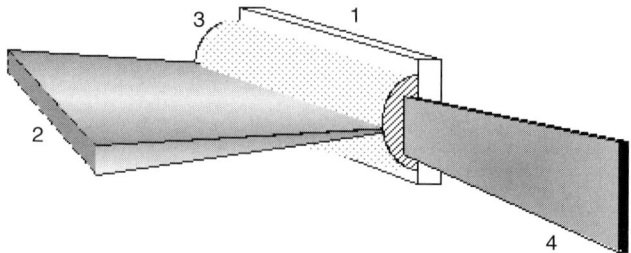

FIG. 1. XUV laser plasma production. 1, solid target; 2, cylindrically focused 1.06- or 0.53-μm laser beam; 3, amplifying plasma column; 4, XUV laser beam.

for C^{5+} (hydrogenic carbon) and 10.6 and 15.4 nm, respectively, for the 5-3 and 4-3 transitions of Al^{10+} (lithium-like aluminium).

The second set of pumping processes is known as the neon-like ion collisional pumping scheme [9–11]. In this case, collisions between neon-like ions and plasma free electrons excite the $2p^6$ ground level to the $2p^5$ $3p$-levels at a great rate. Because the lower $2p^5$ $3s$-levels have a large radiative decay probability to the ground level, $3p$-$3s$ population inversions can occur. As far as amplification takes place between levels of the same main quantum numbers, this pumping scheme needs ions of atomic number Z much larger than that required for pumping by recombination, in order to reach the same laser-line wavelengths. For most practical systems, Z lies between 30 and 40. Since the thermal plasma energy must be high enough to now expel 20 to 30 bound electrons from neutral atoms, the collisional scheme also requires much more power from the driving laser than the recombination scheme does.

Nowadays, using the neon-like scheme is the way to make x-ray lasers suitable for practical applications. Until now, recombination pumping did not yield the large gain-length product necessary to reach this goal. Therefore, we will restrict the material of the next paragraphs to the neon-like lasers. However, remember that other pumping schemes are under investigation. For instance, collisional pumping of nickel-like ions could extend lasing to the shorter wavelength part of soft x-rays [12]. From recent results, a large gain can be obtained near 10 nm with a moderate pumping energy distributed into several pulses, separated by short time intervals [13]. Another example is the demonstration of lasing to the ground state at 13.5 nm with using an experimental arrangement in which the amplifying plasma is produced by a 250 fs UV laser beam focused at the extremity of a LiF microcapillary [14].

The main lasing wavelengths of neon-like ions, including higher and lower Z elements for which a large gain has been demonstrated, are shown in Table I. A neon-like atomic diagram is displayed in the next section.

TABLE I. Main Lasing Lines of Neon-Like Ions[a]

Atomic number	Ion	Wavelengths (nm)			References
22	Ti^{12+}	32.65	45.94	47.21	15
26	Fe^{16+}	25.50	34.04	34.79	16
30	Zn^{20+}	21.22	26.23	26.72	17, 18
32	Ge^{22+}	19.61	23.22	23.63	19, 20
34	Se^{24+}	18.25	20.64	20.97	21
39	Y^{29+}	15.50	15.50	15.71	22
47	Ag^{27+}	12.30	9.94	10.04	23, 24

[a] Wavelengths are given for the $(1/2, 1/2)_{j=0} - (1/2, 1/2)_{j=1}$, $(3/2, 3/2)_{j=2} - (3/2, 1/2)_{j=1}$, $(1/2, 3/2)_{j=2} - (1/2, 1/2)_{j=1}$ transitions, in this order.

7.2.2 Saturated Lasers

For small values of the product of the gain coefficient (G) by the plasma length (L), say, for $G \cdot L < 13$–14 in the case of neon-like plasma lasers, the laser output grows exponentially against L. At larger gain factors ($G \cdot L > 15$) the fast depopulation of the laser upper-level forces the gain coefficient into decreasing and, as a result, the intensity exponential growth terminates. Then the laser reaches the saturation regime in which its efficiency is maximum. XRL saturation was obtained for the first time in 1991, at the 23.6-nm wavelength, with a neon-like germanium laser [25]. In this section we summarize the main experimental conditions of the transition from small gain signal to saturated emission.

Neon-like ion plasmas may provide large gain coefficients from 2 to 6 cm^{-1} or even more. Atomic transitions which supply the lasing lines, reported in Table I, are represented in the diagram of Fig. 2 in the case of neon-like zinc. Numerical codes have been developed for calculating the various processes which lead to population inversion for each of these lines in many elements [26]. They show that, in a given interval of plasma temperature and density, the rates of collisional excitation, populating the $2p^53p$ and $2p^53d$ levels from the ground level, are large compared to those populating the $3s$ levels. Moreover $3d$-$3p$ cascades contribute to the enhancement of $3p$ populations. This is the basis for lasing lines to rise between the $2p^53p$ and $2p^53s$ levels. However important differences arise between the $J = 0$ to $J = 1$ line and the two $J = 2$ to $J = 1$

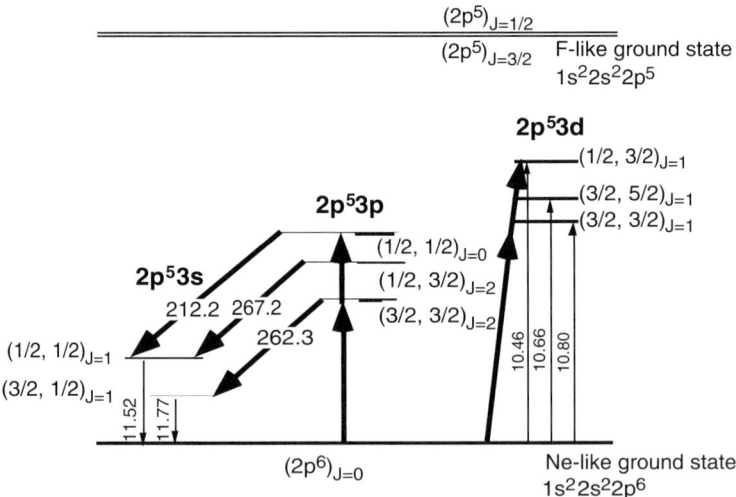

FIG. 2. Main $3p$-$3s$ lasing transitions of neon-like zinc ions. The wavelengths are given in Å.

lines (at 21.2, 26.7, and 26.2 nm, respectively, for the Zn ions represented in Fig. 2). Owing to additional population processes in the second case, the density and temperature required for large gain achievement are not the same for all the lines, namely, about 300 eV and 10^{21} electrons/cm^{-3} for the 21.2-nm line, 500 eV and 5.10^{19} electrons/cm^{-3} for the others. Thus the amplified beams will not propagate in the same plasma region for the three laser lines.

Furthermore the plasma density decreases in the driving laser beam direction, perpendicularly to the target surface. There are density gradients in the transverse direction as well. Correspondingly the refractive index of the medium varies, which involves a three-dimensional deflection of the amplified beams as shown in Fig. 3. The deflection tends to expel the beam from the narrow amplifying zone which is frequently no more than 30 to 100 microns wide. This effect, by limiting the amplifying length and the $G \cdot L$ value, has been the most important obstacle to saturation regime achievement over the years. Deflection can be partially compensated by replacing the plane target of Fig. 1 either with a multiple-target system [27, 28] or with a curved target [29].

In the case of the $J = 0$ to $J = 1$ lines such compensations are not sufficient because the radiation propagates in a narrow region where the density gradient is very steep. These lines were not suitable for XRLs until one observed that to precede the main pulse of the driving laser with a small prepulse enhances their intensity by orders of magnitude [30]. The role of the prepulse is to produce a cold plasma which expands freely during several nanoseconds before the main pulse arrives. In this way, plasma heating takes place in a larger plasma volume where the density gradients are significantly smoothed. The effective amplifying pathlength is considerably increased, which leads the $J = 0$ to $J = 1$ lines to rise as exceptionally intense laser lines. Multiple pulses may also improve the XRL efficiency.

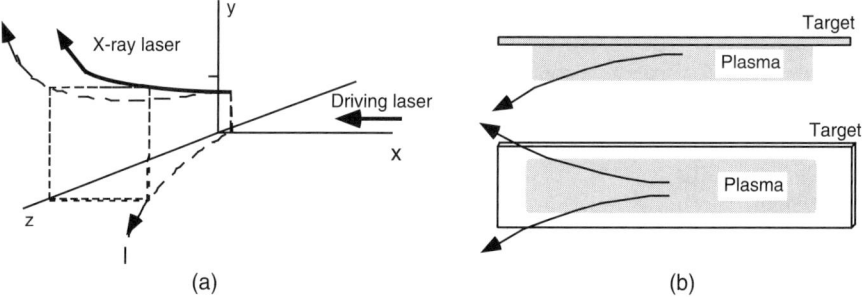

FIG. 3. (a) Sketch of x-ray laser beam refraction due to electron density gradients inside the plasma. (b) The beam leaving the plasma before full amplification owing to the refraction.

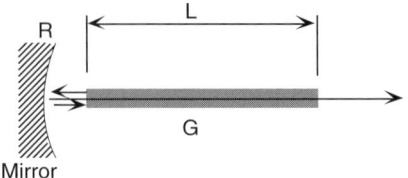

FIG. 4. Sketch of a multilayer curved-mirror, half-cavity used for increasing the x-ray laser gain × length, $G \cdot L$, up to the saturation value.

Moreover, a "half-cavity" consisting of a spherical multilayer mirror put at short distance of the plasma end, as shown in Fig. 4, can increase the $G \cdot L$ value up to laser saturation. After a double-pass through the plasma we have $(G \cdot L)_{\text{eff}} \approx 2G \cdot L - 3$, where the last term on the right roughly accounts for the return-path losses.

Table II displays the output characteristics of the zinc and the yttrium x-ray lasers as examples [31, 32]. The brightnesses reported in the last line of the table are 5 to 6 orders of magnitude larger than the one of the most powerful synchrotron radiation sources. The bandwidth of XRL emission is about 15×10^{-4} nm. XRL beams exhibit coherence features. As an example, Fig. 5 shows a single-shot image of Fresnel fringes produced by the edge of a crosswire located 50 cm from the source, as well as diffraction rings due to small surface defects of a mirror. XRL beams can also acquire a high linear polarization degree by using 45° multilayer mirrors as polarizers [33].

XRL repetition rate, regulated by the cooling time of the driving laser, is three shots per hour for the Zn-laser and is lower for the Y-laser. However, laser technology development can considerably increase the repetition rates. Possibilities also exist to reduce the driving laser energy by using shorter pulses (≈ 100 ps) and new isoelectronic sequences, such as the nickel-like sequence [34].

TABLE II. Comparison between Zinc and Yttrium X-Ray Laser Characteristics

Element	Zn^{20+} (J_{0-1})	Y^{29+} (J_{2-1})
Pump laser energy	0.4 kJ	4 kJ
Wavelength	21.2 nm	15.5 nm
Photons/pulse	$\sim 1 \times 10^{14}$	$\sim 7 \times 10^{14}$
XUV energy	~ 1 mJ	~ 10 mJ
Pulse duration	80 ps	200 ps
Power	~ 12 MW	~ 45 MW
Solid angle	2.5×10^{-5} sr^{-1}	2×10^{-4} sr^{-1}
Brightness (× line width)	1×10^{16} W/cm^2 sr^{-1}	2×10^{15} W/cm^2 sr^{-1}

FIG. 5. CCD picture of an XUV laser beam cross section showing Fresnel fringes due to a crosswire and rings due to diffracting defects at a near mirror surface.

7.2.3 Toward X-Ray Laser Applications

From the previous examples, the main peculiarity of XRLs is the extremely large, quasi-monochromatic-photon number they can supply in short pulses ($t \approx 100$ ps), from a quasi-pinpoint source ($s < 0.01$ mm^2) and within a sharp solid angle. These characteristics award XRL advantages in experiments needing very high monochromatic XUV flux, very short interaction time, and temporal and spatial beam coherence.

An interferometric probe of dense plasmas is a first example of using XRL in a practical application. Plasma regions of much larger density than in the case of current UV interferometry can be diagnosed. For a plasma size of a few hundreds of microns, the probed density can be increased by more than 2 orders of magnitude. The sketch of a successful experiment which combines the yttrium x-ray laser and Mach-Zehnder type interferometer is shown in Fig. 6 [35]. The interferometer optics, including the beamsplitters, is based on multilayer structures coating superpolished substrates. The XRL spatial coherence also makes it possible to resort to a purely reflective bimirror interferometer [36].

A fully different example of XRL utilization, which refers to fast UV-luminescence mechanisms, is shown in Fig. 7 [37]. The XUV laser beam is focused near the surface of a pure cesium iodide crystal. Single shots are sufficient to record luminescence spectra from 200 nm up to the visible light. The spectrum

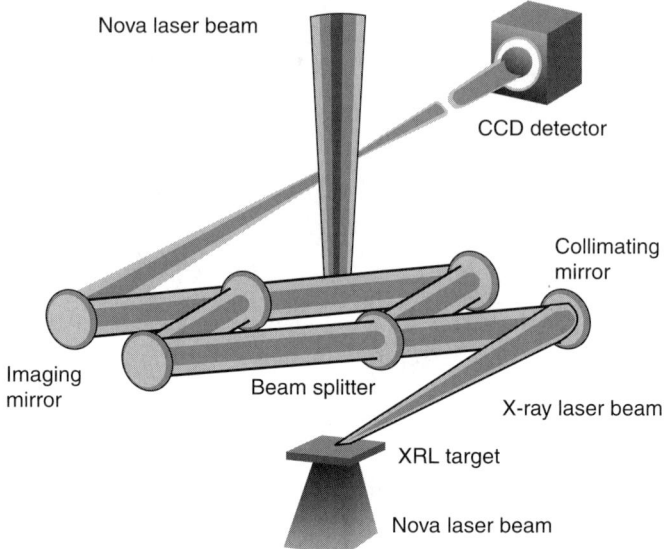

Fig. 6. Interferometric diagnostic of a high-density plasma using an x-ray laser beam. (Courtesy of Luiz Da Silva, Lawrence Livermore National Laboratory.)

excited by the 21.2-nm radiation of the zinc laser is found to be very different from that obtained at much lower excitation intensity with synchrotron radiation, and this difference varies with the XRL intensity. This is expected to provide new information on luminescence center production processes.

Among other possible extensions of current laser methods to the XUV range, high-flux laser–matter interaction, nonlinear optics, cluster characterization,

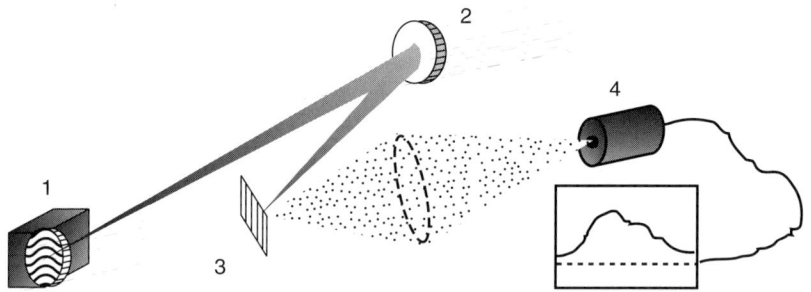

Fig. 7. Sketch of a luminescence experiment with extremely strong soft X-ray excitation. 1, XUV laser; 2, focusing mirror; 3, sample crystal; 4, detector (CCD or streak camera).

study of fractal structures, etc., are all promising topics. Attention is also paid to important technical applications such as microlithography and soft x-ray microscopy.

7.3 High-Order Harmonic Generation

7.3.1 Experiments

High harmonic generation is the nonlinear response of atoms, ions, or even of a steep density-gradient plasma to interaction with an intense electromagnetic field [38]. The experimental production of high harmonics is sketched in Fig. 8. The beam of a short-pulsed laser of frequency ω_L is focused across an atomic jet of rare gas. The laser-radiation harmonics of frequency $q\omega_L$, generated in the direction of the incident beam, are spatially separated by an optical grating. A photon detector is used for performing intensity measurements.

An example of experimental results is displayed in Fig. 9 from data reported in Refs. [39] and [40]. The laser parameters are $\lambda = 1.064$ μm, $\tau_{pulse} = 36$ ps, $P = 1$ GW, and repetition rate $= 10$ Hz. The lens focal length is 200 mm. The figure shows the harmonic photon numbers in argon and xenon, obtained with a laser flux $I_L \approx 3.10^{13}$ W/cm^2 and a gas pressure ≈ 15 Torr. The value of I_L is chosen to fit the saturation intensity of argon, that is to say, the limit at which the atomic medium becomes depleted by ionization.

These results exhibit features of general interest relatively to HHG from rare gases: (1) as a consequence of the atomic dipole interaction parity, one observes only odd order harmonics; (2) the spectra have a characteristic amplitude distribution exhibiting a first steep decrease (respectively up to the 7th and the 5th harmonic for Ar and Xe), a plateau region with a roughly constant harmonic amplitude (respectively up to the 27th and 15th harmonics), then a final steep decrease; (3) the conversion efficiency increases with the Z of the rare gas,

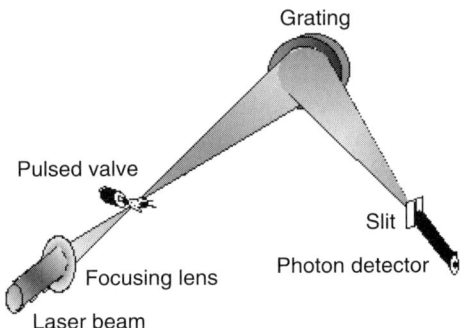

FIG. 8. Principle of a high harmonic generation experiment.

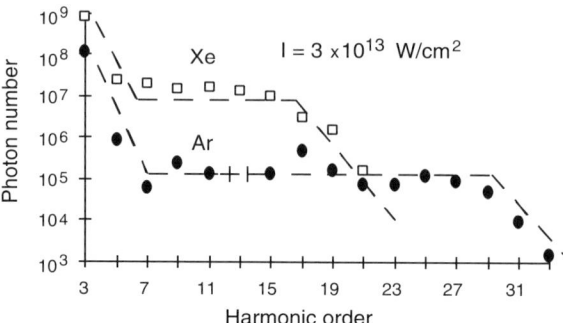

FIG. 9. Intensity of harmonics generated by Xe and Ar [40]. The dashed lines bring out the three typical emission regions, including the horizontal *plateau*.

which results from the corresponding atomic polarizability increase; (4) the photon energy of higher harmonics exceeds the atomic ionization thresholds. Let us mention that the lack of the 13th harmonic of Ar, visible in Fig. 9, is tentatively ascribed by the authors to a resonance with a close discrete level which would enhance ionization to the prejudice of photon emission.

It is also observed that the laser-intensity dependence of the harmonic photon number changes its form when the harmonic order, q, increases. For all rare gases from He to Xe the photon number is found to increase as I_L^q, for $q = 3$ and 5, till the saturation intensity be reached [39]. This law agrees with the lowest order perturbation theory of multiphotonic processes. However, extension of the measurements to higher order harmonics in Ar leads to a breakdown of the I_L^q dependence. For instance, the slopes for the 21st and 33rd harmonics in logarithmic coordinates are respectively ≈ 8 and ≈ 12 instead of 21 and 33. Such discrepancies can be observed for all harmonics lying beyond the "first steep decrease" appearing in Fig. 9 [41]. This fact is considered to be a strong sign that HHG understanding needs a theoretical approach which need not be based on lowest order perturbation calculations (see next section). Other important parameters are gas pressure and focusing geometry. Harmonic photon numbers are nearly proportional to the square of the pressure. This is expected from a coherent emission process. Besides, the confocal parameter $b = 2\pi R^2/\lambda$ (where R is the beam radius and λ the wavelength) is relevant for focusing condition characterization. It is found that harmonic photon numbers are proportional to b^3 [38].

The cutoff energy of HHG (i.e., the maximum harmonic energy accessible with a given atomic gas) is given by $E_{max} = I_p + \alpha U_p$ where I_p is the atom ionization potential and U_p is the "pondemorotive" shift of ionization potential $[U_{p\,(eV)} = 9.33.10^{-14} \times I_{L\,(W/cm^2)} \times \lambda^2_{(mm^2)}]$. This relation was first deduced from single atom calculations which led to $\alpha \approx 3$ [42]. Taking into account the macroscopic propagation of harmonic radiation reduces this value to $\alpha \approx 2$, which is in agreement with experimental measurements [43, 44]. However harmonics of higher order than predicted by the $I_p + \alpha U_p$ rule have been observed recently when using ultrashort 25-fs laser pulses [45]. It is believed that, for equal laser intensity, ultrashort pulses reduce the probability for atoms to be ionized, thus increasing the harmonic conversion efficiency, especially to the highest harmonic orders. The shortest harmonic wavelengths observed to date are 7.1 nm in neon ($q = 35$) and 6.7 nm in helium ($q = 37$) from the 248.6-nm KrF laser, using a 380-fs long pulse and a 4×10^{17} W/cm^2 pulse-averaged intensity [46]. In this experiment, the highest harmonics originated not from neutral atoms but from He$^+$, Ne$^+$, and Ne^{2+} ions, which obviously increases I_p in the $I_p + \alpha U_p$ rule.

Intense laser interaction with solid targets also generates high harmonics [47, 48]. Instead of resulting from the nonlinear atomic polarizability, as in the case of gases, the harmonic emission is then driven by the periodic density perturbations produced by the fast "there and back" electron motion through the steep density gradient of the plasma created close to the solid surface. Owing to the collective character of this process the parity selection rule of single atoms does not apply. Thus odd and even harmonics are emitted with similar intensities. Furthermore, no cutoff energy rule appears in recent calculations [49] nor in experimental observations [50]. Another important characteristic is the solid-target harmonic emission being rather isotropic, whereas gas-target harmonics are tightly concentrated around propagation axis (see later section on Optical Characteristics).

7.3.2 Survey of the Theoretical Background

High harmonics generation is a process in which an atom absorbs q photons of energy ω_L and emits one photon of energy $q\omega_L$. The theory of the whole process involves a microscopic level, relative to the nonlinear behavior of single atoms in strong fields, and a macroscopic level which deals with the coherent propagation of harmonic radiation inside the excited medium. The goal of this section is no more than a very brief introduction to these two aspects of HHG theory. The reader can access original works [40, 42, 49, 51–54].

Calculation of classical trajectories of one electron in the field of a proton, on which superposes an oscillating laser field, provides a simplified but easily understandable view of the physics involved in HHG [55]. An intense laser field increases considerably the eccentricity of an initial periodic Kepler orbit. Moreover, electron trajectories involve fast oscillations at the laser frequency, which contribute to the dipole moment time dependence. Then several types of trajectories appear according to initial conditions. Some trajectories rapidly lead to atom ionization while others remain bound during a long time. Calculation of the time-dependent dipole for a number of trajectories (e.g., 10,000), followed by time averaging their squared Fourrier transform, leads to the spectrum of emitted frequencies. Applied to a classical hydrogen atom, this model remarkably predicts a *plateau* between harmonics 7 and 27 [55b].

Calculation of quantitative conversion efficiencies and many-electron atom treatment calls on the quantum approach. In the case of weak fields ($I_L \approx < 10^{13}$ W/cm^3) where the lowest order perturbation theory can be used, the harmonic spectrum is described by an expansion of the induced polarization in odd powers of the electric field E. The term corresponding to the qth harmonic is

$$P^{(q)} = 2^{1-q} \cdot \chi^{(q)} \cdot E^q, \tag{1}$$

where $\chi^{(q)}$ is the qth-order nonlinear susceptibility. Using the right term of Eq. (1) multiplied by N for a collection of atoms of density N as the source term in propagation equations shows that the harmonic intensity $I^{(q)}$ obeys the following rule [37]:

$$I^{(q)} \alpha |F^{(q)}|^2 \cdot |N \cdot \chi^{(q)}|^2 \cdot I_L^q, \tag{2}$$

which involves the nonlinearity of the macroscopic medium as well as the one of single atoms. In this relation $F^{(q)}$ is the so-called *phase-matching factor* which brings in the difference with the intensity emitted by N independent atoms. Phase mismatch originates, for the most part, from the refractive index difference between fundamental and harmonic radiation and, for another part, from focusing conditions. Weak field calculations show that $|F^{(q)}|^2$ decreases rapidly with q, which contributes to the steep intensity decrease of the first harmonics seen before the plateau in the curves of Fig. 9. The N^2 factor in relation (2) accounts for the coherent character of harmonic emission. This N^2 dependence of harmonic intensities has been confirmed by measuring the number of photons in the 17th harmonic in krypton as a function of the gas pressure [39]. Nevertheless lowest order perturbation theory does not predict the experimentally observed *plateau* of conversion efficiency. The main reason for this failure is in the electrical susceptibilities acquiring a laser intensity dependence under a strong field which is not accounted for in perturbation calculations.

At very large laser intensity a relevant method of HHG calculation is the direct integration of the time-dependent Schrödinger equation [40, 54, 56]. In the case of single-electron atom calculations this equation can be written as

$$i \frac{\partial}{\partial t} \Psi(r \cdot t) = \left(-\frac{1}{2} \frac{d^2}{dr^2} + \frac{l(l+1)}{2r^2} - \frac{1}{r} - E_0 f(t) z \sin(\omega_L t) \right) \Psi(r \cdot t), \tag{3}$$

where E_0 is the laser-peak amplitude and the field being polarized in the z direction; $f(t)$ is the laser-pulse envelope. For many-electron atoms l-dependent effective potentials are separately calculated as functions of the radial coordinate [40]. The time-dependent calculations treat the excitation of a single electron in the mean field of the nucleus and of the remaining electrons, which are assumed

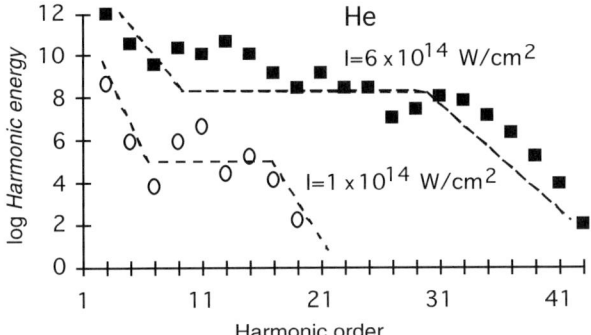

FIG. 10. Calculated harmonics emitted by He for two pump laser intensities [54].

to be frozen in their ground state. The harmonic emission spectrum is obtained by Fourier transform of the time-varying induced dipole:

$$d(\omega) = \frac{1}{T_1 - T_2} \int_{T_1}^{T_2} dt\, e^{-i\omega t} \langle \Psi^*(r \cdot t)|z|\Psi(r \cdot t)\rangle, \tag{4}$$

where the $T_1 - T_2$ interval covers a few cycles (5–10) of the laser pulse. Figure 10, where $|d(\omega)|^2$ is plotted on the vertical axis, presents examples of calculations for helium at two excitation energies. The main experimental features mentioned in the Experiments section are well displayed in these theoretical spectra though propagation effects are not yet included in calculations.

Harmonic propagation calculations in the strong field regime show phase matching to be improved for higher harmonics in comparison with the weak field case. This contributes to the *plateau* formation in the spectra. In addition, propagation smooths the spectral irregularities produced by the resonances in single atom calculations. Another important result is propagation calculations to provide the angular distribution harmonic beams as well as possible spectral shifts and broadening. Initially assumed to be Gaussian, angular profiles develop rings throughout the medium. As a matter of fact such rings are experimentally observed (see Fig. 11 in the next section). Calculations also account for possible spectral shifts and line broadening.

7.3.3 Optical Characteristics

Here we summarize information of interest for source users about the main properties of VUV beams obtained from high-order harmonic generation.

Energy. One can consider 1 to 60 nJ (i.e., $\approx 10^8$ to 6×10^9 photons) the range of energies per pulse supplied by harmonics near the end of the plateau. In each practical case the value depends on the intensity and the wavelength of the laser, on focusing conditions, and on the kind of excited gas. The upper value of the interval has been obtained at $\lambda = 25$ nm, for the 21st harmonic of a 526-nm, 0.65-ps pulse duration laser, generated in helium [57]. The laser intensity was 1.1×10^{16} W/cm^2 and the conversion efficiency, 10^{-7}. Generally speaking, harmonic pulse duration is comprised between the laser pulse duration, τ, and the perturbative limit $\tau/q^{1/2}$ [58]. Current peak powers thus should range from 10 to 100 kW. Let us mention that 10 Hz is a usual repetition rate value.

Angular Distribution. In the case of gas targets, harmonics present a low divergence in the direction of the laser beam. This contrasts with harmonics generated by solid targets, which have a much more isotropic emission. Angular profiles change significantly with the harmonic order. In the weak field limit, the Gaussian-beam cone angle is expected to decrease as $q^{-1/2}$. In fact, harmonics rising away from the *plateau* edge often present irregular profiles with rings and structures. An example of such a profile is given by the 71st harmonic of 1-μm laser in helium, represented in Fig. 11 by data from Ref. [59]. In contrast, beyond the *plateau*, the figure shows the 111th harmonic to exhibit an approximately Gaussian narrow shape. Furthermore the 13th, 15th, and 17th harmonics of a 1-μm laser, generated in Ar, Kr, and Xe, have shown prominent wings, which make their widths comparable to the focused laser profile [60].

Temporal and Spatial Coherence. The temporal—that is, longitudinal—coherence is directly related to the harmonic spectral width $\Delta\lambda$ ($L_{TC} = \lambda^2/\Delta\lambda$). Very short pulses ($\approx 0.1$ ps) may produce relatively large widths (i.e., ≈ 0.1 nm at 30 nm, for instance), which reduce the temporal coherence in comparison with pulse duration around 1 ps.

With regard to the spatial—that is, transverse—coherence, the first experimental data have been obtained only recently by measuring the Young fringe visibility for various harmonic orders [61]. The harmonics were generated in

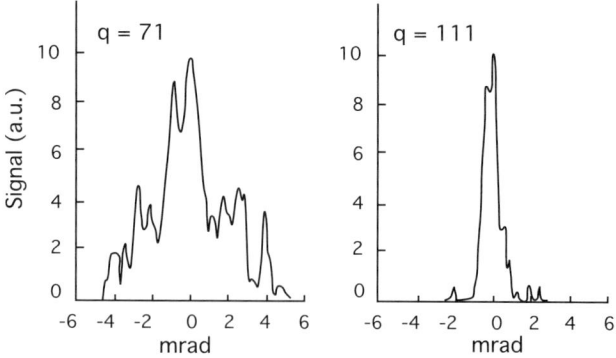

FIG. 11. Angular profiles observed for two high harmonics generated in helium [59].

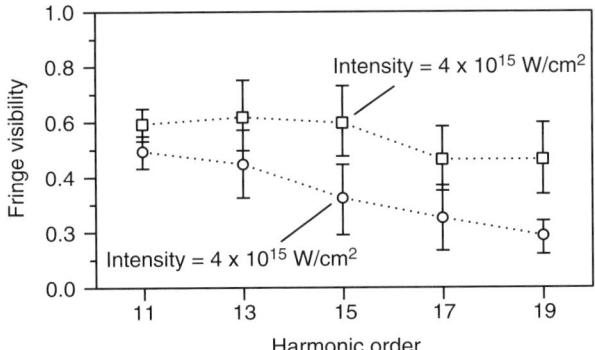

FIG. 12. Fringe contrast versus harmonic order obtained with 50-μm spaced Young slits [61].

helium with a frequency-doubled Nd:glass laser. Figure 12 shows the fringe contrast against the harmonic order obtained with slits of 50-μm spacing and for two laser intensities. From measurements performed by changing the slit spacing, it has been found that the incoherent source diameter is ≈ 15 μm, whereas the size of the harmonic emission at the exit of the gas is estimated to 65 μm.

By way of conclusion about coherent VUV sources, summarizing the main characteristics of x-ray lasers and high harmonic generation shows that, in the present state of the art, (1) the best performances are at the 15 to 30-nm wavelength for XRL against 40 to 50 nm for HHG; (2) peak powers are tens of megawatts for XRL against tens of kilowatts for HHG; (3) the repetition rate is 10^4 to 10^5 larger for HHG; (4) emission duration is 100 to 1000 shorter for HHG; (5) the emission band width is smaller by a factor of ≈ 10 for XRL; (6) the effective incoherent source size is ≈ 5 times smaller for HHG but the number of photons per mode is $\approx 10^3$ larger for XRL; (7) the width of angular distribution is comparable for both sources.

These characteristics should lead to somewhat different applications for each of these two sources.

References

1. Exhaustive information can be found in (a) "International Colloquium on X-ray Lasers" (P. Jaeglé and A. Sureau, eds.) Editions de Physique, Les Ulis, France (1986); (b) "X-Ray Lasers 1990" (G. J. Tallents, ed.), Institute of Physics Conference Series 116, IOP Publishing, Bristol, UK (1990); (c) "X-Ray Lasers 1992" (E. E. Fill, ed.), Institute of Physics Conference Series 125, IOP Publishing, Bristol, UK (1992);

(d) "X-Ray Lasers 1994" (C. C. Eder and D. L. Matthews, eds.), IAIP Conf. Proc. 332, American Institute of Physics, Woodburr, NY, USA (1994); (e) "Proc. Vth Int. Conf. on X-Ray Lasers", Lund, Sweden, in press.
2. J. J. Rocca, M. C. Marconi, F. G. Tomasel, V. N. Shlyaptsev, J. L. A. Chilla, and D. P. Clark, *Proc. SPIE* **2520**, 201 (1995).
3. P. Jaeglé, A. Carillon, P. Dhez, G. Jamelot, A. Sureau, and M. Cukier, *Phys. Lett.* **36A**, 167 (1971).
4. F. E. Irons, and N. J. Peacock, *J. Phys. B* **7**, 1109 (1974).
5. D. Jacoby, G. J. Pert, S. A. Ramsden, L. D. Shorrock, and G. J. Tallents, *Opt. Commun.* **37**, 193 (1981).
6. S. Suckewer, C. H. Skinner, H. M. Milchberg, and D. Voorhees, *Phys. Rev. Lett.* **55**, 1753 (1985).
7. G. Jamelot, P. Jaeglé, A. Carillon, A. Bideau, C. Möller, H. Guennou, and A. Sureau, in "Proc. Conf. on LASERS '81," STS Press, McLeam, VA, pp. 178–183 (1982).
8. P. Jaeglé, G. Jamelot, A. Carillon, A. Klisnick, A. Sureau, and H. Guennou, *J. Opt. Soc. Am. B* **4**, 563 (1987).
9. E. C . Elton, *Appl. Opt.* **14**, 97 (1975).
10. A. V. Vinogradov, I. Sobelman, and E. Yukov, *Sov. J. Quantum Electron.* **6**, 82 (1976).
11. D. L. Matthews, P. L. Hagelstein, M. D. Rosen, M. J. Eckart, N. M. Ceglio, A. U. Hazi, H. Medecki, B. J. MacGowan, J. E. Trebes, B. L. Whitten, E. M. Campbell, C. W. Hatcher, J. H. Scofield, G. Stone, and T. A. Weaver, *Phys. Rev. Lett.* **54**, 110 (1985).
12. H. Daido, Y. Kato, S. Ninomiya, R. Kodama, G. Yuan, Y. Oshikane, M. Tagaki, and H. Takabe, *Phys. Rev. Lett.* **75**, 1074 (1995).
13. S. Sebban, D. Ros, A. G. MacPhee, F. Albert, A. Carillon, P. Jaégle, G. Jamelot, A. Klisnick, C. L. S. Lewis, R. Smith, and C. J. Tallents, "Soft X-Ray Lasers and Applications II," in *Proc. SPIE* 3156 (J. J. Rocca and L. B. Da Silva, eds.), 1997.
14. D. V. Korobkin, C. H. Nam, and S. Suckewer, *Phys. Rev. Lett.* **77**, 5206 (1996).
15. T. Boehly, M. Russotto, R. S. Craxton, R. Epstein, B. Yaakobi, L. B. DaSilva, J. Nilsen, E. A. Chandler, D. J. Fields, B. J. MacGowan, D. L. Matthews, J. H. Scofield, and G. Shimkaveg, *Phys. Rev. A* **42**, 6962 (1990).
16. J. Nilsen, and J. H. Scofield, *Physica Scripta* **49**, 588 (1994).
17. R. C. Elton, "X-ray Lasers," Academic Press, New York, p. 115 (1990).
18. B. Rus, A. Carillon, B. Gauthé, P. Goedtkindt, P. Jaeglé, G. Jamelot, A. Klisnick, A. Sureau, and P. Zeitoun, *J. Opt. Soc. Am. B* **11**, 564 (1994).
19. R. C. Elton, T. N. Lee, and E. A. MacLean, *J. Physique* **C9**, 359 (1987).
20. C. L. S. Lewis, D. Neely, D. O'Neill, J. Uhomoibhi, G. Cairns, A. MacPhee, G. J. Tallents, J. Krishnan, L. Dwivedi, M. H. Key, P. Norreys, R. Kodama, R. E. Burge, G. Slark, M. Brown, G. J.Pert, P. Holden, M. Lightbody, S. A. Ramsden, J. Zhang, C. Smith, P. Jaeglé, G. Jamelot, A. Carillon, A. Klisnick, J. P. Raucourt, J. E. Trebes, M. R. Carter, S. Mrowka, and K. A. Nugent, in "X-Ray Lasers 1992," (E. E. Fill, ed.), IOP Publishing, Bristol, U.K., p. 23 (1992).
21. J. A. Koch, B. J. MacGowan, L. B. DaSilva, D. L. Matthews, J. H. Underwood, P. J. Batson, R. W. Lee, R. A. London, and S. Mrowka, *Phys. Rev. A* **50**, 1877 (1994).
22. C. J. Keane, N. M. Ceglio, B. J. MacGowan, D. L. Matthews, D. G. Nilson, J. E. Trebes, and D. A. Whelan, *J. Phys. B* **22**, 3343 (1989).
23. D. Desenne, L. Berthet, J. L. Bourgade, J. Bruneau, A. Carillon, A. Decoster, A. Dulieu, H. Dumont, S. Jacquemot, P. Jaeglé, G. Jamelot, M. Louis-Jacquet, J. P.

Raucourt, C. Reverdin, J. P. Thébault, and G. Thiell, in "X-Ray Lasers 1990," (G. J. Tallents, ed.), IOP Publishing, Bristol, U.K., p. 351 (1990).
24. D. J. Field, R. S. Walling, G. M. Shimkaveg, B. J. MacGowan, L. B. DaSilva, J. H. Scofield, A. L. Osterheld, T. W. Phillips, M. D. Rosen, D. L. Matthews, W. H. Goldstein, and R. E. Stewart, *Phys. Rev. A* **46**, 1606 (1992).
25. A. Carillon, H. Z. Chen, P. Dhez, L. Dwidedi, J. Jacoby, P. Jaeglé, G. Jamelot, J. Zhang, M. H. Key, A. Kidd, A. Klisnick, R. Kodama, J. Krishnan, C. L. S. Lewis, D. Neely, P. Norreys, D. M. O'Neill, G. J. Pert, S. A. Ramsden, J. P. Raucourt, G. J. Tallents, and J. O. Uhomoibhi, *Phys. Rev. Lett.* **68**, 2917 (1992).
26. See for instance: M. D. Rosen, P. L. Hagelstein, D. L. Matthews, E. M. Campbell, A. U. Hazi, B. L. Whitten, B. MacGowan, R. E. Turner, and R. W. Lee, *Phys. Rev. Lett.* **54**, 106 (1985); A. L. Osterheld, B. K. F. Young, R. S. Walling, W. H. Goldstein, J. Scofield, M. Chen, G. Shimkaveg, M. Carter, R. Shepherd, B. J. MacGowan, L. DaSilva, D. Matthews, S. Maxon, R. London, and R. E. Steward, in "X-Ray Lasers 1992," (E. E. Fill, ed.), IOP Publishing, Bristol, U.K., p. 309 (1992); P. H. Holden, M. T. M. Lightbody, and G. J. Pert, same volume p. 297; S. Jacquemot, in "X-Ray Lasers 1994," (C. C. Eder and P. L. Matthews, eds.), IOP Publishing, Bristol, U.K., p. 279 (1994).
27. S. Wang, Y. Gu, G. Zhou, Y. Ni, S. Yu, S. Fu, C. Mao, Z. Tao, W. Chen, Z. Lin, D. Fan, G. Zhang, J. Sheng, M. Yang, T. Zhang, Y. Shao, H. Peng, and X. He, *Chinese Phys Lett.* **8**, 618 (1991).
28. D. Neely, C. L. S. Lewis, D. N. O'Neill, J. Uhomoibhi, M. H. Key, S. J. Rose, G. J. Tallents, and S. A. Ramsden, *Opt. Commun.* **87**, 231 (1992).
29. R. Kodama, D. Neely, Y. Kato, H. Daido, K. Murai, G. Huan, A. MacPhee, and C. L. S. Lewis, *Phys. Rev. Lett.* **73**, 3215 (1994).
30. T. Boehly, M. Russotto, R. S. Craxton, R. Epstein, B. Yaakobi, L. B. Da Silva, J. Nilsen, E. A. Candler, D. J. Fields, B. J. MacGowan, D. L. Matthews, J. H. Scofield, and G. Shimkaveg, *Phys. Rev. A* **42**, 6962 (1990).
31. P. Jaeglé, S. Sebban, A. Carillon, G. Jamelot, A. Klisnick, P. Zeitoun, B. Rus, F. Albert, and D. Ros, in "Proc. Vth Int. Conf. on X-Ray Lasers," Lund, Sweden, in press.
32. L. B. Da Silva, B. J. MacGowan, S Mrowka, J. A. Koch, R. A. London, D. L. Matthews, J. H. Underwood, *Optics Letters* **18**, 1174 (1993).
33. B. Rus, C. L. S Lewis, G. F. Cairns, P. Dhez, P. Jaeglé, M. H. Key, D. Neely, A. G. MacPhee, S. A. Ramsden, C. G. Smith, A. Sureau, *Phys. Rev. A.* **51**, 2316 (1995).
34. H. Daido, Y. Kato, S. Ninomiya, R. Kodama, G. Yuan, Y. Oshikane, M. Tagaki, and H. Takabe, *Phys. Rev. Lett.* **75**, 1074 (1995).
35. L. B. DaSilva, T. W. Barbee, R. Cauble, P. Celliers, D. Ciarlo, S. Libby, R. A. London, D. Matthews, S. Mrowka, J. C. Moreno, D. Ress, J. E. Trebbes, A. S. Wan, and F. Weber, *Phys. Rev. Lett.* **74**, 3991 (1995).
36. F. Albert, D. Joyeux, P. Jaeglé, A Carillon, J. P. Chauvineau, G. Jamelot, A. Klisnick, J. C. Lagron, D. Phalippou, D. Ros, S. Sebban, and P. Zeitoun, *Optics Communications* **143**, 184 (1997).
37. P. Jaeglé, S. Sebban, A. Carillon, G. Jamelot, A. Klisnick, P. Zeitoun, B. Rus, M. Nantel, F. Albert, and D. Ros, *J. Appl. Phys.* **81**, 2406 (1997).
38. See, for instance, A. McPherson, G. Gibson, H. Jara, U. Johann, T. S. Luk, I. A. McIntyre, K. Boyer, and C. K. Rhodes, *J. Opt. Soc. Am. B.* **4**, 595 (1987).
39. X. F. Li, A. L'Huillier, M. Ferray, L. A. Lompré, and G. Mainfray, *Phys. Rev. A* **39**, 5751 (1989).

40. A. L'Huillier, K. J. Shafer, and K. C. Kulander, *J. Phys. B: Atom. Molec. Opt. Phys.* **24**, 3315 (1991).
41. A. Lompré, A. L'Huillier, P. Monot, M. Ferray, G. Mainfray, and C. Manus, *J. Opt. Soc. Am.* **7**, 754 (1990).
42. L. Krause, K. J. Schafer, and K. C. Kulander, *Phys. Rev. Lett.* **68**, 3535 (1992).
43. A. L'Huillier, M. Lewensyein, P. Salières, P. Balcou, M. Yu. Ivanov, J. Larsson, and C. G. Wahlström, *Phys. Rev. A* **48**, R3433 (1993).
44. J. J. Macklin, J. D. Kmetec, and C. L. Gordon, *Phys. Rev. Lett.* **70**, 766 (1993).
45. J. Zhou, J. Peatross, M. M. Murname, and H. C. Kapteyn, *Phys. Rev. Lett.* **76**, 752 (1996).
46. S. G. Preston, A. Sanpera, M. Zepf, W. J. Blyth, C. G. Smith, J. S. Wark, M. H. Key, K. Burnett, M. Nakai, D. Neely, and A.A. Offenberger, *Phys. Rev. A* **53**, R31 (1996).
47. R. L. Carman, C. K. Rhodes, and R. F. Benjamin, *Phys. Rev A* **24**, 2649 (1981).
48. D. von der Linde, T. Engers, G. Jenke, and P. Agostini, *Phys. Rev. A* **52**, R25 (1995).
49. P. Gibbon, *Phys Rev. Lett.* **76**, 50 (1996).
50. P. A. Norreys, M. Zepf, S. Moustsaizis, A. P. Fews, J. Zhang, P. Lee, M. Bakarezos, C. N. Danson, A. Dyson, P. Gibbon, P. Loukakos, D. Neely, F. N. Walsh, J. S. Wark, and A. E. Dangor, *Phys. Rev. Lett.* **76**, 1832 (1996).
51. Bo Gao and A. F. Starace, *Phys. Rev. A* **39**, 4550 (1989).
52. Liwen Pan, K. T. Taylor, and C. W. Clarke, *Phys. Rev. A* **39**, 4894 (1989).
53. A. L'Huillier, P. Balcou, S. Candel, K. J. Schafer, and K. C. Kulander, *Phys. Rev. A.* **46**, 2778 (1992).
54. J. L. Krause, K. J. Schafer, and K. C. Kulander, *Phys. Rev. A.* **45**, 4998 (1992).
55. See, for instance, (a) J. G. Leopold and I. C. Percival, *J. Phys. B: Atom. Molec. Phys.* **12**, 709 (1979); (b) G. Bandarage, A. Maquet, and J. Cooper, *Phys. Rev. A* **41**, 1744 (1990).
56. K C. Kulander and B. W. Shore, *Phys. Rev. Lett.* **62**, 524 (1989) and *J. Opt. Soc. Am. B* **7**, 502 (1990).
57. T. Ditmire, J. K. Crane, H. Nguyen, L. B. Da Silva, and M. D. Perry, *Phys. Rev. A* **51**, R902 (1995).
58. D. Joyeux, P. Jaeglé, and A. L'Huillier, in "Trends in Optics," Vol. 3 (A. Consortini, ed.), Academic Press, New York, p. 371 (1996).
59. J. W. G. Tisch, R. A. Smith, J. E. Muffet, M. Ciarrocca, J. P. Marangos, and M. H. R. Hutchinson, *Phys. Rev. A* **49**, R28 (1994).
60. J. Peatross and D. D. Meyerhofer, *Phys. Rev. A* **51**, R906 (1995).
61. T. Ditmire, E. T. Gumbrell, R. A. Smith, J. W. G. Tisch, D. D. Meyerhofer, and M. H. R. Hutchinson, *Phys. Rev. Lett.* **77**, 4756 (1996).

8. RADIOMETRIC CHARACTERIZATION OF VUV SOURCES

Michael Kühne

Physikalisch-Technische Bundesanstalt
Berlin, Germany

8.1 Quantitative VUV Radiometry by Use of Source Standards

In the field of vacuum ultraviolet (VUV) radiation physics a major objective is the study of the interaction of VUV radiation with matter or the investigation of VUV radiation emitted by matter in the course of other interaction processes. In both cases quantitative measurements of the VUV radiation are generally required. To perform and evaluate these measurements is the objective of VUV radiometry. For commonly used radiometric quantities, see Table I. A more complete listing can be found in [1].

> **Warning:** This is a short review and cannot provide complete information. The information contained in this chapter and the references may not be sufficient to achieve the performance or to reproduce the results described. Also it is not within the scope of this chapter to provide information concerning safety aspects. It is the responsibility of the reader to consult and establish appropriate safety procedures and determine the applicability of regulatory limitations prior to using the information provided in this chapter or in its references.

There are fundamentally two different ways to determine the radiometric properties listed in Table I:

1. To measure the radiometric property with an absolute detector standard (with sufficient spectral resolving power if spectral resolution is required)
2. To compare the radiometric property of the source under investigation with another source of known radiometric properties (radiometric source standard) utilizing a spectrometer with sufficient spectral resolving power (if spectral resolution is required).

The disadvantage of method 1 in the VUV spectral range is that in most practical cases the absolute detector cannot provide the necessary spectral resolving power so that the detector needs to be combined with a VUV monochromator. The spectral transmission of this monochromator needs to be determined with sufficient accuracy. This is often quite difficult as the spectral transmission may depend on the way the optical components are illuminated, which might well be different for the calibration of the spectrometer and the intended characterization of the radiation source.

TABLE I. Frequently Used Radiometric Units

Symbol	Unit	Quantity	Dimension
Q	Radiant energy	Quantity of radiation	J
ϕ	Radiant power (also named radiant flux)	Radiation energy per unit time	W
$\phi_\lambda(\lambda)$	Spectral radiant power	Radiant power emitted by a source at the wavelength λ per unit wavelength	W m^{-1}
$L_\lambda(\lambda)$	Spectral radiance	Spectral radiant power emitted by a surface element of a source per solid angle	W m^{-3} sr^{-1}
$I(\lambda)$	Radiant intensity	Radiant power emitted by a source per solid angle	W sr^{-1}
$I_\lambda(\lambda)$	Spectral concentration of radiant intensity	Spectral radiant power emitted by a source per solid angle	W m^{-1} sr^{-1}
$E_\lambda(\lambda)$	Spectral irradiance	Spectral radiant power emitted by a source passing through an area in a defined distance from the source	W m^{-3}

Note: Spectral units can be also expressed per unit photon energy, e.g., spectral radiant power $\Phi_E(E)$, dimension W/eV, or in terms of photons, e.g., spectral photon flux $\Phi_E(E)$, dimension photons/s.

Another difficulty in the VUV spectral domain is the change of the instruments spectral transmission due to ageing of the optical components under irradiation in particular in the presence of hydrocarbons.

For these reasons, method 2 is generally employed for characterizing radiometric source properties. Application of method 2 requires the availability of suitable VUV radiometric source standards. At the highest level there is the primary standard. It signifies the direct and independent realization of the definition of a unit. At the same time it represents the realization with the highest precision. Such primary standards are usually maintained at national metrological institutions and serve as national standards. Secondary standards are linked to these primary standards by intercomparisons. Their realization of the unit may be based on different physical principles than the primary standard, only sufficient reproducibility is required.

8.2 Primary Standards

8.2.1 Blackbody Radiators

The classical primary standard in the field of radiometry is the blackbody radiator. Its spectral radiation emission is determined by Planck's radiation law and is a function of the temperature T and wavelength λ only. For a detailed description of blackbody radiators see Quinn [2]. Indeed the realization of the International Temperature Scale of 1990 (ITS-90) is based above 1235 K on Planck's law [3].

Unfortunately, however, the classical blackbody radiator is only of very limited use to VUV radiometry because for technical reasons such blackbodies can be realized only to temperatures up to about 3300 K. The maximum radiation emission is therefore in the infrared region and no suitable radiation emission occurs at wavelengths significantly shorter than 200 nm.

For the field of VUV radiometry therefore other sources had to be developed and investigated for their suitability to serve as primary standards. The first candidates were sources like wall stabilized arcs. These stationery noble gas plasma sources can achieve temperatures of 15,000 K under conditions close to local thermodynamic equilibrium (LTE). At this temperature the maximum radiation emission lies at about 200 nm and blackbody radiation would be usable down to about 40 nm. The radiation emission of such arc plasmas is composed of continuum radiation superposed with emission lines. When suitable doping gases such as H, C, or N are added in the proper concentration the spectral emission of the resonance lines of these atoms becomes optically thick. That is when the product of the spectral absorption coefficient $a(\lambda, T)$ and the thickness of the emitting layer l becomes large compared to 1 ($al \gg 1$). In this case the spectral radiance $L_\lambda(\lambda, T)$ of the plasma at the wavelength of the resonance line approaches the blackbody spectral radiance $L_\lambda^{BB}(\lambda, T)$:

$$L_\lambda(\lambda, T) = \{1 - \exp[-a(\lambda, T)l]\} L_\lambda^{BB}(\lambda, T) \approx L_\lambda^{BB}(\lambda, T).$$

Such radiation sources have been realized, for example, by Boldt [4], Stuck and Wende [5], Key and Preston [6], and Kaase [7]. To prevent the reabsorption of the resonance line radiation by cold boundary layers, either an outer gas layer containing no trace gases or a differential pumping system has been used. An example of such a differentially pumped wall stabilized arc is described by Grützmacher and Wende [8]. A schematic of such an arc source taken from Kühne and Wende [9] is given in Fig. 1. The operating pressure lies between 0.5 and 2 bar. A three-stage differential pumping system is used to reduce the pressure to high vacuum level. This source can be utilized as a radiometric source standard down to wavelengths just below 100 nm. The uncertainty of the spectral radiance is mainly determined by the uncertainty of the plasma temperature T. At $T = 15,000$ K a temperature uncertainty $\Delta T/T = 0.01$ causes an uncertainty $\Delta L_\lambda / L_\lambda \cong 0.05$ at 200 nm.

8.2.2 Continuum Emission from a Hydrogen Plasma

For wall stabilized arcs the continuum emission of most gases can be calculated from atomic data but only with significant uncertainties. The optically thin continuum emission of hydrogen under LTE conditions, however, can be

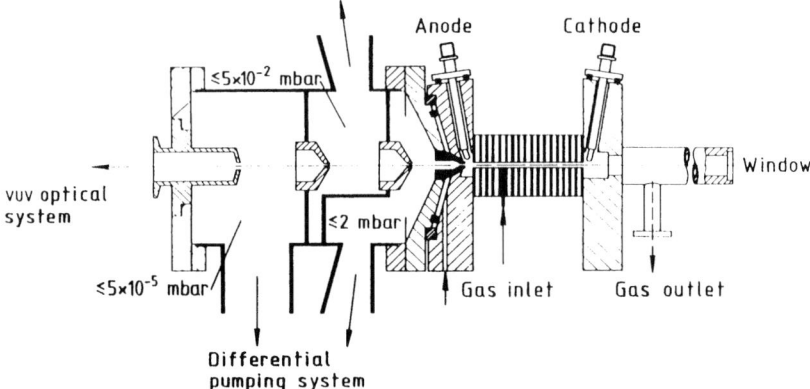

FIG. 1. Wall stabilized high-pressure arc with three-stage differential pumping system for self-absorption-free observation of spectral lines from optically thick layers. [Reprinted with permission from M. Kühne and B. Wende, *J. Phys. E.: Sci. Instrum.* **13**, 637–647 (1985). Copyright © 1985 Institute of Physics.]

determined from fundamental physics based on the knowledge of the plasma temperature and its electron density. On this basis, Ott *et al.* [10] developed a source standard for wavelengths above the hydrogen resonance line (124 to 360 nm), utilizing the radiation of the Balmer continuum with uncertainties of 5% above 140 nm, 9% at 130 nm, and 14% at 124 nm of the spectral radiance. The method has been extended to the Lyman continuum (53 to 92 nm) by Behringer and Thoma [11] with an uncertainty of 15% for the spectral radiance. To avoid reabsorption of the radiation by cold hydrogen, a helium gas layer has been used that is transparent down to 53 nm in combination with a differential pumping system.

Summarizing the experience using arc plasma sources as primary source standards it must be said that the result was somewhat disappointing. In no way uncertainties in the percent range or less could be achieved. Below 53 nm no practical primary source standards were available at all. This unsatisfactory situation remained until electron storage rings as dedicated sources of synchrotron radiation became available for VUV radiometry.

8.2.3 Electron Storage Ring as a Radiometric Source Standard

In the 1940s it became evident that high-energy electrons accelerated on circular paths would be emitters of electromagnetic radiation covering the full range from the far infrared to the x-ray region [12–14]. Optical radiation from an electron synchrotron was first observed in 1947 [15]. The use of synchrotrons as fundamental radiometric standards was pointed out by Tomboulian and Hartmann in 1956 [16]. With the increasing number of electron synchrotrons the systematic use of synchrotron radiation began in the mid 1960s; see, for example, Codling and Madden [17], Lembke and Labs [18], Pitz [19], and Key and Ward [20].

For detailed description of synchrotron radiation and its application to VUV physics see Chapter 1 of this book. This chapter discusses only the specific properties of synchrotron radiation sources with respect to their application as primary source standards.

A major problem with electron synchrotrons as radiation sources was the constant change of electron energy and the loss of electrons during the acceleration cycle, which prevented the practical use of synchrotrons as primary source standards. This situation changed with the development of electron storage rings where electrons with practically constant energy could be stored with beam lifetimes in the order of several hours.

According to Schwinger [14] the spectral radiant power of an electron storage ring emitted in an aperture A is given by

$$\Phi_\lambda^{SR}(\lambda) = \Phi_\lambda^{SR\|} + \Phi_\lambda^{SR\perp}$$

$$= \frac{2e\rho^2 Jb}{3\varepsilon_0 \lambda^4 \gamma^4 d^{SR}} \left[\int_{\psi_0 - a/2d^{SR}}^{\psi_0 + a/2d^{SR}} [1 + (\gamma\psi)^2]^2 K_{2/3}^2(\xi)\, d\psi \right.$$

$$\left. + \int_{\psi_0 - a/2d^{SR}}^{\psi_0 + a/2d^{SR}} [1 + (\gamma\psi)^2](\gamma\psi)^2 K_{1/3}^2(\xi)\, d\psi \right]$$

with

$$\gamma = \frac{W}{m_0 c^2}; \qquad \xi = \frac{2\pi\rho}{3\gamma^3 \lambda}[1 + (\gamma\psi)^2]^{3/2}; \qquad \rho = \frac{W}{ecB}.$$

Unlike thermal radiation synchrotron radiation is completely polarized. There are two components: $\Phi_\lambda^{SR\|}(\lambda)$ with the electrical field strength parallel to the electron orbit plane and $\Phi_\lambda^{SR\perp}(\lambda)$ with electrical field strength perpendicular to the electron orbit plane. Between the two components a fixed phase shift of π exists. The radius of curvature of the circular path at the tangent point of observation is denoted ρ; and W, e, and m_0 are the energy, charge, and rest mass, respectively, of the electrons in the beam which form a current J; B is the magnetic inductance at the point of observation; c is the speed of light; and $K_{1/3}$ and $K_{2/3}$ are Bessel functions of the second kind. The definitions of a, b, d^{SR} and ψ_0 are given in Fig. 2.

The preceding relations hold true strictly only for electrons moving on an ideal circular path. Due to the criteria for stable acceleration and containment of the electrons in an electron storage ring, the electrons are located in bunches in which they oscillate around the circular orbit. These oscillations in the bunches can be described by horizontal and vertical halfwidths of σ_x and σ_y and corresponding horizontal and vertical divergence σ'_x and σ'_y. The product of $\sigma_x \sigma'_x$ and $\sigma_y \sigma'_y$ is constant. While the horizontal bunch width and divergence have no influence on the spectral radiant power emitted into aperture A, the vertical halfwidth σ_y and the vertical divergence σ'_y can have a measurable influence, particularly at very short wavelengths and for small vertical apertures. Both parameters can be combined into an effective vertical halfwidth σ_{y*}, which combines the effects of σ_y and σ'_y; see Arnold and Ulm [21].

To use an electron storage ring as a primary source standard, provisions must be taken to measure the following parameters with sufficient accuracy.

1. *Electron storage ring*: Electron energy W, magnetic inductance B at the tangent point under observation, electron current J stored in the ring at the time of observation, and effective vertical beam size σ_{y*}.
2. *Radiometric laboratory*: Height a, width b (or alternatively the radius r in the case of a circular aperture), and distance d^{SR} of the aperture A from the tangent point of observation as well as the angle ψ_0 between the center of the aperture A and the electron orbit plane.

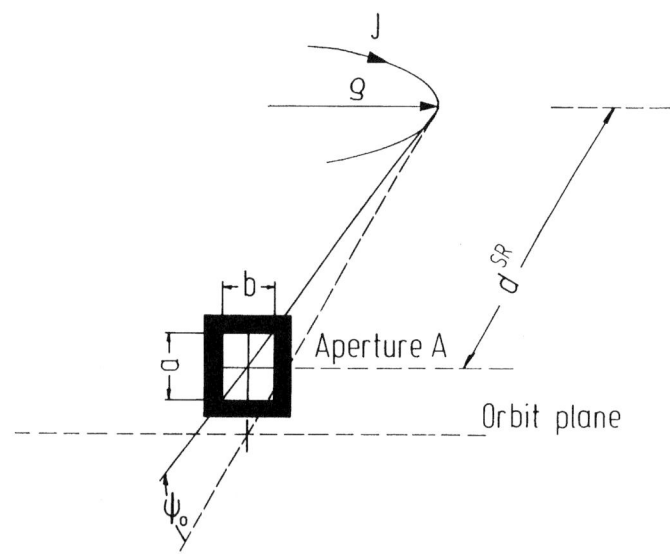

FIG. 2. Spectral radiant power $\Phi_\lambda^{SR}(\lambda)$ of an electron storage ring passing through an aperture stop A with area $a \times b$ at a distance d^{SR} from the tangent point, where ψ_0 is the angle of the center of A with the respect to the plane of the electron orbit, as seen from the tangent point. [Reprinted with permission from J. Fischer, M. Kühne, and B. Wende, *Appl. Opt.* **23**, 4252–4260 (1984). Copyright © 1984 Optical Society of America.]

Worldwide several electron storage rings like SURF II (USA), TERAS (Japan), VEPP-2M (Russia), and BESSY (Germany) have been investigated and used as primary radiometric source standards. As an example for the achievable high accuracy, the uncertainties for the spectral photon flux of BESSY into a given aperture at photon energies between 1 eV and 15 keV, [21–23] are shown in Table II. These very low uncertainties combined with the possibility of varying the spectral photon flux by 9 orders of magnitude from a single stored electron to currents of hundreds of milliamperes, make electron storage rings by far the best primary source standards for the VUV region today.

8.3 Radiometric Characterization of Secondary Source Standards

Radiometric source standards are mainly used for the radiometric characteriztion of other radiation sources by comparison or for the determination of the spectral sensitivity of radiometric or spectroscopic instrumentation. If a characterization at the highest metrological level is required in general a primary

TABLE II. Uncertainties (1σ Level) Due to Storage Ring Parameters and Geometrical Quantities for Photon Fluxes $\Phi(E)$ Realized with Bending-Magnet Radiation of the Standard Source BESSY I from the Near-IR (Photon Energy 1 eV) to the X-Ray Region (15 keV)

	Relative uncertainties $\Delta\Phi(E)/\Phi(E)$		
Source of uncertainty	1@eV (%)	5@keV (%)	15@keV (%)
Electron energy $W = (851.94 \pm 0.06)$ McV (measured by resonant spin depolarization)	0.0047	0.093	0.28
Electron current $I = (10 \pm 0.002)$ mA (measured by two dc transformers)	0.020	0.020	0.020
Extension divergence $\Sigma y = (33.5 \pm 4.3)$ μrad of the electron beam	0.0014	0.14	0.11
Magnetic induction $B = (1.592\,60 \pm 0.000\,12)$ T (measured by an NMR probe)	0.0050	0.045	0.14
Distance $D = (15\,837 \pm 2)$ mm^a $= @(7468@\pm@2)$ mm^b	0.025	0.037	0.033
Radius of aperture $r = (2.5010 \pm 0.0002)$ mm^a $= @(2.501@\pm@0.001)$ mm^b	0.016	0.048	0.042
Emission angle $\Psi = (0 \pm 15)$ μrad^a $= @0@\pm@9)$ μrad^b	0.0011	0.040	0.030
Total rms uncertainty $\Delta\Phi(E)/\Phi(E)$	0.037	0.19	0.34

^{a} For 1 eV.
^{b} For ≥ 5 keV.
Reprinted with permission from B. Wende, *Metrologia* **32**, 419–424 (1995/96). Copyright © 1996 Bureau International des Poids et Mesures, Sévres.

radiometric source standard should be employed to minimize the overall uncertainty. In such a case an electron storage ring primary source standard offers the smallest achievable uncertainties. However, due to the high cost of building and operating such radiation sources the availability of such primary standards will always be very limited. Aside from these access limitations there are other reasons why it can be preferable not to use an electron storage ring but to utilize a secondary source standard that has been calibrated against an electron ring primary standard:

1. Electron storage rings are radiation sources that cover the full electromagnetic spectrum from the far IR to the short-wavelength VUV or the x-ray region depending on the storage ring parameters, with a maximum radiation emission in the short-wavelength VUV or even in the x-ray region. If the relative spectral power distribution of the source or the spectral sensitivity of the instrument to be characterized is significantly different from the spectral radiation distribution of the storage ring, than severe problems from higher orders of the monochromator grating or from stray light can occur. In such a case the advantage of the small uncertainty of the primary standard is lost as other uncertainties become dominant.

2. Electron storage rings require UHV conditions to achieve the necessary electron-beam lifetimes. In addition the radiometric beamlines at the ring require very clean hydrocarbon free vacuum conditions; otherwise the optical components will be contaminated and rapidly degrade under the intense VUV radiation flux. Sources or instrumentation that cannot meet these requirements cannot be calibrated at storage rings.

3. Because storage rings are typically multiuser systems, in general strict beam time scheduling is required, in turn requiring well-in-advance-scheduling of the calibration tests. This is a serious drawback for instrumentation like satellite spectrometers where, on one hand, schedule slips often occur, but where, on the other hand, tight overall deadlines have to be observed.

4. The instrumentation to be characterized may require special environmental conditions, like cleanroom requirements, which cannot be met at the storage site or it might be just too delicate or too bulky to be transported to the storage ring site.

5. The characterization intended does not need the ultimate accuracy so that for cost savings a secondary source standard can be utilized and the calibration can be performed at the users own laboratory.

For one or more of these reasons it can be very desirable to use secondary source standards calibrated against an electron storage ring primary standard. An additional advantage is that for the calibration of that secondary source standard the instrumentation at the storage ring site can be optimized in order to minimize the effects of stray light and higher order contributions by use of the suitable optical components taking into account the particular spectral radiant power distribution of the storage ring used.

8.3.1 General Concept

The general concept for the radiometric characterization of a secondary source standard is that the determination (calibration) of the radiometric properties of interest, for example, spectral radiant power is performed by comparison to the primary standard. No absolute measurements are required as the absolute value is obtained from the ratio of the detector signals and the absolute value of the primary standard.

While the details of the layout of the instrumentation will depend on the particular source standard, the radiometric property of interest, the wavelength range, require spectral resolution and many other factors, some features can be described that are typical for most instrumentations. In the VUV spectral range the losses of mirrors and gratings are in general quite significant, so the general rule is to minimize the number of optical elements. A basic design would consist of an imaging mirror focusing the radiation emitted from the sources onto the entrance slit of a monochromator. The monochromator contains a figured grating acting as disperser and focusing mirror simultaneously. Behind the exit slit of the monochromator a photoelectric detector is located converting the photons received into an electric signal which is read out for further processing.

The basic assumption is that the ratio of the detector signals obtained when alternatively observing the two sources will be equal to the ratio of the radiometric property that is to be determined. For example, the spectral radiance $L_\lambda^{SE}(\lambda)$ of a secondary standard is determined from the ratio of the detector currents of the primary standard $i^{PR}(\lambda)$ and secondary standard $i^{SE}(\lambda)$ and the known absolute spectral radiance of the primary standard $L_\lambda^{PR}(\lambda)$:

$$L_\lambda^{SE}(\lambda) = \frac{i^{SE}(\lambda)}{i^{PR}(\lambda)} L_\lambda^{PR}(\lambda).$$

However, this equation will be correct only under ideal conditions that are very difficult to meet in practice. It assumes, for example, in the case of spectral radiance the following to be true:

1. The standards have emitting areas of sufficient size to illuminate homogeneously the entrance slit area of the monochromator. Size of source effects can be neglected.
2. The distances from the sources to the entrance slit are equal (the slit area corresponds to the same source areas).

3. The same areas are illuminated on the optical elements and on the detector under the same conditions.
4. The polarization of the radiation of both sources is the same.
5. No nonlinearity exists for the detector.

In an ideal case the two sources to be compared would be placed on a rail system that would move them and position them alternatively at the same location so that their images can be placed identically onto the entrance slit. With the proper optical apertures and monochromator slit widths we can then be assured that the illumination of all optical components and the detector is identical. However, due to the size of many sources this solutions can not often be realized-certainly not in the case of a storage ring serving as the primary standard.

The next best solution is to place both sources at the same distance but at different locations from the entrance slit and rotate the imaging mirror so that either source can be imaged onto the entrance slit of the monchromator. If enough space is available in the laboratory that certainly is the way to go. However, at synchrotron radiation laboratories this approach will in general not be possible. Due to the packed space in such laboratories and the geometrical constrictions given by the use of several beamlines often quite close together it is often necessary to place the secondary standard at a distance to the imaging mirror that is significantly smaller than the distance from the tangent point to the mirror. This basically has two consequences:

1. Because the distance from the mirror to the sources will be different, so will be the distance from the mirror to the images of the sources. That will in general require that either the mirror or the monochromator be moved along the optical axis to bring the entrance slit to the position of the image. A solution must be found to perform this movement under UHV conditions, which are required for operation at storage rings.
2. Due to the laws of optics it is not possible to illuminate identical areas on the mirror and the grating for both sources. If the same area is illuminated on the mirror then the radiation from the secondary standard will illuminate a smaller area on the grating. In general, the inhomogeneity in diffraction efficiency of a grating will be larger than that of the reflectance of a mirror, making it desirable to have the illuminated areas on the grating of approximately the same size. This will also provide the same illuminated area on the detector, which is also important because of local efficiency variations.

When using an electron storage ring as a primary standard, keep in mind that synchrotron radiation is fully polarized (see Chapter 1) with the status of polarization depending on the angle in respect to the orbit plane. Most secondary standards in the VUV emit unpolarized or differently polarized radiation as synchrotron radiation. Because the efficiency of gratings and mirrors will in general be polarization dependent, the polarization properties of the instrumentation must be determined and taken into account. In the VUV good polarizers are hard to come by or not available at all. As a consequence the polarization characteristics of the instrumentation has to be determined using the calculable degree of polarization of the synchrotron radiation and different orientations of the calibration instrumentation in respect to the electron orbit plane (see, e.g., Fischer *et al.* [24] for details).

The electron beam emitting the synchrotron radiation has an elliptical cross section and a Gaussian spatial electron density distribution. Electron storage rings are therefore in general unsuitable to serve directly as primary standards of spectral radiance. In most cases the spectral radiant power passing through an aperture of precisely known size and location is calculated and then the spectral radiant power of the secondary standard determined by comparison. This spectral radiant power is then converted into other radiometric properties, such as spectral concentration of radiant intensity or spectral radiance from the geometrical data of the instrumentation, like the entrance slit area.

In the following sections instrumentation developed and operated at the laboratory for VUV radiometry of the Physikalisch-Technische Bundesanstalt at the Berlin electron storage ring BESSY is described. The instrumentation was set up for the calibration of secondary source standards using BESSY as a primary source standard. The instrumentation was designed to take into account and allow corrections for the different degrees of polarization of the radiation from the storage ring and from the secondary source standards. If such instrumentation were used to compare laboratory sources with secondary source standards these particular features would not be required in most cases. To illustrate the performance achieved by the instrumentation, examples for secondary source standard calibrations are given. To optimize the performance and to minimize the effects of stray light and higher diffraction orders, three different set of instruments were built.

8.3.2 Instrumentation for Soft X-Ray Source Comparisons

The wavelength range of 0.6 to 6 nm (200 to 2000 eV) is the traditional grazing incidence domain. The short-wavelength cutoff is given by the efficiency limits of classical reflection gratings. For shorter wavelengths in general crystal spectrometers are required. This wavelength range includes the so-called "water window" between 2.4 and 4.4 nm, which is of importance for soft x-ray microscopy, and the range from 0.6 to 2 nm, which is important to proximity lithography.

The motivation for the development of this instrumentation was therefore to characterize soft x-ray sources with an application potential for soft x-ray microscopy or lithography. As the optical requirements in this field call for point-like sources of high soft x-ray radiation output, the radiometric property to be characterized is the spectral concentration of radiant intensity [W m^{-1} sr^{-1}] or the spectral concentration of radiant intensity time integrated per pulse [J m^{-1} sr^{-1}] often also expressed in [ph m^{-1} sr^{-1}]. Typical sources for this application are pulsed plasma sources such as laser-produced plasmas or plasma foci.

To perform a comparison of the spectral concentration of radiant intensity of the electron storage ring BESSY with that of a point-like plasma source, the following boundary conditions for the optical design of the instrumentation apply:

1. The radiation from both sources emitted in a well-defined solid angle has to be focused into the entrance slit of a monochromator. The image sizes of both sources at the location of the entrance slit must be small enough so as not to prevent any radiation from entering the entrance slit. This requires an imaging mirror with sufficient demagnification. A further requirement is that the image of the source in the entrance slit plane be sufficiently small to achieve sufficient spectral resolution.
2. No stigmatic imaging is required because it is the spectral concentration of radiant intensity [W m^{-1} sr^{-1}] that is the radiometric property under investigation and not the spectral radiance [W m^{-3} sr^{-1}].
3. Because pulsed soft x-ray plasma sources are very bright, only a small solid angle of observation is needed.

As a result of applying these three criteria a concave gold-coated mirror illuminated at an angle incidence of 88.5 deg was chosen for the imaging optics [25]. With a radius of curvature of 22 m the resulting meridional and saggital focal lengths are about 300 mm and more than 400 m (which practically, is infinity) respectively. The mirror can be rotated around its horizontal axis to allow alternatively the imaging of the electron storage ring tangent point or the plasma source into the entrance slit of the monchromator (Fig. 3). Due to the small meridional focal length the line images from either the tangent point of the storage ring or from the source under investigation are formed nearly in the same image plane so that the same monochromator entrance slit position can be used for both sources, see [25] for details.

For the operation of BESSY with an effective vertical beam size of ≈500 μm the demagnified image has a vertical height of ≈15 μm. Considering the small difference between the image plane and the entrance slit plane (the entrance slit plane was located halfway between the image plane of the tangent point and the image plane of the plasma source) an effective vertical image size of 27 μm was to be expected. This was confirmed by measurements using an entrance slit width of 20 μm, which resulted in collection of 90% of the total radiation. A

Rowland circle monochromator with a concave grating having a 5-m curvature was used for spectral dispersion. The laminar ion-etched and gold-coated grating had a line density of 1200 lines/mm and was used under a fixed angle of incidence of 88 deg. Different entrance slits with widths ranging from 20 μm to 1 mm could be exchanged under vacuum. No mechanical Rowland circle was realized but the exit slit was moved by two computer-controlled linear translation stages. The exit slits with widths ranging from 36 μm up to 1.5 mm could also be exchanged under vacuum. A mechanical guiding system kept the exit slit planes always perpendicular to the axis defined by the center of the grating and the center of the exit slit. The same system also kept the open electron multiplier in position behind the exit slit so that the illuminated area did not move on the detector during a wavelength scan. The travel range of the exit slit extends from zero order up to 6 nm in first order. The imaging mirror and the monochromator components are mounted on a granite bench and are enclosed by a UHV vacuum tank. The vacuum tank is decoupled from the granite bench and its spectrometer components using membrane bellows. This way the optical alignment is not disturbed by the vacuum tank due to pump down or venting. Great care was

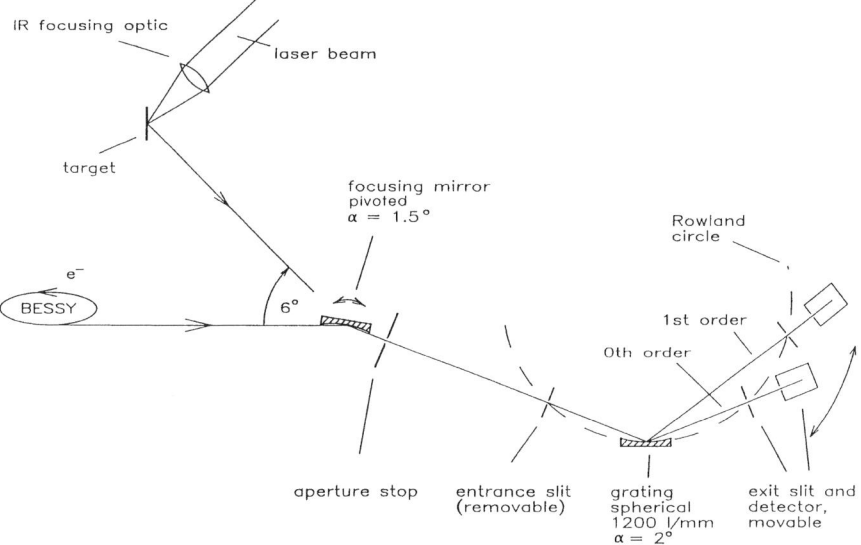

FIG. 3. Instrumentation for the radiometric characterization of soft x-ray sources by comparison with the calculable synchrotron radiation emission of the electron storage ring BESSY in the spectral region from 0.6 to 6 nm (200 to 2000 eV). The vertical plane is shown. Distances and angles are not to scale. The focusing mirror can be rotated around a horizontal axis perpendicular to the figure plane. [Reprinted with permission from R. Thornagel and M. Kühne, *Rev. Sci. Instrum.* **60**, 1920–1923 (1989). Copyright © 1989 American Institute of Physics.]

taken to minimize the amount of hydrocarbons present in the vacuum system, which was operated at a total pressure of 2×10^{-7} Pa.

Due to the different degree of polarization of the radiation emitted from the storage ring or from the plasma source, it is necessary to determine the polarization characteristic of the mirror/spectrometer combination. This determination can be performed using the polarized synchrotron radiation To do so, the complete mirror/spectrometer system, which is mounted on the same granite bench, can be rotated by 90 deg around a horizontal axis lying in the synchrotron radiation plane. The rotation can be performed without braking the vacuum by using a differentially pumped hollow shaft. The evaluation of the polarization characteristic is then performed with the mirror and the grating plane parallel (0 deg position, s polarization) or perpendicular (90 deg position, p polarization) to the electron orbit plane using the linearly polarized synchrotron radiation in the electron orbit plane. No difference was found between the s- and p-polarization measurements within the measurement uncertainty ($\approx 2\%$ over most of the spectral range), which can be attributed to the large angles of incidence used on the mirror and the grating.

Because the synchrotron radiation spectrum of BESSY extends to well below 0.2 nm this optical instrumentation using an imaging mirror and a grating under grazing incidence could well be sensitive to higher diffraction order contamination. To evaluate this problem, an energy-resolving Si(Li) detector was mounted behind the exit slit and the spectrum recorded for different storage ring energies (Fig. 4). At BESSY's standard operating energy of 756 MeV, almost no higher orders were present below a wavelength of 1.65 nm (see Fig. 5). For observations at longer wavelengths the storage ring needs to be operated at 570 MeV. The instrumentation and primary standard related minimum achievable uncertainty in the determination of the spectral concentration of radiant intensity was evaluated to 6% for continuum sources and 9% for line emission sources [26].

A laser-produced carbon plasma was the first plasma source investigated due to the high interest in such a source for soft x-ray microscopy. Many soft x-ray microscopes use Fresnel zone plates for imaging. Such zone plates suffer from strong chromatic aberrations. If a source emitting a continuous spectrum is used for illumination, then, in general, a predispersing element is needed that significantly reduces the optical throughput. Under proper laser operating conditions it should be possible to produce intense line radiation from ionized carbon with lines in the water window that are well separated. The use of such radiation in connection with multilayer optics could provide intense monochromatic radiation. Because the details of the above experiment can be found in [27] only a summary is provided here.

Pulses of 100 ps with energies of up to 300 mJ from a mode-locked Nd:YAG system operating at a repetition rate of 10 Hz were focused to a diameter of

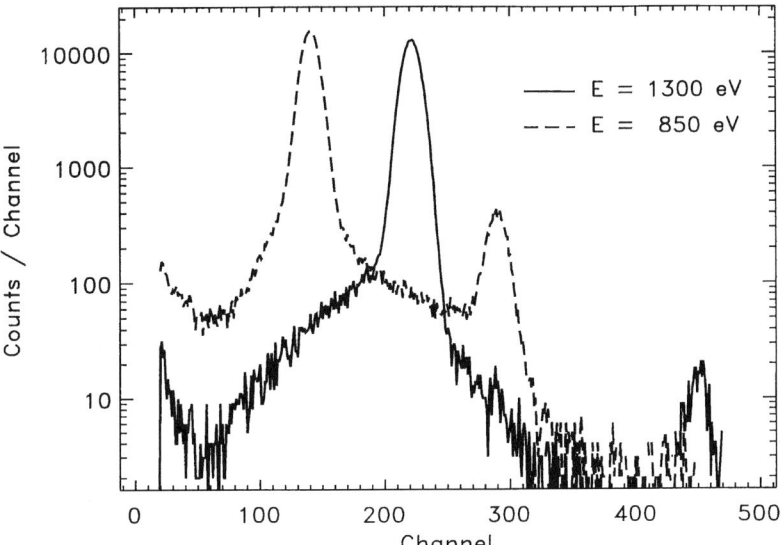

FIG. 4. Spectrum recorded with a Si(Li) detector at a fixed monochromator setting of $E = 850$ eV (dashed line) and $E = 1300$ eV (solid line) at a storage ring energy of 756 MeV. The spectral lines corresponding to the first- and second-order photons at the double channnel number can be seen. [Reprinted with permission from R. Thornagel and M. Kühne, *Rev. Sci. Instrum.* **60**, 1920–1923 (1989). Copyright © 1989 American Institute of Physics.]

≈50 μm on a plane carbon target. The exit slit width was selected so that the full line profile lay inside the bandwidth of the monochromator. The line integrated emission per laser pulse was determined by comparison with the calculable spectral concentration of radiant intensity of BESSY. Table III shows the results. In the strongest line, C V $1s^2$-$1s2p$ at 308 eV, about 5.5×10^{12} photons/sr/laser pulse are emitted. The observed spectrum (Fig. 6) consists of a clean emission spectrum from hydrogen and helium-like carbon ions. In the vicinity of the strongest line the ratio between the line emission and the continuum is at least 500:1. An overall conversion efficiency of laser radiation to single-line soft x-ray radiation of close to 1% has been achieved. The overall soft x-ray conversion efficiency was determined to be about 2.5%.

8.3.3 Instrumentation for EUV Source Comparisons

For the spectral region of 5 to 100 nm, special instrumentation for source comparisons was built at the radiometric laboratory of PTB at BESSY [24]. The equipment consisted of an ellipsoidal mirror and a toroidal grating monochromator (TGM) (see Fig. 7). The angle of incidence on the ellipsoidal mirror is set to 86 deg and a TGM with an angle of 162 deg between the arms is used. The TGM is equipped with three ion-etched gratings with a line density of 200, 600, and 1800 lines/mm to cover the wavelengths ranges of 36 to 100 nm, 12 to 36 nm, and 4 to 12 nm, respectively. The short-wavelength cutoff is around 5 nm due to the angle of incidence of the grating. In the early years the system was used up to the 100-nm wavelength. However, since the completion of a normal incidence calibration station (see Section 8.3.4) it is not used above 60 nm because of the presence of higher diffraction orders.

The center of the mirror is located 15 m off the tangent point. Because of the spatial constraints of the laboratory, the source under investigation had to be placed within a distance of only 5 m away from the mirror. The mirror can be rotated around a vertical axis to focus the radiation either from the tangent point or from the source under investigation into the entrance slit of the monchromator using the same angle of incidence. The entrance slit size is chosen so that no radiation is cut off by the slit jaws.

To compensate for the different distances of the sources and the corresponding different image distances, the TGM can be moved under UHV along its optical entrance axis. A common aperture stop A limits the solid angles from

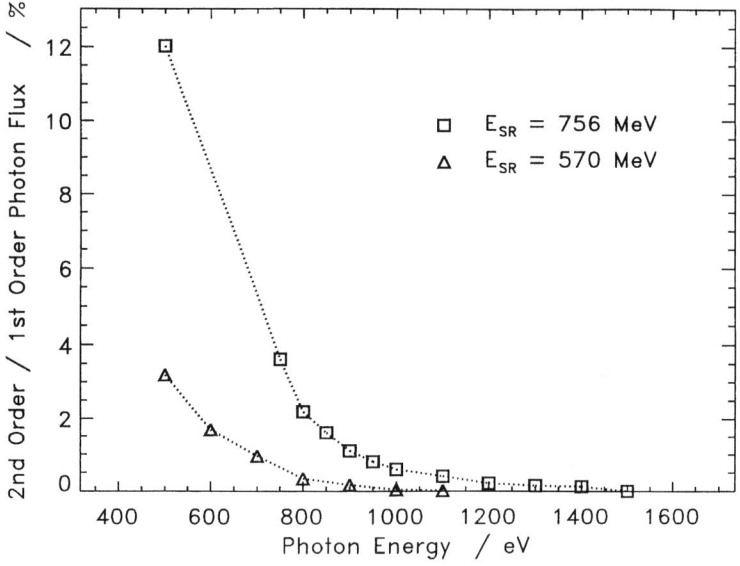

FIG. 5. Ratio of second- to first-order photon flux behind the exit slit at a storage ring energy of 756 and 570 MeV, respectively. The low-energy limit was due to the low-energy cutoff of the Si(Li) detector that was used. [Reprinted with permission from R. Thornagel and M. Kühne, *Rev. Sci. Instrum.* **60**, 1920–1923 (1989). Copyright © 1989 American Institute of Physics.]

TABLE III. Radiation Emission $\int I(t)\, dt$ in Photons per Steradian per Pulse of the Lines of C V and C VI at a Laser Energy of 215 mJ on the Target, Radiated Energy per Pulse W_{ph} into 2π Under Assumption of an Isotropic Distribution, and Conversion Factor k from Laser Energy on the Target into X-Ray Radiation into 2π

Ion	Transition	$h\nu$ (eV)	$\int I(t)\, dt$ ($\times 10^{10}$ ph/sr)	W_{ph} (mJ)	k (%)
C V	$1s^2$–$1s2p$	308	549	1.79	0.84
	$1s^2$–$1s3p$	354	114	0.43	0.20
	$1s^2$–$1s4p$	371	64	0.25	0.16
C VI	$1s$–$2p$	367	456	1.78	0.83
	$1s$–$3p$	435	153	0.71	0.33
	$1s$–$4p$	459	55	0.21	0.13
	$1s$–$5p$	470	25	0.12	0.06
Sum				5.35	2.5

Reprinted with permission from M. Kühne and R. Thornagel, in "X-Ray Microscopy III" (A. G. Michette, G. R. Morrison, and C. J. Buckley, eds.), Springer Series in Optical Science, Springer Verlag, Berlin, **67**, p. 39–92 (1992). Copyright © 1992 Springer Verlag Berlin Heidelberg.

both sources. To determine the polarization characteristics of the TGM the monochromator can be rotated around its optical entrance axis to allow measurements to be made with the monochromator plane parallel and perpendicular to the electron orbital plane. The influence on the polarization caused by the ellipsoidal mirror is small due to the large angle of incidence and can be calculated from the optical properties of the kanigen-coated mirror; see [24] for details. Another major limitation of the achievable calibration uncertainty stems from the amount of higher diffraction orders present in the source spectra. For the quantitative determination of the higher orders present in the synchrotron radiation spectrum, a reflectometer containing a 1000 lines/mm free-standing transmission grating was installed behind the exit slit [28]. The number of higher orders present in the TGM synchrotron radiation spectra is shown in Fig. 8. By operating the electron storage ring at a suitable low electron energy and by selecting the appropriate grating and absorption edge filter, the amount of higher orders present in the synchrotron radiation spectrum can be kept at or below 5% for all wavelengths below 40 nm. Sources that have been radiometrically characterized include laser-produced tungsten plasma sources (7–100 nm) [29], hollow cathode sources (13–60 nm) [30], and Penning sources (10–17 nm) [31].

A special type of source standard had to be built and investigated for the radiometric calibration of the solar telescope-spectrometer CDS (Coronal Diagnostics Spectrometer), with a spectral range of 15–80 nm and its sister system SUMER (Solar Ultraviolet Measurements of Emitted Radiation), with a spectral

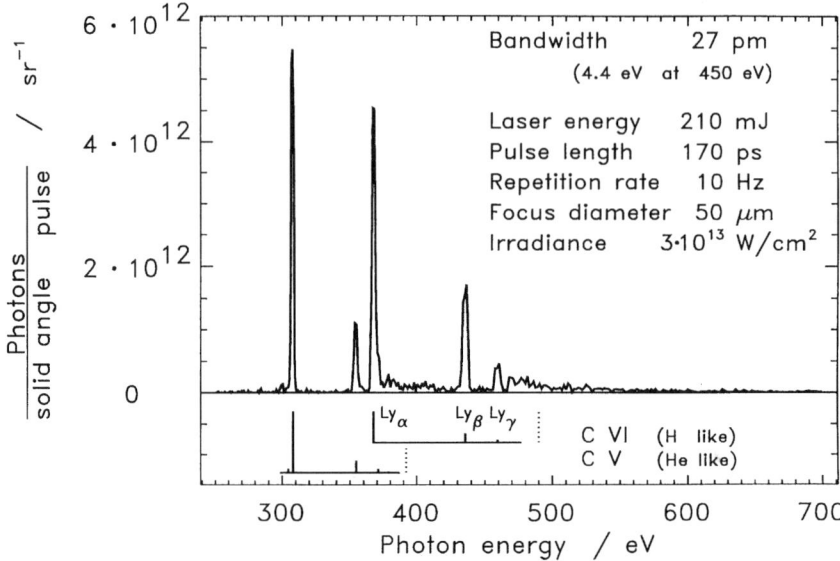

FIG. 6. Spectrum of a laser-produced plasma. $\int_t I_{\Delta E}$ is the radiant emission in photons/sr within the bandwith of the monochromator ($\Delta E = 2.0$ eV at 300 eV, $\Delta E = 4.4$ eV at $E = 450$ eV), time integrated over one pulse. The expected lines and the ionisation energies (dotted) of C V and C VI are shown. [Reprinted with permission from M. Kühne and R. ThornageL, in "X-Ray Microscopy III" (A. G. Michette, G. R. Morrison, and C. J. Buckley, eds.), Springer Series in Optical Sciences, Vol. 67, pp. 39–42, Springer Verlag, Berlin (1992). Copyright © 1992 Springer Verlag Berlin Heidelberg.]

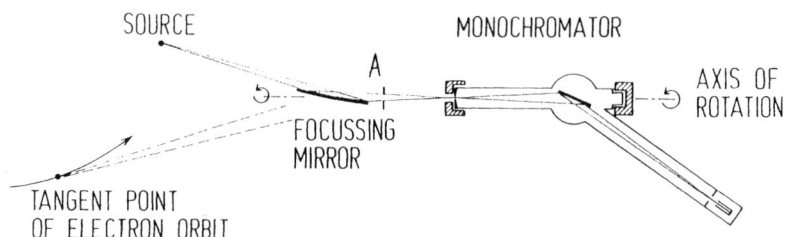

FIG. 7. Instrumentation for the radiometric characterization of EUV sources by comparison with the calculable radiation emission from the electron storage ring BESSY in the spectral region from 5 to 100 nm. [Reprinted with permission from J. Fischer, M. Kühne, and B. Wende, *Appl. Opt.* **23**, 4252–4260 (1984). Copyright © 1984 Optical Society of America.]

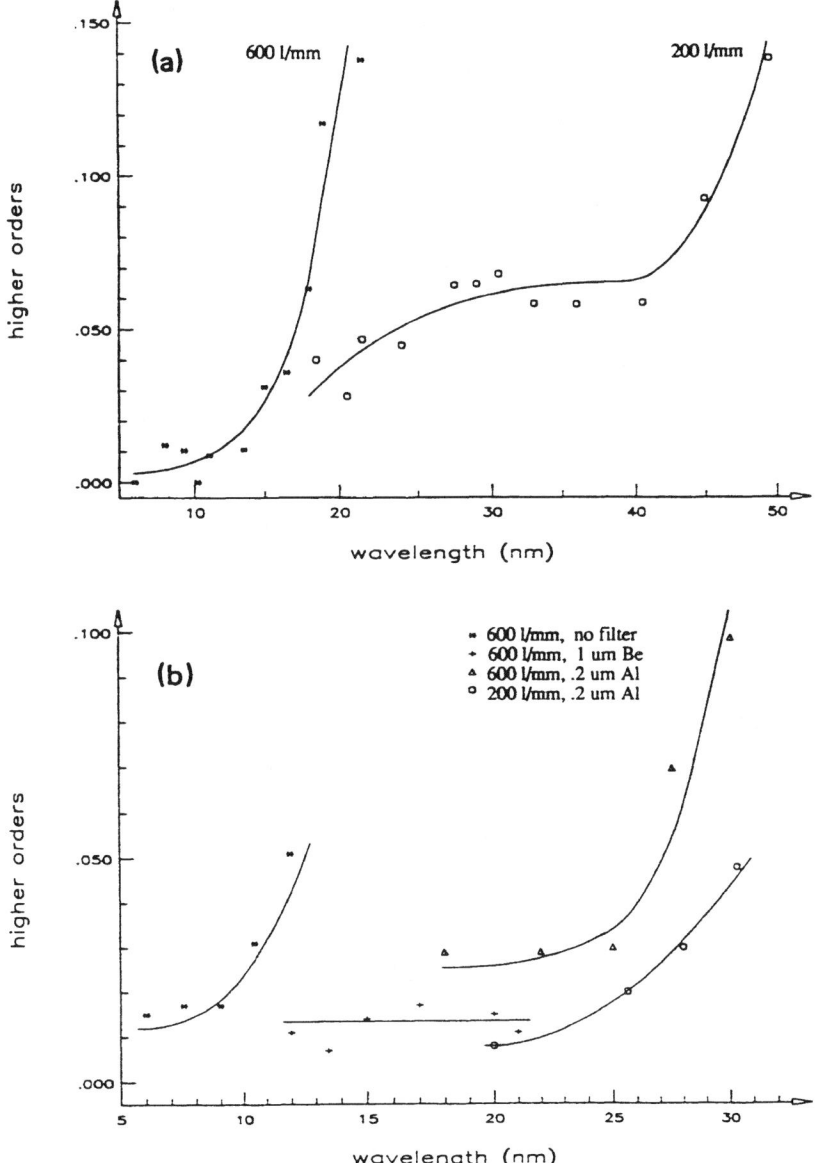

FIG. 8. Higher order contribution in the synchrotron radiation spectrum of the TGM (a) at electron beam energy of 337 MeV and (b) at electron beam energy of 756 MeV. Displayed is the fraction of the total detector signal that is caused by second or higher orders. [Reprinted with permission from M. Kühne and P. Müller, *Rev. Sci. Instrum.* **60**, 2101–2104 (1989). Copyright © 1989 American Institute of Physics.]

range of 50–160 nm. Both were employed on the SOHO satellite [32–34], which was successfully launched on December 5, 1995.

For the radiometric calibration of these solar telescope-spectrometer systems the radiometric source standards had to be placed at infinity to simulate the sun and produce a focused image at the respective focal plane. Because this condition cannot be fulfilled in the laboratory for obvious reasons, collimating optics for the source standards were needed to provide a parallel beam simulating an infinite object distance. For that purpose a high-current hollow cathode source [30, 35] had to be combined with a type II Wolter telescope for the calibration of CDS and with a normal incidence concave mirror for the SUMER system.

The CDS calibration source utilized a Wolter telescope supplied by NASA that had been build as a spare for the SERTS project (a solar telescope to be flown on a sounding rocket). It is illuminated from the back by a hollow cathode source with a flux limiting aperture of 0.6 mm. It has a focal length of 2120 mm and produces a parallel beam with a diameter of 5 mm. The hollow cathode was operated at a current of 2 A and a voltage drop across the electrodes of 400 V. An aluminium cathode was used with either He, Ne, or Ar as a buffer gas. A compact two-stage differential pumping system allowed observation of the VUV radiation under UHV conditions.

For calibration, the TGM was moved along its optical axis so that the ellipsoidal mirror focused the parallel beam from the source into the monochromator entrance slit. To minimize the higher orders in the recorded synchrotron radiation spectrum, the storage ring was operated at a reduced electron energy of 340 MeV. In addition, at specific wavelength ranges absorption edge filters were used (Al: 20–32 nm; Mg: 25–41 nm; Sn: 53–59 nm). The measurements were obtained with the 200 line/mm grating (22–60 nm) and the 600 line/mm grating (15–22 nm). For a list of calibrated lines see Table IV. The lines above 60 nm have been calibrated using the instrumentation described in Section 8.3.4. The uncertainties (1 σ level) of the photon flux in the emission lines are between 7 and 8%. The contributions to the uncertainty for wavelengths of 20 and 40 nm are detailed in Table V.

8.3.4 Instrumentation for VUV and UV Source Comparisons

For comparison of VUV and UV radiation sources with the calculable spectral radiant power of the electron storage ring BESSY in the spectral range from 40 to 400 nm, instrumentation was build based on a normal incidence imaging mirror and a normal incidence monochromator. A detailed description of the instrumentation is found in [36]. The 1-m, 15-deg, McPherson-type monochromator and the 1-m focal length concave mirror are mounted on a frame that can be rotated around a vertical axis (Fig. 9). This allows the instrumentation alternatively to view different sources including the tangent point of the storage ring. UHV valves and a small vacuum lock are installed to allow the vacuum to be broken and to change the connection to a beam line leading to a different source without venting either the instrumentation or the corresponding beam line. The monochromator can be moved under UHV conditions ≈400 mm along its optical axis to compensate for object distances from 4 m to infinity (sources with collimated beams). As explained in Section 8.3.2 it is not possible with sources at different object distances to illuminate the same area on the grating and the mirror. Because, in general, the spatial variation for the grating efficiency exceeds the variation of reflectance from a mirror, a solid angle limiting aperture stop was installed in front of the concave mirror at such a distance that it provided equal area of illumination on

TABLE IV. Photon Flux in Selected Emission Lines of the CDS Calibration Source

Wavelength (nm)	Atom/ion	Photon (flux/s^{-1})	Relative uncertainty (%) (1σ)
16.01 to 16.17	Al IV/Ne	5.43×10^4	8
16.95 to 17.56	Al III/Ne	1.22×10^5	8
16.01 to 16.17	Al IV/Ar	4.86×10^4	8
16.95 to 17.56	Al III/Ar	8.77×10^4	8
20.43 to 20.89	Ne IV	3.43×10^5	8
21.26	Ne IV	1.23×10^5	8
21.54 to 21.88	Ne IV	4.81×10^5	8
22.26 to 22.36	Ne IV	2.03×10^5	8
26.71 to 26.77	Ne III	1.01×10^6	8
28.25 to 28.39	Ne III	2.61×10^6	7
30.11	Ne III	1.55×10^6	7
30.38	He II	4.89×10^7	7
30.86	Ne III	6.21×10^5	7
31.31 to 31.39	Ne III	1.96×10^6	7
37.93	Ne III	1.09×10^7	7
40.59 to 40.71	Ne II	2.01×10^7	7
44.50 to 44.78	Ne III	3.89×10^7	7
46.07 to 46.24	Ne II	1.57×10^8	7
48.81 to 49.11	Ne III	1.75×10^7	7
53.70	He I	7.16×10^6	7
58.43	He I	1.79×10^8	7
71.81 to 74.53	Ar II	2.54×10^7	7
73.59	Ne I	1.00×10^8	7
74.37	Ne I	5.88×10^7	7
76.92	Ar III	2.25×10^6	7

Source parameters: Discharge current, 2 A; discharge voltage, 400 V; diameter of collimated beam, 5 mm.
Reprinted with permission from J. Hollandt, M. Kühne, M. C. E. Huber, and B. Wende, *Astron. Astrophys. Suppl. Ser.* **115**, 561–572 (1996). Copyright © 1996 Astronomy & Astrophysics.

TABLE V. Contributions to the Relative Uncertainty of the Photon Flux in Selected Emission Lines of the CDS Calibration Source

Parameter	Relative uncertainty (%) (1σ)	
	20@nm	40@nm
Spectral photon flux of synchrotron radiation	1.1	0.8
Higher diffraction orders in synchrotron radiation	0.6	0.6
Local efficiency variation of optical components and detector	3	3
Linearity of detector	0.6	0.6
Wavelength calibration	0.6	0.6
Polarizing properties of the ellipsoidal mirror and the monochromator-detector system and degree of polarization of synchrotron radiation	3.2	3.2
Signal-to-noise ratio of detector from the emission lines of the CDS calbration source	3	—
Long-term stability of the CDS calibration source	5	5
Overall systematic relative uncertainty (addition in quadrature)	7.5	6.8

Reprinted with permission from R. A. Harrison, B. J. Kent, E. C. Sawyer, J. Hollandt, M. Kühne, W. Paustian, B. Wende, and M. C. E. Huber, *Metrologia* **32**, 647–652 (1995/96). Copyright © 1996 Bureau International des Poids et Mesures, Sévres.

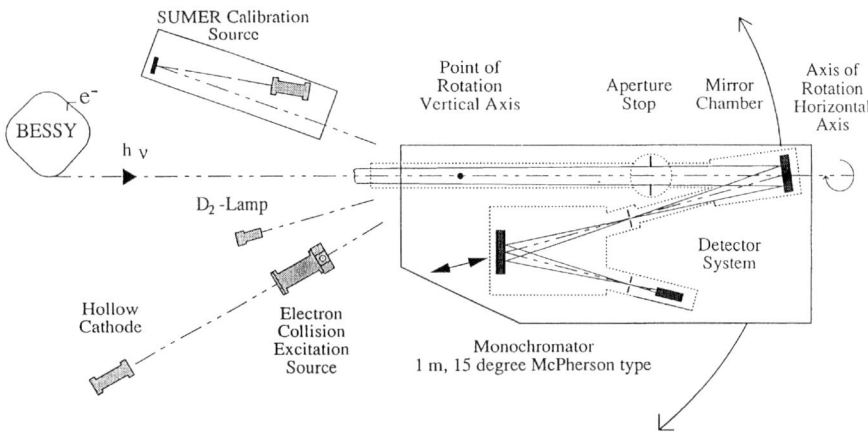

FIG. 9. Schematic drawing of the instrumentation for the chracterization of VUV and UV radiation sources. Shown are the synchrotron radiation beam line and different VUV sources. [Reprinted with permission from J. Hollandt, W. Jans, M. Kühne, F. Lindenlauf, and B. Wende, *Rev. Sci. Instrum.* **63**, 1278–1281 (1992). Copyright © 1992 American Institute of Physics.]

the grating for both sources. Exactly equal areas are obtained with a distance of 23.8 m from the tangent point to the aperture stop and, respectively, 5 m for the source under investigation.

To allow for the proper corrections due to the differently polarized radiation from the different sources, the complete instrumentation can be rotated around the horizontal axis. The polarization characteristics of the instrumentation are determined by measurements both with the grating normal parallel and perpendicular to the electron orbit plane. For details see [36]. To detect the radiation an open electron multiplier was used with the first dynode operated as a photo cathode. The photon flux can be registered alternatively in the photon current or photon counting mode. To maximize the spectral responsivity, imaging mirrors coated with Os, Au, and SiC can be installed. The differently coated mirrors are also important for the reduction of higher order contamination of the recorded spectra [35] (see Fig. 10).

Using the emission lines from a hollow cathode source, the ratios for first- and second-order diffraction efficiencies have been determined for the gratings in use. As an example the second-order to first-order ratio of a 1200 lines/mm grating blazed for maximum efficiency at 45 nm is shown in Fig. 11. The details

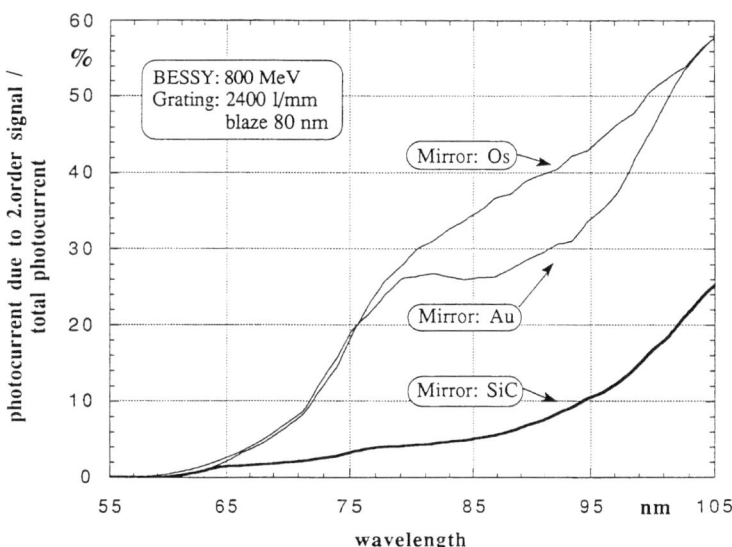

FIG. 10. Fraction of the total detected synchrotron radiation signal that is due to second-order radiation for differently coated imaging mirrors. [Reprinted with permission from J. Hollandt, M. Kühne, and B. Wende, *Appl. Opt.* **33**, 68–74 (1994). Copyright © 1994 Optical Society of America.]

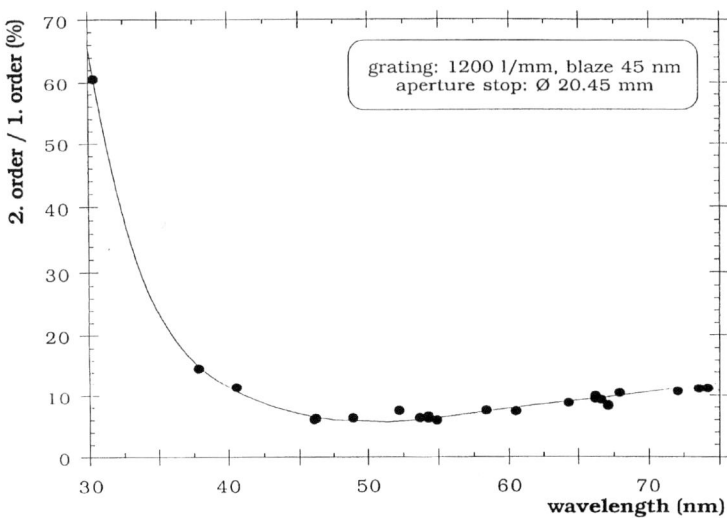

FIG. 11. Efficiency ratio of second to first diffraction order of a reflection concave grating obtained with a hollow cathode line radiation source. [Reprinted with permission from J. Hollandt, W. Jans, M. Kühne, F. Lindenlauf, and B. Wende, *Rev. Sci. Instrum.* **63**, 1278–1281 (1992). Copyright © 1992 American Institute of Physics.]

TABLE VI. Contributions to the Relative Uncertainty ($\sqrt{3}\sigma$ Level) of the Photoemission Cross Section σ (λ = 91.98 nm) for the Ar II $3s3p^6\ ^2S_{1/2}$–$3s^23p^5\ ^2P_{3/2}$ Transition at 2-keV Excitation Energy

Source of uncertainty	Relative uncertainty of photoemission cross section $\Delta\sigma/\sigma$ (%)
Target gas density n	
Temperature	0.2
Pressure	0.7
Pressure profile	0.9
Energy of the exciting electrons E	0.4
Current of the exciting electron i	0.7
Observed length of the electron beam l	0.3
Solid angle Ω	0.5
Counting statistics	0.6
Background correction	0.3
Subtotal in quadrature	1.7
Spectral responsivity of the monochromator-detector system with Si C mirror $\overline{s(\lambda)}$	4.1
Sum in quadrature	4.4

Reprinted with permission from W. Jans, B. Möbus, M. Kühne, G. Ulm, A. Werner, and K.-H. Schartner, *Appl. Opt.* **34**, 3671–3680 (1995). Copyright © 1995 Optical Society of America.

of the procedure are described in [36]. Gratings with ruling densities of 1200 and 2400 lines/mm can be installed using kinematic grating holders.

The instrumentation has been used to characterize the radiometric properties of source standards, among them high-current hollow cathodes [35], a radiometric source standard for the calibration of the solar VUV telescope SUMER on SOHO [32–34], and an electron-beam excitation source [37, 38].

Electron-beam excitation sources (see Chapter 4 for details), are among the secondary source standards with the highest reproducibility. The achievable uncertainties for the absolute emission cross section of the electron-impact-induced Ar II $3s3p^6\,^2S_{1/2} - 3s^23p^5\,^2P_{3/2}$ transition is shown in Table VI. It demonstrates what uncertainties are obtainable for radiometric calibrations of suitable UV sources. Please note that the uncertainty is quoted on a $\sqrt{3}\sigma$ level. A list of 18 electron-impact-induced emission cross sections with 1σ uncertainties ranging from 2.8 to 5% can be found in [38].

References

1. International Union of Pure and Applied Physics. Document UIP 20 (1978).
2. T. J. Quinn, "Temperature," Academic Press, London (1990).
3. H. Preston-Thomas, *Metrologia* **27**, 3–10 (1990).
4. G. Boldt, *Space Sci. Rev.* **11**, 728–772 (1970).
5. D. Stuck and B. Wende, *J. Opt. Soc. Am.* **62**, 96–100 (1972).
6. P. J. Key and R. C. Preston, *Appl. Opt.* **16**, 2477–2485 (1977).
7. H. Kaase, *Optik* **59**, 1–12 (1981).
8. K. Grützmacher and B. Wende, *Phys. Rev. A* **18**, 2140–2149 (1978).
9. M. Kühne and B. Wende, *J. Phys. E: Sci. Instrum.* **18**, 637–647 (1985).
10. W. R. Ott, K. Behringer and G. Gieres, *Appl. Opt.* **14**, 2121–2128 (1975).
11. K. Behringer and P. Thoma, *Appl. Opt.* **18**, 2586–2594 (1979).
12. D. Ivanenko and I. Pomeranschuk, *Phys. Rev.* **65**, 343 (1944).
13. J. Schwinger, *Phys. Rev.* **70**, 798–799 (1946).
14. J. Schwinger, *Phys. Rev.* **75**, 1912–1925 (1949).
15. F. R. Elder, R. V. Langmuir and H. C. Pollock, *Phys. Rev.* **74**, 52–56 (1948).
16. D. H. Tomboulian and P. L. Hartman, *Phys. Rev.* **102**, 1423–1447 (1956).
17. K. Codling and R. P Madden, *J. Appl. Phys.* **36**, 380–387 (1965).
18. D. Lemke and D. Labs, *Appl. Opt.* **6**, 1043–1048 (1967).
19. E. Pitz, *Appl. Opt.* **8**, 255–259 (1969).
20. P. J. Key and T. H. Ward, *Metrologia* **14**, 17–29 (1978).
21. D. Arnold and G. Ulm, *Rev. Sci. Instrum.* **63**, 1539–1542, (1992).
22. B. Wende, *Metrologia* **32**, 419–424 (1996).
23. R. Thornagel, J. Fischer, R. Friedrich, M. Stock, G. Ulm, B. Wende, *Metrologia* **32**, 459–462 (1996).
24. J. Fischer, M. Kühne and B. Wende, *Appl. Opt.* **23**, 4252–4260 (1984).
25. R. Thornagel and M. Kühne, *Rev. Sci. Instrum.* **60**, 1920–1923 (1989).
26. R. Thornagel, Thesis, Technische Universität Berlin (1990).

27. M. Kühne and R. Thornagel, in "X-ray microscopy III" (A. G. Michette, G. R. Morrison, and C. J. Buckley, eds.), Springer Series in Optical Sciences, Vol. 67, pp. 39–42, Springer Verlag, Berlin (1992).
28. M. Kühne and P. Müller, *Rev. Sci. Instrum.* **60**, 2101–2104 (1989).
29. J. Fischer, M. Kühne, and B. Wende, *Metrologia* **23**, 179–186 (1986/87).
30. K. Danzmann, M. Günther, J. Fischer, M. Kock, and M. Kühne, *Appl. Opt.* **27**, 4947–4951 (1988).
31. C. Heise, J. Hollandt, R. Kling, M. Kock, and M. Kühne, *Appl. Opt.* **33**, 5111–5117 (1994).
32. J. Hollandt, M. C. E. Huber, and M. Kühne, *Metrologia* **30**, 381–388 (1993).
33. R. A. Harrison, B. J. Kent, E. C. Sawyer, J. Hollandt, M. Kühne, W. Paustian, B. Wende, and M. C. E. Huber, *Metrologia* **32**, 647–652 (1995/96).
34. J. Hollandt, M. Kühne, M. C. E. Huber, and B. Wende, *Astron. Astrophys. Suppl. Ser.* **115**, 561–572 (1996).
35. J. Hollandt, M. Kühne, and B. Wende, *Appl. Opt.* **33**, 68–74 (1994).
36. J. Hollandt, W. Jans, M. Kühne, F. Lindenlauf, and B. Wende, *Rev. Sci. Instrum.* **63**, 1278–1281 (1992).
37. W. Jans, B. Möbus, M. Kühne, G. Ulm, A. Werner, and K.-H. Schartner, *Appl. Opt.* **34**, 3671–3680 (1995).
38. W. Jans, B. Möbus, M. Kühne, G. Ulm, A. Werner, and K.-H. Schartner, *Phys. Rev. A* **55**, 1890–1898 (1997).

9. IMAGING PROPERTIES AND ABERRATIONS OF SPHERICAL OPTICS AND NONSPHERICAL OPTICS

James H. Underwood

Center for X-Ray Optics, Lawrence Berkeley National Laboratory
Berkeley, California

9.1 Need for Mirror Optics in Vacuum and Extreme Ultraviolet Spectroscopy

For spectroscopy in any spectral region optical components to disperse the radiation into a spectrum are required. Whereas in the visible, UV, and IR regions both prisms and gratings can be used, at wavelengths ranging from the x-ray and soft x-ray regions to the vacuum ultraviolet (VUV) region, there are no suitable materials for prisms, and thus only gratings are available. In the x-ray region natural and artificially grown crystals can also be used. These dispersive elements are treated in Chapter 19. A spectroscopic instrument such as a spectrograph or monochromator also requires optical components to deflect and focus the radiation and to form images. However, at short wavelengths lenses are not available because all materials become highly absorbing. Lithium fluoride 1–2 mm thick transmits down to 1040 Å (Fig. 1) and can be used as vacuum windows, but this is too thin to be useful as a lens. Below this wavelength, therefore, mirrors must be used.

Mirrors must be made from (or coated with) a reflective material to achieve a reflectivity high enough to construct a spectroscopic system of reasonable efficiency. Although multilayer coatings can be used for some specific applications (see Chapter 14), metallic coatings giving broadband reflectivity are most commonly used. The normal incidence reflectivity of metals decreases considerably in the VUV and operation of mirrors at short wavelengths requires small glancing angles (see Chapters 10 and 13). The best reflectors are the heavy metals such as gold and platinum. Figure 1 shows the reflectivity of platinum at several glancing angles. If 10% is defined as a minimum acceptable reflectivity, then a normal incidence Pt mirror will operate down to 400 Å, a 10-deg glancing mirror down to about 50 Å, and a 1-deg mirror down to less than 10 Å. For this reason, spectroscopic systems for soft x-rays and EUV are usually constructed using glancing incidence mirrors and gratings.

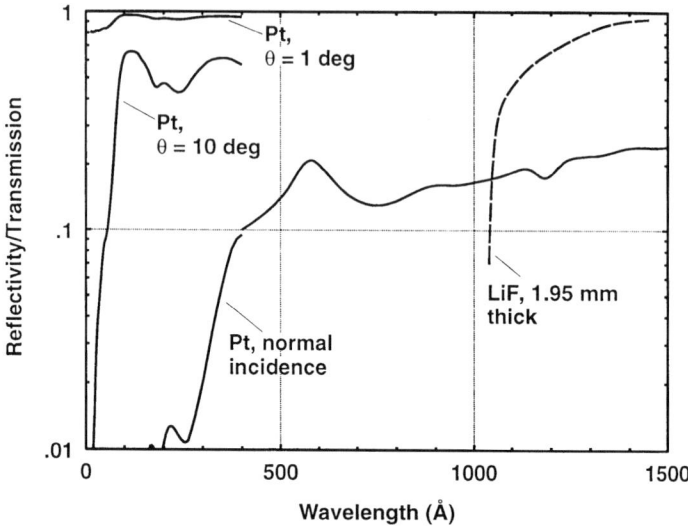

FIG. 1. Optical materials for the far and extreme UV. The dashed curve is the transmission of a lithium fluoride window 1.95 mm thick. The three solid curves represent the reflectivity of a Pt mirror at normal incidence, 10 deg, and 1 deg glancing incidence. The figure demonstrates the need for glancing incidence optics at wavelengths below about 400 Å.

9.1.1 Design of EUV Spectroscopy Systems

The first stage in the design of an optical system is the specification of its purpose and the performance (wavelength coverage, spectral, spatial and time resolution, throughput, etc.) required of it. The next is to choose candidate configurations that may meet these goals and to lay them out in the physical space available using first-order, or *paraxial*, geometrical optics theory. This establishes object and image distances, focal lengths, spectral dispersions, etc., and enable a first calculation of the instrument performance to be made. Once a satisfactory paraxial design has been established, the performance can be evaluated with objects and images of finite size. At this point image defects or *aberrations* will be encountered and the effect of these aberrations on the performance must be evaluated using higher order geometrical optics theory. This process is iterative; if the final predicted performance does not meet specifications, or the optics needed cannot be made (with the available funding!), the earlier steps are repeated.

X-ray/EUV systems use no lenses, so the complete theory of geometrical optics is not needed. In addition, spectroscopic systems do not normally require high-resolution imaging over an extended object field, and the treatment of aberrations can be simplified. In what follows we have abstracted only those aspects of geometrical optics relevant to spectroscopy, or those that are needed for a more complete understanding.

9.2 The Law of Reflection

The basic geometrical optics law governing mirror optics is the *law of reflection*. It states that (1) the incident ray, the reflected ray, and the normal to the surface at the point of incidence are all in the same plane, and (2) that the angle of incidence i is equal to the angle of reflection i'. These conditions are summarized in Fig. 2. The law of reflection can be expressed as a single vector equation:

$$\mathbf{i'} = \mathbf{i} - 2\mathbf{n}(\mathbf{n} \cdot \mathbf{i}), \qquad (1)$$

where $\mathbf{n}$ is a unit vector normal to the reflecting surface and $\mathbf{i}, \mathbf{i'}$ are unit vectors along the incident and reflected rays, respectively. This equation is useful in ray tracing. In ordinary optical practice the angle of incidence i is measured from the surface normal to the incident ray. In the x-ray/EUV regions it is customary to use the angle $\theta = \pi/2 - i$ measured from the tangent to the surface; the glancing (or grazing) angle. The law of reflection is valid for all wavelengths. Thus an optical system using only mirrors is *achromatic*; there is no variation of imaging properties, such as focal length, with wavelength.

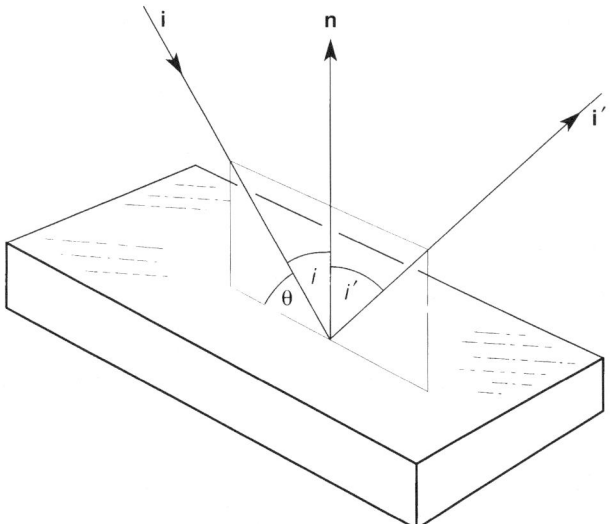

FIG. 2. The law of reflection, $\mathbf{i}$, $\mathbf{n}$, and $\mathbf{i'}$ are in the same plane, and $\angle i' = \angle i$.

9.3 Paraxial Optics

The first-order approximation to the geometrical optics of an imaging system treats only *paraxial* rays, that is, rays that are everywhere close to the optical axis and that make small angles $\alpha, \beta, \ldots$, with the axis, so that everywhere the sines of the angles can be approximated by the angles themselves, $\sin \alpha \approx \alpha$, etc. Higher order approximations include higher order terms in these expansions.

9.3.3 Axisymmetric Systems and the Thin Lens Formula

Most optics textbooks deal primarily with *axisymmetrical* systems of lenses and mirrors (cameras, telescopes, etc.), in which the imaging properties of the system are invariant to rotations of the optical system around a single common axis (the *optical axis*). Most x-ray/EUV systems are not symmetrical. However, we first review the symmetrical case in order to pursue its extension to the nonsymmetrical case.

Figure 3 shows schematically a symmetrical system of several elements. These may be lenses or mirrors. An ideal optical system carries out a one-to-one mapping of the object, a three-dimensional array of point sources, into another array of points which constitute the image. If in addition there is a one-to-one correspondence between straight lines in the object space and straight lines in the image space, then there will be correspondence between object and image planes. Corresponding elements are called *conjugate* elements. This *collinear mapping* allows definition of the cardinal points of the optical system: the focal

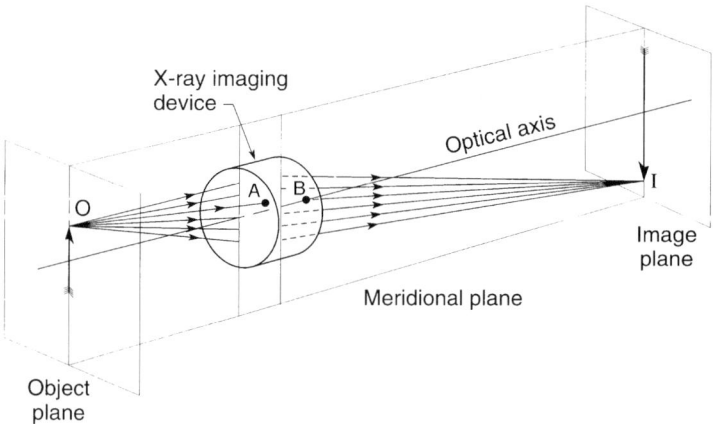

FIG. 3. An axisymmetrical optical system. The meridian plane is defined by the object point *O*, the image point *I*, and the optical axis. All rays are in the meridian plane except for *OABI*, which is a general or *skew ray*.

points, principal points, and nodal points. Under these assumptions we may derive the Gaussian equation for focusing by a thin lens:

$$\frac{1}{u} + \frac{1}{v} = \frac{1}{f}, \qquad (2)$$

where u is the object distance and v is the image distance, measured from the principal plane of the lens and f is the *focal length*. The transverse *magnification* M of the system (the ratio of the length of a line in the image plane to the length of its conjugate in the image plane) is

$$M = -v/u. \qquad (3)$$

The front and rear focal lengths of a lens or mirror can be derived in the paraxial approximation; for the reflecting systems of interest in x-ray/EUV spectroscopy these are the same. A spherical mirror of radius R at normal incidence has a focal length given by $f = R/2$. The case of glancing incidence is discussed in Section 9.5.1.

9.3.2 Sign Conventions

The preceding equations require corresponding sign conventions. Using the term *lens* to refer to any optical element or system that can be represented in the paraxial approximation by a thin lens:

1. Light is assumed to travel from left to right.
2. Object distances (u) are positive when the object is to the left of the lens and negative when the object is to the right of the lens. Thus objects to the right are virtual objects.
3. Image distances (v) are positive when the image is to the right of the lens and negative when the image is to the left of the lens. Thus images to the left are virtual images.
4. The focal length is positive for a converging lens and negative for a diverging lens.
5. Transverse directions are positive when measured upward from the axis and negative when measured downward.

For mirrors the sign conventions for use with these formulas are essentially the same, if the following interpretations are made: Item 1 remains the same. For items 2 and 3 replace "lens" with "center of the mirror." In item 4, for concave mirrors R is positive, for convex mirrors R is negative. Finally, item 5 remains the same. These sign conventions also apply to glancing incidence optics. Note that a single concave mirror *inverts* the image whether at glancing or normal incidence.

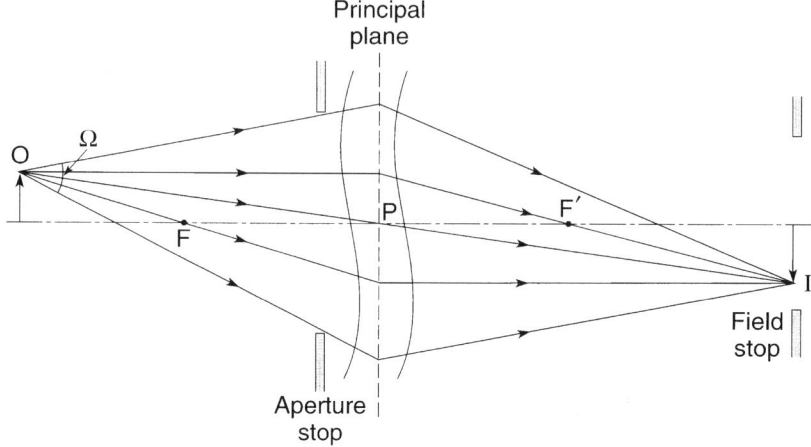

FIG. 4. Aperture and field stops. The aperture stop defines the solid angle Ω accepted by the system and the field stop defines the width w of the object field that can be viewed.

9.4 Geometrical Optics with Finite Apertures and Objects

Paraxial optics assumes objects made up of source points on or close to the axis, and rays that make small angles with the axis. A real system must image objects of finite extent with elements of finite size and, as a result, image defects, or *aberrations*, will arise.

9.4.1 Stops and Pupils

In a real system the most extreme rays will make relatively large angles with the axis and be far from it at some points. The paths of these rays are determined by the boundaries of *stops*. Figure 4, which depicts a "thin lens" system, shows the two principal kinds. An *aperture stop* determines the amount of light that can traverse the system, and the brightness of the image, by limiting the solid angle of the ray bundle diverging from O. It may be an actual diaphragm placed in the optical train or it may be formed by the periphery of the optical components themselves. A *field stop* limits the size of the image, and thus the size of the object, that is viewed. It is usually formed by the boundaries of the film or detector in the focal plane.

In Fig. 4 the aperture stop is on the object side of the optical system and is said to be in object space. In this position it forms the *entrance pupil* of the system. The aperture-limiting stop may also be situated on the opposite side of the system—in image space, in which case it forms the *exit pupil* (Fig. 5). In

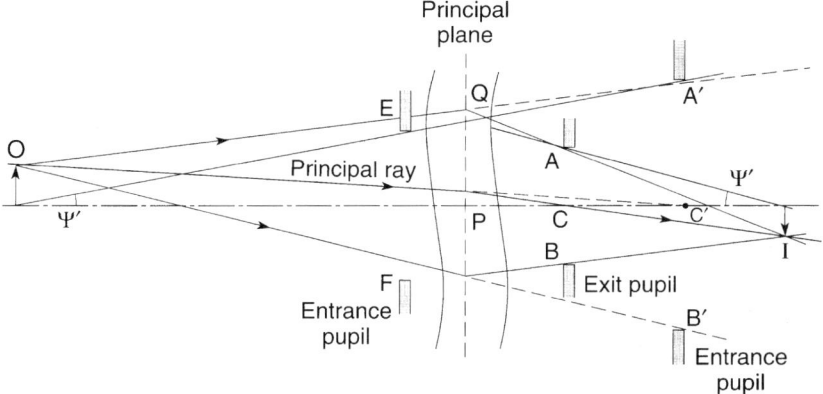

FIG. 5. Entrance pupils, exit pupils, and vignetting. ACB is the exit pupil and $A'C'B'$, its image, is the entrance pupil. EF is an additional stop, which also acts as an entrance pupil. The overlapping of EF and $A'C'B'$ cuts off or *vignettes* rays such as OQ and reduces the intensity of the image at the edge of the field.

either case, a corresponding exit or entrance pupil may be found by constructing the image of the real (physical) pupil. The image $A'B'C'$ of the exit pupil ACB lies to the right of the system in an imaginary extension of object space. It is equivalent to the exit pupil, as the bundle of rays passing through it subtends the same solid angle as that defined by the real exit pupil. By a similar construction the virtual exit pupil may be found. If the physical pupils in the system are not carefully placed, the real pupils and their images may interact to decrease the intensity of the image at the edges of the field. For example, the ray OQ is cut off by the entrance pupil EF; otherwise it would pass through the exit pupil at A. This effect is called *vignetting*.

A ray that passes through the center C' of the entrance pupil will also pass through the center C of the exit pupil and is termed a *principal ray*.

9.4.2 Ray and Wavefront Aberrations

Nonparaxial rays emanating from an object point O will not in general pass through the image plane at the Gaussian point I. This "error" in focusing decreases the image sharpness and is termed *aberration*. The imaging qualities of an optical system are determined to a large extent by the amount and type of inherent aberrations.

Aberrations may be described either in terms of their effect on rays (*ray aberration*) or on wavefronts (*wave aberration*). We define a *homocentric pencil of rays* as a bundle of rays that passes through some common point. Because rays are perpendicular to the wavefronts, the wavefronts in homocentric pencils are spherical, centered on the apex. Any lack of homocentricity, that is, aberration, may thus be expressed as a departure from sphericity of the corresponding wavefront. Although there is no wavelength associated with these surfaces, they are called *geometrical wavefronts* and are a good approximation to physical wavefronts except in the neighborhood of geometrical shadow boundaries: Diffraction presents the ultimate limit to the resolution of a system. However, at the short wavelengths of x-rays and EUV, diffraction effects are minimal, and the geometrical ray or wavefront approximation is particularly good.

9.4.3 Aberrations of an Axisymmetrical Optical System

Although most x-ray/EUV optical systems are not axisymmetric, we begin with a treatment of the symmetrical system because (1) the theory and terminology of aberrations has been developed most extensively for the symmetrical system and an understanding of this is crucial to our topic; and (2) there are a few x-ray optical systems which are in fact symmetrical, and to which the standard theory of aberrations (see, e.g., [1, 2]) can be directly applied. These include the various types of Wolter glancing incidence optics and the multilayer coated Schwarzchild optics and telescopes. We adopt a selective treatment because not all aberrations are equally important in spectroscopy.

In Fig. 6, let O be a point in the object plane and OA a general ray entering the system. Let x, y be Cartesian coordinates in the object plane. No generality is lost by assuming that O is on the y-axis; y is then designated the field coordinate. (If the object is at infinity, y is replaced by the off-axis angle α.) After reflection at the surfaces, the ray OA will leave the exit pupil at the point P and finally intersect the receiving plane (the plane for which the aberrations are to be found; it need not necessarily be the Gaussian image plane) at I. Let the principal ray from O intersect the receiving plane at I_g. Then the vector II_g is the transverse

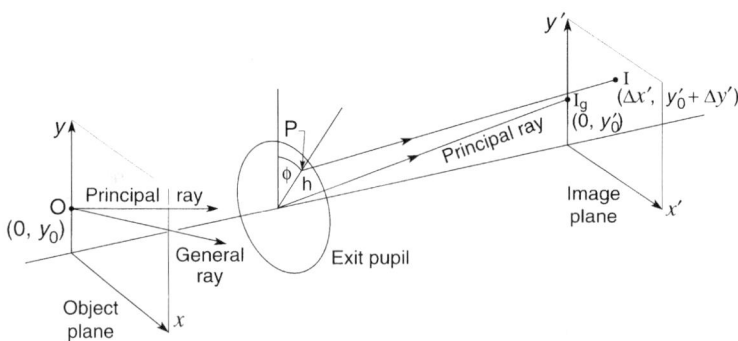

FIG. 6. Coordinate system for the classification of the ray aberrations of an axisymmetrical optical system.

ray aberration. With respect to coordinates x', y' in the image plane, I_g is at point $(0, y_0')$ and I at $(\Delta x', y_0' + \Delta y')$. For any system, the aberration components or displacements $\Delta x'$, $\Delta y'$ depend only on the positions of the points O and P. If the position of P in the pupil is denoted by polar coordinates h and ϕ, then $\Delta x'$ and $\Delta y'$ can be expressed as a power series:

$$\Delta x' = \sum_{j=0}^{\infty} \sum_{k=0}^{\infty} a_{jk} y^j h^k F_{jk}(\phi),$$

$$\Delta y' = \sum_{j=0}^{\infty} \sum_{k=0}^{\infty} b_{jk} y^j h^k G_{jk}(\phi).$$

(4)

Each term in either of these series denotes a particular aberration. In a symmetrical system only particular combinations of j and k values are allowed. This can be seen as follows: Suppose that the coordinate axes in Fig. 6 are rotated 180 deg, so that y becomes $-y$ and h becomes $-h$. It is clear that $\Delta x'$ and $\Delta y'$ must also change signs. Since the a_{jk} (or the b_{jk}) are independent, the condition $j = k = $ odd must be satisfied. The terms in the power series expansion for $\Delta x'$ or $\Delta y'$ can then be arranged schematically as shown in Table I (following [3]). In this table the bold entries represent the aberrations of a symmetrical system. The top left diagonal (**h** to **y**) contains the first-order or Gaussian terms. These are not, strictly speaking, aberrations; they may be made to vanish by choosing the receiving plane to be the Gaussian image plane. However, it is worth bearing in mind that the term in h is a "defocusing" term, which may be used to balance other aberrations, spherical aberration in particular. The center diagonal (**h**^{3} to **y**^{3}) represents the third-order or Seidel terms, and the third diagonal (**h**^{5} to **y**^{5}), the fifth-order terms. Some of the rows and columns represent aberrations with common characteristics and are specifically named. Terms in h only (first column) are called *spherical aberration* terms, those linear in y (second column) *linear coma* terms, and those in y only (first row) *distortion* terms.

TABLE I. Terms in the Expansion for Ray Aberrations

	y^0	y^1	y^2	y^3	y^4	y^5
h^0	1	**y**	y^2	**y**^{3}	y^4	**y**^{5}
h^1	**h**	yh	**y**^{2}**h**	$y^3 h$	**y**^{4}**h**	
h^2	h^2	yh^2	$y^2 h^2$	**y**^{3}**h**^{2}		
h^3	**h**^{3}	yh^3	**y**^{2}**h**^{3}			
h^4	h^4	yh^4				
h^5	**h**^{5}			+Higher terms		

9.4.4 The Primary Aberrations

If terms of third order ($j + k = 3$) only are retained, we can show that

$$\Delta x' = a_{03} h^3 \sin \phi + a_{12} y h^2 \sin 2\phi + a_{21} y^2 h \sin \phi,$$
$$\Delta y' = a_{03} h^3 \cos \phi + a_{12} y h^2 (2 + \cos 2\phi) + b_{21} y^2 h \cos \phi + b_{30} y^3. \quad (5)$$

The five independent coefficients in these equations are dependent on the parameters of the particular optical systems under consideration and can be identified with the five primary or Seidel aberrations.

9.4.4.1 Spherical Aberration.
If all coefficients are zero except for a_{03}, then Eq. (5) becomes:

$$\Delta x' = a_{03} h^3 \sin \phi,$$
$$\Delta y' = a_{03} h^3 \cos \phi. \quad (6)$$

These are the parametric equations of a circle; as the point P traces a *zonal family of rays* (i.e., as it describes a circle in the exit pupil), a corresponding single circle in the receiving plane is traced out. This corresponding zonal family of rays is focused at successively greater distances behind the Gaussian plane (if a_{03} is positive) or in front of it (if a_{03} is negative), and a point object is imaged into a circular patch (Fig. 7). This is *spherical aberration*. In the Gaussian plane, the radius of the aberration patch is determined by the *marginal rays* having the greatest h value, $h = h_m$. By shifting the receiving plane, its size can be reduced; its minimum size (*circle of least confusion*) is four times smaller than its size at the Gaussian plane. Spherical aberration can also be reduced or eliminated by figuring the optical surfaces to a shape other than a sphere (*aspherization*). Spherical aberration is clearly the only aberration that remains if the object point is on-axis ($y = 0$).

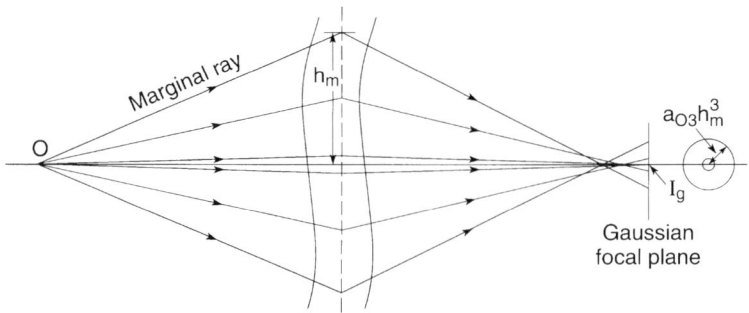

FIG. 7. Third-order spherical aberration. Rays passing through the system at progressively larger values of h are focused progressively closer to the principal plane of the optical system.

9.4.4.2 Coma and the Abbé Sine Condition.
If a_{12} is the only nonzero term, we obtain

$$\Delta x' = a_{12} y h^2 \sin 2\phi,$$
$$\Delta y' = a_{12} y h^2 (2 + \cos 2\phi). \qquad (7)$$

As the point P traces a circle in the pupil, the ray traces a double circle in the receiving plane. This circle expands and becomes more offset from the Gaussian point in the y-direction as h increases, so that the aberration figure in bounded on two sides by lines intersecting at 60 deg and on the third side by an arc of a circle formed by the marginal rays (Fig. 8). This "comet-like" figure accounts for the name *coma* given to this aberration. Coma is, as we shall see, a variation of magnification with aperture. Its effects cannot be reduced by shifting the receiving plane, or by aspherization. Coma is corrected by satisfying the Abbé sine condition, which can be developed as follows.

If a symmetrical system forms a stigmatic image A' of an axial point A, that is, it is free from spherical aberration for the conjugates A, A', then it is possible to derive, from Fermat's principle, conditions for the stigmatic imaging of points close to A. One of these is illustrated in Fig. 9. Let any ray from A intersect the

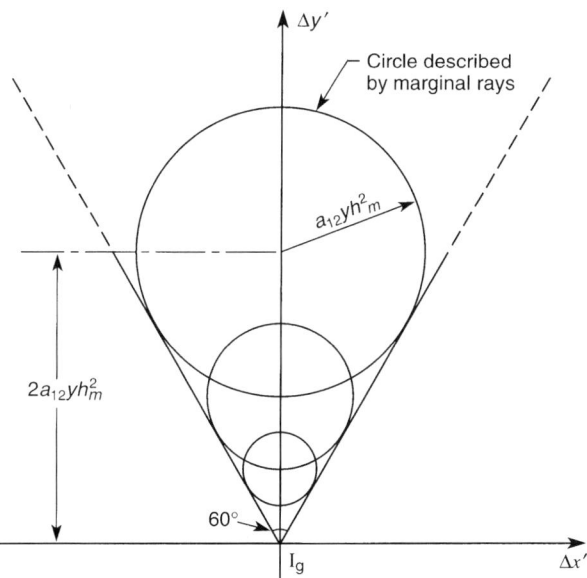

FIG. 8. Transverse ray aberration for third-order coma. The classic "comet-like" aberration figure is bounded on two sides by lines intersecting at 60 deg and on the third side by an arc of a circle formed by the marginal rays.

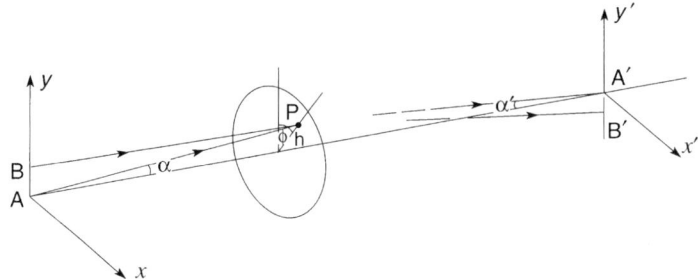

FIG. 9. Illustrating the Abbé sine condition.

entrance pupil at P; let α be the angle between AP and the axis and α' the angle between the corresponding ray through A' and the axis. Then a nearby point B in the x-y plane will be imaged stigmatically at B' if, for all possible positions of P,

$$\sin \alpha / \sin \alpha' = \text{constant}. \tag{8}$$

The relation of Eq. (8) is called the Abbé sine condition. Interestingly, it relates the stigmatic imaging of off-axis points to quantities defined entirely for on-axis points. However, satisfaction of the sine condition only ensures stigmatism to first order, that is, it ensures that those terms independent of the field coordinate y and those linear in y—the spherical aberration and linear coma terms—will be absent from the aberration expansions Eqs. (4) (higher powers of y may still remain). A system free from spherical aberration, which also satisfies the sine condition, is said to be *aplanatic*. If the object plane is at infinity, the sine condition takes the special form:

$$h/\sin \alpha' = \text{constant}. \tag{9}$$

It is clear that the constant in Eq. (8) is the system magnification M. Freedom from comatic aberration thus requires that the magnification remain constant for all ray paths through the system. The sine condition is also equivalent to the statement that the *principal surface*—the surface on which the rays in object space, if traced forward, intersect the corresponding rays in image space, if traced backward—should be spherical. This is important for the understanding of coma in glancing incidence systems.

9.4.4.3 Astigmatism and Curvature of Field.

The two aberrations represented by a_{21} and b_{21} are usually considered together. For general values of these two coefficients the equations

$$\begin{aligned} \Delta x' &= a_{21} y^2 h \sin \phi \\ \Delta y' &= b_{21} y^2 h \cos \phi \end{aligned} \tag{10}$$

evidently represent an ellipse in the Gaussian plane. Further insight may be gained by making the substitutions:

$$A = \frac{b_{21} - a_{21}}{2}, \quad C = \frac{3a_{21} - b_{21}}{2}. \quad (11)$$

Noting that any term with a simple linear dependence on the pupil coordinate h is equivalent to a shift of focal plane, we see that the equations

$$\Delta x' = (A + C - k)y^2 h \sin \phi$$
$$\Delta y' = (2A + C - k)y^2 h \cos \phi \quad (12)$$

represent the displacements on surface shifted by an amount $\Delta z = ky^2$ from the Gaussian plane, where k is an additional parameter. For the k values

$$k_T = 3A + C, \quad k_S = A + C, \quad (13)$$

the aberration figures degenerate into two straight lines of length $4Ay^2 h_m$. These are, respectively, the tangential and sagittal focal lines; they are formed on the tangential and sagittal focal surfaces. All tangential rays (those for which $\phi = 0$ or 180 deg) come to a focus on the tangential surface; sagittal rays (those for which $\phi = 90$ or 270 deg) focus on the sagittal surface (Fig. 10). The two surfaces are paraboloidal, separated by the distance or *astigmatic difference* $2Ay^2$, which increases as the square of the field coordinate. Their curvatures at the common point of intersection with the optical axis are

$$c_T = 2(3A + C), \quad c_S = 2(A + C). \quad (14)$$

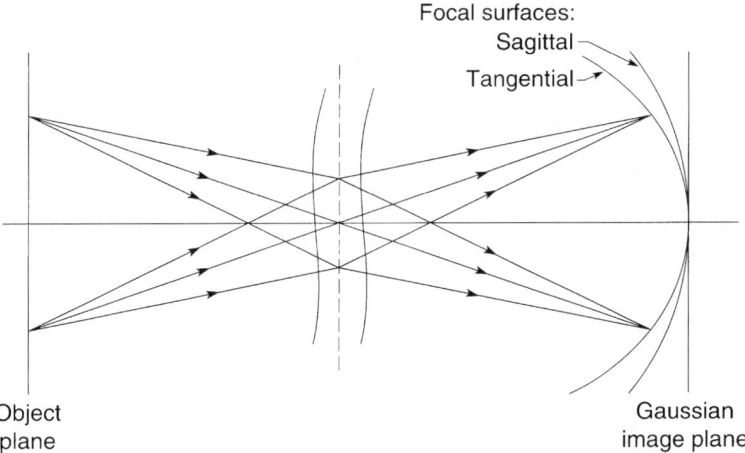

FIG. 10. Astigmatism and curvature of field. Formation of astigmatic images of a plane object on the tangential and sagittal image surfaces. Tangential rays only are shown.

Evidently, if $C = 0$ these two surfaces still remain distinct. The aberration governed by A alone is called *astigmatism*. On the other hand, if $A = 0$, the two surfaces become one, with polar curvature

$$c = 2C, \qquad (15)$$

and both tangential and sagittal rays focus on a curved surface. This aberration is therefore called *curvature of field*; the curvature c is the *Petzval curvature*. Even if $C = 0$, the presence of the astigmatic coefficient A means that the best image is formed on a curved surface, halfway between the two astigmatic focal surfaces.

9.4.4.4 Distortion.
The final primary aberration is represented by

$$\Delta x' = 0,$$
$$\Delta y' = b_{30} y^3. \qquad (16)$$

Because there is no dependence on the pupil coordinate, these equations represent point foci that have been displaced by an amount proportional to y^3, that is, a *distortion*. Two kinds of distortion are possible; if the coefficient $b_{30} > 0$, *pincushion distortion* results; if $b_{30} < 0$, the result is *barrel distortion*. Distortion is rarely important in UV spectroscopic systems.

9.4.5 Secondary Aberrations

The higher order aberrations ($j + k > 3$) are rarely encountered in spectroscopic systems, although in Wolter-type optics (see Section 9.6) the higher order spherical aberration terms in the first column of Table I are not insignificant.

9.5 Nonaxisymmetrical Systems

9.5.1 Mirrors at Glancing Incidence

Figure 11 is a representation of a glancing incidence optical system, which will rarely be axisymmetric. Although there is no recognizable axis of symmetry, the equivalent of an optical axis can be defined. We can define a point O_0 as the center of the object field, and the aperture limiting the illumination of the mirror M_1 (the aperture stop) will usually be of some regular shape such as an ellipse or rectangle, for which a center C_1 can be found. We call this point the center of the mirror. The ray that passes from the center of the object field through the center of the mirror and on to the image plane (and which thus defines the center I_0 of the image field) is the closest thing to an optical axis. Although it consists of line segments rather than a single straight line, the same definition of paraxiality can be applied as for a symmetrical system; the paraxial rays are never far from this axis and make only small angles with it.

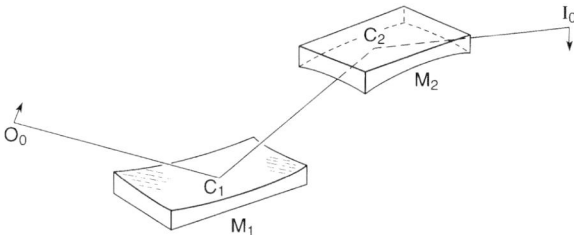

FIG. 11. An x-ray optical system of two mirrors, having a plane of symmetry. The normals to M_1 and M_2 are coplanar and are contained in the meridan plane. The line $O_0 C_1 C_2 I_0$ defines an "optical axis" for such a system.

We see, however, that this definition of the optical axis of a nonsymmetrical system introduces a new consideration; not all the rays striking the mirror are equivalent. In particular, we may define two kinds of rays. The plane defined by the axis and the normal to the mirror is called the *tangential* (or *meridional*) plane, and rays whose points of incidence on the mirror are contained in this plane are termed tangential (or meridional) rays. The plane containing the optical axis which is perpendicular to the meridional plane is termed the *sagittal* plane, and rays whose points of incidence are in this plane are termed sagittal rays. Now there are two sets of paraxial focusing conditions, one for the tangential and one for the sagittal rays. For a spherical mirror these can be derived geometrically (see e.g., [4]) or by using Fermat's principle; we quote them here without derivation:

$$\frac{1}{u} + \frac{1}{v_T} = \frac{1}{f_T} = \frac{2}{R \sin \theta}, \tag{17a}$$

$$\frac{1}{u} + \frac{1}{v_S} = \frac{1}{f_S} = \frac{2 \sin \theta}{R}, \tag{17b}$$

where v_T, v_S are the tangential and sagittal image distances, and f_T, f_S, the tangential and sagittal focal lengths. By analogy with the equivalent, well-known set of equations for the case of refraction at a single spherical surface we call Eqs. (17) the Coddington equations for a spherical reflector. Equations (17) together with their sign conventions (Section 9.3.2), are the most fundamental and useful equations in the design of glancing incidence systems.

The strongly different focal lengths for tangential and sagittal rays result in extreme astigmatism. It is convenient to treat it here rather than under "aberrations," because then the remaining aberrations can be discussed only in the tangential plane.

9.5.2 Astigmatism of a Single Spherical Mirror

Consider a fixed image distance u. At normal incidence, $\theta = \pi/2$, the two images are coincident, However, as θ decreases toward glancing incidence the images formed by the sagittal and focal ray bundles diverge, since f_T becomes progressively smaller while f_S becomes progressively larger—the *astigmatic difference* (Section 9.4.4.3) increases rapidly. In fact, at some value of θ, f_S will become equal to u and the sagittal image will be formed at infinity. Further decrease of θ leads to the sagittal image becoming virtual. Consider a particular case, a mirror of radius $+10$ m used at an angle of 1 deg, suitable for the reflection of 10 keV x-rays from platinum. We find that $f_T = 0.087$ m while $f_S = 286.5$ m. For an object point distant $u = 1$ m from the mirror, the image formed by the tangential rays is distant $v_T = +0.095$ m from the mirror (real), whereas the sagittal image is formed at $v_S = -1.0035$ m and is thus virtual. Essentially, two *line* images are formed by the mirror at distances v_T and v_S (Fig. 12). Thus the single spherical mirror at glancing incidence does not form *stigmatic* images (Gk.: *stigma*, a spot) of a point source but strongly *astigmatic* ones.

Some special cases of Eqs. (17) are of particular interest as discussed in the following subsections.

9.5.2.1 Normal Incidence: $\theta = \pi/2$.

The same sign conventions can be used, with care, for this case. Both equations become identical, $f_T = f_S = f = R/2$ as we saw in Section 9.3.1, and the tangential and sagittal images coincide. Even small departures from normal incidence, however, will lead to finite astigmatism.

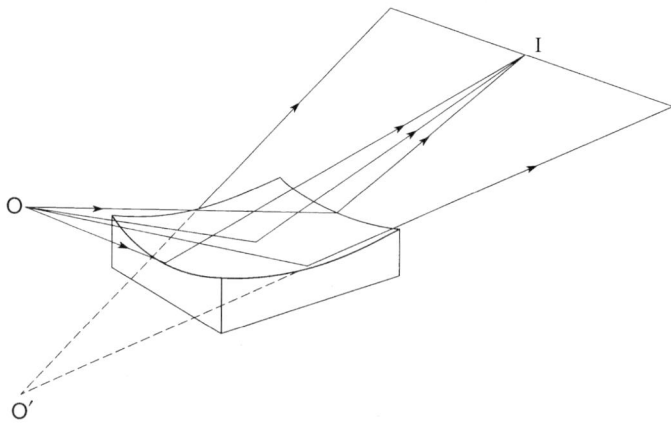

FIG. 12. Astigmatic image formation by a single spherical mirror at glancing incidence. The focus for tangential rays is at I, while the focus for sagittal rays is at O', and may be virtual as shown here. Thus the image of O is a line focus.

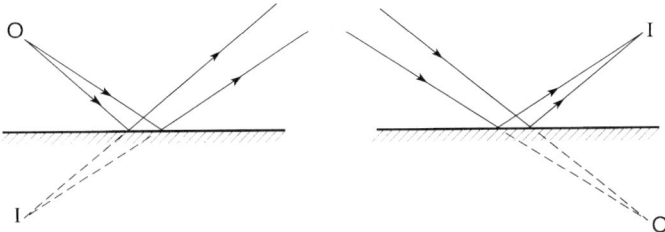

FIG. 13. A plane mirror forms inverted images at unit magnification. (a) A virtual image is formed of real objects. (b) A real image is formed of virtual objects.

9.5.2.2 Plane Mirrors: $R = \infty$. We see that $v_T = v_S = -u$ for all values of θ. Thus a plane mirror forms inverted, stigmatic images at unit magnification, virtual images of real objects, and real images of virtual objects (Fig. 13). The distance d between corresponding points on the object and image is dependent on the glancing angle, clearly, $d = 2r \sin \theta$, where r is the distance between the point of incidence on the mirror and the object point.

Plane mirrors are used extensively in x-ray/EUV spectroscopy to deflect a beam of radiation (i.e., to turn it through an angle). The purpose of this deflection may be connected with the optical design of the spectroscopic instrument, or it may be done to achieve a physical effect such as filtration, order sorting, or linear polarization. In Fig. 14a a two-mirror system, which returns the beam to its original direction, is shown. This system causes a lateral translation of the beam as shown, but restores the original image orientation. In Fig. 14b a three-mirror system is shown; this system returns the beam to its original direction without lateral deflection, but inverts the image. In general, a plane mirror system with an odd number of mirrors will invert the image, while an even number of mirrors will not.

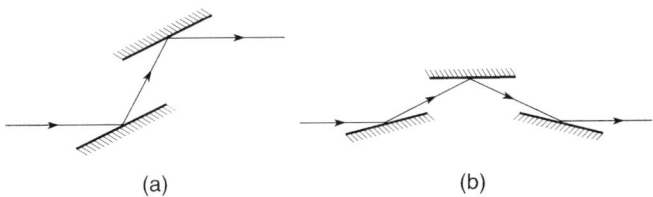

FIG. 14. (a) System of two plane mirrors ("periscope" system). This returns the beam to its original direction and gives an erect image, but causes a lateral translation of the beam. (b) Three-mirror system. This system returns the beam to its original direction without lateral translation, but inverts the image. Three mirrors is the minimum for a system that neither translates nor deviates the beam.

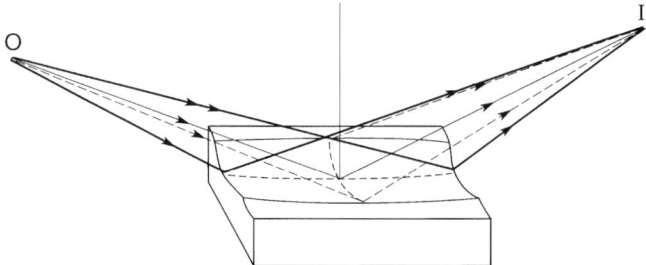

FIG. 15. Correction of the astigmatism of a single mirror by means of a toroid.

9.5.3 Correction of the Astigmatism of a Glancing Incidence Spherical Mirror

While the tangential line image from a single glancing incidence spherical mirror can be ideal for imaging a small or "point" source onto the slit of a spectrograph, the severe astigmatism is intolerable in applications such as microscopes. Two methods for accomplishing correction are discussed next.

9.5.3.1 Toroidal Mirrors. A surface can be constructed, which presents a different value of the radius of curvature R to sagittal rays than it does to tangential rays. Let the radius of curvature in the meridional section be R_M, and in the perpendicular (equatorial) direction R_E. This surface is an element of a toroid. Specifically, let $R_E = R_M \sin^2 \theta$. Substituting into Eqs. (17), we find that $f_S = f_T = 2/R_M \sin \theta$, and both tangential and sagittal images coincide (Fig. 15).

In practice, this approach has two problems. The scheme works for any pair of object/image conjugates but only for a specific glancing angle. As this angle is varied, the tangential and sagittal images move in opposite directions, as for a spherical mirror. This gives the toroid unpleasant properties in cases where varying angles of incidence are encountered. The other difficulty is the practical one of making the mirror. To return to the example of Section 9.3.4, we find that for $R_M = +10$ m, $\theta = 1$ deg, R_E must be $+3$ mm. Thus the mirror shape required resembles the bore of a gently curved capillary tube. Such a mirror is very difficult to manufacture to exacting optical tolerances, especially, as we shall see later, if the meridional section must depart from a circular shape to reduce aberrations. However, toroidal or tube mirrors have been used in x-ray telescopes, where object and image distance are large and allow reasonable equatorial radii on the mirrors.

9.5.3.2 Kirkpatrick–Baez Optics. In the example of Section 9.5.2, we see that the spherical mirror has a very weak effect in the sagittal direction. The sagittal image is only 3.5 mm further from the mirror than the source. Thus this virtual image can be used as the source point for a second spherical mirror, following the first and orthogonal to it, so that the sagittal rays from the first mirror

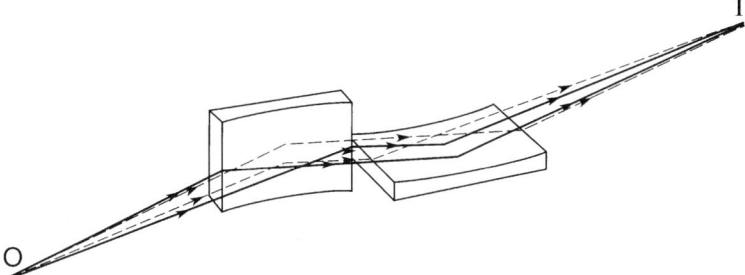

FIG. 16. Principle of the Kirkpatrick–Baez mirror system. The rays shown in bold are the tangential rays for the first mirror and are strongly focused by it, but weakly focused by the second mirror, for which they are sagittal rays. The opposite is true for the dashed rays.

become the tangential rays for the second and are strongly focused by it. This is the principle of the Kirkpatrick and Baez mirror system [5], shown in Fig. 16.

Let distances and radii associated with the first mirror and second mirrors be denoted $u_{T1}, R_1, \ldots, u_{T2}, R_2, \ldots$, respectively, and let the separation between the mirror centers be s. Then we have the following conditions:

$$u_{T2} = -v_{S1} + s; \quad v_{T2} = v_{T1} - s. \quad (18)$$

Then from Eqs. (17) we may find the radius R_2 of the second mirror. For the example of Section 9.5.2, let $s = 0.03$ m and $\theta = 1$ deg, then $R_2 = 7.07$ m. The original tangential image from the first mirror is displaced by the addition of the second mirror. However, on calculating its new position as the sagittal image formed by the second mirror we find only an insignificant shift. Thus a cylindrical mirror (equivalent to a toroidal mirror with $R_E = \infty$) can be used in place of either mirror. Cylindrical mirrors have practical advantages over spherical mirrors, as we shall see later.

9.5.4 Aberrations of Nonsymmetrical Systems

The lack of rotational symmetry in glancing incidence x-ray optical systems leads to aberrations additional to the primary aberrations. Examination of only rays in the *plane* of symmetry (see Fig. 11), that is, meridional rays, gives considerable insight. This is especially true with systems composed of spherical mirrors, such as the Kirkpatrick–Baez system, because the focusing in the sagittal plane of each mirror is very weak. As in Section 9.4.3 we take the reference "axis" in object space to be a ray from an arbitrary object point O_0 (the center of field) to an arbitrary pupil point P_0 (preferably the center). In image space, the corresponding ray, $O_0 I_0$ after reflection at the mirror surfaces, serves as the reference line. Paraxial rays are then defined to be those making a small angle with O_0P_0 and are everywhere a small distance from it. The definition of the Gaussian image point I_g follows as before; the Gaussian image plane is the plane perpendicular to O_0I_0 through the Gaussian image point, with I_0 the center of field in image space. The principal ray from a general object point O is the ray that passes through P_0.

The field coordinate y is now the distance O_0O and the pupil coordinate h the distance PP_0 (or an equivalent quantity, such as the angle P_0P_0 between the principal ray and the aberrant ray). With these definitions, the displacement $\Delta y' = II_g$ can be expressed as a double power series in y and h, analogous to Eq. (4):

$$\Delta y' = \sum_{j=0}^{\infty} \sum_{k=0}^{\infty} a_{jk} y^j h^k; \tag{19}$$

a_{00} is zero by the definition of ray aberration; if y and h are zero, so is the aberration. The Gaussian terms $a_{01}h$ and $a_{10}y$ have the same meaning as for the axisymmetric case, namely, focal shift and magnification. As in the symmetric case, we interpret terms in h alone as spherical aberration, terms in y alone as distortion, and terms linear in y as coma (Table I). In the nonsymmetric case, terms where $j + k$ is even do not necessarily vanish, and new aberrations, such as $a_{11}yh$ and $a_{02}h^2$, appear.

9.5.4.1 Aberrations of a Mirror of Circular Cross Section.
The aberrations of a mirror having a circular meridional cross section have been treated by a number of authors, including Kirkpatrick and Baez [5], Dyson [6], Montel [7, 8], and McGee and Burrows [9].

9.5.4.2 Rowland Circle and Obliquity of Field.
The term $a_{11}yh$, being linear in h, represents a focal plane shift. This shift, being linear with y, can also be regarded as a form of coma, and defines a focal surface that is oblique to the Gaussian focal plane. If a_{11} is positive, this surface inclined away from the principal ray. This aberration—*obliquity of field*—is peculiar to nonsymmetric systems. Its origin can be seen from Eq. (17a) for tangential imaging. Setting $u = \infty$, $v_T = r$, we obtain the equation

$$r = (R \sin \theta)/2. \tag{20}$$

This equation is the equation of a circle of radius $R/4$ in polar coordinates (r, θ) with the origin on the circumference. Tangential rays from an infinitely distant object point are brought to a focus on this circle, the *focal circle* (Fig. 17). Alternatively, if we set $v_T = u = r$, we obtain:

$$r = R \sin \theta, \tag{21}$$

which is a circle of radius $R/2$. Thus, conjugate points of unit magnification lie on this circle, termed the *Rowland circle* (Fig. 17), and extended objects lying along the Rowland circle will be imaged into extended images along the

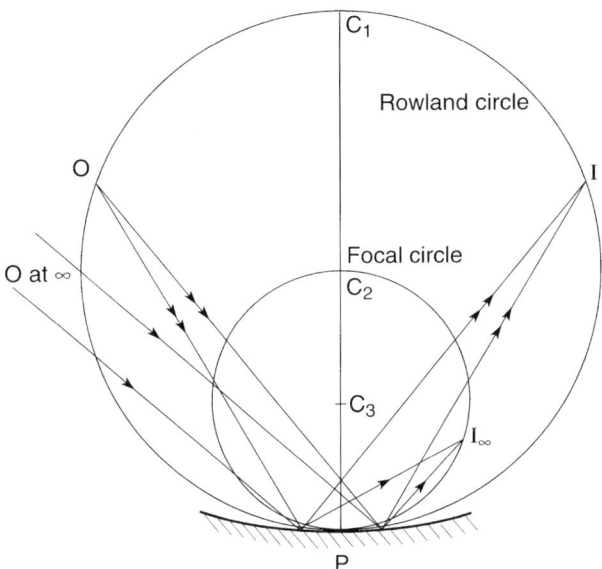

FIG. 17. Focusing conditions for a mirror of circular cross section (sphere or cylinder) of radius R. Points O on the Rowland circle of radius $R/2$ are refocused to points I on the Rowland circle. Object points at infinity are focused to image points I_∞ on the focal circle, of radius $R/4$.

Rowland circle. Object points that are outside the Rowland circle will be imaged between these two circles. An object vector **dE**, which is perpendicular to the optical axis will be imaged into the vector **dE'**, which is strongly inclined to the axis, as shown in Fig. 18. This illustrates the *obliquity of field* aberration for the case of unit magnification. For magnifications other than unity (objects or images far from the Rowland circle) this aberration is much worse. Pattee [10] showed that if the object vector **dE** is at an angle γ to the axis it is imaged into the vector **dE'** at angle γ' to the axis, where;

$$\cot \gamma' = (M + 1) \cot \theta - M \cot \gamma, \qquad (22)$$

which for $\gamma = 90$ deg becomes:

$$\tan \gamma' = \frac{\tan \theta}{M + 1}, \qquad (23)$$

or for small θ, large M,

$$\gamma' = \theta/M. \qquad (24)$$

So, if the example mirror of Section 9.5.2 were to be used as a 10× magnifier, it would have a focal plane making an angle of 0.1 deg to the optical axis.

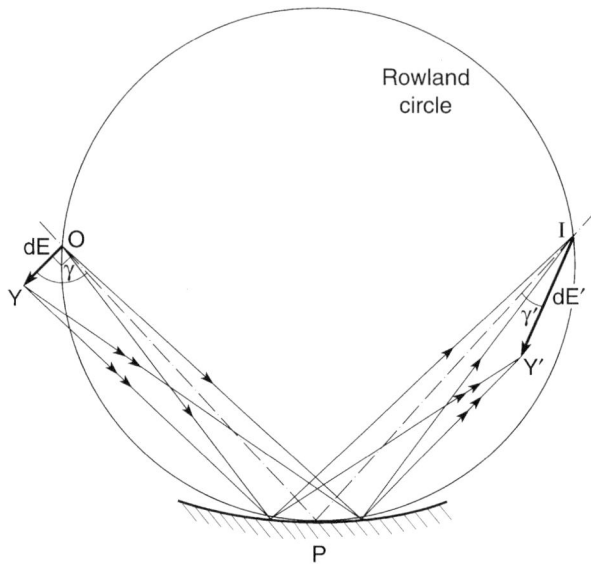

FIG. 18. Points outside the Rowland circle, such as Y, are focused to image points such as Y', inside the Rowland circle. Thus, the vector $\mathbf{dE}$, perpendicular to the axis, will be imaged into the vector $\mathbf{dE}'$ at an angle to the axis.

9.5.4.3 Spherical Aberration Terms.
The spherical aberration of a mirror at glancing incidence can be expressed

$$\Delta y' = a_{00} + a_{01}h + a_{02}h^2 + a_{03}h^3 + \cdots. \tag{25}$$

We encounter the new term $a_{02}h^2$. Because, like Seidel coma (Section 9.4.4.2), its dependence on the aperture coordinate is quadratic, this aberration is frequently called "coma" or "the coma-like aberration." However, there is no dependence on the field coordinate, and this terminology is confusing. The quadratic term (the "primary spherical aberration" for this case) and all other terms in Eq. (25) can be made zero by the usual methods of spherical aberration correction, such as aspherization. These do not work for true coma terms, as we shall see.

Figure 19 shows schematically the tangential focusing of rays from the on-axis point object O_0 by a circular mirror operating as a demagnifier ($M = -0.5$). The Gaussian image point is indicated by I_g. Successive infinitesimal pencils of rays $O_0 A \ldots O_0 B$, form images of O_0 at the corresponding points $I_A \ldots I_B$, which can be computed from Eqs. (17a and b). The locus $I_A I_g I_B$ of all such foci is called the *caustic curve*, or *caustic*. We note that in this particular case all rays from the mirror intersect the Gaussian plane on the side of I_g toward the mirror (negative values of y'). This effect creates the characteristic

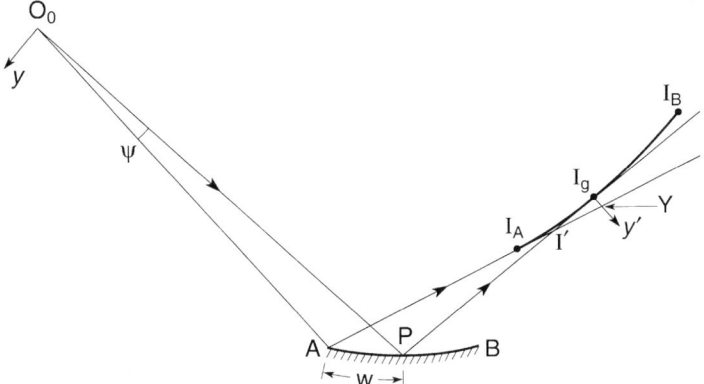

FIG. 19. Spherical aberration of a mirror of circular meridional section. A ray bundle through the pole P forms an image of the object point O_0 at the Gaussian point I_g. Ray bundles through points on the mirror between A and B form images of O_0 whose locus forms the caustic $I_A I_g I_B$. For clarity the ray $O_0 B I_B$ is not shown.

flared image of a spherical mirror used at nonunit magnification, which is sharp at the origin of y' (the geometrical optics intensity from the density of rays is infinite there), and fades away in the $-y'$ direction. The curve $I_A I_g I_B$ is a segment of the complete caustic of the hemispherical mirror. Many authors (e.g., [11]) have derived expressions describing its shape, and for the spherical aberration of spherical reflectors at glancing incidence. The shape of the caustic is dependent on the magnification. Figure 20 shows the shapes for $M = 0$ (object at infinity), $M = -0.5$, and $M = -1$. These curves are for a mirror with $R = 1$ m and $\theta = 1$ deg, and for clarity the figures are anamorphic. Note that for $M < -1$ the caustic is reversed, that is, the sharp side is toward the mirror and the flare is toward the $+y'$ direction.

Expressions for the spherical aberration terms of a mirror of circular cross section can be derived by first principles using the light path function or equivalent methods. A useful expression was developed by Kirkpatrick and Baez [5]. In Fig. 19, an infinitesimal ray bundle from O_0 through mirror point A, with aperture coordinate ψ, intersects the Gaussian plane at Y and the principal ray at I'. The distance $I_G I'$ is the longitudinal spherical aberration and $\Delta y = YI_G$ the transverse spherical aberration. Using some approximations, Kirkpatrick and Baez showed that

$$\Delta y' = -\frac{3\mu a u \psi}{\mu + 1} \left[\frac{\mu^2 + 2a\mu - 1}{\mu + 1 - 2a\mu} \right], \tag{26}$$

where $a = \psi/\theta$ and $\mu = v/u = -M$ [this change of sign of M, and that on the right-hand side of Eq. (24) are required to satisfy the same sign convention (v)

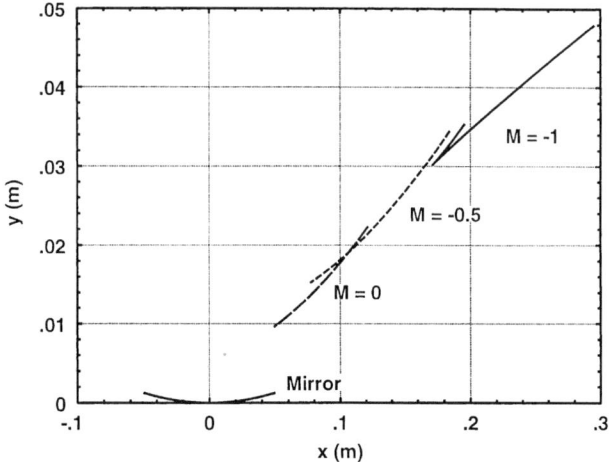

FIG. 20. Caustic curves for a mirror with $R = 1$ m, $\theta = 1$ deg at various values of the magnification M.

of Section 9.3.2 regarding M]. With further approximation (small ψ) we obtain

$$\Delta y' = -\frac{3R\mu}{2}\left[\frac{\mu - 1}{\mu}\psi^2 + \frac{2\mu}{\theta(\mu + 1)}\psi^3\right], \qquad (27)$$

which gives expressions for the "primary" and "secondary" spherical aberration terms in Eq. (23).

Figure 21 is a comparison of the transverse aberration calculated from Eq. (27) with the results of exact ray tracing for $R = 10$ m, $\theta = 1$ deg. For $M = 0$, $\Delta y' = (3/2)\psi^2$, which is essentially a perfect match to the ray-tracing results. At other magnifications the match is sufficiently good for most practical purposes. The reversal of the caustic for $M < -1$ is clearly seen.

The other aberrations of the nonsymmetrical system that appear in the expansion of Eqs. (17) depend on the first or higher powers of the field coordinate y. Apart from field obliquity, these are normally not important in EUV spectroscopic systems, which do not usually require extended images. The object is usually a small source such as a plasma, synchrotron radiation source, or entrance slit, which is defined to be "on-axis". Then y does not attain significant values, and aberrations dependent on the field are insignificant. However, we treat them briefly for completeness.

9.5.4.4 Coma.
Coma is a variation of magnification over the pupil. In the glancing incidence case (linear) coma is represented by the terms in the second column of Table I. Hence obliquity of field can be regarded as a form of coma.

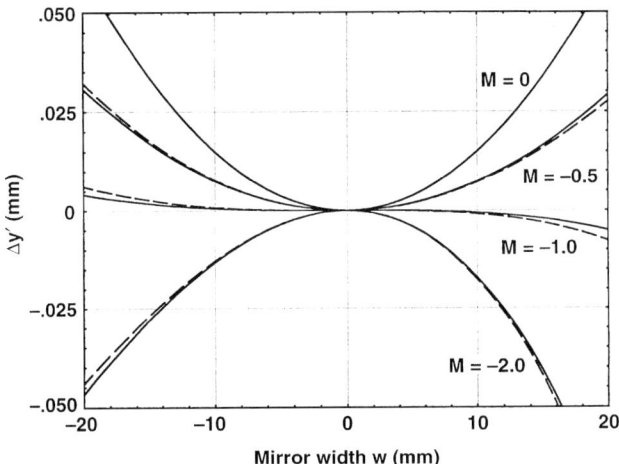

FIG. 21. A comparison of the transverse aberration calculated from Eq. (27) with the results of exact ray tracing for a spherical mirror with $R = 10$ m, and $\theta = 1$ deg.

Both aberrations are illustrated in Fig. 22. The paraxial imaging of object OY by a point A on the mirror gives an image IY_A with a certain magnification M_A and obliquity θ/M_A, whereas the image IY_B produced by a mirror point B has both a different magnification M_B and a different obliquity. This results from the large

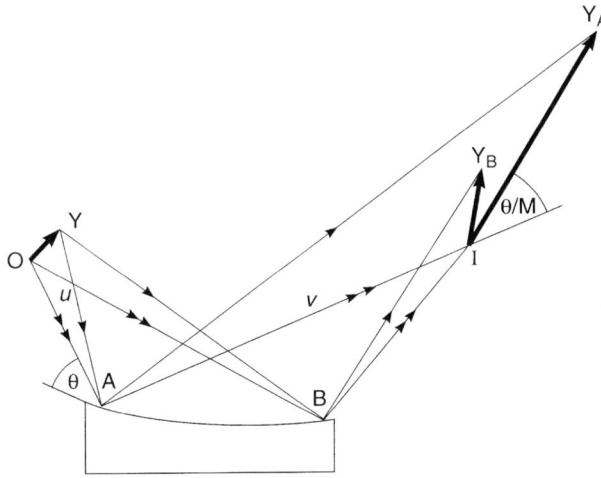

FIG. 22. Obliquity of field can be regarded as a form of coma. The paraxial imaging of object OY by a point A on the mirror gives an image IY_A with a certain magnification M_A and obliquity θ/M_A, whereas the image IY_B produced by a mirror point B has both a different magnification M_B and a different obliquity.

change in both θ and v/u across the mirror. Coma correction is discussed later.

9.5.4.5 Astigmatism and Field Curvature.
Astigmatism does not appear as a term in Eqs. (17) since it requires rays out of the tangential plane. Essentially, it arises because there are two paraxial focal lengths, one for tangential and one for sagittal rays (Section 9.5.2). Hence there are two focal surfaces as depicted in Fig. 9. After field obliquity and coma have been corrected (see later discussion) the best focal surface may have some residual curvature. This is normally insignificant in spectroscopic applications.

9.5.4.6 Distortion.
Distortion is represented by the row of terms in y alone in Table I. These are also unimportant in most spectroscopic applications, because the field y is usually small and a faithful image representation of the object is usually of little importance. In the Kirkpatrick–Baez mirror arrangement, note that there is a strong anamorphism (different magnifications in the two perpendicular directions) arising from the fact that the two mirrors divide the object–image distance in different ratios.

9.5.5 Correction of the Aberrations of a Circular Mirror

9.5.5.1 Obliquity of Field.
This effect comes about because each object point is imaged by a pencil of rays that makes a slightly different angle with the mirror and thus undergoes a different magnification (Figs. 17 and 22). Field obliquity is the most significant aberration problem for the construction of Kirkpatrick–Baez microscopes and other optical systems that require high spatial resolution. Although in the Rowland circle spectrograph the film or other detector is made to conform to the curved focal plane, it is impractical in many cases to set the detector at such a small angle. Other methods of correction have been devised; Dyson [6] showed that if two spherical mirrors are arranged so that the normals at their centers are in the same plane and pointing in the same direction (i.e., opposite to the configuration depicted in Fig. 11), simultaneous correction of field obliquity, coma, and the quadratic ("primary") spherical aberration can be achieved. Dyson also showed that an aperture stop introduced at a distance of $2/3f$ from the center of the mirror on the image side, an object normal to the principal ray will give an image that is also normal to it. At the same time the angular aperture of the beam and the image intensity are decreased. Neither of these two methods has enjoyed wide application in spectroscopy.

9.5.5.2 Spherical Aberration.
From Eq. (27) and Fig. 21 we see that spherical aberration can be decreased by decreasing the mirror aperture and by choosing unit magnification, for which the "quadratic" spherical aberration disappears. However, a small aperture mirror at unit magnification is frequently far from optimum for a spectroscopic system requiring high throughput. Also, a

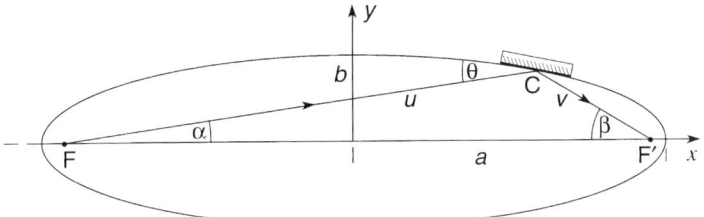

FIG. 23. Focussing from point to point by an ellipse, where F and F' are the foci of the ellipse. The other parameters are explained in the text.

decreased aperture increases the width of the diffraction pattern, so that the resolution cannot be improved indefinitely by this means.

It is clear from Fig. 19 that there is a focal plane location I' for which the focal patch is smaller than it is at the Gaussian focus. Thus even for a spherical mirror the spherical aberration may be alleviated somewhat by shifting the focal plane. Also, spherical aberration can be corrected by aspherizing the mirror, which means distorting the meridian cross section into a curve other than a circle. The curve required for point-to-point focusing is an ellipse (Fig. 23). If one conjugate is at infinity, the curve required is a parabola, and for a virtual image a hyperbola. These are internally reflecting (concave) surfaces. Since most spectroscopic applications have real objects at finite distances and require converging optics, we derive the parameters of the ellipse here.

Referring to Fig. 23, if the object is distance u, the image distance v, and the glancing angle θ at the mirror center C, we wish to find the semi-axes a and b in the ellipse equation:

$$\frac{x^2}{a^2} + \frac{y^2}{b^2} = 1. \tag{28}$$

Noting that the focal properties of the ellipse give

$$a = (u + v)/2 \qquad FF' = 2\sqrt{a^2 - b^2}, \tag{29}$$

where FF' is the intrafocal distance, the law of cosines can be applied to the triangle FCF' to find b

$$b = \sqrt{\frac{uv}{2}[1 - \cos(\pi - 2\theta)]}. \tag{30}$$

In a spectroscopic system consisting of a spherical (circular) mirror and a grating, it is also possible to use the line spacing variation of the grating to correct the terms in Eq. (23). This technique, analogous to the use of a phase corrector plate in visible light optics, is discussed in Vol. 32, Chapter 3.

9.5.5.3 Coma. "Aspherizing" the circular mirror does not reduce coma, which is even more severe for an elliptical mirror. Coma correction is achieved by satisfying the Abbé sine condition, that is, by ensuring that the magnification stays constant for all paths across the aperture. This can be achieved by using systems with multiple mirrors, in particular, pairs of mirrors as depicted in Fig. 11. This method of correction is discussed in more detail in Section 9.6, which deals with toroidal and Wolter optics. We note that the system depicted in Fig. 11 is not corrected for astigmatism, and so the equivalent Kirkpatrick–Baez imaging system consists of *four* mirrors (two acomatic pairs).

9.5.5.4 Astigmatism.   This topic was treated in Section 9.5.3.

9.5.5.5 Curvature of Field and Distortion.   The use of an aperture stop to erect the image onto the Gaussian plane results in a focal surface that is flat only to a first approximation. This residual curvature of field could, in theory, be corrected by distorting the film or detector surface into a curve. The optimum curve may be found by ray tracing [6]. There is also some residual distortion.

9.5.6 Ray Tracing

9.5.6.1 Representation and Reduction of Aberrations.   In real systems, aberrations of all types and orders will be mixed, but in many cases one or more particular types will dominate. In the optical design these must be evaluated and reduced to acceptable levels. The small number of purely reflecting surfaces makes the design of x-ray and EUV systems a much simpler task than it is for multielement lens systems designed for ultimate performance in visible light. It is possible to optimize a system by studying the aberrations analytically, as we did in Section 9.4. This approach is preferable because it provides the most insight. Alternatively, ray tracing can be used; an appropriate number of rays is traced through the system to yield the displacements $\Delta x'$, $\Delta y'$ directly. Computer ray tracing yields a large amount of data very rapidly, and an appropriate method of presentation must be chosen to allow evaluation of the predominant aberrations. One method of approach is represented by the spot diagram. A relatively large number of general rays from a point source is traced, and their intersections with the image plane are represented as separate points in a two-dimensional plot representing the aberrated image of the source. Normally, the rays are chosen to be uniformly spaced in the pupil, so that the density of points in a particular region of the spot diagram represents the light intensity in that region. Although spot diagrams do not show directly which aberrations are most important, it is possible to draw inferences from the general shape and symmetry of the pattern. They are usually computed for a variety of y values or for an extended source. The information thus generated may be too unwieldy for assimilation and evaluation, and it is often convenient to reduce the information contained in the spot diagram to one number, or figure of merit. This may, for instance, be the full extent of the spot diagram in either the tangential or sagittal direction. Alternatively, the irradiance curve of the image of a point source (the point spread function) may be derived from the spot diagram and its full width at half maximum measured. The same quantity may be obtained for a line source (the line spread function). In x-ray optics, the rms blur circle radius σ has been found useful as a figure of merit. If N uniformly spaced rays are traced through the pupil, this quantity is then defined by

$$\sigma = \frac{1}{N} \sum_{i=1}^{N} (\Delta x_i'^2 + \Delta y_i'^2)^{1/2}, \tag{31}$$

where $\Delta x_i'$, $\Delta y_i'$ are the displacements for the ith aberrant ray. The origin for the measurement of these quantities need not be the Gaussian image; by this definition the same value of σ will be obtained if another origin is chosen.

Many software packages for ray tracing are available. One very useful set of programs is the SHADOW package developed by Cerrina and coworkers [12]. It was specifically developed for calculations of synchrotron radiation beamlines and spectroscopic systems for EUV and x-rays, and so contains routines for specifying sources (bending magnet, undulator, etc.) as well as mirrors, gratings, crystals, and multilayers.

9.6 Toroidal and Wolter Optics

From Fig. 23, it is clear that if the figure is rotated about the ellipse axis FF' the mirror surface becomes an ellipsoid of revolution that focuses all rays from F to F'. However, for an off-axis point the single ellipsoid suffers from severe aberrations. The reason is clear if a small element of the ellipsoid around C is considered to be, in the limit, a toroid whose major tangential radius R_M is given by Eq. (17a) and whose minor radius $R_E = R_M/\sin^2 \theta$. The ray through C from an off-axis point will encounter a different glancing angle at C and hence this element of the mirror will have different focal lengths for tangential and sagittal rays. When all rays from O are traced through an infinitesimal annulus of the ellipsoid to the focal plane the result is a circular pattern, which is actually a part of the coma pattern of Fig. 8. An azimuthal segment of the ellipsoid will produce a corresponding segment of this circle. The reason for this strong coma is clear; the principal surface for a single mirror is the mirror surface itself. To satisfy the sine condition, this surface should be close to spherical, and normal to the rays. However, for a glancing incidence ellipsoid, the mirror surface and, hence, the principal surface lies almost parallel to the rays. It is evident that this deficiency cannot be corrected by altering the shape of a single mirror. In fact, coma cannot be well corrected in any glancing incidence system consisting of an odd number of mirrors.

Suppose that a second surface of revolution is adjoined to the paraboloid in the manner shown in Fig. 24. We can see that now the principal surface is more

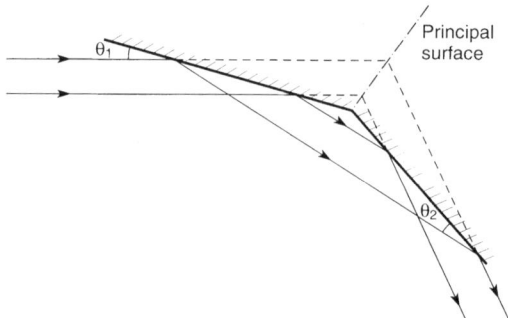

FIG. 24. Correction of coma by a two-mirror system. The combination of the two reflectors gives a principal surface that is approximately perpendicular to the rays and thus allows at least approximate satisfaction of the Abbé sine condition. Application of this principle to mirrors that are surfaces of revolution are shown in Fig. 25.

or less perpendicular to the rays leaving the region of the joint. Wolter [13] showed that certain combinations of conic sections of revolution yield systems, which are to a good approximation, aplanatic. These are depicted in Fig. 25. All are shown with one infinite conjugate; for finite conjugates each paraboloidal element is replaced by a glancing incidence ellipsoid.

Wolter's type I system is composed of a concave (internally reflecting) paraboloid followed by a concave hyperboloid (Fig. 25b). The focus of the paraboloid and that of one branch of the hyperboloid (whose surface plays no part in the reflection of rays) are arranged to coincide at F'. From the properties of a hyperbola, it can be seen that rays entering the system parallel to the axis will converge to the other focus of the hyperboloid at F and thus there will be no spherical aberration. Telescopes of Wolter's type I have been used extensively in x-ray astronomy, and their image characteristics have been investigated in detail.

Wolter's type II system, depicted in Fig. 25c, again consists of a paraboloid of revolution followed by a hyperboloid, but in this case the second element is convex, or externally reflecting. This system is the exact glancing incidence analog of the Cassegrain telescope, and has the same "telephoto" property: The focal length is greater than the total physical length of the system. Practical use of the type II system requires relatively large glancing angles of about 10 deg and for this reason has been found more useful in extreme ultraviolet, rather than x-ray, applications.

Finally, a Wolter's type III system consists of a *convex* paraboloid followed by a concave *ellipsoid*. No practical use has yet been found for this system, and its image characteristics have not been examined in detail.

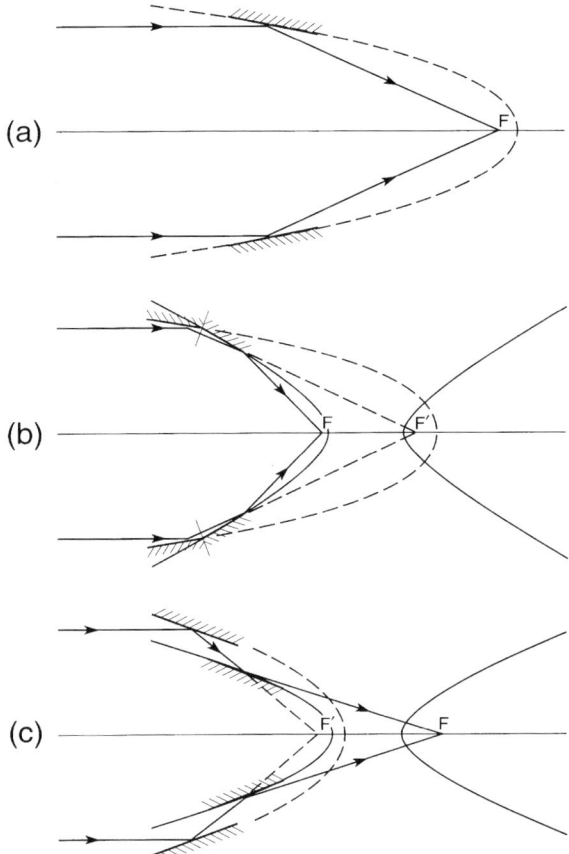

FIG. 25. (a) Focussing of parallel rays by a single paraboloid. (b) Wolter's type I optical system is composed of a concave (internally reflecting) paraboloid followed by a concave hyperboloid. (c) Wolter's type II system is composed of a concave paraboloid followed by a *convex* hyperboloid. A third system (type III), not illustrated here, consists of a *convex* paraboloid followed by a concave *ellipsoid*.

9.7 Fabrication of Mirrors

9.7.1 Conventional Techniques

Most mirrors for EUV/x-ray spectroscopy, in particular spheres and planes, are made by conventional optical techniques of grinding, polishing, and interferometric testing to the required figure. For applications requiring mimimal scattering a low value of rms surface roughness must in addition be achieved.

Polishing technologies developed to produce low-scatter mirrors for the visible and other spectral regions can produce mirrors of a few ångstroms roughness, which is adequate for most spectroscopic applications. For most applications a glass or fused silica material is adequate, but in some applications, such as synchrotron radiation beamlines and others where a high heat load may be encountered, other materials such as single-crystal silicon, silicon carbide, and metals may be employed for their advantageous properties such as high thermal conductivity, low coefficient of thermal expansion, or machinability.

The random nature of conventional grinding and polishing techniques naturally generates planes and spheres, but the manufacture of surfaces of revolution such as cylinders, toroids, ellipsoids, and Wolter-style optics requires a different polishing technology. In this case the optical element must be rotated about an axis during the polishing process, as in a lathe. In fact, such optics have been made, in metals and other materials, by turning with a diamond tool in a precision lathe. These methods allow the alternative of making a negative mold or "mandrel" of the desired optical element and then replicating the desired surface by electroforming or casting in epoxy. It is difficult, with these manufacturing techniques, to produce a surface with the same low-roughness, low-scatter properties possible with conventional polishing. Furthermore, since the optics are of unusual shapes that are not amenable to interferometric testing with ordinary optical shop equipment, it is difficult to achieve close tolerances on the surface figure on this kind of optics. It is therefore preferable to use planes or spheres wherever possible.

9.7.2 Figure Errors

Minimizing aberrations in the geometrical optics design will be ineffective if the surfaces are not manufactured to a high degree of accuracy. Errors in the figure will, in the geometrical optics approximation, cause the rays to intercept the focal plane at some distance from the Gaussian focus, that is, there will be transverse ray errors akin to aberrations. In visible light technology, figure errors are frequently specified in terms of the number of wavelengths of departure from the true figure permitted over the full aperture, or in some equivalent unit such as fringes/inch. Such a specification is equivalent to specifying an angle, and because x-ray and EUV optics are now routinely measured using instruments such as the long trace profiler, it is now customary to specify figure errors in terms of the allowable deviation of the slope of the actual surface from that of the ideal one. As estimate of the slope error can be made from the required spatial resolution of the optical element and the distance v from the element to the image. For example, if we require a resolution of 1 μm and the distance from the mirror to the focus is 100 mm, no ray should make an angle of more than 5 μrad with its ideal path. This means that no element of the surface should make an angle of more than half this amount with the ideal surface, that is, the slope errors should be less than 2.5 μrad. Of course, the period of the errors and their power spectral distribution is important, but this topic is beyond the scope of this article.

9.7.3 Mirrors Generated by Bending

It is possible to make an excellent mirror by starting with a high-quality sphere or plane, and then bending it to a desired curve. The use of a bent optical flat as a glancing incidence optical element to achieve refocusing of x-rays was first discussed by Ehrenberg [14]. By bending a flat into a section of a right circular cylinder, he brought CuK radiation (1.54 Å) to a fine line focus. The technique was used by Franks [15] and by Franks and Breakwell [16] to construct low-angle scattering cameras. In this method, the flat is bent by the application of equal and opposite end couples (the so-called four-point bending method), so that the reflecting surface has a circular cross section. Of course, for freedom from spherical aberration we require a conic section as described in Section 9.5.2.2. Because two mirrors arranged in the Kirkpatrick–Baez geometry can be used to achieve two-dimensional (tangential and sagittal) focusing, we need not make surfaces of revolution but only elliptic (parabolic, hyperbolic, etc.) cylinders, for which the equatorial radius in infinite. Such surfaces can be formed by elastic bending of a mirror. Underwood [17] described how an initially flat strip mirror may be bent to generate such a curve, or to match any sufficiently well-behaved curve, to a high degree of accuracy. The desired match is obtained by applying the optimum combination of end couples and by varying the cross-sectional moment of inertia along the length of the beam.

Referring again to Fig. 23, suppose we want to bend a uniform glass beam so as to match the section of the elliptic cylinder that lies between $x = x_0 + 1$. The differential equation for an elastically bent beam is

$$EI \frac{d^2y}{dx^2} = M(x), \tag{32}$$

where x = distance measured along the beam, y = vertical deflection of the beam, E = Young's modulus, I = moment of inertia of the beam cross section, and $M(x)$ = bending moment as a function of length.

The ellipse is described by Eq. (26), with a and b given by Eqs. (27) and (28). The slope of the curve at point (x, y) is given by

$$\frac{dy}{dx} = -\frac{b^2}{a^2}\frac{x}{y} = -\frac{bx}{a^2}\left[1 - \frac{x^2}{a^2}\right]^{-1/2}. \tag{33}$$

When the deflections are small, dy/dx is small, d^2y/dx^2 is approximately equal to the curvature $1/\rho$ of the beam, and we can write

$$\frac{1}{\rho} \approx \frac{d^2y}{dx^2} = -\frac{b}{a^2}\left[\frac{x^2}{a^2}\left(1-\frac{x^2}{a^2}\right)^{-3/2} + \left(1-\frac{x^2}{a^2}\right)^{-1/2}\right]. \quad (34)$$

To take a concrete example, we examine the case where $u = 100$ (length units), $v = 10$, $\theta = 1$ deg, and the section of the ellipse under consideration is five units in length, centered at $x = 45$ from the origin of coordinates; this is a 10× demagnifying mirror. The curvature of the ideal ellipse is plotted in Fig. 26. Also shown in this figure are the curvatures of beams bent with end moments (or couples) M_1 and M_2 chosen in various ways to approximate the ideal curve. The dashed-dotted line represents a beam bent by the four-point bending method, that is, with equal end couples $M_1 = M_2$. This gives a curve of constant curvature—a circle—matching the ellipse curvature at a single point. A

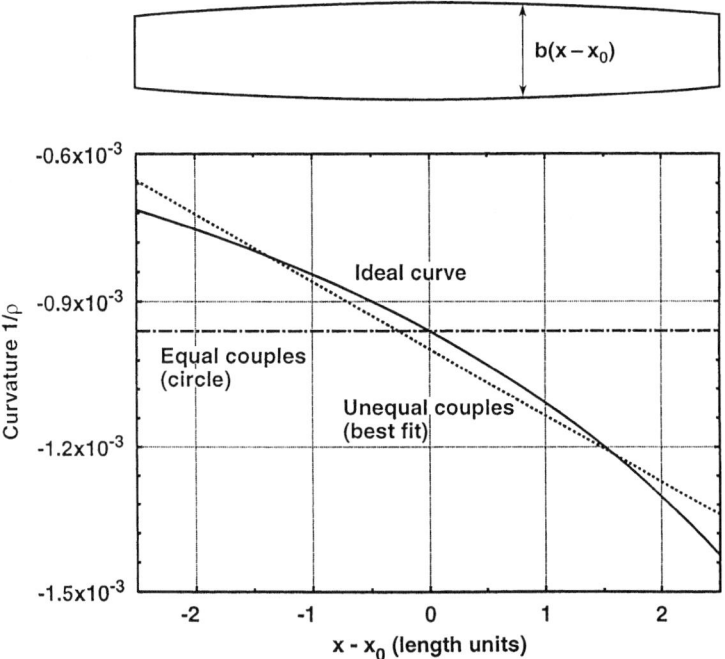

FIG. 26. Comparison of the curvature of an ideal ellipse (solid curve) with a circular mirror obtained by bending with equal end couples (dashed-dotted line) and the best fitting approximation curve obtained by bending with unequal end couples (dotted curve). For this example, correction of the curvature by profiling the mirror results in the "torpedo-shaped" profile, approximately 30% wider in the center than at the ends, is shown above the graph.

better fit to the ideal curvature is obtained by applying finite but unequal end couples, $M_1 \neq M_2$ (dotted curve). It is represented by

$$\frac{d^2y}{dx^2} = \frac{1}{EI(x)} \left[\frac{(M_2 - M_1)}{l}(x - x_0) + M_1 \right], \tag{35}$$

where M_1 and M_2 are the bending moments at $x = x_0$ and $x = x_0 + 1$ respectively.

Variational methods can be used to find the values of M_1 and M_2 that give the best match to the ellipse [17]. This fit will clearly give a reduction of spherical aberration over the circular mirror and an image quality that will be adequate for many applications. However, it is possible to fit the desired ellipse even more closely by varying I along the beam, that is, by making I a function $I(x)$ of x as we have done in Eq. (35). Equation (32) remains valid if the variation of I with x is not too rapid. If the beam is first bent to the approximate shape by some suitable choice of unequal end couples we need only vary I a small amount to make up the difference between the ideal curve and the dotted curve in Fig. 26. The solution for the variation of $I(x)$ is then simply obtained by equating the right-hand sides of Eqs. (34) and (35). For a beam of rectangular cross section $I = bd^3/12$, where b = breadth, and d = depth, the variation of I can be simply accomplished by varying b. For the example this results in a "torpedo-shaped" profile approximately 30% wider in the center than at the ends (Fig. 26).

This method can be applied to the other conics or to *any* well-behaved curve for which values d^2y/dx^2 can be computed. If d is not too large and variations of $I(x)$ are comparatively gentle, a beam bent in this way will match the idea curve exactly. However, lengthwise variations in E or d, and the failure to match the desired $b(x)$ curve accurately, will introduce errors, as will waves or bends in the material used to make the beam. We note, however, that *linear* errors in all these quantities may be taken out by slightly adjusting the values of the end couples. An advantage of a variable figure mirror is that the focal length can be altered to bring the focus to diffent positions.

9.7.4 Use of Bent Mirrors in Spectroscopic Applications

An example of the spectrographic use of an elliptical mirror produced by bending is given by Behring et al. [18]. The mirror was used to obtain spatial resolved spectra of laser-heated plasmas in the 6- to 370-Å spectral region. It is interesting that the spatial resolution was limited in this experiment, not by the elliptical mirror figure, but by the field obliquity. Figure 27 illustrates the use of a bent mirror to correct the astigmatism in a fixed deviation monochromator using a spherical grating, a concept introduced by Hettrick and Underwood [19]. We note that this system is essentially a Kirkpatrick–Baez imaging system with a grating ruled on one of the elements. This monochromator is incorporated in

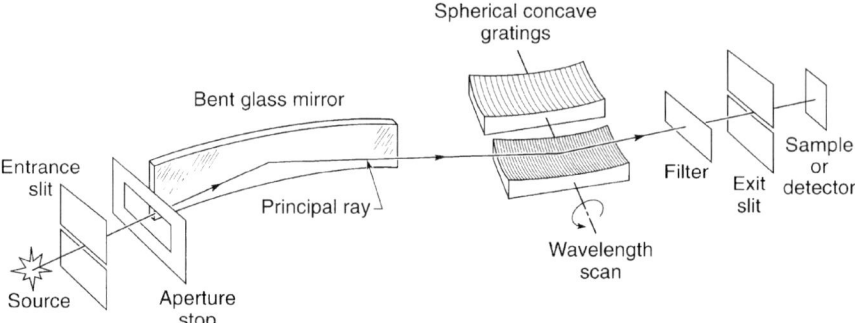

FIG. 27. Use of a premirror to correct the astigmatism of a monochromator with a spherical grating. The variable curvature mirror may be adjusted to form an image of the source in the horizontal direction on the slit or on the sample or detector at any distance behind the slit.

the EUV reflectometer using a laser plasma source described elsewhere [20]. Other examples of the use of glancing incidence mirrors in x-ray spectrographs, monochromators, and instrumentation such as synchrotron radiation beamlines are presented in Chapter 2.

References

1. W. T. Welford, "Aberrations of the Symmetrical Optical System," Academic Press, New York (1974).
2. H. A. Buchdahl, "An Introduction to Hamiltonian Optics," Cambridge University Press, Cambridge, MA (1970).
3. A. Maréchal, "Traite d'Optique Instrumentale—Tome 1—Imagerie Geometrique, Aberrations," Editions de la Revue d'Optique theorique et instrumentale, Paris (1952).
4. R. A. Houston, "A Treatise on Light," Longmans, London (1938).
5. P. Kirkpatrick and A. V. Baez, *J. Opt. Soc. Am.* **38**, 766–774 (1948).
6. J. Dyson, *Proc. Phys. Soc.* **B65**, 580–589 (1952).
7. M. Montel, *Rev. Opt. Theor. Instrum.* **32**, 585 (1953).
8. M. Montel, "X-Ray Microscopy and Microradiography," Academic Press, New York, p. 177 (1957).
9. J. F. McGee and J. W. Burrows, *Proc. SPIE* **106**, 107–111 (1977).
10. H. H. Pattee, Ph.D. thesis, Stranford University (1953).
11. D. Korsch, "Reflective Optics," Academic Press, Sand Diego, CA (1991).
12. C. Welnak, G. J. Chen, and F. Cerrina, *Nucl. Instrum. Methods* **A347**, 344–347 (1994). The SHADOW software is available via the World Wide Web at http://shadow.xraylith.wisc.edu/shadow/.
13. H. Wolter, *Ann. der Phys.* **10**, 42 (1952).
14. W. Ehrenberg, *J. Opt. Soc. Am.* **39**, 741 (1949).
15. A Franks, *Brit. J. Appl. Phys.* **9**, 349 (1958).

16. A. Franks and P. R. Breakwell, *J. Appl. Cryst.* **7**, 122 (1974).
17. J. H. Underwood, *Space Sci. Instrum.* **3**, 259–270 (1977).
18. W. E. Behring, J. H. Underwood, C. M. Brown, U. Feldman, J. F. Seely, F. J. Marshall, and M. C. Richardson, *Appl. Opt.* **27**, 2762–2767 (1988).
19. M. C. Hettrick and J. H. Underwood, *Appl. Opt.* **25**, 4228–4231 (1986).
20. E. M. Gullikson, J. H. Underwood, P. C. Batson, and V. Nikitin, *J. X-Ray Sci. Technol.* **3**, 382–299 (1993).

10. REFLECTOMETERS

W. R. Hunter
SFA Inc.
Largo, Maryland

10.1 Introduction

In this chapter, a reflectometer is defined as a mechanism that carries a sample and detector, both of which can be maneuvered to permit measurement of the radiant intensity specularly reflected or scattered from the sample. The mechanism must be supplied by a radiation source, for example, a monochromator. Generally, measurements of reflectance require knowledge of the angular position of both sample and detector with respect to the incident radiation; hence, a reflectometer might also be classified as a goniometer.

The angle of incidence, ϕ, at which a reflectance measurement is made is defined as the angle between the incident beam and the surface normal. These two directions also define the plane of incidence. In x-ray measurements the angle of the incident beam is usually specified as the angle between the beam and the surface, the glancing angle θ, which is the complement of ϕ. X-ray terminology refers to θ–2θ scans because the detector is at an angle 2θ to intercept the beam reflected from the sample at an angle θ.

In principle, measurements can be made at all angles of incidence except normal incidence, $\phi = 0$ deg where the detector obscures the incident beam, and $\phi = 90$ deg where the incident radiation is parallel to the surface and, therefore, not reflected.

In an ideal reflectometer one has a parallel radiation beam of uniform radiance and a detector of uniform sensitivity. Monochromatic vacuum ultraviolet (VUV) radiation beams usually diverge from a source, for example, the exit slit of a monochromator. Thus, the angle of incidence is uncertain by an amount equal to the divergence angle. Not only do monochromatic VUV beams diverge (or converge), they are seldom uniform in intensity over their cross sections because the optical element dispersing the radiation is not uniformly efficient across its surface. To compound the problem, detectors are seldom uniform in their response over their sensitive area. Consequently, the angle of incidence spread caused by the divergence can be weighted by both beam and detector non-uniformity. Furthermore, both the divergence and direction of the beam may change with wavelength because some monochromators move the grating with respect to the exit slit when changing wavelength. Generally this change is small but it can be large enough to invalidate angle-of-incidence settings.

So far no mention has been made of the linearity of detector response to radiation intensity, or of the effect of stray light, on reflectance measurements. Manufacturers usually specify a range over which the detector response is linear. If the range is large, checking linearity can present problems beyond the scope of this chapter. A linear response is tacitly assumed in all that follows.

Stray light is an insidious problem that cannot be treated generally. Some comments on the subject are given later. Faced with this litany of imperfections, the experimentalist must approach the "simple" measurement of reflectance with some caution. With careful attention to detail most of these imperfections can be overcome.

10.2 Specular Reflectance Measurements

Specular reflectance is defined as the ratio of the intensity of a radiation beam after reflection from a nonscattering surface to the intensity of the beam incident on that surface; that is, the intensity of the reflected beam *relative* to the incident beam intensity. Defining the incident intensity as I_0 and the reflected intensity as I, the reflectance is $R = I/I_0$, a decimal value often presented as a percent. Sometimes this ratio is referred to as the *absolute reflectance*, perhaps to distinguish it from measurements in which only I is available. One infers that from I only, *relative reflectance* values are obtained. In view of the definition of R, the terms *absolute* and *relative* have little or no meaning.

10.2.1 Measurements at a Fixed Angle of Incidence

Reflectometers can be divided into two classes, those measuring reflectance at a fixed angle of incidence, and those capable of measuring reflectance at different angles of incidence. Consider first reflectance measurements at a fixed angle close enough to normal incidence so that the result is essentially independent of polarization. Figure 1 illustrates one way in which such reflectance measurements can be done. Mirror M_2 is the mirror whose reflectance, R_2, is to

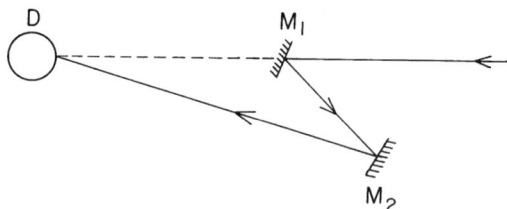

FIG. 1. Arrangement for measuring reflectance at a fixed angle of incidence. M_1 is a mirror of known reflectance, M_2 is the mirror to be measured, and D is the detector. M_1 is removed for measuring I_0.

be measured, and M_1 is another mirror whose reflectance, R_1, is known. The detector, D, is fixed and I_0 is obtained by removing mirror M_1 from the beam so that the beam directly illuminates D. Intensity I is obtained with M_1 and M_2 in the positions shown so that $I = I_0 \times R_1 \times R_2$. If the two mirrors are at the same angle of incidence, $I/(I_0 \times R_1)$ gives R_2. The fact that neither mirror is used exactly at normal incidence is not significant because the change in R from normal to the small angle of incidence used is practically negligible. Under these conditions the effective distance of the detector from the slit changes so errors could arise if the beam divergence causes the detector to be overfilled when this distance is greatest. Although this scheme is adaptable to any fixed angle of incidence, at angles much larger than normal the reflectance can change so rapidly with angle of incidence that sufficiently accurate angular settings of the two mirrors are extremely difficult. Furthermore, for large ϕ the result is not independent of polarization and the measurements should be made in two orientations of the plane of incidence, as mentioned below.

An alternative way to measure reflectance at a fixed angle of incidence near normal incidence uses the photoyield from two identical surfaces. LeBlanc et al. [1] used an experimental arrangement similar to that shown in Fig. 1 but without a detector. Mirror M_1 was set to an angle of incidence of 15 deg and the radiation reflected from it was incident on M_2 at the same angle. Because the photoyield is proportional to the absorptance A ($A = 1 - R - T$, where T is the transmittance through the mirror surface into the material), the photocurrent from the first mirror is $i_1 = I_0 \times A_1$ and that from the second mirror is $i_2 = I_0 \times R_1 \times A_2$. The ratio of the photoyields, i_2/i_1, is the reflectance because $A_1 = A_2$. There is no explicit measurement of I_0. To use this method successfully, one must ensure that M_1 and M_2 are not contaminated to the extent that the photoyield of one is affected more than the other and that neither emitting surface is overfilled.

In principle, the method used by LeBlanc et al. could also be used to measure reflectance at larger angles of incidence if the two samples could be rotated so as to preserve the angle of incidence on both and if the reflectometer azimuth could be rotated to assess polarization.

10.2.2 Measurements at Oblique Incidence

For convenience, the intensity of a reflected beam is divided into two components that are linearly polarized, R_s perpendicular to, and R_p parallel to, the plane of incidence. The intensities of these components are not the same for nonpolarized incident radiation, therefore, if the incident radiation is partially linearly polarized reflectance measurements should be made in two orientations perpendicular to each other, preferably parallel to and perpendicular to the plane of incidence, to evaluate the effect of polarization.

Figure 2 [2] is a schematic diagram of a reflectometer capable of measuring reflectances at different angles of incidence and often used in the determination of optical constants in the VUV. The primary conditions that the instrument must meet are (1) both the mirror, M, and detector rotate about a common axis, O, which lies in the reflecting surface of the mirror; and (2) the optical axis of the illuminating beam XX' is centered on O. If these two conditions are met, XX' is also the optical axis of the reflectometer. The reflectometer radius is ℓ, the distance from the point of reflection to the detector surface; N is the mirror normal, ϕ is the angle of incidence, and ϕ' is the angle of reflection. The geometry is such that the effective distance of the detector from the source is fixed, and I_0 is measured by removing the mirror from the beam. The detector must be able to rotate from the I_0 position to the smallest angle of incidence desired, but preferably through a full 360 deg, and the mirror through 180 deg, to avoid some systematic errors that are discussed later.

In earlier times the detector–sample motions were often mechanically coupled, for example, via angular drives in the speed ratio of 2/1, so that a

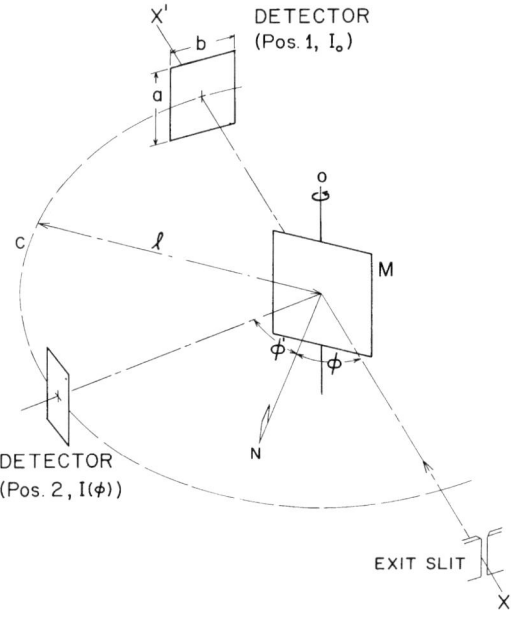

FIG. 2. Schematic diagram of a reflectometer that can measure reflectance at different angles of incidence, where O is the axis of rotation of both the mirror to be measured, M, and the detector; ϕ and ϕ' are the angles of incidence and reflection, respectively; and N is the mirror normal. The detector is shown in two positions; for measuring I_0 (position 1) and I (position 2). The radius of the detector is given by ℓ, and XX' is the reflectometer optical axis [2].

continuous recording of reflectance versus angle of incidence (R vs ϕ) could be obtained at a given wavelength. In the absence of coupling, R vs ϕ data could still be obtained at discrete angles by setting the mirror at ϕ and adjusting the detector to ϕ'.

Mechanical coupling had an advantage in that when the reflected signal was small and noisy there was no question that the mirror and detector were properly aligned. Under the same conditions, aligning of uncoupled motions could be quite uncertain. A number of schemes have been devised for angular drives with a 2/1 speed ratio [3–6] but they are not described here because most modern reflectometers are computer controlled and angular tracking is programmed into the software, thus providing "electronic" coupling.

10.2.3 Design of Reflectometers

10.2.3.1 Systematic Errors in Reflectance Measurements.

Hunter [2] and Draper and Murphy [4] have discussed the cause of errors in measuring reflectances and how to avoid them. The following discussion is adapted from Hunter's paper. If the surface of the mirror is displaced from the common rotational axis one has a mechanical displacement error, as illustrated in Fig. 3 [2]. The radius of the small central circle is the displacement of the mirror from the axis of rotation. The designations M and m represent the mirror at two different angles of incidence and A and a are the corresponding positions where the reflected beam should intersect the detector circle if there were no error. When the mirror is set for an angle of incidence ϕ, the incident beam is reflected from the mirror to A' if error is present. This point corresponds to a detector angle of $\phi' + \delta$ rather than ϕ', hence, while the reflected intensity is $I(\phi)$ (the mirror is properly set), it is being measured at $\phi' + \delta$. Similarly, if for the same angle of incidence the mirror is displaced diametrically across the small circle, the reflected intensity would be measured at a detector angle of $\phi' - \delta$. The same is true for mirror position m and the detector circle locations a and a'. If the mirror and detector motions are not coupled, a small displacement error usually causes no problems because the mirror is at the correct angle of incidence and the detector can be rotated to A'. Actually the displacement error causes the distance of the detector from the source to change slightly; however, this change is usually negligible. If the motions are coupled, however, the net effect is that the reflected beam drifts off the center of the detector as the angle of incidence increases. As shown in the figure, it appears to "lead" the detector as the angle of incidence increases, and it "lags" if the mirror is displaced diametrically across the small circle. In the absence of other errors the leading, or lagging, of the light patch relative to the detector makes the displacement error fairly easy to detect. If one illuminates the reflectometer with visible radiation and scans in angles of incidence from large to small angles with the detector and mirror motions coupled, the light patch leads the detector if the mirror is behind the axis of rotation (as shown in Fig. 3). If the mirror is in front of the axis of rotation, the opposite is the case. These relative motions hold true regardless of the side of the reflectometer's optical axis on which the observations are being made. When designing a reflectometer, provision should be made to move the mirror parallel to its normal to bring the mirror surface and axis of rotation into coincidence. This is an adjustment used in the initial setup of the reflectometer and should seldom need changing.

A mechanical deviation error occurs when the common axes of rotation of the mirror and detector are not centered on the optical axis of the beam, as shown in

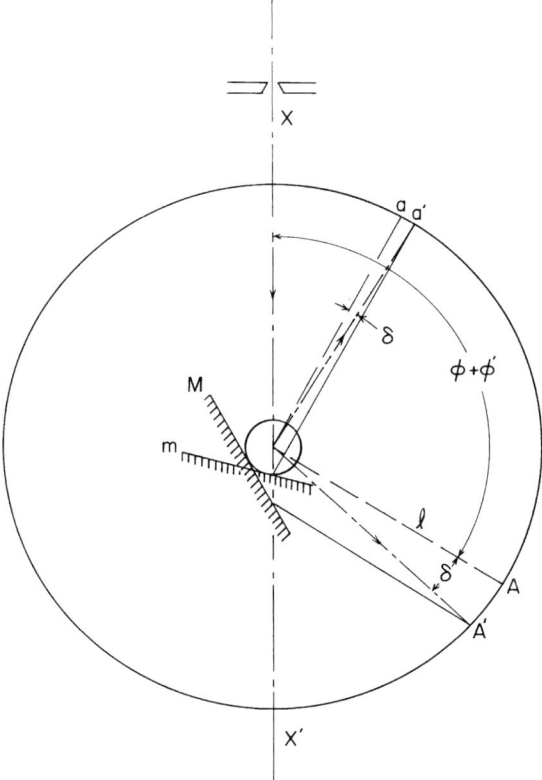

FIG. 3. Schematic diagram showing the mechanical displacement error. The diameter of the small circle is the displacement error; M and m represent the displaced mirror at two positions; A and a are the positions of the reflected ray if the displacement is 0; A' and a' are the positions of the ray reflected from the displaced mirror; ℓ is the reflectometer radius; ϕ and ϕ' are the angles of incidence and reflection, respectively; and XX' is the reflectometer optical axis. For a discussion of δ see the text [2].

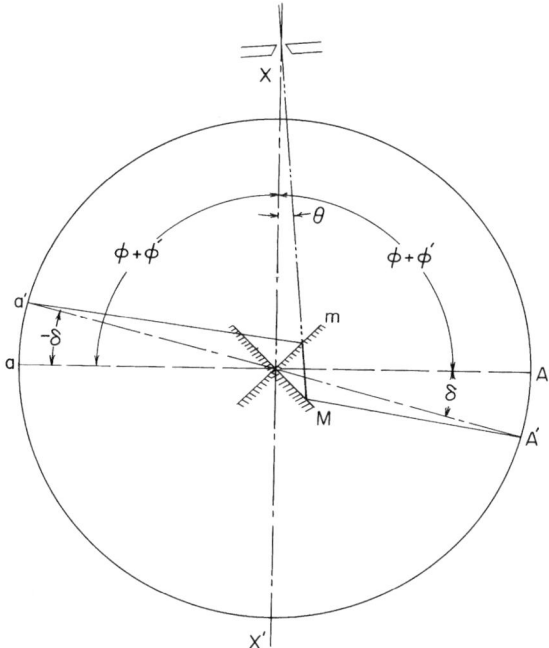

FIG. 4. Schematic diagram showing the mechanical deviation error, where XX' is the optical axis of the reflectometer which deviates from the optical axis of the incident beam by the angle θ. Otherwise the symbols are the same as in Fig. 3 [2].

Fig 4 [2]. The net effect is that the angle of incidence is either slightly more or less than expected. If the detector is then rotated to collect the beam, the intensity measurement is obtained at an angle of incidence of $\phi \pm \delta$. The mechanical deviation error cannot be corrected by translating the mirror along its normal, as was the case with the displacement error. As the figure shows, if the deviation angle causes a positive error on one side of the beam, the error is negative on the other side. If the deviation is not large, that is, the deviated beam does not fall outside the sensitive area of the detector, the reflectance can be measured on either side of the radiation beam and the signals for the corresponding angles averaged, a procedure that eliminates the effect of the error caused by mechanical deviation.

In principle, the mechanical deviation error can be detected in the same manner as the displacement error in the absence of other errors. As the reflectometer is scanned in angle of incidence, the light patch lags, or leads, by a *constant amount* on both sides of the optical axis. If the beam lies on the other side of the optical axis, the relative motions are reversed. The mechanical deviation error can be eliminated if either the optical axis of the beam or reflectometer can be shifted to make the two coincide. A device for controlling the direction of the beam from a VUV monochromator has been described by Hunter and Chaimson [7].

Nonparallelism of the incident beam, up to 3–4 deg of divergence, usually produces a negligible error [8] because the reflectance changes slowly enough that the average over the divergence is very close to the actual value at the selected angle. The error may not be negligible, however, if beam, or detector, or both are nonuniform.

Nonuniformity of the beam and detector can also present problems. On reflection in the mirror, the beam is reversed in the plane of incidence with respect to its orientation when I_0 is measured. The situation is shown in Fig. 5 where the beam is shaded to illustrate its nonuniformity and an arrow drawn on the detector to show its orientation. Both beam and detector nonuniformities can affect the apparent angle of incidence to cause errors in the measurement. The effect of nonuniformity can be alleviated to a certain extent by (1) stopping down the beam until the change in intensity across its width is small and (2) selecting a portion of the detector with small nonuniformity.

Errors can also be caused by fluctuations of the incident intensity. Fast fluctuations can be averaged by adjusting the time constants of the system, but long-term drifts must be treated separately. One approach is to make measurements in a timed sequence that averages the effects of long-term drifts.

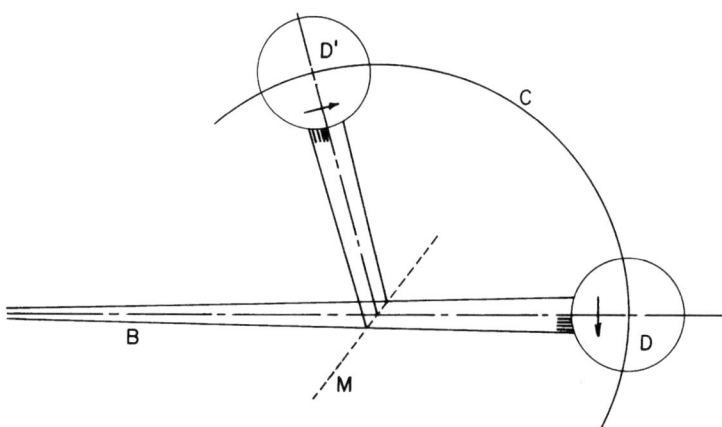

FIG. 5. Schematic diagram showing the effects of nonuniform beam and detector, where B represents the incident beam, M the mirror, and D the detector. The shading on one side of the beam shows how the position of the beam on the detector is reversed after reflection. The arrow on the detector shows its orientation at the I_0 measurement and at other angles of incidence.

If reflectance measurements of a curved surface are required, one must calculate the image dimensions using the simple laws of mirror optics to ensure that the image size does not exceed the size of the sensitive area of the detector. The object distance is the distance from the slit to the optic so the image distance will fall either between optic and detector, on the detector, or past the detector if the optic is concave. If convex, the image distance is virtual. If the image size is larger than the detector, considerable error may result. Krumrey et al. [9] have discussed the measurement of reflectance of curved surfaces in the soft x-ray region.

Stray Light. One of the most common causes of errors in reflectance measurements, and usually the most difficult to detect, is the presence of stray light, that is, radiation of wavelengths other than those intended to be in the monochromator passband. Stray light can arise from many sources: scattering from the walls of the monochromator, light leaks in either the monochromator or reflectometer housing, etc. These sources can be eliminated by proper baffling and construction; however, the dispersing source itself can be a major source of stray light. Because no amount of baffling is effective in eliminating this source, it becomes necessary to correct the reflectance values.

With a line source a correction can be made by measuring I_0 and $I(\phi)$ at the wavelength of interest and then repeating the measurement, at a wavelength setting where only the background exists but close to the line of interest, to obtain $I_0(s)$ and $I(\phi)_s$. From these data a fairly accurate reflectance value is given by

$$R = [I(\phi) - I(\phi)_s]/[I_0 - I_0(s)].$$

If the source emits a continuum, such as synchrotron radiation, such simple corrections are not possible. One can reduce the stray light by using holographic gratings, but the problem remains of reducing, or eliminating, the higher grating orders. Unbacked metal film transmission filters and reflecting filters [10] can help in reducing, but cannot eliminate, higher order intensities. If the mode of operation of the dispersing instrument is flexible enough to permit obtaining the wavelength of interest at different diffracting angles, the grating can be set closer to normal incidence, which may help in suppressing higher orders. Hunter and Rife [11] have discussed the suppression of higher orders by using appropriate coatings on the monochromator optics.

10.2.3.2 Instrumental Requirements. In view of earlier discussions in this chapter, it is recommended that the reflectometer mechanism for measuring specular reflectance as a function of angle of incidence have at least the following capabilities:

1. An adjustment to obtain coaxial rotation of the mirror and detector, for example, by moving the mirror parallel to its normal
2. The ability to measure reflectance on either side of the reflectometer optical axis

3. The ability to rotate the entire reflectometer about its optical axis to correct for polarization
4. The ability to do $\theta/2\theta$ tracking and to rotate the detector and mirror independently

In addition, an I_0 detector that constantly monitors the monochromatic beam intensity after the exit slit, by abstracting part of the beam, will aid greatly in nullifying the effect of a slowly drifting incident beam intensity. Some synchrotron radiation establishments route a signal proportional to the beam current in the ring to beamline users. Such a signal tracks the exponential decay of the ring current, but it may not be proportional to the monochromatic beam as the monochromator is scanned. It is not a substitute for an I_0 monitor.

A very important requirement is the accuracy with which the sample and detector angles can be set. The author strives to have the angular setting repeatable to 0.1 deg. Such accuracy usually requires a direct drive for both sample and detector. In high vacuum instruments direct drives require either a shaft with an O-ring seal or internal driving motors. In ultrahigh vacuum instruments O-ring seals and internal drive motors may not be acceptable and one resorts to rotary motions with spring-loaded Teflon seals that are differentially pumped as direct drives for the sample and detector. If the geometry is such that the rotary motion cannot be used as a direct drive, the author has found studded pulleys driving perforated metal belts under tension to be satisfactory for coupling rotary motions to sample and detector motions [12]. Coupling using gears is not as satisfactory because of the backlash, which may be overcome by always driving in the same direction, and because of gear errors.

Further requirements may be imposed depending on the uses to which the reflectometer will be put and the limitations inherent in VUV systems. For example, one of the limitations of VUV systems is the small beam area, dictated by the geometry of the system and the necessity to maintain a small beam to ensure uniformity. A large optic can only be measured at one spot unless the reflectometer has some means for moving the optic so the surface can be scanned. One solution is to vent the reflectometer, move the optic so that a new location can be measured, pump down, then make the measurement, which is extremely wasteful of time and energy. If this sort of measurement must be made, the reflectometer must be equipped with suitable motions for manipulating the optic being measured. More on this subject late.

10.2.3.3 Reflectometers for Specular Reflectance Measurements at Different Angles of Incidence.
Purcell [13] has made perhaps the simplest, perhaps the most unusual, VUV reflectometer. It is not contained in a vacuum system but makes use of atmospheric windows for VUV radiation. Consequently, its spectral range is limited to approximately 1216 Å. The source is a hydrogen arc and the detector a thermoluminescent phosphor ($CaSO_4$: Mn). The

phosphor is attached to a nickel mesh that can be heated by passing an electric current through it. When illuminated by VUV radiation, the phosphor stores energy proportional to the dose and releases the energy as visible radiation when heated. This detector is insensitive to visible or UV radiation. A total air path of 5 cm is used. A pair of slits 0.25 mm wide and about 1 cm apart define a beam that is about 1 mm wide at the detector distance. The arrangement is essentially that of Fig. 2. Phosphor exposure times were 1 min. After exposure the phosphor was heated while in front of a photomultiplier. The integrated response of the multiplier is proportional to the intensity of the VUV radiation. The ratio of the response at angle ϕ to the response at 90 deg is the reflectance at that angle. All the adjustments can be made visually after interposing a glass filter in front of the source since the phosphor is completely insensitive to visible or near UV; consequently, it is not necessary to darken the room during exposure of the phosphor. However, there is no absorption window in water vapor at these wavelengths so it is important to make measurements when the humidity is low, or to reduce it as much as possible when measurements are being made.

Socker et al. [14] has used a slightly different version of this scheme for measuring the reflectance of Al/MgF$_2$ coatings at 1216 Å at atmospheric pressure.

High Vacuum Reflectometers. Vilesov et al. [15] have developed a reflectometer capable of measuring reflectance at different angles of incidence using a detector scheme similar to that of LeBlanc et al. [1]. The sample is enclosed in an evacuated sphere and is connected, electrically, to an electrometer. The inside of the sphere is coated with aquadag so as to have a uniform photoemission over its entire surface and is held at a potential of -315 V. When VUV radiation impinges on the sample, the current collected by the sample and measured with the electrometer is $I_1 = I_0 \times A \times R$ where I_0 is the incident intensity, A is the photoyield of the aquadag, and R is the reflectance of the sample. If the sample is removed from the beam so that the beam is incident directly on the aquadag, the current collected by the sample is $I_2 = I_0 \times A$, and the ratio of the two currents yields R. Vilesov et al. point out that the following conditions must be met: (1) The photoyield is equal for all points on the sphere, (2) the intensity does not change between measurements of I_1 and I_2, (3) the fraction of radiation reflected by the sphere and absorbed by the sample can be neglected, and (4) the radiation always falls on the sphere at the same angle of incidence. Condition 4 is achieved by placing the sample at the center of the sphere. The authors claim that their aquadag coatings were uniform to within $\pm 1\%$, which they attribute to the fact that in the VUV photoelectrons are emitted from the bulk, rather than the surface of the sample; therefore, slight surface contamination has little effect.

Vilesov et al. also point out that the experimental arrangement described here can be used to measure diffuse reflection in the VUV (see discussion later).

A reflectometer used by Hunter [2], originally designed by F. S. Johnson, consists of a steel cylinder separating two flat plates. Each cylinder end and each plate has matching V-grooves for steel balls so that the plates can rotate when the reflectometer is evacuated. The plates extend into the cylinder and have O-ring grooves that seal against the polished cylinder wall. One plate carries the detector, which is off-center, and the other the sample. Each plate is, in essence, a worm wheel with teeth around the periphery. The teeth on the detector plate have twice the pitch of those on the sample plate so that $\theta/2\theta$ scans can be made. A sample holder is mounted eccentrically on a removable flange so that the sample can be rotated out of the beam for an I_0 measurement. An adjustable stop ensures that the mirror is in the correct position when it is rotated back into the beam. Two pins in the bottom plate of the reflectometer orient the flange with respect to the optical axis so that it can be removed and replaced with no loss in alignment. The top of the mirror is pressed against a straight bar and the bottom against a screw. Adjusting the screw makes the mirror surface parallel to the axis of rotation. Another adjustment allows the mirror to be moved parallel to its normal so the mirror surface can be made coincident with the axis of rotation.

A reflectometer described by Madden and Canfield [3] was designed to be part of a reflectometer-evaporator. The sample holder could be rotated to cover a hole between the evaporator and reflectometer chambers so that evaporations and reflectance measurements could be made without exposing the mirror to air. The angle divider was such that measurements could be made only on one side of the optical axis. Other than that, the reflectometer contained all the other necessary adjustments.

Michels *et al.* [16] have designed and built a high vacuum reflectometer, intended primarily to measure diffraction grating efficiency. Because gratings do not have uniform efficiency across their aperture it is necessary to manoeuver the grating to sample different parts of the surface. The grating is held in a trolley that is translated along a track having the same radius of curvature as the grating. The track can be rotated to select the angle of incidence and the grating can be tilted. Thus the entire grating surface can be scanned and an efficiency map constructed from the results. External stepping motors drive rotational, tilting, and translational motions through O-ring seals. This instrument has the ability to change the radius of the reflectometer, that is, change the detector–sample distance to facilitate measurement of curved optics. Hunter and Prinz [17] have reported measuring concave gratings at grazing incidence using this instrument.

All the high vacuum reflectometers described thus far place the detector inside the vacuum enclosure. Smith [18] designed a reflectometer with the detector outside of the vacuum enclosure; a schematic diagram of his arrangement is shown in Fig. 6. He used a Pyrex rod bent to resemble a question mark, which he refers to as a light pipe, *LP*. A flat surface on the end of the rod, at

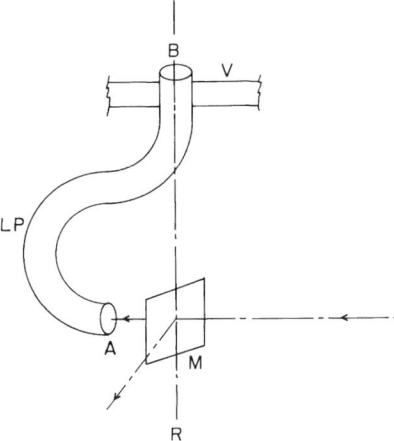

FIG. 6. Schematic diagram of the light pipe detector arrangement used by Smith [18].

position A, held sodium salicylate perpendicular to the beam and sufficiently far from the mirror to permit rotation of the mirror. The light pipe then curved such that the straight portion came througth the vacuum system, V, at the center of rotation, B. An O-ring seal was used. The photomultiplier was supported close to the end of the light pipe in air and isolated from extraneous light.

Ultrahigh Vacuum Reflectometers. With the advent of synchrotron radiation there has been a great interest in studying the properties, including reflectance, of pristine surfaces or surfaces contaminated in a controlled manner. Designing a reflectometer for UHV use, however, presents an order of magnitude increase in complexity because of the problems of rotating the detector and sample to obtain R vs ϕ data, and of rotating the components to account for polarization. Feuerbacher *et al.* [19] have described a UHV reflectometer that has the required rotational degrees of freedom and can achieve pressures of 5×10^{-10} Torr. Rotation of the components is done using a wobble stick type of rotary motion that drives two wheels, one holding the detector and the other the sample. Rather than a continuous angular motion for sample and detector, they have chosen discrete steps of 7.5 deg. These steps are obtained by using wheel and detent mechanisms for both mirror and detector. The justification for discrete motion is that it is easier to obtain an accurate angle setting using fixed angle positions inside the chamber than by measuring the rotation of directly coupled feedthroughs, especially wobble sticks. Discrete angular steps have a disadvantage, however, when measuring optical constants by the R vs ϕ method or the angular response of multilayer coatings or gratings. Maximum accuracy requires reflectance measurements within a certain angular range that changes with changing n, k. If the angular positions are fixed, angles giving maximum accuracy may not always coincide with the fixed angles. Furthermore, if the reflectometer is also to be used in measuring scattering as a function of angle, fixed angular positions of the detector may be inconvenient.

Hogrefe et al. [20] have described a UHV reflectometer that employs continuous, angular motions of sample and detector. Their vacuum vessel is a cylindrical tank. Radiation enters through the center of the cylindrical wall and passes through, or close to, the cylinder axis. Sample and detector angular motions are brought into the tank through the center of the two opposing flat surfaces. The drives are direct and use two stages of differential pumping. A computer controls the angular motions with a reputed accuracy of 0.01 deg. Aligning the sample and detector shafts for both concentricity and coincidence requires locating the supporting flanges accurately on the flat end plates. The authors want coincidence to <50 µm and concentricity to <0.01 deg. They report a pressure of 2×10^{-10} Torr in the chamber.

Hunter and Rife [12] have described a reflectometer suitable for UHV use. The instrument is intended to measure specular reflectance and in-plane and out-of-plane scattering, and has a provision for rotating the sample about its normal. The entire mechanism is contained in a stainless steel cylindrical chamber 18 in. in diameter and 18 in. long. Sealing at either end of the cylinder is accomplished through the use of 18-in. Wheeler flanges using copper wire O-rings. A number of ports are located on the cylindrical surface midway between the ends. Two of them, 8-in. flanges diametrically opposed, serve to admit radiation and align the instrument. During operation the flange opposite the radiation entrance has a transfer chamber attached to permit interchanging samples in the reflectometer without breaking the vacuum. Fig. 7 is a cross-sectional drawing of the mechanism. Three aluminum coaxial cylinders are supported by stainless steel ball bearing races. The races are lubricated with MoS_2. The inner cylinder carries the sample, the middle cylinder controls the altitude of the detector to permit out-of-plane scattering measurements, and the outermost cylinder carries the detector and is mounted in a support attached to a 10-in.-diameter flange on one of the 18-in. Wheeler flanges of the UHV chamber. A support of this type ensures, within the accuracy of the bearings and machining of the cylinders, coincident sample and detector axes of rotation. In addition, supporting all rotary motions from one location eliminates misalignment of the axes should the vacuum vessel flex on evacuation. The original mechanism used worm wheels and worms, driven through wobble sticks, for the rotary motion, and an angular precision of 0.05 deg was expected if the mechanism was always driven in the same direction. But no suitable lubricant could be found for long-term lubrication. Neither MoS_2 nor ion-implanted WS_2 were useful; within a few tens of hours of operation the worms were badly worn. The wobble sticks and worm gears were abandoned in favor of differentially pumped direct drives and metal belt

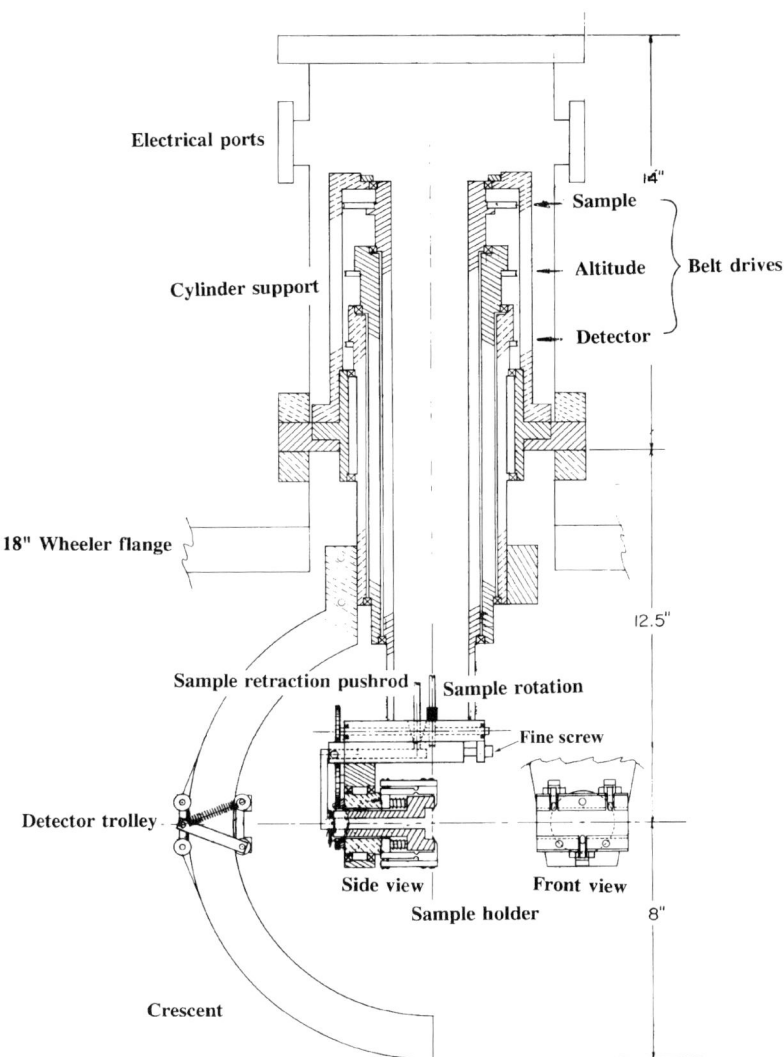

FIG. 7. A diagram of the ultrahigh vacuum reflectometer reported by Hunter and Rife [12]. Three coaxial cylinders are supported by ball bearings mounted in the worm support, which is, in turn, supported on a 10 in. flange attached to the 18 in. Wheeler flange. The worm gears and wheels have been replaced by belt drives directly driven by differentially pumped rotary motions that come through the uppermost flange. The worm gear enclosure is a vacuum enclosure and has an array of ports around the top for electrical leads. The fine screw translates the sample holder parallel to the sample normal. The detector trolley carries the detector along the crescent when out-of-plane measurements are to be made. (From [12] with permission from Elsevier Science, Netherlands.)

couplings, as mentioned earlier. This substitution eliminated the problem of sliding friction in a UHV environment.

If the UHV reflectometer must measure the sample at different positions on its surface, the design problem becomes much more difficult than for a high vacuum instrument wherein O-ring seals can be used for the motional feedthroughs. Windt and Catura [21] have designed a UHV reflectometer with movable internal stages to measure the uniformity of multilayer coatings on large optics for use in space instruments. Internal stepping motors are used, obviating the need for mechanical motions through the vacuum wall. A diagram of their reflectometer is shown in Fig. 8. Subsequently Windt and Waskiewicz [22] have described a UHV reflectometer capable of a number of motions that allow the investigation of the uniformity of surface coatings. Their instrument, built by Thermionics Northwest, is shown in Figs. 9a and b. Figure 9a is a

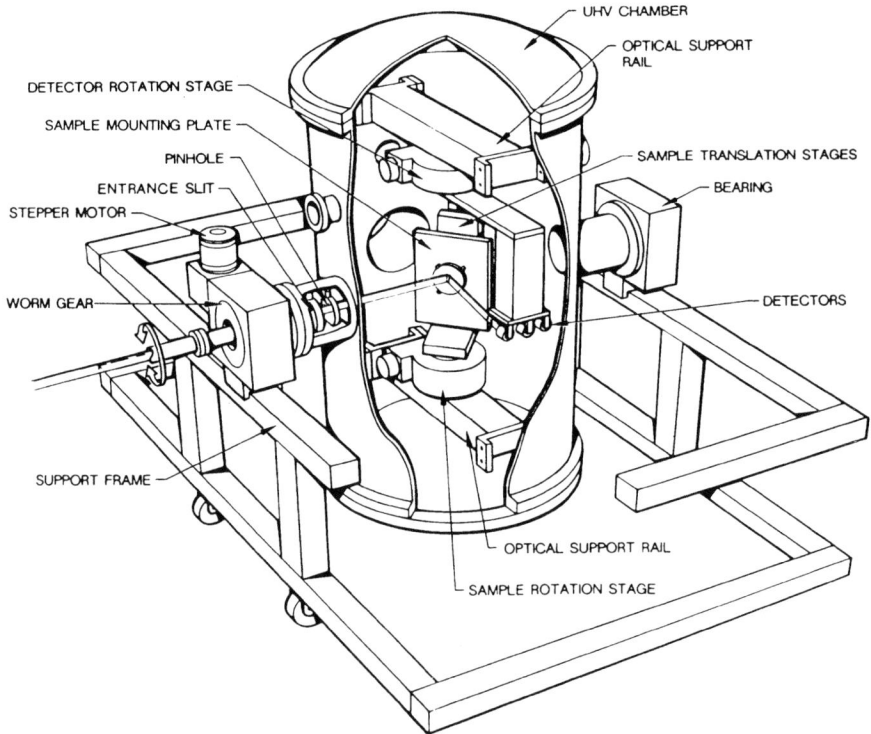

FIG. 8. A diagram of the reflectometer reported by Windt and Catura [21]. The sample rotation stage can also be translated to survey the sample surface. The detector rotation stage can accommodate a number of detectors for various dynamic ranges. The entire reflectometer can be rotated about its optical axis for evaluating polarization. (From [21] with permission from SPIE.)

SPECULAR REFLECTANCE MEASUREMENTS 199

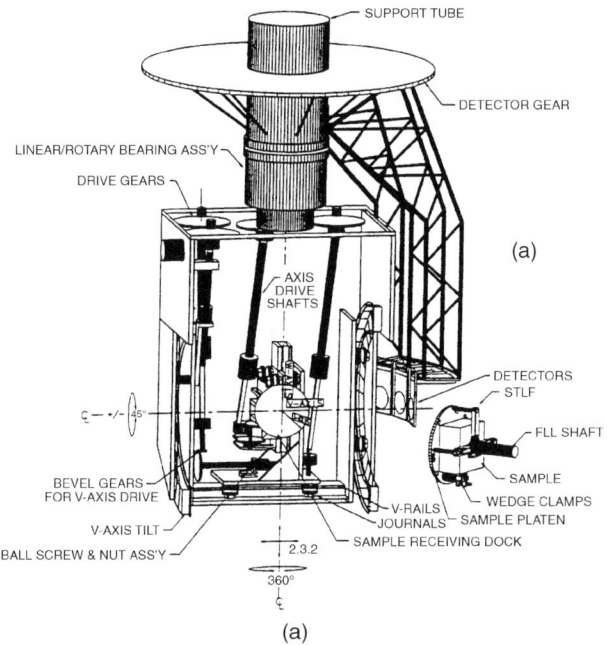

(a)

(b)

FIG. 9. (a) The reflectometer mechanism of Windt and Waskiewicz [22]. The entire assembly is mounted from the 24-in. diameter top flange of a vacuum chamber. The detector and sample assemblies rotate independently. A number of detectors can be mounted. The sample stage can move in x, y, and z directions as well as tilt and rotate so that the entire surface of a sample can be measured. (b) Photograph of the mechanism of part a. (courtesy of Thermionics Northwest from [22].)

schematic diagram of the reflectometer mechanism. The detector gear drives a framework that can carry a number of detectors. The sample has x, y, and z motions as well as tilt and rotation. Samples can be inserted into the reflectometer through a special airlock that can be pumped independently of the main vacuum vessel. Figure 9b is a photograph of the mechanism.

10.3 Diffuse Reflectance

Diffuse reflectance values are obtained from photometric measurements of radiation scattered by a surface in a nonspecular direction. In the visible spectral region, one can collect all the radiation scattered into the hemisphere, via an integrating sphere, at the detector to obtain total integrated scatter (TIS). Such an arrangement is possible because the coating in the integrating sphere has a diffuse reflectance close to unity over most of the visible region. In the VUV spectral region coating reflectances are much smaller so that the integrating sphere is not a useful approach. One must measure differential scatter at a number of scattering angles and integrate over the hemisphere into which the radiation is scattered. In making scattering measurements it is important to know the distance from the scattering surface to the detector, and to know the dimensions (area) of the detector sensitive surface so the solid angle acceptance can be calculated, values not so important for purely specular reflectance measurements.

If the roughness of the surface is isotropic, the scattering should also be isotropic, and integrating the measured differential scatter in the plane of incidence should give the TIS. Most of the reflectometers described in this chapter can be used for scattering measurements if the detector can be moved independently of the sample. An anisotropic scattering surface requires that the sample be rotated about its normal or that provision be made to move the detector out of the plane of incidence.

For polished surfaces the scattered intensity may decrease very rapidly away from the specular direction, consequently a detector that can measure specular reflection may not be sensitive enough to measure scattered radiation. Conversely, a detector sensitive enough for scattering usually cannot handle the large intensities associated with the specular and direct beams. Thus two detectors may be required. If so they must be calibrated against each other at an intensity compatible with both.

The use of very sensitive detectors, for example, photon counters, can cause additional problems for the unwary. If the detector is sensitive enough to count a few photons per second, a channeltron for example, it may detect photons, singly or multiply scattered, from any scattering surface in the reflectometer. Figure 10 shows the scattering signal in one of the author's reflectometers [2]. A

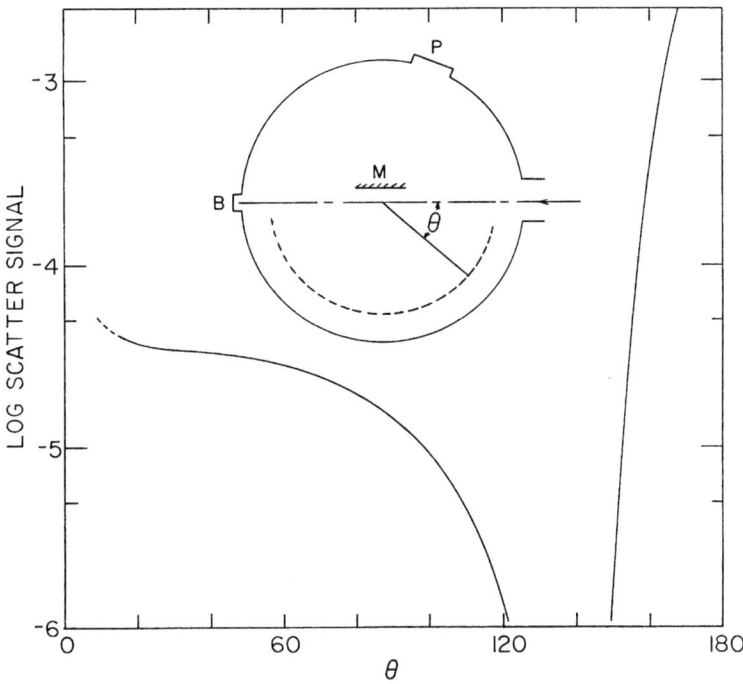

FIG. 10. Single and multiple scattering in a reflectometer. The radiation (584 Å) enters from the right and impinges on a brass plug. The mirror, M, is retracted, and θ is the angle between the radiation beam and the detector. The dashed line shows the detector path, which has a radius of about 3.5 in. The curve shows the signal (normalized count rate) as a function of angle. For small angles the detector sees radiation scattered directly from the brass plug. At larger angles the brass plug can no longer be seen and the signal is that scattered from the chamber wall after scattering from the brass plug and the slit jaws. At large angles, where the scatter signal is rising rapidly, the detector is beginning to see scattering from the slit jaws.

schematic diagram of the reflectometer is shown for reference. The radiation, 584 Å from a helium glow discharge, enters from the right and impinges on a brass plug, B, that plugs a port used for alignment. The end of B does not intrude as far as the reflectometer wall; it is recessed about 2 cm into the port and, in addition, has a slanted surface not normal to the incident beam. For the measurement shown, mirror, M is retracted from the beam. At an angle near 0 deg the detector is looking toward B and detects radiation scattered by B. As the detector angle is increased, the scattering signal drops as B becomes occulted, and reaches a very small value at an angle of about 130 deg where the detector is looking directly at the unilluminated mirror, which subtends a large solid angle. Also at this angle the scattering signal is secondary scattering of radiation originally scattered by B and the slit jaws that is occulted by M. At an angle of about 150 deg the detector begins to see scattering from the slit jaws, which increases very rapidly as the detector moves toward the radiation beam.

Adequate trapping of the specularly reflected radiation beam is a requirement, otherwise the scattered intensity versus scattering angle will have a misleading asymmetry. The author has experimented with different types of traps and found a polished wedge to be very good. Figure 11 shows the measured counts per second from a stainless steel wedge of an included 30 deg angle. The intensity of the beam illuminating the wedge, again 584 Å, was approximately 6×10^5 counts/s. The abscissa shows the positions of the beam on the wedge; for positions 1–5 and 7–11 the beam fell outside of the wedge on finely machined stainless steel. For position 6 the beam was entirely within the wedge. The wedge surfaces themselves were polished.

If the scattering surface is to be investigated at different angles of incidence, the wedge trap must be moved to intercept the specularly reflected beam. Rather than design a mechanism to enable the wedge to follow the reflected beam, the author designed a continuous wedge in the form of two aluminum rings that just fit inside the reflectometer. One edge of each ring was bevelled, then buffed to a high polish. Inside the reflectometer the bevelled surfaces of the rings were put

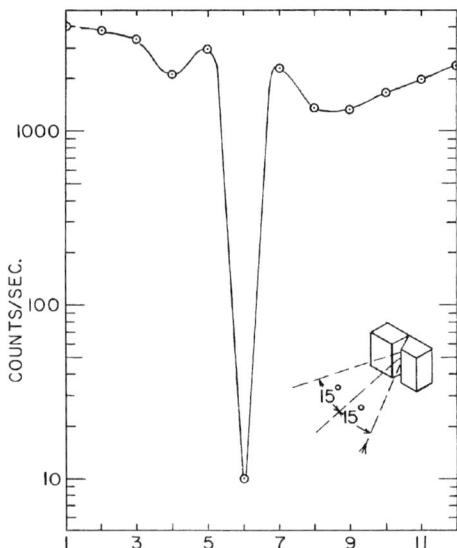

FIG. 11. Measured attenuation of a wedge radiation trap with an included 30 deg angle. The positions from 1–5 and 7–12 show the scattered signal from the flat portion of the finely machined stainless steel wedge. Postion 6 is the signal when the radiation falls entirely within the wedge. The wedge jaws were buffed to a high polish.

in contact, thus forming a continuous wedge around the reflectometer except for diametrically opposed holes; one for the entrance of the beam and the other for alignment.

Although wedges reduce the intensity of the radiation beam by a large factor, they undoubtedly increase the number of photoelectrons in the reflectometer. Channeltrons are good detectors of photoelectrons and require adequate electric shielding to prevent their detection.

Johnston et al. [23] have used the reflectometer of Madden and Canfield [3] equipped with a Bendix M306 magnetic multiplier used in the current measuring mode to measure scattering from substrates and evaporated gold and aluminum films. Mattson [24] and Jark and Stöhr [25] have described instruments for measuring scattering.

10.4 Reflectometry Facilities

In closing it should be mentioned that the recent development of multilayer reflecting coatings for use in synchrotron radiation research, space research, and soft x-ray lithography has prompted the establishment of at least three reflectometry facilities. One is at the National Institute of Standards and Technology (NIST) [26, 27], another at the University of Padova in Italy [28], and one at the Advanced Light Source [29] at the Lawrence Berkeley Laboratory.

References

1. L. J. LeBlanc, J. S. Farrell, and D. W. Juenker, *J. Opt. Soc. Am.* **54**, 956–957 (1964).
2. W. R. Hunter, *Appl. Opt.* **6**, 2140–2150 (1967).
3. R. P. Madden and L. R. Canfield, *J. Opt. Soc. Am.* **51**, 838–845 (1961).
4. J. C. Draper and E. Murphy, *J. Phys. E.* **3**, 633–635 (1970).
5. R. C. Vehse, J. C. Sutherland, and E. T. Arakawa, *Rev. Sci. Instrum.* **39**, 268–269 (1968).
6. L. C. Emerson, J. T. Cox, G. L. Ostrom, L. R. Painter, and G. H. Cunningham, *Rev. Sci. Instrum.* **47**, 1065–1068 (1976).
7. W. R. Hunter and R. K. Chaimson, *Appl. Opt.* **13**, 2913–2918 (1974).
8. W. R. Hunter, *J. Opt. Soc. Am.* **55**, 1197–1204 (1965).
9. M. Krumrey, M. Kühne, P. Müller, and F. Scholze, *Proc. SPIE* **1547**, 136–143 (1991).
10. W. R. Hunter and J. P. Long, *Appl. Opt.* **33**, 1264–1269 (1994).
11. W. R. Hunter and J. C. Rife, *Appl. Opt.* **23**, 293–299 (1984).
12. W. R. Hunter and J. C. Rife, *Nucl. Instrum. Methods* **A246**, 465–468 (1986).
13. J. D. Purcell, *J. Opt. Soc. Am.* **43**, 1166–1169 (1953).
14. D. Socker, Naval Research Laboratory, Washington, DC, private communication.
15. F. I. Vilesov, A. A. Azgrubskii, and M. M. Kirillova, *Opt. Spectrosc.* **23**, 79–82 (1967).

16. D. J. Michels, T. L. Mikes, and W. R. Hunter, *Appl. Opt.* **13**, 1223–1229 (1974).
17. W. R. Hunter and D. K. Prinz, *Appl. Opt.* **16**, 3171–3175 (1977).
18. A. Smith, *J. Opt. Soc. am.* **50**, 862–864 (1960).
19. B. Feuerbacher, M. Skilbowski, and R. P. Godwin, *Rev. Sci. Instrum.* **40**, 305–306 (1969).
20. H. Hogrefe, D. Giesenberg, R.-P. Haelbich, and C. Kunz, *Nucl. Instrum. Methods* **208**, 415–418 (1983).
21. D. L. Windt and R. C. Catura, *Proc. SPIE* **984**, 82–88 (1988).
22. D. L. Windt and W. K. Wakiewicz, *Proc. SPIE* **1547**, 144–158 (1991).
23. R. G. Johnston, L. R. Canfield, and R. P. Madden, *Appl. Opt.* **6**, 719–722 (1967).
24. L. Mattsson, *Proc. SPIE* **525**, 189–196 (1985).
25. W. Jark and J. Stöhr, *Nucl. Instrum. Methods* **A266**, 654–658 (1988).
26. C. Tarrio, R. N. Watts, T. B. Lucartorto, M. Haass, T. A. Calcott, and J. Jia, *J. X-Ray Sci. Technol.* **4**, 96–101 (1994).
27. C. Tarrio, R. N. Watts, T. B. Lucartorto, M. Haass, T. A. Calcott, and J. Jia, *Proc. SPIE* **2011**, 534–539 (1994).
28. P. Villoresi, G. Naletto, P. Nicolosi, G. Tondello, and E. Jannitti, *Proc. SPIE* **1742**, 314–323 (1992).
29. J. H. Underwood, E. M. Gullikson, M. Koike, P. J. Batson, P. E. Denham, K. D. Franck, R. E. Tackaberry, and W. R. Steele, *Rev. Sci. Instrum.* **67**, 1–5 (1996).

11. REFLECTANCE SPECTRA OF SINGLE MATERIALS

W. R. Hunter
SFA Inc.
Largo, Maryland

11.1 Introduction

Rather than show published reflectance spectra of single materials, usually measured close to normal incidence and over limited spectral ranges, spectra calculated from published optical constants [1] and from other sources are included in this chapter. The advantage of calculated spectra is that they can be shown over a wide wavelength range and calculated at any angle of incidence. The disadvantage is that if the optical constants are not correct, neither are the spectra. The optical constants in [1], carefully chosen by the various contributors to those volumes, are certainly accurate enough to show correct spectral features and to enable calculation of fairly accurate reflectance values. Rather than show curves for all the materials in [1], only materials that have proven useful as reflecting coatings or components of multilayers (ML) have been selected.

The spectra shown here are calculated for angles of incidence of 0 and 89 deg, corresponding to normal and grazing incidence, respectively. The wavelength scale is logarithmic and the reflectance scale is linear in order to give a clearer picture of the behavior of the reflectance at the short wavelengths. Curves for a 35-deg angle of incidence, the nominal angle of incidence for the commonly used Seya–Namioka monochromator, are not shown because they are usually similar in contour and value to the normal incidence values. If curves for other angles of incidence are required, it is a simple matter to set up a computer program and use the n, k values from [1] to do the calculation.

Locations of binding energies are shown where there appears to be a corresponding feature in the reflectance curves. These energies are taken from Siegbahn *et al.* [2].

Most of these materials can be sputtered, however, the advantage of sputtering over thermal deposition, or vice versa, is not generally known. Some metals, for example, Pt, appear to have the same reflectance values whether sputtered or thermally evaporated, but the same may not be true of Al or ZnS. Conventional deposition methods used for making single-layer coatings are mentioned and the behavior of the reflectance with time and environment is briefly discussed if known. For some materials, such as aluminum, it is severe, for others nonexistent. A few comments are also made on the effect of a space environment because many of these materials are used in rocket and satellite spectrographs.

Short-term exposures (about 1 week) during shuttle missions have been discussed by Gull et al. [3] and long-term exposures (6 years) on the LDEF experiment have been discussed by Herzig et al. [4].

11.2 Aluminum

Figure 1 shows the calculated reflectance spectrum of aluminum (solid lines). Aluminum is a quasi-free-electron metal that has its plasma oscillation at about 840 Å. To wavelengths longer than 840 Å, the normal incidence reflectance values are large, about 90% or more. At wavelengths less than 1000 Å the reflectance drops sharply and is effectively zero at 500 Å and less. At grazing incidence the reflectance values are quite large, 98 to 99% and remain so to wavelengths as short as about 500 Å. At 840 Å, where the normal incidence reflectance values are rapidly decreasing, the grazing incidence values appear to reach their maximum value. Shortward of 500 Å, the reflectance values gradually decrease until at the $L_{2,3}$ soft x-ray edge they drop rapidly to about 50%. To even shorterwavelengths the reflectance values again increase and remain

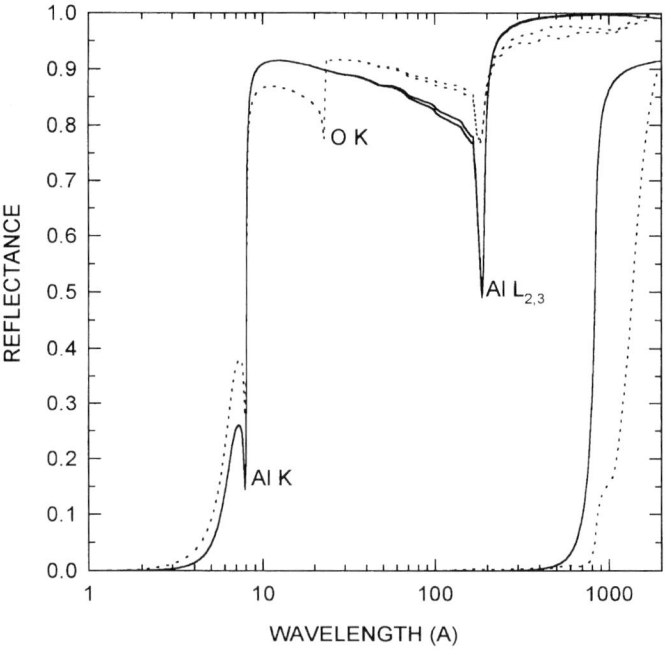

FIG. 1. Reflectance spectra of Al and oxidized Al at 0 and 89 deg.

large until the wavelength reaches the vicinity of the K edge, 8 Å, where the reflectance drops rapidly.

The values shown by the solid lines in Fig. 1 can never be achieved in earth-bound laboratories because the aluminum oxidizes so readily and the presence of the oxide layer drastically modifies the reflectance spectrum of a pristine aluminum surface. Figure 1 also shows the calculated reflectance spectrum of oxidized aluminum (dashed lines) where the oxide thickness is that of the natural oxide, nominally 30 Å. The most dramatic change is in the normal incidence reflectance, which drops to rather small values between 2000 and 1000 Å. For example, the normal incidence reflectance of unoxidized aluminum at 1000 Å is approximately 87%, but the formation of 30 Å of oxide reduces that value to about 15%. At an 89-deg angle of incidence the reflectance values are depressed slightly from about 1800 Å to about 200 Å, elevated slightly from the $L_{2,3}$ edge to about 23 Å, and again somewhat depressed from 23 Å to the Al K edge at about 8 Å. Generally the effect of oxidation on the grazing incidence reflectance is small.

The reflectance of Al films is quite sensitive to evaporation conditions. Fast evaporations (1–2 s to visual opacity) from filaments at pressures of 10^{-5} to 10^{-6} are required for highest reflectance [5]. At lower pressures slower evaporations can achieve the same reflectance values [6].

Herzig *et al.* [4] found that the reflectance of naturally oxidized Al in space decreased at wavelengths shorter than 2500 Å, possibly because of additional oxidation engendered by oxygen atoms at orbital altitude.

11.3 Silicon

Figure 2 shows the reflectance spectrum of silicon (solid lines). Silicon is another quasi-free-electron material with a plasma oscillation at about 760 Å. Its normal incidence reflectance spectrum is quite similar to that of aluminum, although in the spectral region of large normal incidence reflectance values, these values reach only about 70%. At grazing incidence the reflectance values are large, and reach a maximum at about the plasma oscillation. To shorter wavelengths they decrease slowly until in the region of the $L_{2,3}$ edge they drop rapidly to a value between 30 and 40%. At wavelengths shorter than the $L_{2,3}$ edge the grazing incidence reflectance values are again large and remain so until the K edge is reached at 6.7 Å.

At wavelengths longer than the plasma oscillation, the grazing incidence reflectance of silicon begins to show a separation of the polarization components. The perpendicular component remains large but the parallel component decreases.

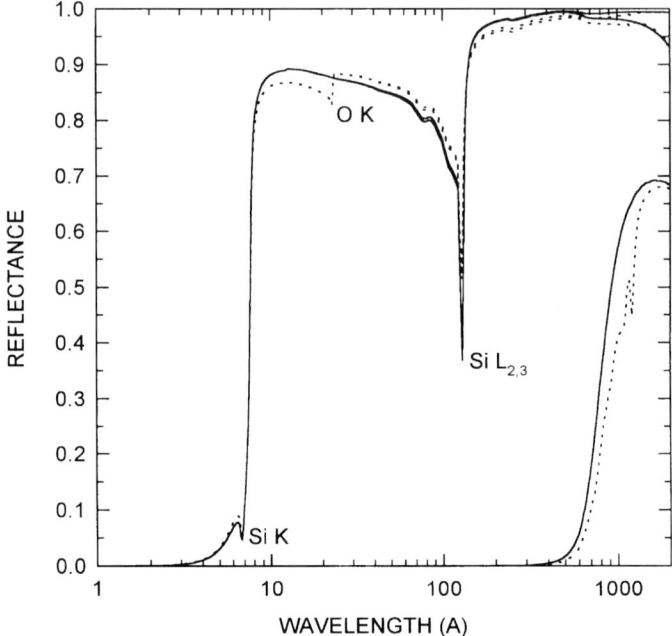

FIG. 2. Reflectance spectra of Si and oxidized Si at 0 and 89 deg.

As with aluminum, an oxide forms on silicon when it is exposed to air. However, the rate at which it forms appears to be much slower, and an equilibrium thickness may be approximately 10 Å. Figure 2 also shows how an oxide film 30 Å thick modifies the reflectance spectrum of silicon (dashed lines). The behavior of oxidized Si is quite similar to that of Al. The slight reflectance peak at about 1200 Å is characteristic of SiO_2 (see Fig. 17 later in this chapter).

Evaporation by electron-beam gun is the conventional method of deposition. Thermal evaporation from tungsten does not work well because molten Si disolves the tungsten. Si is most useful as a ML component.

11.4 Beryllium

Beryllium is used primarily as a component in multilayers. The normal incidence reflectance shown in Fig. 3 is for unoxidized Be because optical constants of BeO for wavelengths longer than 240 Å are not available. As with Al and Si, Be is a quasi-free-electron material with a plasma oscillation at about 640 Å. Consequently, the normal incidence reflectance values that are between 60

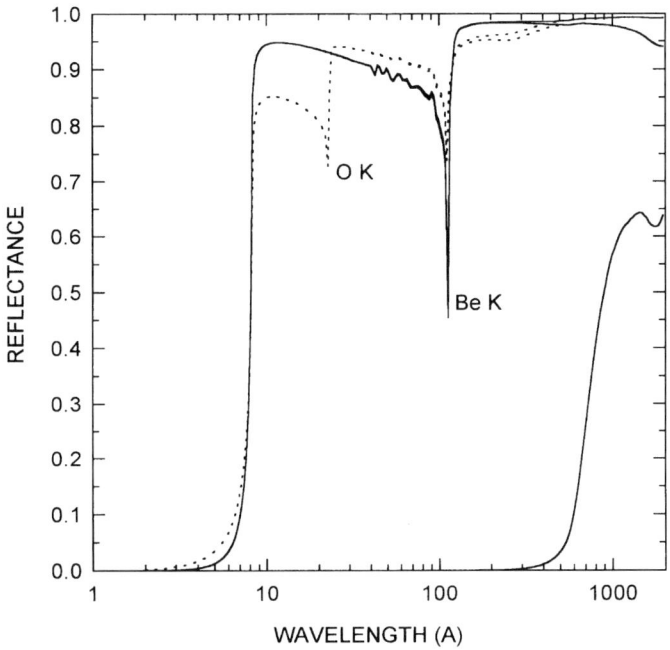

FIG. 3. Reflectance spectra of Be and oxidized Be at 0 and 89 deg.

and 65% from 2000 and 1000 Å drop rapidly to small values when the wavelength is less than 640 Å. Probably any oxide growth will lower these values. At grazing incidence the solid lines show the reflectance of unoxidized Be. The values are large and there is a sharp minima at the Be K edge. The dashed lines show the grazing incidence reflectance of Be with 30 Å of oxide. Addition of the oxide introduces another sharp minimum at the O K edge. An oxide thickness of 30 Å was chosen because Skulina *et al.* [7] found 30 Å of BeO on a sputtered Be film 5000 Å thick. They did not discuss the stability of the oxide layer.

Beryllium can be evaporated by an electron-beam gun or it can be sputtered. Thermal evaporation from tungsten filaments or boats does not work well because molten beryllium dissolves tungsten.

11.5 Gold

Figure 4 shows the reflectance spectra of gold at 0 and 89 deg. Gold has appreciable normal incidence reflectance values to wavelengths as short as 400 Å. A number of features seen in the spectrum are caused by interband transitions. The grazing incidence spectrum shows the separation of polarization

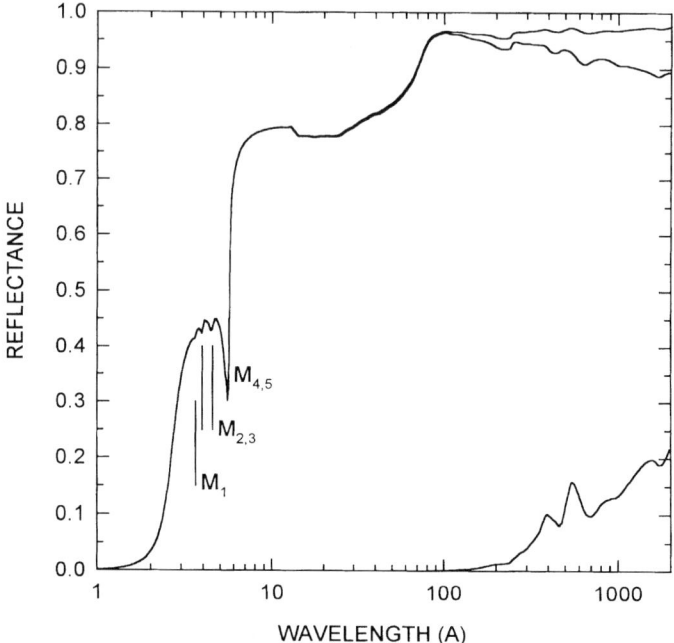

FIG. 4. Reflectance spectra of Au at 0 and 89 deg.

components to wavelengths as short as a few hundred angstroms. There is a maximum value at about 100 Å, and then the reflectance begins to decrease to shorter wavelengths. The sudden decrease between 5 and 6 Å is due to the $M_{4,5}$ edge of gold. The very small oscillations at shorter wavelengths are due to other M edges of gold.

The reflectance of gold is not very sensitive to evaporation conditions [8]. Evaporation from boats works quite well but tends to throw microscopically small globules of gold onto the substrate. Bach [9] evaporates gold from filaments to avoid globules. Gold shows no appreciable aging on exposure to clean air. Herzig et al. [4] reported a loss in reflectance of Au which they attributed to sputtering and roughening by oxygen atoms in orbit.

11.6 Silver

Figure 5 shows the reflectance spectra of silver at 0 and 89 deg. The normal incidence reflectance (NIR) is less than 10% for wavelengths as long as 1500 Å, which is considerably less than some of the platinum metals. At grazing incidence the reflectance values are large down to 100 Å and, to shorter

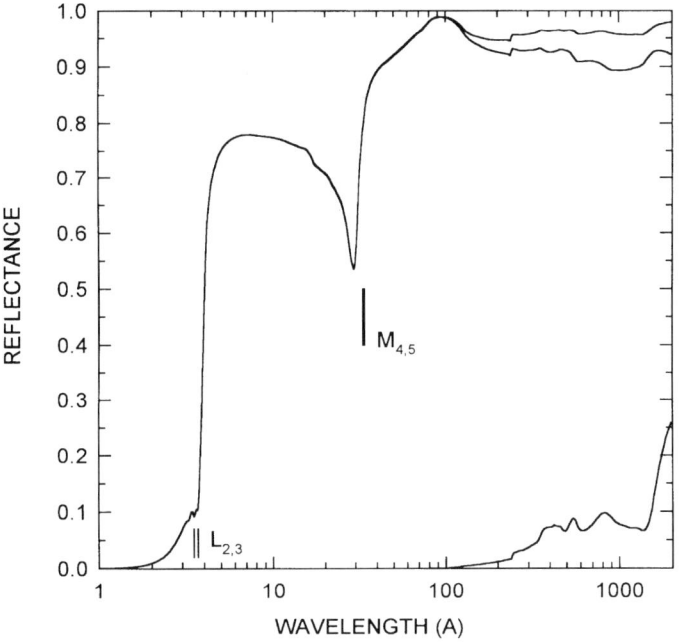

FIG. 5. Reflectance spectra of Ag at 0 and 89 deg.

wavelengths, decrease to a minimum of about 60% at the $M_{4,5}$ edge. To even shorter wavelengths the reflectance increases slowly to about 80% then drops suddenly at the $L_{2,3}$ edge. Compared to other materials, silver is not a very useful reflecting coating at normal incidence. Silver has very little loss in reflectance in the 1000- to 2000-Å spectral region on exposure to clean air [10]. However, silver tarnishes, that is, silver sulfide forms on its surface if it is exposed to an atmosphere containing traces of sulfur compounds for any length of time.

Silver is usually evaporated out of tungsten boats onto room temperature substrates.

11.7 Copper

Copper also has fairly small NIR values, as shown in Fig. 6. At grazing incidence, however, it has a fairly high reflectance, above 90% from 2000 Å to about 20 Å where there is a sharp drop due to the $L_{2,3}$ edge. Copper should be a useful reflecting coating for the grazing incidence spectral region and has been used in synchrotron radiation instrumentation by Rehn and Jones [11].

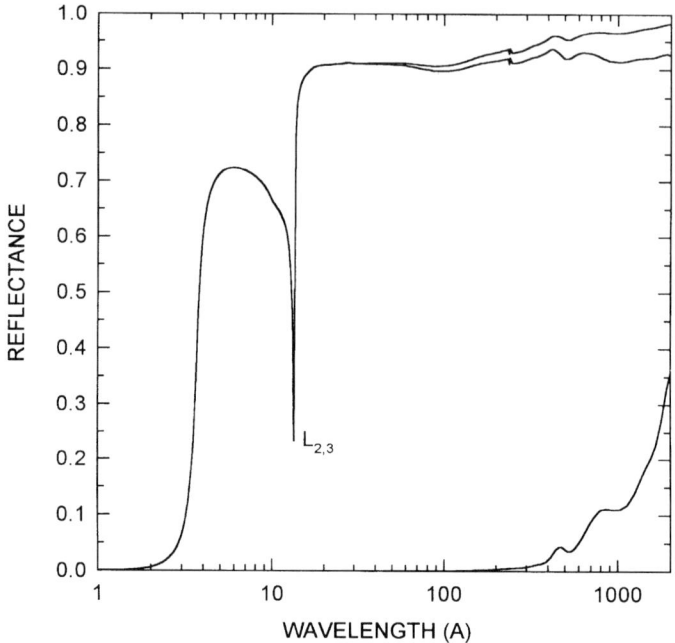

FIG. 6. Reflectance spectra of Cu at 0 and 89 deg.

Copper shows a loss in reflectance in the 1000- to 2000-Å spectral region because of oxidation. Moreover, this loss is continuous because the oxide does not form a stable barrier to diffusion of oxygen to the metal [10].

Copper is evaporated out of tungsten boats onto room temperature substrates.

11.8 Platinum

Platinum is a useful coating for normal incidence to wavelengths as short as 300 Å, and has been used extensively in space research as well as in the laboratory. At grazing incidence it has a high reflectance to wavelengths as short as 100 Å. To shorter wavelengths, the reflectance drops slowly but remains above 80% to wavelengths as short as 8 Å. The $M_{4,5}$ edge occurs at about 5.8 Å. The reflectance spectra of platinum at 0- and 89-deg angles of incidence are shown in Fig. 7.

Under normal conditions in the laboratory, platinum shows no loss in reflectance on exposure to air. Gull et al. [3] have flown platinum samples, and samples of other members of the platinum group, in the space shuttle and found

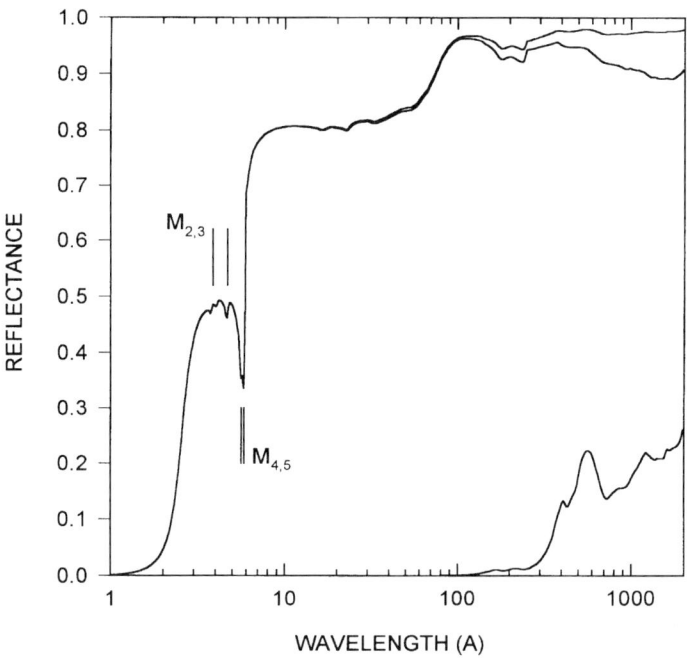

FIG. 7. Reflectance spectra of Pt at 0 and 89 deg.

that platinum appeared not to be much affected by the atomic oxygen encountered at that altitude. Herzig *et al.* [4] found essentially no change in the reflectance of Pt samples recovered from LDEF.

Vacuum deposition of platinum can be done by thermal evaporation, using an electron-beam gun. Best adherence and reflectance are obtained when it is deposited onto substrates heated to between 300 and 400°C. For normal incidence use the coatings on glass substrates should not be opaquely reflecting for highest reflectance. Calculations show that a thickness of 400 Å is generally optimum [12]. Sputtering also produces good platinum films but the substrate temperature required for good adhesion is not known. Electrolytic and electroless deposition can also be used to produce smooth films about 1000 Å thick. How the reflectances of such films compare with those produced *in vacuo* is not known.

11.9 Iridium

The normal incidence reflectance values of iridium are shown in Fig. 8. They are larger than those of platinum. At grazing incidence its reflectance spectrum closely resembles that of platinum.

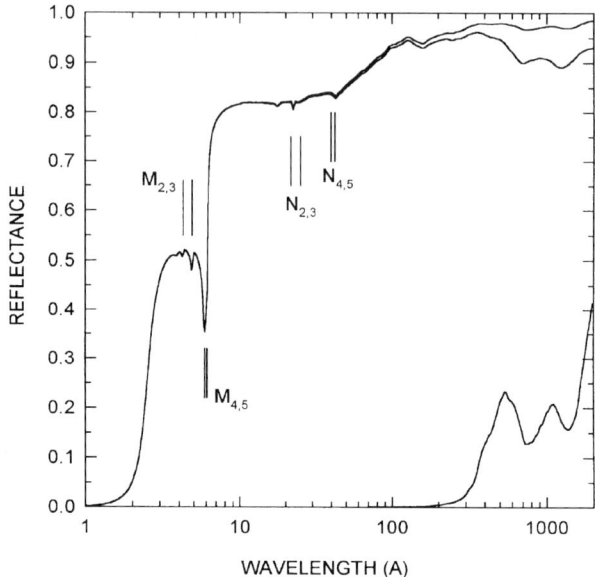

FIG. 8. Reflectance spectra of Ir at 0 and 89 deg.

Iridium shows no signs of aging when exposed to normal laboratory air, but it is affected by the space environment encountered in shuttle flights. Gull *et al.* [3] included samples of iridium on the shuttle flight mentioned above and found that it is attacked by atomic oxygen, resulting in loss in reflectance and crazing of the coating. Herzig *et al.* [4] found a small loss in reflectance.

An electron-beam gun is required for thermal evaporation of iridium. As with platinum, best adherence and reflectance are obtained with substrates heated to between 300 and 400°C. On glass substrates, iridium films show highest reflectance when they are not opaque [13].

11.10 Osmium

Figure 9 shows the reflectance spectra of osmium. At normal incidence osmium has some of the highest reflectance values of any single material, approximately 30% at about 500 Å. At grazing incidence its reflectance spectrum resembles that of Pt.

Osmium usually shows very little loss on exposure to air in the laboratory; however, there have been reports of catastrophic degradation of some osmium coatings during storage in the laboratory. Gull *et al.* [3] found that osmium coatings sometimes completely disappear when exposed to the environment that

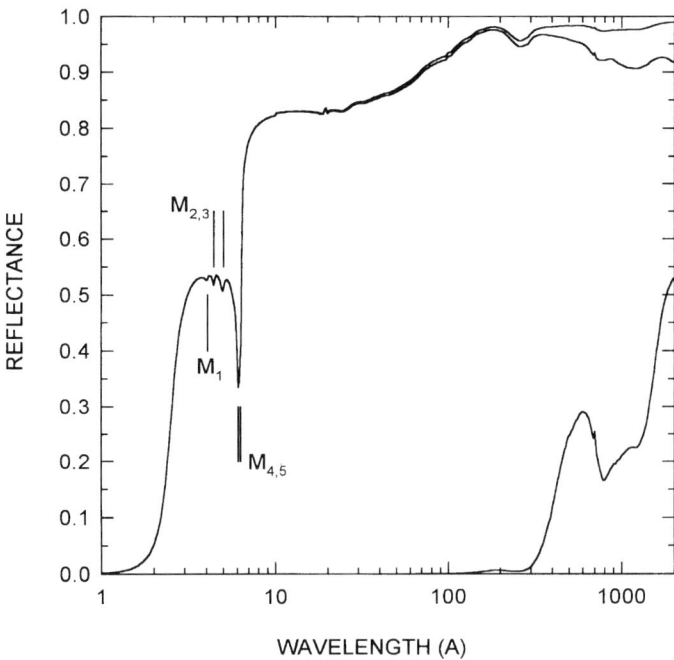

FIG. 9. Reflectance spectra of Os at 0 and 89 deg.

exists during shuttle flights. Herzig et al. [4] also found that the Os coating had completely disappeared; the reflectance spectrum of the "osmium" sample after the flight was that of Cr, the binding layer between the Os coating and the substrate.

Osmium must be evaporated using an electron-beam gun. Best adherence is obtained with a substrate heated to between 300 and 400°C, but the reflectance appears to be independent of substrate temperature. As with platinum, osmium films on glass substrates have highest reflectance when not opaque [14]. No information is available on sputtered or chemically deposited osmium.

11.11 Rhodium

Rhodium is extremely resistant to chemical and mechanical attack. Because of this resistance it has been used as a coating for searchlight mirrors and in other applications where the coating is subject to hostile conditions. For large metal mirrors electrodeposition is used to make the coating.

The normal incidence reflectance values of rhodium are not very large and are characterized mainly by the reflectance minimum at about 1216 Å. At grazing

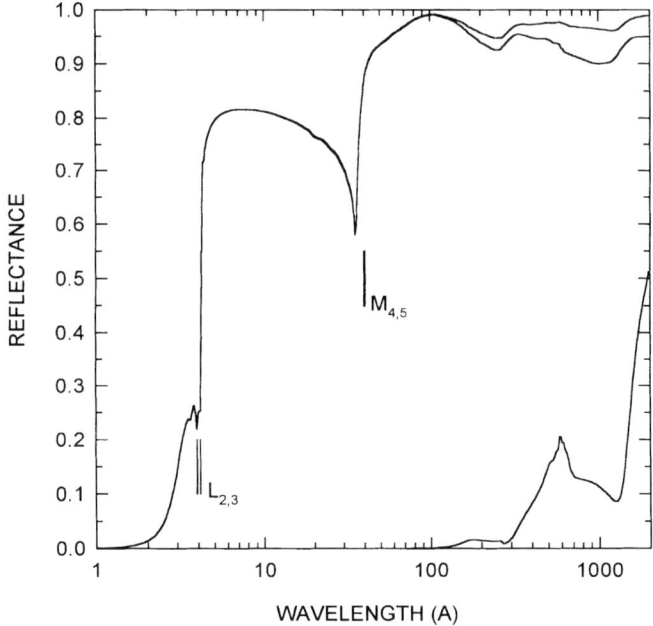

FIG. 10. Reflectance spectra of Rh at 0 and 89 deg.

incidence its reflectance values are large but there is a rapid drop in reflectance at about 40 Å due to the $M_{4,5}$ edge. The rapid drop at about 4 Å is due to the $L_{2,3}$ edge. Rhodium does not appear to be as useful as other members of the platinum group. The reflectance spectra of rhodium are shown in Fig. 10.

Rhodium shows no loss in reflectance on exposure to normal laboratory air and is fairly easy to evaporate using an electron-beam gun. Best adherence is obtained by using hot substrates, 300 to 400°C. On glass substrates, highest reflectance is obtained when the film is not opaque but the maximum values are not much larger than for opaque films except at the reflectance minimum at 1216 Å where the reflectance can be increase from 9% for opaque coatings to almost 17% for a coating about 100 Å thick [15]. No information is available on sputtered or electroless deposited rhodium.

11.12 Tungsten

Figure 11 shows the reflectance spectra of tungsten. With the exception of aluminum and silicon, tungsten has the largest reflectance values at normal incidence of any of the coating materials measured, reaching 35% at 600 Å. At

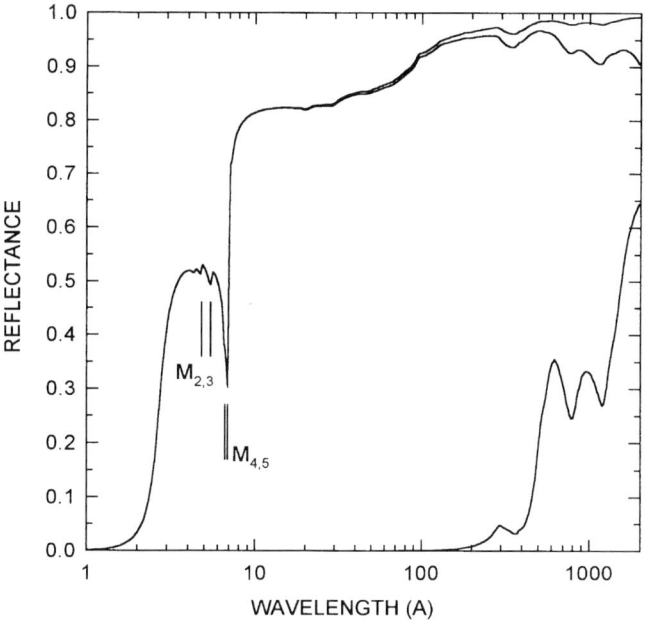

FIG. 11. Reflectance spectra of W at 0 and 89 deg.

grazing incidence its reflectance spectrum resembles that of platinum and has reflectance values in excess of 80% to wavelengths as short as 10 Å.

On initial exposure to dry air, tungsten coatings show a loss in reflectance of approximately 1 to 1.5% caused by the formation of an oxide film. It is estimated by Cox et al. [16] that the thickness of oxide is about 10 Å, and that it is a stable film. On exposure to air outside of a desiccator, the reflectance drops even more. Cox et al. attribute this additional increase to the presence of water vapor.

Tungsten requires an electron-beam gun for evaporation and has highest reflectance when deposited on a hot (300–400°C) substrate.

11.13 Nickel

Nickel is not a useful reflecting coating for normal incidence but is a good ML component. At grazing incidence, however, it has uniformly large reflectance values (>90%) to wavelengths as short as 15 Å where the reflectance begins to drop rapidly because of the $L_{2,3}$ edge, as shown in Fig. 12. In this respect, its grazing incidence reflectance spectrum resembles that of copper, as

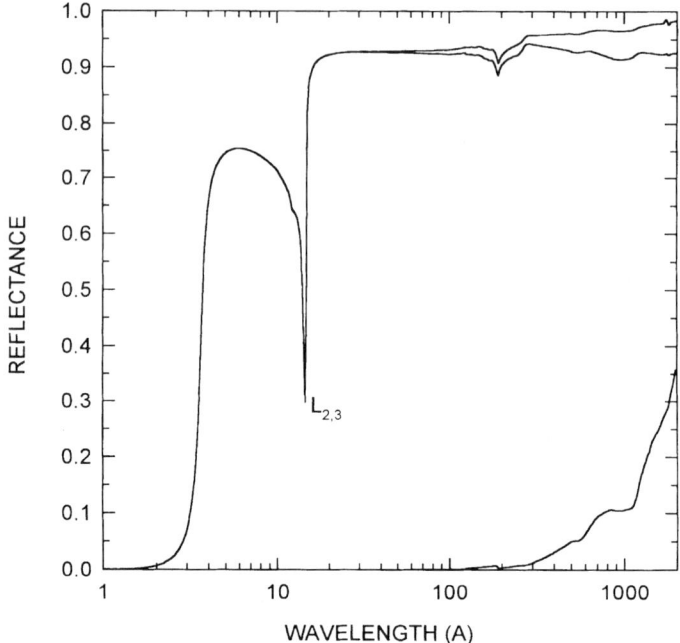

FIG. 12. Reflectance spectra of Ni at 0 and 89 deg.

can be seen in Fig. 6. There are no reports on the aging properties of nickel films.

Nickel is best evaporated using an electron-beam gun. There are no reports on the dependence of reflectance on evaporation conditions and substrate temperatures.

11.14 Chromium

Figure 13 shows the reflectance spectra of chromium. Between 240 and 310 Å optical constants are not currently available. Chromium appears not to be very useful for a normal incidence reflector but has fairly large grazing incidence reflectance values down to about 25 Å. Cr is usually evaporated from a tungsten boat, and is used to improve adherence of gold and other metals to glass substrates. Very strong stresses are set up in Cr films during deposition and if the deposition is carried close to visual opacity the stresses are great enough to tear the film from the substrate.

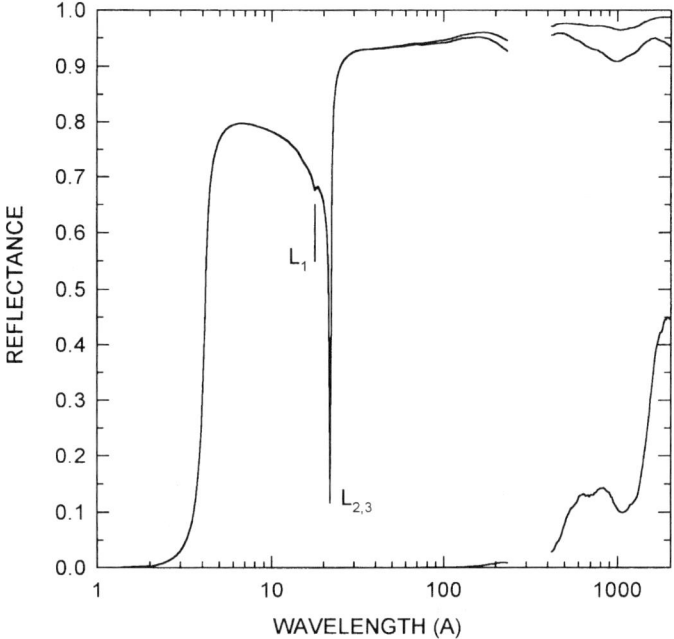

FIG. 13. Reflectance spectra of Cr at 0 and 89 deg.

11.15 Molybdenum

Figure 14 shows the reflectance spectra of Mo. At normal incidence there is a peak between 700 and 800 Å that could be useful for a narrowband reflector, otherwise NIR is best achieved using other metals. At grazing incidence the reflectance values are large from 2000 Å to about 50 or 60 Å where there is a drop due to the $M_{4,5}$ edge. Slightly above 5 Å the reflectance drops rapidly due to the $L_{2,3}$ edge.

Molybdenum is seldom used as a reflecting coating, it finds much greater use as a component for multilayer coatings. It does oxidize but optical constants for the oxide are not available over a very wide range of wavelengths.

11.16 Tantalum

The reflectance spectra of Ta are shown in Fig. 15. The break in the curves is due to a lack of optical constants. Tantalum has a useful NIR spectrum from 2000 to about 700 Å. At grazing incidence the reflectance is fairly constant from 2000 to about 200 Å, then begins to decrease until the $M_{4,5}$ edge is reached. To

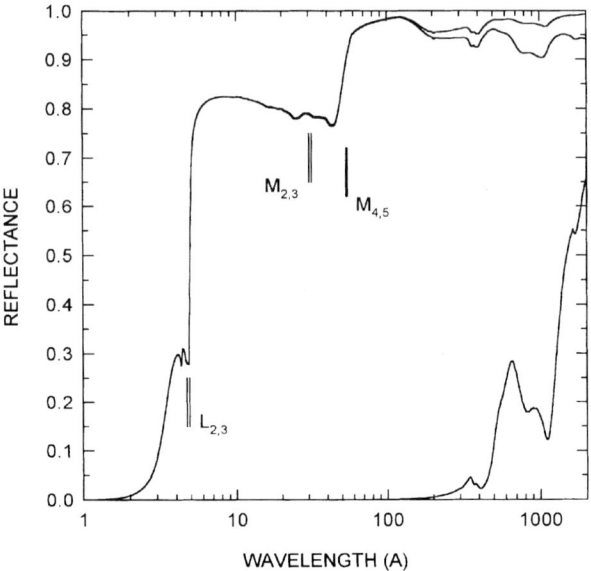

FIG. 14. Reflectance spectra of Mo at 0 and 89 deg.

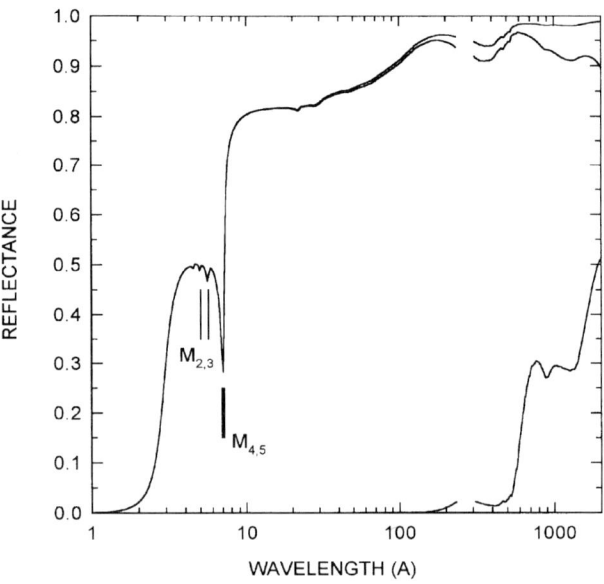

FIG. 15. Reflectance spectra of Ta at 0 and 89 deg.

shorter wavelengths there is a semi-plateau but the reflectance drops to very small values at wavelengths less than about 3 Å.

11.17 Zinc Sulfide

Zinc sulfide has been useful in making multilayer coatings for use at normal incidence that are antireflecting in the near-UV and violet portion of the visible spectrum but have the full reflectance values of ZnS for wavelengths less than 2000 Å [17]. Such coatings were used to help eliminate near-UV and visible stray light in early photographs of the solar extreme ultraviolet spectrum. The reflectance spectra of ZnS are shown in Fig. 16. There appears to be no loss in reflectance of ZnS coatings on exposure to air unless exposed to UV radiation at the same time, in which case the ZnS is converted to ZnO [18].

The evaporation conditions for ZnS have been described by Cox and Hass [19].

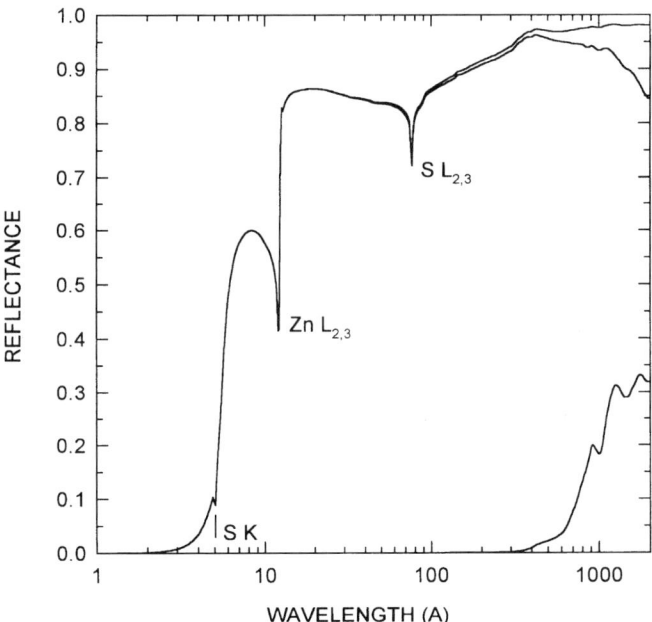

FIG. 16. Reflectance spectra of ZnS at 0 and 89 deg.

11.18 SiO$_2$

The reflectance spectra of SiO$_2$ are shown in Fig. 17. From about 700 to 1200 Å there are reflectance features associated with the structure of the material. These features are especially pronounced when measuring crystalline SiO$_2$, and less so for fused silica. The measured index of refraction at the largest peak is about 2.8. As a normal incidence reflector it is not useful because of its nonuniform reflectance. At grazing incidence the reflectance values are large down to about 25 Å. To shorter wavelengths there is a small plateau until the reflectance drops severely due to the Si K edge.

11.19 Al$_2$O$_3$

Aluminum oxide is not useful as a normal incidence reflector. At grazing incidence the reflectance values are fairly large down to about 8 Å where the reflectance drops rapidly due to the Al K edge. Between 2000 and 8 Å there are two absorption edges that will introduce small discontinuities in reflectance if the material is used as a mirror. Fig. 18 shows the reflectance spectra of aluminum oxide.

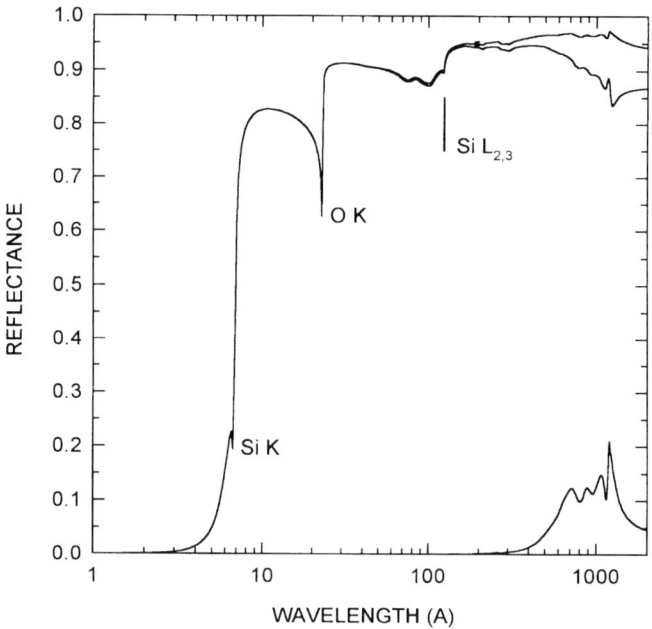

FIG. 17. Reflectance spectra of SiO$_2$ at 0 and 89 deg.

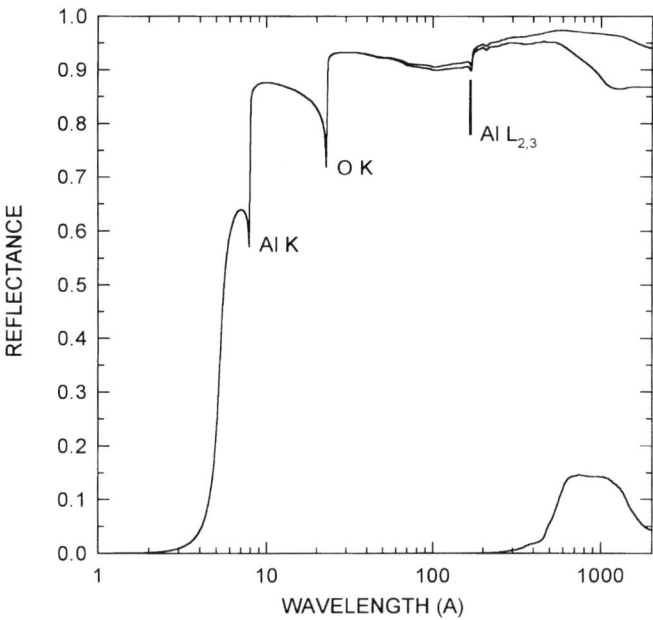

FIG. 18. Reflectance spectra of Al_2O_3 at 0 and 89 deg.

11.20 Carbon

Carbon is not widely used as a reflecting coating and, in fact, is usually considered a contaminant on other coatings. Carbon contamination is especially severe in instruments for use with synchrotron radiation, and extensive literature exists on the subject of removing carbon contamination from reflecting coatings (see Chapter 9, Vol. 32).

Figure 19 shows the reflectance spectra of carbon. At normal incidence it is not useful but at grazing incidence it has fairly large reflectance values to wavelengths as short as 50 Å, and from about 30 to 8 Å so the reflectance values are large enough to be useful. However, in the vicinity of the K edge, at 44 Å, the reflectance drops drastically. This drop appears on carbon-contaminated reflecting surfaces and makes measurements in the region around 44 Å rather difficult because of the loss of beam intensity. In the wavelength regions below and above the K edge, carbon coatings of the proper thickness can increase the reflectance; however, the first indication of carbon contamination is a loss in reflectance so carbon coatings are seldom allowed to thicken to the point where the reflectance is increased.

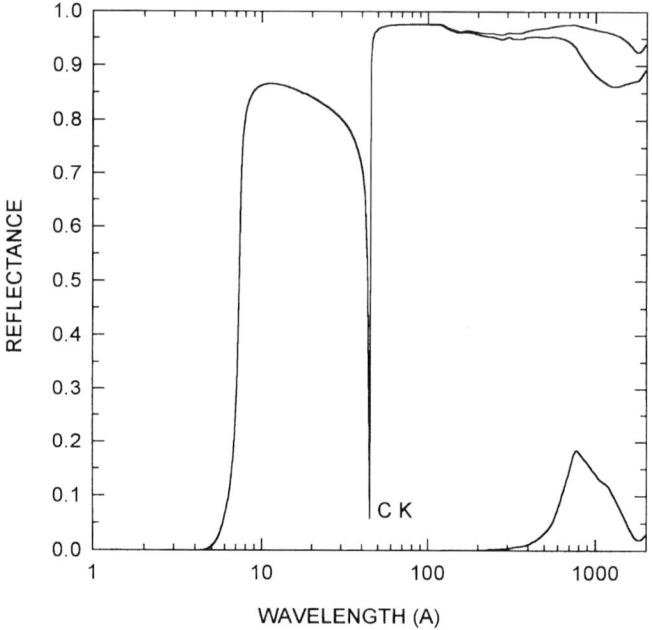

FIG. 19. Reflectance spectra of C at 0 and 89 deg.

Carbon can be evaporated using an electron-beam gun and is sometimes deposited from a carbon arc in vacuum. No studies have been done on the effect of evaporation conditions on the reflectance of carbon, nor on its aging.

11.21 Silicon Carbide

Although carbon by itself is not a welcome material in vacuum systems, SiC is an extremely useful reflecting material. Its reflection spectra are shown in Fig. 20. It has comparatively large reflectance values from about 600 to 2000 Å and, in grazing incidence, from 2000 to about 7 Å. The grazing incidence spectrum is not smooth because of the presence of the Si $L_{2,3}$ and C K edges.

SiC is considered a good candidate for both synchrotron radiation experiments and extreme ultraviolet spectrographs for use in space research because it is light, has good thermal conductance, and is almost indestructable mechanically. It is, however, attacked by orbital atomic oxygen [4], which reduces the NIR to less than 5% from about 600 to 1300 Å. Seely et al. [20] have exposed SiC to atomic oxygen in laboratory experiments and found similar losses in reflectance.

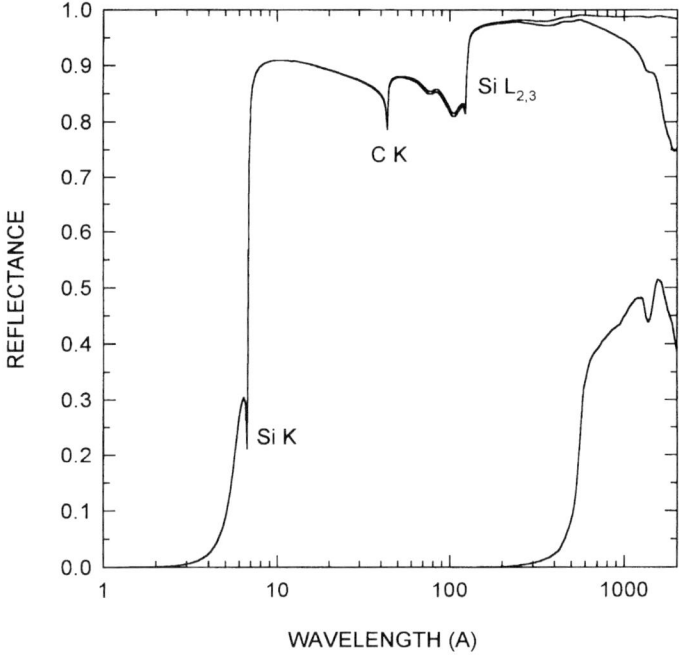

FIG. 20. Reflectance spectra of SiC at 0 and 89 deg.

References

1. E. D. Palik, ed., "Handbook of Optical Constants of Solids," Academic Press, New York (1985) and E. D. Palik, ed., "Handbook of Optical Constants of Solids, II," Academic Press, New York (1991).
2. K. Siegbahn, C. Nording, A. Fahlman, R. Nordberg, K. Hamrin, J. Hedman, G. Johansson, T. Bergmark, S.-E. Karlsson, I. Lindgren, and B. Lindberg, in "ESCA Atomic, Molecular, and Solid State Structure Studied by Means of Electron Spectroscopy," Almquist & Wiksell, Uppsala, Sweden (1967).
3. T. R. Gull, H. Herzig, J. F. Osantowski, and A. R. Toft, *Appl. Opt.* **24**, 2660–2665 (1985).
4. H. Herzig, A. R. Toft, and C. M. Fleetwood, Jr., *Appl. Opt.* **32**, 1796–1804 (1993).
5. G. Haas, W. R. Hunter, and R. Tousey, *J. Opt. Soc. Am.* **46**, 1009–1012 (1956).
6. E. T. Hutcheson, G. Hass, and J. K. Coulter, *Opt. Commun.* **3**, 213 (1971), and B. P. Feuerbacher and W. Steinmann, *Opt. Commun.* **1**, 81 (1969).
7. K. M. Skulina, C. S. Alford, R. M. Bionta, D. M. Makowiecki, E. M. Gullikson, R. Soufli, J. B. Kortright, and J. H. Underwood, *Appl. Opt.* **34**, 3727–3730 (1995).
8. L. R. Canfield, G. Hass, and W. R. Hunter, *J. Phys.* **25**, 124–129 (1964).
9. B. W. Bach, Hyperfine Inc., Boulder, CO, private communication.
10. L. R. Canfield and G. Hass, *J. Opt. Soc. Am.* **55**, 61–64 (1965).

11. V. Rehn and V. O. Jones, *Opt. Eng.* **17**, 504 (1978).
12. W. R. Hunter, D. W. Angel, and G. Hass, *J. Opt. Soc. Am.* **69**, 1695–1699 (1979).
13. G. Hass, G. F. Jacobus, and W. R. Hunter, *J. Opt. Soc. Am.* **57**, 758–762 (1967).
14. J. T. Cox, G. Hass, J. B. Ramsey, and W. R. Hunter, *J. Opt. Soc. Am.* **63**, 435–438 (1973).
15. J. T. Cox, G. Hass, and W. R. Hunter, *J. Opt. Soc. Am.* **61**, 360–364 (1971).
16. J. T. Cox, G. Hass, J. B. Ramsey, and W. R. Hunter, *J. Opt. Soc. Am.* **62**, 781–785 (1972).
17. G. Hass and R. Tousey, *J. Opt. Soc. Am.* **49**, 593 (1959).
18. G. Hass, J. B. Heaney, W. R. Hunter, and D. W. Angel, *Appl. Opt.* **19**, 2480–2481 (1980).
19. J. T. Cox and G. Hass, *J. Opt. Soc. Am.* **48**, 677 (1958).
20. J. F. Seely, G. E. Holland, W. R. Hunter, R. P. McCoy, K. F. Dymond, and M. Corson, *Appl. Opt.* **32**, 1805–1810 (1993).

12. POLARIZATION

W. R. Hunter
SFA Inc.
Largo, Maryland

12.1 Introduction

This chapter is concerned with devices that control the state of polarization of radiation in the vacuum ultraviolet (VUV) region. The state of polarization can be controlled by transmission or reflection, and depends on the optical constants of the polarizing material and the angle of incidence of the radiation. The optical constants are n, the index of refraction, and k, the extinction coefficient. They are the two components of the *complex index of refraction*, given by $N = n + ik$. Note that n is never zero, but for a nonabsorbing medium, $k = 0$. The angle of incidence is the angle between the direction of propagation of the incident radiation and the surface normal. These two directions also define the plane of incidence.

Within the family of transmission polarizers one has refraction and double refraction, for which $k = 0$ or is extremely small, and selective absorption (dichroism) for which $k \neq 0$. Transmission polarizers can produce either linear polarization, for example, a Rochon or Nicol prism, or, in the case of quarter-wave plates, elliptical or circular polarization. They are limited in their wavelength range by their fundamental absorption, which occurs at different wavelengths for the different transmitting crystals in the VUV (see Chapter 16). Polarization by dichroic action has been reported for the soft x-ray region but has not seen much use in the VUV so far.

Polarization produced by reflection invokes a tacit assumption that the surface is smooth; roughness is small compared with the wavelength because, as the surface roughens randomly, reflected radiation is depolarized. Using interference to control polarization on reflection is well understood but as of this writing not many attempts have been made in this direction in the VUV or soft x-ray regions. Generally polarization on reflection is elliptical; however, under certain conditions circular polarization can be obtained by reflection. An excellent treatment of polarization in general has been given by Bennett and Bennett [1] and J. M. Bennett [2].

12.2 Linear Polarization

Radiation is considered to be linearly polarized if the electric vector vibrates primarily in a plane with any components not in the plane having a very small amplitude. In formal descriptions of linear polarization it is convenient to consider polarized radiation as the resultant of two equal linearly polarized components, perpendicular to each other and with the same phase. If we label the components as I_s and I_p then linear polarization is often specified by the degree of polarization which is given by

$$p = (I_s - I_p)/(I_s + I_p),$$

where p can vary between the limits of $+1$ for $I_p = 0$ and -1 for $I_s = 0$. The signs here represent a convention and not a fundamental property. Percent polarization is merely $p \times 100$. Polarization may also be specified by the ratio of the less intense component to the more intense component, the extinction ratio [1, 2]. A more useful specification, in the author's opinion, is the ratio of the more intense component to the less intense component, the component ratio. Figure 1 compares the component ratio to the degree of polarization as a function of the value of I_s when $I_s \geq 0.5$. If $I_s = 0.5$, the light is not polarized. As I_s increases, the polarization increases ever more rapidly; $p = 99\%$ when $R_s/R_p = 100$. For larger values of polarization the degree of polarization is not

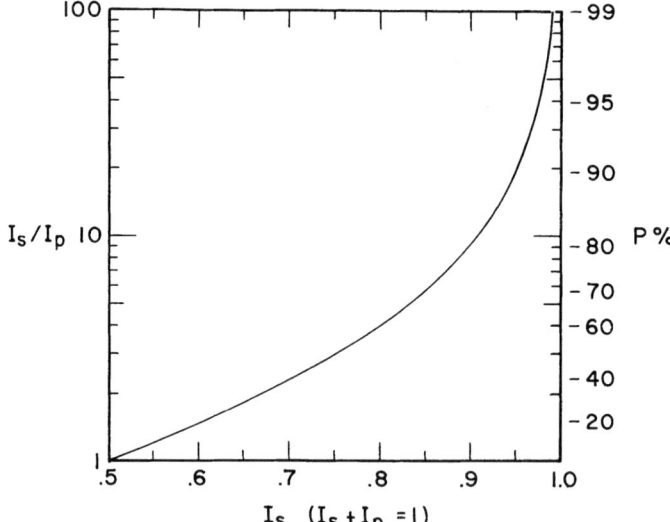

FIG. 1. Comparison of the degree of polarization (right ordinate) and the component ratio (left ordinate) for nonpolarized radiation ($I_s = 0.5$) to complete polarization ($I_s = 1$).

very informative. For example, if the component ratio increases from 100 to 1000, the degree of polarization has increased to 99.9%, an increase of 0.9%. The same curve is obtained if $I_p > I_s$ and by substituting I_p for the abscissa and I_p/I_s for the ordinate.

12.2.1 Transmitting Polarizers

Transmitting polarizers made of birefringent dielectrics can be used to within a few hundred angstroms, or less, of their cutoff wavelength. Wollaston and double Rochon prism polarizers made of MgF_2 have been reported by Johnson [3] and by Steinmetz et al. [4]. Both are beamsplitting prisms; that is, they produce two plane polarized beams, or rays, at right angles to each other and they represent two examples of polarization by double refraction. A Wollaston prism deviates both rays exiting from it but the Rochon only deviates the extraordinary ray. Thus, the Rochon prism arrangement may be more convenient for most purposes. These prisms, and probably most crystalline or glassy prisms, are subject to radiation damage in the presence of intense electric fields. Garton [5] has observed darkening inside Rochon prisms constructed of MgF_2 for use in the VUV, possibly caused by radiation damage.

Pile-of-plates polarizers have been in use for many years in the infrared region to obtain linear polarization and are examples of polarization by refraction. They are made by stacking plates of a transmitting material such that the angles of incidence on all the plates are equal and approximately at the Brewster angle. Because there is a slight displacement of the beam for a pile leaning in one orientation, the plates are usually divided into two groups of equal numbers and stacked leaning in the opposite directions, thus avoiding displacement. At the Brewster angle the component parallel to the plane of incidence, I_p, has zero reflectance and is transmitted without loss (if there is no absorption in the plate). The perpendicular component, I_s, is only partially transmitted. Therefore, the more plates used the more complete the polarization. The degree of polarization for a pile-of-plates polarizer with the plates at the Brewster angle is given by [1, 2]

$$p = (1 - A)/(1 + A),$$

where

$$A = [2n^2/(n^4 + 1)]^m,$$

where n is the index of refraction of the plates and m is the number of plates. This result assumes multiple reflections within plates but not between them. Figure 2 shows the degree of polarization for such a polarizer containing 2 (single plate), 4, and up to 12 surfaces (six plates). It is apparent that in the infrared where index values are large the pile-of-plates polarizer can be quite useful with just a few plates. In the VUV where index values rarely exceed 2.0 the number of plates must be increased to increase the polarization.

Walker [6] constructed a pile-of-plates polarizer from cleaved LiF plates held such that the angle of incidence was 60 deg. He tried both four- and six-plate combinations. Using both a polarizer and an analyzer, he found the angular extinction on rotation of the analyzer was a very good approximation to a $\cos^2$ curve and concluded that the radiation from the polarizer was indeed highly polarized. Assuming an index value of 1.6 for Walker's LiF plates, a six-plate polarizer would provide a degree of polarization of 0.98 and, for four plates, 0.9155. From Fig. 1 the corresponding component ratios are 80 and 10, respectively.

Hinson [7] used a pile-of-plates analyzer consisting of eight LiF plates to measure the polarization of a normal incidence VUV monochromator and Schellman *et al.* [8] used a pile-of-plates polarizer to measure the birefringence of quartz in the Schumann region.

At wavelengths less than the transmission limit of LiF (≈ 1050 Å) there are a number of metals that have transmission windows. It is possible that thin films of these materials might have useful polarizing capabilities when illuminated at

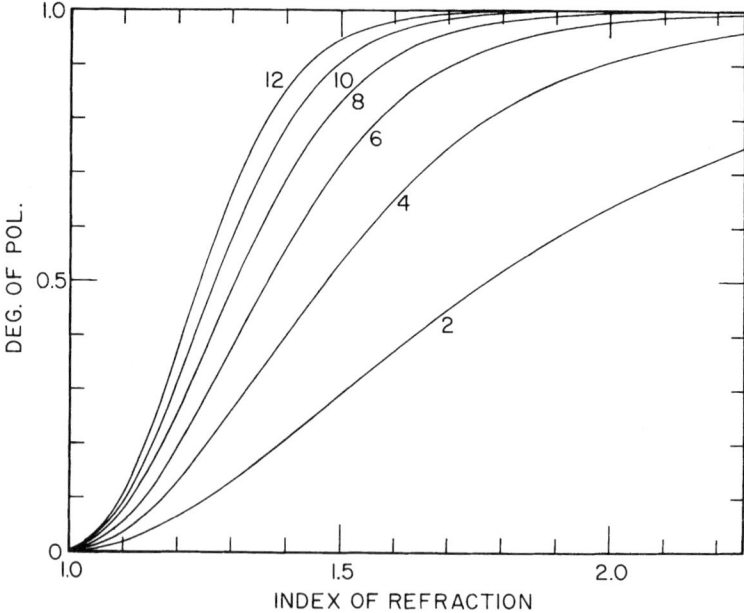

FIG. 2. The degree of polarization vs index of refraction for a pile-of-plates polarizer at the Brewster angle. The numbers indicate the number of surfaces (twice the number of plates).

oblique incidence, so Hunter [9] investigated the use of thin, unbacked films as transmission polarizers and found that a single aluminum film, without oxide, is a rather poor transmission polarizer; the maximum polarization occurs at about 700 Å and the component ratio is about 3, while the transmittance is approximately 11%. The effect of the natural oxide (30 Å thick) is to emphasize interference effects and polarization, but only at the longer wavelengths (>500 Å) where the differences between the indices of aluminum and its oxide are becoming appreciable.

If unbacked multilayers of aluminum and its oxide, or other metals, could be made, the effect would be to increase the component ratio. Figure 3 shows a calculation comparing the component ratios of multilayers of aluminum/ aluminum oxide arranged in the order oxide/metal/oxide/metal/.../oxide at 700 Å as a function of angle of incidence. If one considers a period to be metal/ oxide, then three periods on an unbacked substrate of oxide 30 Å thick has an component ratio of about 24 but a transmittance of about 2.7×10^{-4}%. Two periods provide a component ratio of approximately 9 and a transmittance of

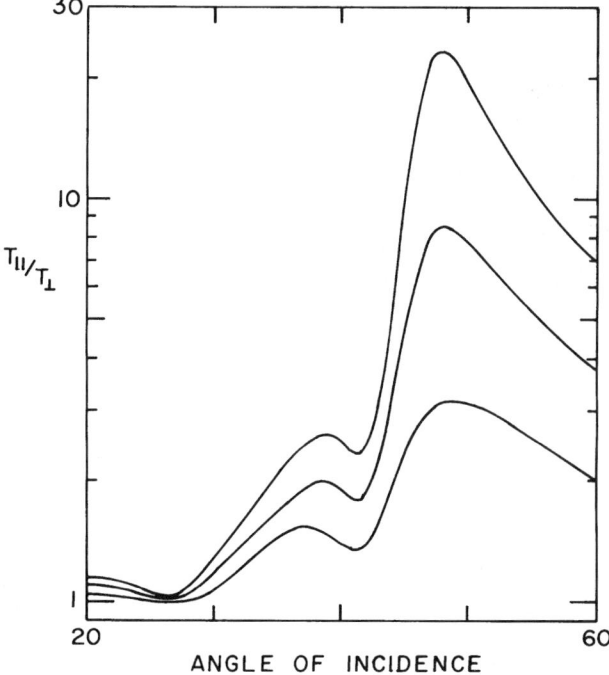

FIG. 3. Calculated polarization on transmission through a film consisting of one, two, and three periods of aluminum/oxide on a substrate of oxide as a function of angle of incidence. The aluminum thicknesses are 500 Å and the oxide thicknesses are 30 Å.

0.012%. Thus the gain in polarization by increasing the number of periods is offset by the loss in transmission. It may be possible to use unbacked metal films as transmission polarizers at certain specific wavelengths or over very small wavelength regions, but the use of unbacked metal films as general-purpose polarizers appears to be unrewarding.

Dichroic Polarizers. Schellman et al. [10] have obtained linear polarization by transmission through very thin sections of crystalline calcite. They found that the extraordinary ray was transmitted to wavelengths as short as 1650 Å, whereas the ordinary ray was strongly absorbed at wavelengths less than 2000 Å. Their thin sections of crystal ranged from 0.05 to 0.17 mm thick.

12.2.2 Reflecting Polarizers

The conventional method of treating polarization on reflection is to resolve nonpolarized radiation (electric vectors in all possible directions perpendicular to the direction of propagation) into two equal components, perpendicular and parallel to the plane of incidence. This is equivalent to a linearly polarized beam of radiation with the plane of polarization inclined at 45 deg to the plane of incidence. The angle of inclination is referred to as the azimuthal angle, Ψ, of the incident wave.

The reflected radiation intensity consists of the s component (R_s) polarized perpendicular to the plane of incidence and the p component (R_p) polarized parallel to the plane of incidence. These two components are considered to be independent and can be calculated as follows [11]:

$$R_s = [(a - \cos\phi)^2 + b^2]/[(a + \cos\phi)^2 + b^2],$$
$$R_p = R_s[(a - \sin\phi\tan\phi)^2 + b^2]/[(a + \sin\phi\tan\phi)^2 + b^2],$$

and the net, or average, reflectance is

$$R = (R_s + R_p)/2,$$

where

$$a^2 = 0.5 \cdot \{[(n^2k^2 - \sin^2\phi)^2 + 4n^4k^2]^{0.5} + n^2k^2 - \sin^2\phi\},$$
$$b^2 = 0.5 \cdot \{[(n^2k^2 - \sin^2\phi)^2 + 4n^2k^2]^{0.5} - n^2k^2 - \sin^2\phi\}$$

where n and k are the optical constants and ϕ is the angle of incidence.

The value of R_s increases monotonically from normal to grazing incidence but R_p has a minimum between these two angles. The value of the s component is always greater than that of the p component except in the case of total reflection when the two components are equal.

On reflection, the phase of each component changes with respect to the phase of the incident radiation. These phase changes can be calculated, but more important to this chapter is the phase difference between the components, Δ, which can be calculated as follows [11]:

$$\tan \Delta = [2b \sin \phi \tan \phi]/[\sin^2 \phi \tan^2 \phi - (a^2 + b^2)],$$

where a, b, and ϕ have the same meaning as in the preceding formulas.

There are two points of interest:

1. If $\phi = 45$ deg, $\tan \phi = 1$ and $\sin \phi = \cos \phi$; therefore one has the relation $R_p = R_s^2$. If the reflectance is written in amplitude form, one has $[r_s e^{-\delta(s)}]^2 = r_p e^{-\delta(p)}$, so that the phase change of the parallel component is twice that of the perpendicular component. These relations were first pointed out by Abeles [12] in 1950.
2. Berning [13] points out that if $\Psi \neq 45$ deg, then the net reflectance is

$$R = R_p \cos^2 \Psi + R_s \sin^2 \Psi,$$

and if the incident radiation is elliptically polarized with relative amplitudes w_s and w_p, referred to the direction of the s and p components, then

$$R = (w_s^2 R_s + w_p^2 R_p)/(w_s^2 + w_p^2).$$

If the reflecting material is nonabsorbing ($k = 0$), and the index is greater than unity, the phase difference, Δ, between the two components is π from normal incidence to the Brewster angle, $\tan \phi = n$ where ϕ is the angle of incidence and n is the index of refraction, where it changes abruptly to zero, that is, the first derivative of Δ is not continuous. Δ retains this new value from the Brewster angle to grazing incidence. The minimum value of R_p, which is zero, occurs at the Brewster angle. A true Brewster angle ($R_p = 0$) occurs only for non-absorbing materials.

If the material is absorbing ($k > 0$) R_p never reaches zero. The angle at which the minimum R_p occurs is sometimes referred to as the pseudo–Brewster angle and always occurs at an angle of incidence larger than $\tan \phi = n$, where n is the real part of the complex index of refraction. Δ is π at normal incidence and changes to zero at grazing incidence with no discontinuities in the first derivative. The angle of incidence at which the phase difference takes on the value of $\pi/2$ is defined as the principal angle of incidence. $\Delta = \pi/2$ is one of two conditions required for circular polarization. The other is that the components be equal. They are not if the incident radiation is nonpolarized so that circular polarization by reflection from the surface of an absorbing material requires $\Psi \neq 45$ deg. These conditions are discussed later in the section on circular polarization by reflection.

Both Humphreys-Owen [14] and Damany [15] have arrived at formulae giving the location of the pseudo–Brewster angle and, in addition, Damany has also arrived at a formula for the location of the angle of maximum polarization.

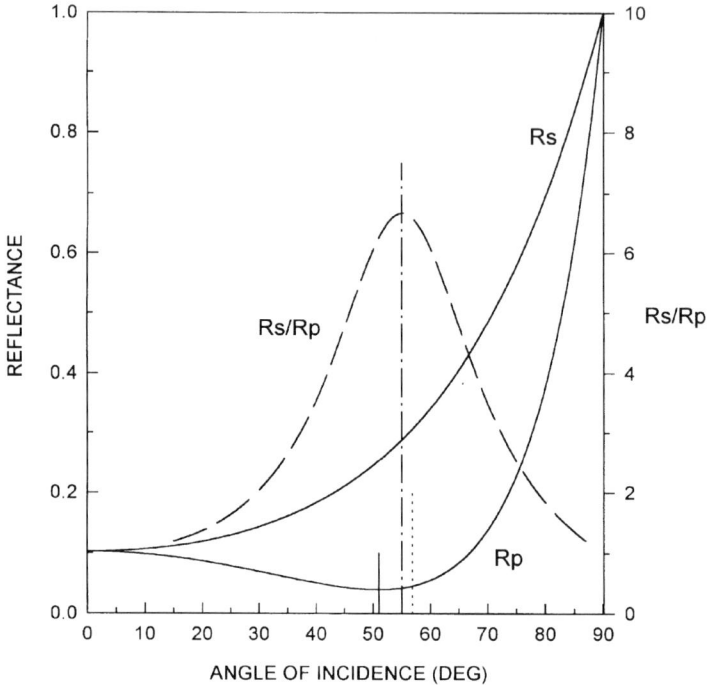

FIG. 4. Calculated reflectance of gold at 500 Å as a function of angle of incidence showing the behavior of the R_s and R_p components and the polarization. The short vertical line indicates the minimum value of R_p (the pseudo–Brewster angle), the dashed-dotted line the maximum polarization, and the dotted line the principal angle of incidence [16].

The conditions mentioned earlier form the basis for ellipsometry. With an ellipsometer one measures the phase difference between the two components, Δ, and the angle through which the plane of vibration of the electric vector has been rotated on reflection. From these measurements the optical constants of a single material can be derived. In the visible region, ellipsometry goes far beyond just determining n, k of a single material; it can be used to determine n, k of layers, the layer thickness, etc. In the VUV the difficulty of measuring phase differences limits the usefulness of ellipsometry.

Figure 4 [16] shows the calculated values of R_s and R_p and the polarization as a function of angle of incidence for a surface with the optical constants $N = 1.16 + i0.71$. These optical constants are approximately those of gold at 500 Å [17]. At normal incidence the reflectance for both components has the value of 10.2%. As the angle of incidence increases, the value of R_p decreases and reaches a minimum, which, for these optical constants, occurs at 51 deg and has a value of approximately 4%. Its location is indicated by the short vertical line. To larger angles of incidence R_p increases monotonically to unity at 90 deg. In Fig. 4 the polarization maximum occurs at 55 deg, shown by the long dashed-dotted vertical line, and the principal angle of incidence is the vertical dotted line at about 57 deg.

The behavior illustrated in Fig. 4 is typical of any isotropic absorbing material that is thick enough not to show interference effects from a second surface. If the material has a very large reflectance at normal incidence, the two components can be approximately equal from 0 deg to fairly large angles of incidence (80 deg or so) before the p component decreases to its minimum.

Figure 5 [16] shows the polarization to be expected from some values of n, k usually found in the VUV. The n values for each curve are shown on the abscissa and the k values on the ordinate. The curves are on a semi-log plot; each rectangle covers a decade. The value of P_{max} generally lies between 1 and 10 but two values of P_{max} exceed 10 and one is greater than 100. The general trend is that, for a given k, P_{max} increases as n increases, and for a given n, P_{max} decreases as k increases.

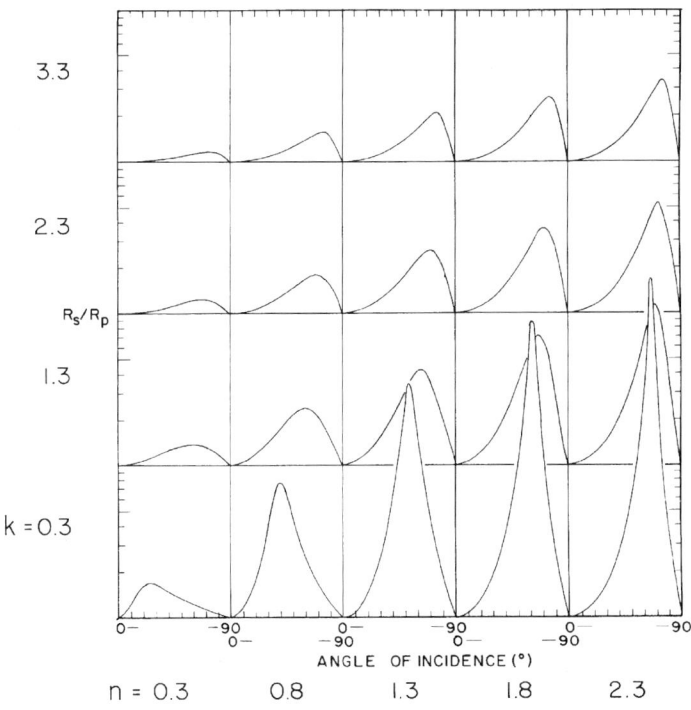

FIG. 5. The calculated polarization to be expected for n, k values in the VUV spectral range. The polarization increases as the n/k ratio increases [16].

The proportionality between n/k and the maximum polarization is not unexpected. For the extreme case of a pure dielectric, $k = 0$, the ratio n/k is infinite, and at the Brewster angle the maximum value of the polarization is also infinite because $R_p = 0$. This behavior has also been pointed out by Sasaki and Fukutani [18].

Hunter [16] calculated the values of the maximum polarization, the corresponding values of R_s, and the angle of incidence for a number of materials that might be useful as polarizers. Figures 6 [16] and 7 [16] illustrate the behavior to be expected of a dielectric material (crystalline MgF_2) and an absorbing material (gold), respectively. The optical constants for the calculation for MgF_2 were taken from Hanson et al. [19] for wavelengths less than 1000 Å and from Hass [20] for wavelengths longer than 1100 Å. Values between 1000 and 1100 Å were interpolated. From 1100 Å to longer wavelengths, the polarization is infinite and the value of ϕ is the Brewster angle. At 1100 Å, R_s has a value of about 50%, which slowly decreases to just over 10% at 2000 Å. At the shorter wavelengths, the polarization ranges from about 50 to 4, R_s ranges from 20 to 30% from 1000 Å to about 600 Å, then decreases rapidly to shorter wavelengths, and ϕ, which has an average value of about 50 deg between 1000 and 600 Å, suddenly decreases to the vicinity of 45 deg.

In contrast to MgF_2, gold is absorbing over the entire wavelength range from 300 to 2000 Å. Consequently, there is no true Brewster angle for this material, and the polarization is always less than 7.5. The optical constants used in this calculation were measured by Canfield et al. [17]. As with MgF_2, the angle for maximum polarization for gold also approaches 45 deg at the short wavelengths. This trend is characteristic of all materials close to the soft x-ray region. The index of refraction approaches unity and the extinction coefficient becomes small. Thus the pseudo–Brewster angle approaches 45 deg and all materials have the angle of maximum polarization at 45 deg. This condition ($n \approx 1$) extends into the x-ray region with the result that x-rays scattered at 90 deg to the primary beam are completely plane polarized, as observed by Barkla [21] and reported in 1906. Under these conditions the ratio of the two components can be very large; however, the s component, the larger of the two, is itself very small; on the order of 1% or 0.1%. As a consequence, a multireflecting polarizer at the short wavelengths may be very inefficient.

Studies of polarization due to single materials have also been made by Stephan et al. [22] who calculated the polarization to be expected from Pyrex, CaF_2, LiF, and corundum from 200 to 2000 Å. Cazaux [23] has studied the polarization to be expected from crystalline graphite.

Figures 6 and 7 show that the polarization obtained from a dielectric at the Brewster angle far exceeds anything available from a metallic surface whatever the angle of incidence. Thus, a fairly good single-reflection polarizer can be made using a dielectric at its Brewster angle for wavelengths longer than its

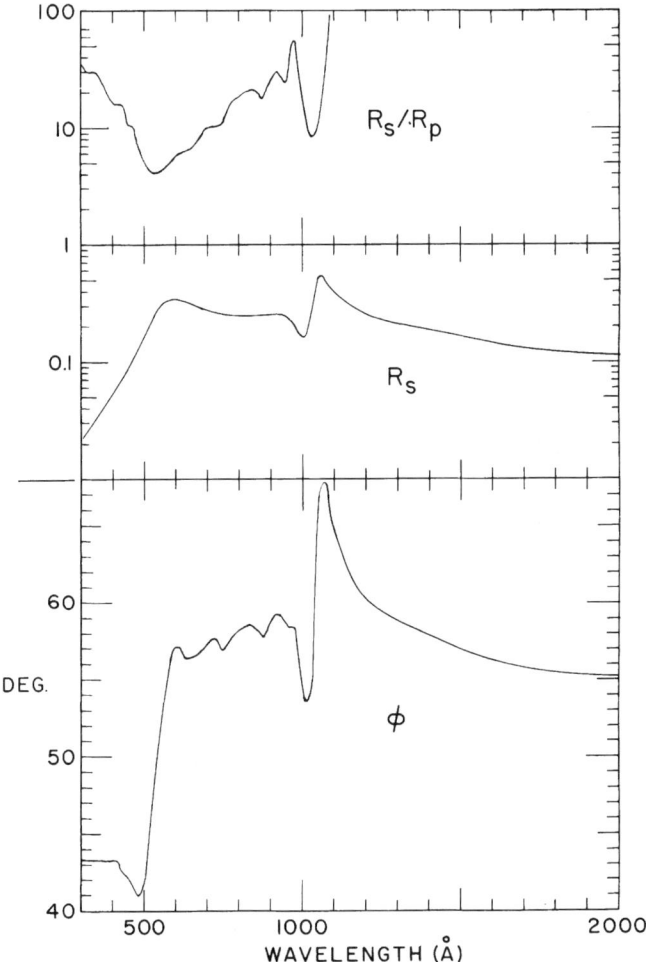

FIG. 6. The calculated polarization (R_s/R_p), R_s, the pseudo–Brewster angles ($\lambda <$ 1100 Å), and Brewster angles ($\lambda >$ 1100 Å) for crystalline MgF_2 from 300 to 2000 Å. From about 1100 Å to longer wavelengths $R_s\backslash R_p$ is infinite [16].

cutoff wavelength. It is somewhat inconvenient to use single-reflection polarizers that divert the radiation through double the angle of incidence, especially since the angle must be changed with wavelength to stay at the Brewster angle. Although two reflections can restore the direction of the radiation beam, the beam will still be displaced. To avoid such an inconvenience, three- and four-mirror polarizers have been developed that neither deviate nor displace the beam. Figure 8 [16] shows schematic diagrams of these polarizers. The planes of

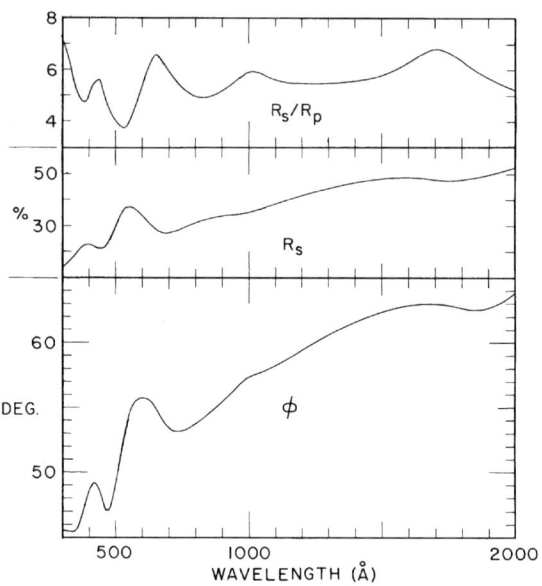

Fig. 7. The calculated polarization (R_s/R_p), R_s, the pseudo–Brewster angles for gold from 300 to 2000 Å [16].

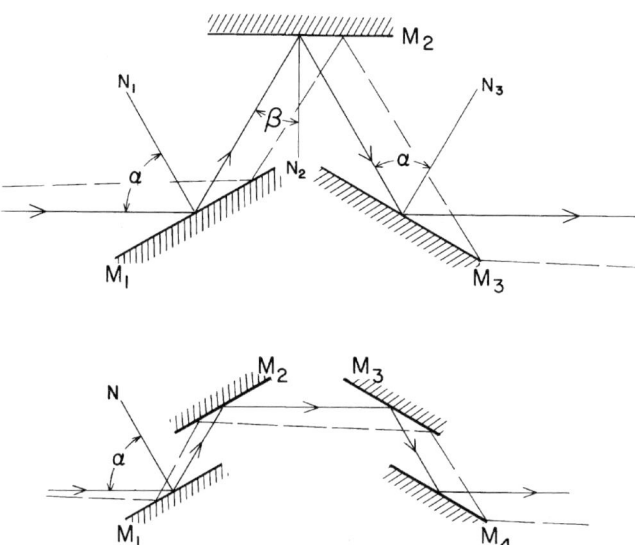

Fig. 8. Schematic diagrams of three- and four-mirror polarizers that do not deviate the beam [16].

incidence of the mirrors in each polarizer are parallel. For the four-mirror polarizer the angles of incidence are all equal. For the three-mirror polarizer only the first and third angles of incidence are equal. The relation between the first (α) and second (β) angles of incidence is given by $2\alpha - \beta = 90$ deg or, equivalently, the glancing angle on the middle mirror is twice that on the first and third mirrors. The solid lines represent the central ray and the dashed lines an oblique ray. The oblique rays show that the angles of incidence change in the same sense. That is, if an oblique ray is incident on the first mirror at a smaller angle than the central ray, the angles of incidence on the other mirrors are also smaller than that of the central ray. They also show that the three-mirror polarizer inverts the beam but the four-mirror polarizer does not. In effect, the three-mirror polarizer is an image rotator; the image rotates at twice the angular speed of the polarizer.

One of the fundamental differences between the three- and four-mirror polarizers is that the four-mirror polarizer can have all four mirrors at the optimum angle for polarization, but the three-mirror polarizer can have at most two mirrors set at the optimum angle. Hence, one can obtain a higher degree of polarization from a four-mirror polarizer than from a three-mirror polarizer; however, the throughput of the four-mirror polarizer is less than that of the three-mirror polarizer.

If the mirrors of the four-mirror polarizer are set so that the central ray is at maximum polarization, the polarization will decrease for all other rays. The extent of the decrease is controlled by the acceptance angle of the four-mirror polarizer, which, in turn, is governed by the smallest of the mirrors in the train. It is not as easy to see how the polarization changes within the acceptance angle of the three-mirror polarizer because all the angles of incidence are not equal. One must calculate the change. Hunter [16] has made such a calculation giving the polarization, throughput, and angular limit within which the polarization change is 5% or less than that of the central ray from 300 to 2000 Å for a number of coating materials. His results are shown in extensive tables to which the reader is referred [16].

Three-mirror polarizers have been reported by Hancock and Samson [24], Winter et al. [25], Robin et al. [26], Hamm et al. [27], Horton et al. [28], Matsui and Walker [29], Hass and Hunter [30], Remneva et al. [31], Saito et al. [32], and Koide et al. [33]. Robin et al. used micaceous biotite at its Brewster angle (61-deg angle of incidence) as the first mirror in their three-mirror polarizer and Al/MgF$_2$ mirrors as the final two mirrors. They found that the polarizer performed very well from about 1100 to 6000 Å. Matsui and Walker used biotite as the first and third mirrors and the second mirror surface was Al/MgF$_2$. Hamm et al. used mirrors coated with gold or silver and used the polarizer to study the polarization of gratings in the VUV. Horton et al. used gold-coated mirrors in their three-mirror polarizer. Remneva et al. studied the polarizing properties of

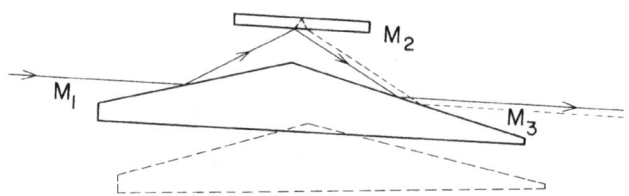

FIG. 9. Schematic diagram of a three-mirror polarizer. Mirrors M_1 and M_3 are part of a prism that can be removed from the beam to permit the use of nonpolarized radiation. Mirror M_2 is a MgF_2 plate at approximately the Brewster angle. The MgF_2 should be wedge shaped to avoid the reflection from the second surface [30].

silver, copper, SiO_2, zinc sulfide, aluminum oxide, magnesium fluoride, gold, and indium and chose gold as the most satisfactory mirror coating. Hass and Hunter used Al/MgF_2 mirrors as their first and third mirrors and an uncoated MgF_2 plate, approximately at its Brewster angle, as their second mirror. Their first and third mirrors were on a prism that could be removed from the radiation beam, thus they could select either polarized radiation or radiation directly from the monochromator. Figure 9 [30] shows a schematic diagram of their arrangement. The angles of incidence on the prism faces are 75 deg and that on the MgF_2 plate 60 deg, which is approximately the Brewster angle for MgF_2 at 1200 Å. The prism faces were simultaneously coated with Al/MgF_2. It was found that the second surface of the MgF_2 plate also contributed a polarized beam that was slightly displaced from the first. A plate with a wedge can shift the second beam so it can be occulted.

Rosenbaum et al. [34] have used a four-mirror polarizer with gold mirror coatings to measure the polarization of synchrotron radiation dispersed by a normal incidence monochromator. They report component ratios of between 300 and 400 for wavelengths between 600 and 950 Å. The shortest wavelength they measured was about 500 Å where the ratio was on the order of 100.

The reflecting polarizers reported to date have mirrors fixed in orientation. Because the angle of incidence for optimum polarization changes with wavelength, it seems reasonable to design an apparatus in which the angles of incidence can be changed. This is not a straightforward task for a three-mirror polarizer, but, Watanabe [35] has designed a device that is capable of the motions required for an adjustable three-mirror polarizer. A schematic diagram of his mechanism is shown in Fig. 10. The optical axis is designated by XX'. Two mirrors, M_1 and M_3, pivot about axes that lie in their front surfaces, that are parallel, and that intersect the optical axis at right angles. Mirror M_2 moves parallel to its normal, the y direction, which is perpendicular to the optical axis. The rotations of M_1 and M_3 are constrained by rods, rigidly attached to the mirrors, that have sliding pivots, P_1 and P_2, on the plate to which M_2 is attached. Thus when M_2 moves up or down, M_1 must rotate such that the beam reflected

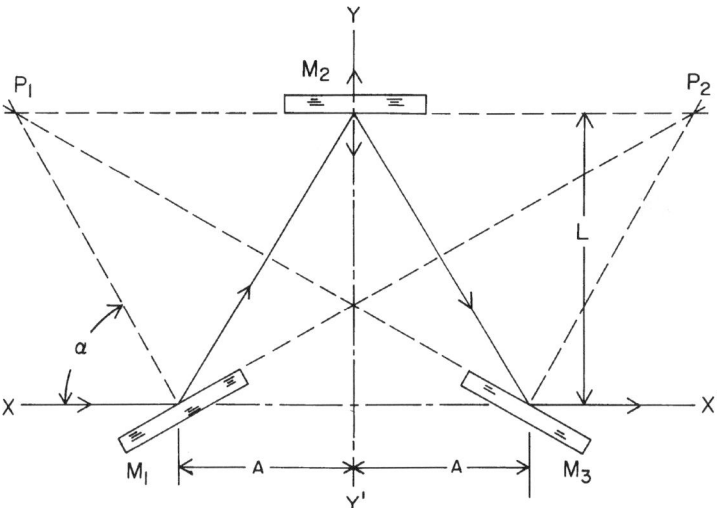

FIG. 10. A three mirror reflector designed by Watanabe that permits the angles of incidence on the mirrors to be changed without deviating the beam. (Adapted from [35] with permission from the Institute of Physics.)

from it always impinges on the center of M_2. Mirror M_3 is constrained to rotate in the same manner. The angle of incidence, α, is related to the movement of M_2, L, by the relation,

$$L = A \cot(2\alpha - 90 \text{ deg}),$$

where A is half the distance between the first and third mirrors. Watanabe designed his apparatus so that it could be rotated about the optical axis, which would allow the relative polarization of the beam to be measured.

An adjustable four-mirror polarizer is somewhat easier to design. Hunter and Rife [36] have designed an adjustable four-mirror polarizer for use with synchrotron radiation that functions as shown in the schematic diagram of Fig. 11. The optical axis (beam path) is shown by the dashed line. Mirror pairs, M_1–M_2 and M_3–M_4, are supported on cams with circular edges. The cams pivot about the axes designated by the small circles centered on the front surfaces of M_1 and M_4. These axes are parallel to each other and perpendicular to the optical axis. Thin metal bands are fastened to the circular edges of the cams and to a common plunger so that the mirror pairs can be adjusted simultaneously to the same angle of incidence. Radiation is reflected from the first small mirror to the first large mirror, thence to the second large mirror and, finally, to the second small mirror. The beam is neither deviated nor displaced, although the pathlength is increased slightly as the angle of incidence is increased. As the wavelength changes, the

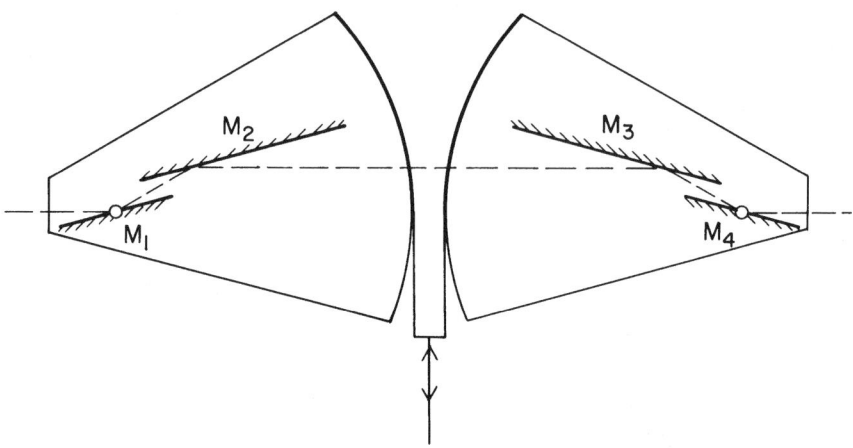

FIG. 11. A design for an adjustable four-mirror polarizer.

mirror pairs are rotated to the angle for optimum polarization. The tables in [16] show that the angle for maximum polarization does not exceed 58 deg from 300 to 2000 Å, therefore, the cams need not rotate more than about 60 deg.

The NRL four-mirror polarizer is also capable of shifting the mirror assemblies completely out of the beam should the need arise. The complexity of the design required for an UHV polarizer ruled out the ability to rotate the polarizer about the optical axis.

A four-mirror polarizer can also serve as a higher order suppressor for wavelengths less than about 300 Å. This is done by adjusting the glancing angles such that the desired wavelength is being reflected at an angle just smaller than the critical angle. The critical angle of the higher order is larger so the reflecting surfaces have small reflectance values for the higher orders. Gluskin et al. [37] have used a similar arrangement for higher order suppression consisting of only two mirrors that rotated together to produce a beam of constant deviation but that was displaced from the incident beam.

12.2.3 Interference Polarizers

Polarization by interference (on reflection) occurs in the VUV with some fairly large component ratios. Although the throughputs of the polarizers are usually rather small, an exception is the Al/MgF_2 coating at large angles of incidence. Figure 12 [38] shows both the calculated and measured reflectance of Al/MgF_2 coatings for different thicknesses of MgF_2 at 1216 Å. Only the coating with the 250-Å-thick MgF_2 coating shows useful polarization, and that at about an 85-deg angle of incidence. The component ratio is about 17 and the

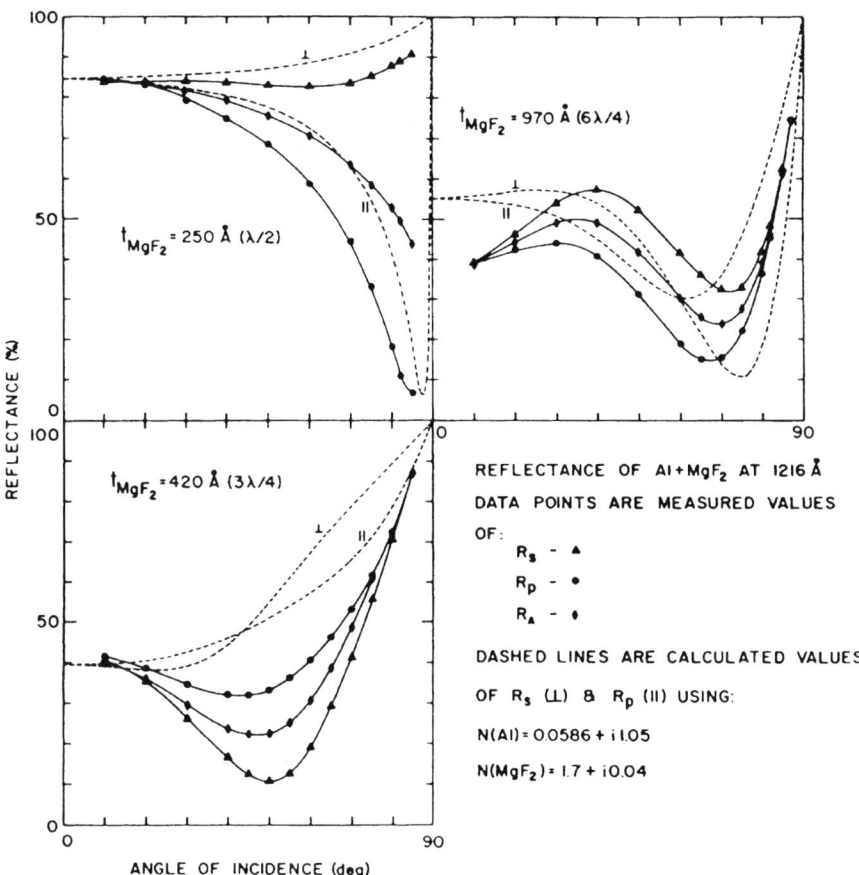

FIG. 12. Reflectance of Al + MgF$_2$ for different thicknesses of MgF$_2$ using polarized and nonpolarized radiation at 1216 Å. The solid curves are measured values and the dashed curves are calculated [38].

reflectance for the perpendicular component is about 90%. A properly oriented four-mirror polarizer comprised of such coatings would produce a very large degree of polarization at 1216 Å. The calculations do not match the measurements very well for reasons unknown. Perhaps the optical constants used for the MgF$_2$ film are not really characteristic of this particular coating. Calculations and measurements show that this amount of polarization does not persist at wavelengths removed from 1216 Å; consequently, a polarizer made from these coatings will have a rather small wavelength range. More recently Kim *et al.* [39] have designed an MgF$_2$/Al/MgF$_2$ reflecting layer structure for use as a polarizer at 1216 Å. The angle of incidence is 45 deg.

TABLE I. Polarization Maxima and Minima and Throughput for an Oxidized Aluminum Film 1000 Å Thick on Fused Silica with Oxide Thickness of 30 Å

	Maxima			Minima			
λ (Å)	R_s/R_p	R_s	α	R_s/R_p	R_p/R_s	R_p	α
700	1.75	0.051	19.0	0.315	3.17	0.025	26.0
600	21.7	0.061	34.6	0.00752	133.0	0.013	28.6
500	176.0	0.095	49.6	0.0689	14.5	0.012	44.0
400	51.3	0.017	49.4	0.0283	35.3	0.0019	47.2
300	67.5	0.031	44.8				
200	765.0	0.00083	43.8				
100	818.0	0.0028	44.2				
50	75,290.0	0.000084	44.6				

At somewhat shorter wavelengths (<800 Å) calculations indicate that aged aluminum films on glass substrates should have some surprisingly large component ratios caused by interference between the rays reflected from the vacuum–layer and layer–substrate interfaces. Although the component ratios are large, the throughputs are rather small. As an example, Table I lists the component ratios, throughput (value of R_s or R_p) and the angles at which the greatest maximum or smallest minimum occurs for an oxidized aluminum film 1000 Å thick and for wavelengths of 600 to 50 Å. Those columns under *Maxima* are for the ratio R_s/R_p and under *Minima* the inverse. For example, at 600 Å the maximum component ratio is 21.7 with an R_s value of 6.1% at an angle of incidence of 34.6 deg. The minimum R_s/R_p is 0.00752 ($R_p/R_s = 133$) with an R_p value of 1.3% at an angle of incidence of 28.6 deg. Using different aluminum film thicknesses, the maxima and minima occur at different angles.

Interference dominates the performance of the coating until the wavelength decreases to less than 400 Å, and either R_s/R_p or R_p/R_s can be maximum, depending on the aluminum film thickness. At about 300 Å, interference effects become ineffective, R_s/R_p is always greater than 1.0 at all angles; hence, there are no listings under minima for $\lambda \leq 300$ Å. The maximum ratio (R_s/R_p) moves toward 45 deg for the simple reason that all of the indices of refraction tend toward the value of unity.

12.2.4 Pseudo-Brewster Angle Polarizers

Radiation reflected from layered synthetic microstructures (LSMs) is also polarized, partly due to interference, and partly because of the approach of the indices of refraction of the layers to unity. If the index ≈ 1, then the pseudo–Brewster angle is ≈ 45 deg, and the value of R_p is very small. It is in the vicinity of the pseudo–Brewster angle where the maximum polarization occurs. Khandar et al. [40] have discussed the subject and give an illustration of the polarization to be expected from a multilayer consisting of Si/Hf with an unspecified number of periods, adjusted for maximum reflectance at 304 Å. The angle of incidence was 45 deg and the value of R_s is about 55% while that of R_p is about 2%. In this case the resonance within the layers keeps the value of R_s large (see Chapter 14), while the low value of R_p is due to the indices of refraction being approximately unity and the polarizer being operated at approximately the Brewster angle.

Yanagihara et al. [41] describe a multilayer polarizer for use over the energy range from 80 to 150 eV that shows similar behavior. They used two multilayer mirrors obtained by halving a silicon wafer substrate coated with a 21-period C/Ru multilayer. The multilayer was designed for maximum reflectance, 34%, at 99 eV and a 45-deg angle of incidence. The two multilayers were mounted on an apparatus resembling a double crystal x-ray monochromator; that is, the multilayers could be translated and rotated so that the emergent beam direction remained fixed and the angles of incidence on each multilayer were always equal. As the energy is changed, the angle of incidence on the multilayers is also changed so that the multilayers are at maximum reflectance [R_s (max)]. An illustration of the change in reflectance vs energy is given in this volume (see Figs. 20a and b in Chapter 16). The multilayers are operating at maximum polarization efficiency and reflectance only at 99 eV where the pseudo–Brewster angle is ≈45 deg. However, even at other energies, R_s was sufficiently greater than R_p so that useful polarization was obtained. The throughput of the two-mirror polarizer ranged from about 3% at 80 eV to a maximum of about 12% at 99 eV and decreased to about 4% at 150 eV. Corrsponding values for R_p ranged from 0.1% to a maximum of 0.8% and then decreased to about 0.6%. The polarization was never less than 75% and rose to 99% at 99 eV. Yanagihara et al. also operated the polarizer from 160 to 300 eV and found no measurable throughput, thus the polarizer also suppresses the second harmonic when used in the 80 to 150 eV ranges.

12.2.5 Analyzers for Linear Polarization

Rabinovitch et al. [42] have used a single-reflection analyzer, a mirror at a 45-deg angle of incidence, to measure the polarization of VUV radiation emerging from a monochromator. They give two analyses to find the polarization from reflectance measurements using the single-reflection analyzer. The first of these analyses assumes the electric vector to be either parallel, I_p, or perpendicular, I_s, to the slit and defines the polarization as $I_p/I_s = g$. A measurement of the reflected intensity with the plane of incidence of the analyzer perpendicular to the slit gives

$$R_1 = (I_p R_s + I_s R_p)/(I_p + I_s),$$

where R_s and R_p are the reflectances of the analyzer mirror at a 45-deg angle of incidence. Another measurement with the plane of incidence of the analyzer parallel to the slit gives

$$R_2 = (I_p R_p + I_s R_s)/(I_p + I_s).$$

Dividing by I_s and rearranging,

$$R_1 = (R_p + gR_s)/(1 + g),$$
$$R_2 = (R_p g + R_s)/(1 + g).$$

Rabinovitch et al. used the Abeles relationship given earlier, $(R_s^2) = R_p$, in the two preceding equations above to obtain

$$g = \frac{R_2[1 + 4(R_1 + R_2)]^{0.5} - (R_2 + 2R_1)}{R_1[1 + 4(R_1 + R_2)]^{0.5} - (R_1 + 2R_2)},$$

which gives g in terms of R_1 and R_2 only and it is not necessary to know the optical properties of the analyzer mirror. If the radiation incident on the mirror is not parallel, which is usually the case, the Abeles relationship is not valid; however, if the divergence is not more than ± 2 deg, nonparallelism is a very minor factor.

Rabinovitch et al. [42] have also considered the case where the maximum of the electric vector is neither perpendicular nor parallel to the slit. The reader is referred to their paper for the details. Hamm et al. [27] have also published an analysis of polarization measurements without restricting themselves to a single reflection.

The accuracy with which the polarization can be measured is largely dependent on the accuracy with which R_1 and R_2 can be measured. Intuitively, we would expect a single mirror analyzer to be most accurate at the angle of incidence where P is greatest, and this will be true if the errors in measuring the reflectances are ignored. If, however, R_1 or R_2 is small at the angle of maximum P, the error in measuring the reflectance will cause the value of g to be shifted. To calculate the amount of this shift, an error model must be chosen. Hunter [16] has done such an analysis and published figures showing the errors to be expected with single and multiple reflection analyzers.

Mirrors inclined at 45 deg have been used by Samson [43, 44] for analyzers and by Gluskin [45] as polarimeters.

12.3 Circular Polarization

12.3.1 Circular Polarization on Reflection

Circular polarization of unpolarized radiation by reflection requires that the two reflected components be equal and have a phase difference of $\pi/2$. Such a phase difference is not hard to achieve but equalizing the reflected components can be difficult. If, however, one can use total internal reflection, both components are equal and it remains only to make their phase difference $\pi/2$.

According to Jenkins and White [46], Fresnel used a glass rhomb, as shown in Fig. 13a to change linearly polarized light to circularly polarized light. The angle of incidence of the light on the rhomb face is 0 deg and it exits at the same angle. The azimuth of the linearly polarized light is set at 45 deg with respect to the planes of incidence within the rhomb. At internal angles of incidence of 54 deg, the phase change was 45 deg, so after two reflections a phase difference of $\pi/2$ was achieved and the totally reflected light was circularly polarized. The same effect can be achieved without deviating the beam by using a double rhomb; adding a second rhomb as shown in Fig. 13b. In this case the phase difference on each reflection must be only 22.5 deg so that after four reflections the phase difference is $\pi/2$. One objection to the double Fresnel rhomb is that it can be quite long if the internal angles of incidence are large. One way of circumventing the length is to use triple total internal reflections. An undeviated circularly polarized beam can be obtained using three reflections, as shown in Fig. 13c (Fig. 13d is discussed later). As with the three-mirror polarizer previously discussed, the three reflections do not occur at the same angles of incidence so the phase changes are not the same. Whatever the phase changes,

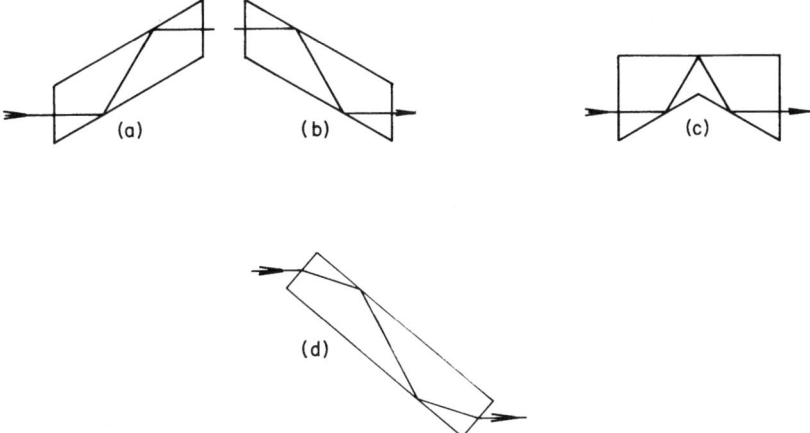

FIG. 13. The single Fresnel rhomb (a) has two total internal reflections. For circular polarization the phase differences on each reflection must be 45 deg. If it is coupled with another single rhomb (b) it becomes a double Fresnel rhomb and the phase difference on each reflection must be 22.5 deg for circular polarization. Circular polarization can also be achieved with the three-reflection device (c) if the sum of the phase differences adds up to 90 deg. By allowing the radiation to be incident obliquely on a rhomb (d) the error in phase difference as a function of wavelength can be reduced.

they must add up to $\pi/2$ to obtain circular polarization. One further point is that if the polarizer must be rotated to determine the purity of circular polarization, the double Fresnel rhomb does not rotate the beam but the three-mirror polarizer does.

LiF is transparent in the VUV to wavelengths as short as about 1100 Å and could be used to make rhombs, as could any other nonbirefringent, transparent material, for example, CaF_2, BaF_2, and fused silica. However, the index of refraction of the rhomb material must be such that the proper phase difference (PD) can be achieved. In calculating the PD the inverse of the index must be used because the radiation goes from a more dense to a less dense medium. For indices less than unity the PD is π at normal incidence until the Brewster angle is reached, at which point it abruptly becomes zero. It remains at zero until the critical angle, ϕ_C, is reached. To larger angles the PD rises abruptly to a maximum, then decreases to zero at grazing incidence. Stratton [47] gives a formula for calculating the angle at which the maximum PD occurs, $\sin^2 \alpha = 2n^2/(1 + n^2)$, where n is the inverse of the index and α is the angle of incidence. Figure 14 shows the behavior of the PD for a number of indices of

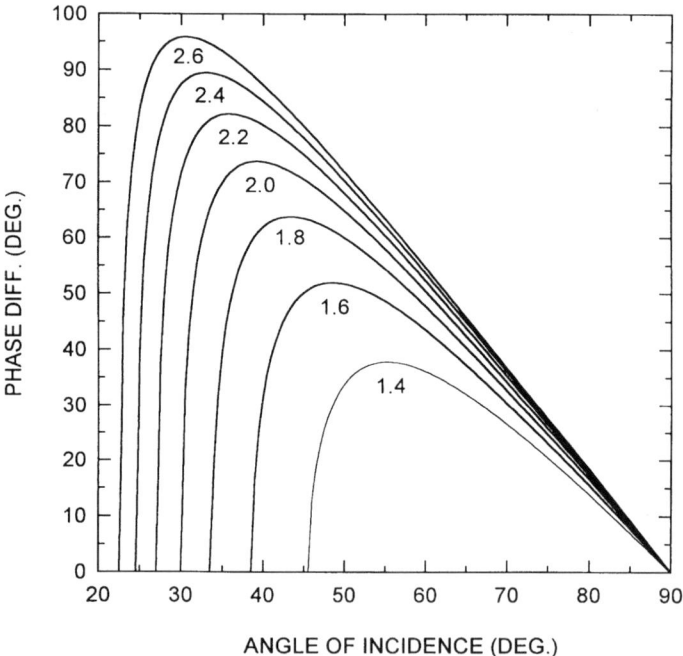

FIG. 14. Calculated phase difference vs angle of incidence for different values of the index of refraction. The position of the Brewster angle is not shown.

refraction ranging from 1.4 (0.714...) to 2.6 (0.3486...) at angles larger than ϕ_C. The largest index value, 2.6, was chosen to illustrate that the PD can reach, and exceed, the value of $\pi/2$. Thus, under certain conditions there can be three principal angles of incidence—one at the Brewster angle and two between ϕ_C and grazing incidence. In the VUV the index values of dielectrics in their transmission windows are unlikely to be as large as 2.6 or less than 1.4.

Figure 14 shows that the PD is double valued with respect to the angle of incidence, thus all the indices shown in the figure have two angles of incidence at which the PD is 22.5 deg. Using these PD values, one can draw contours of constant PD = 22.5 deg in the angle of incidence–index of refraction plane for index values often encountered in the VUV, as shown in Fig. 15. The solid line represents the PD of 22.5 deg and the dashed lines indicate PDs of 21.5 and 23.5 deg, a PD change of ± 1 deg, which gives some idea of the tolerance of the PD with respect to angle of incidence and index of refraction. For large indices and small angles the three curves are practically indistinguishable so tolerance is extremely tight; the slightest error in angle results in a large change of PD. If the angles are large, however, the tolerance is much looser. For example, Fig. 15

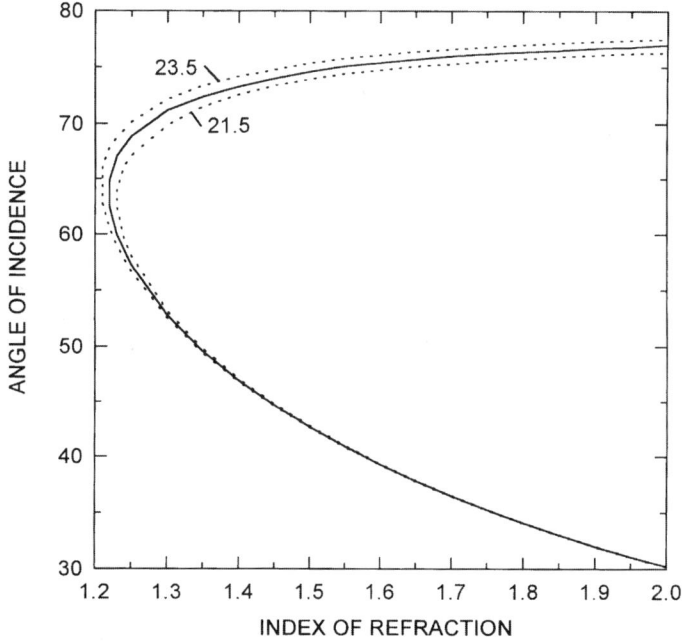

FIG. 15. A contour showing the constant phase difference of 22.5 deg (solid line) as a function of angle of incidence and index of refraction, and contours for ± 1 deg (dashed lines).

shows that at about a 70-deg angle of incidence, an error of about ±1 deg causes an error in PD of about ±1.5 deg. But how does the error in PD for one reflection affect the PD error in a quarter-wave polarizer? The answer depends on the number of reflections in the polarizer. If a ray enters the double rhomb at an angle, δ, slightly deviated from the optical axis, it exits at the same angle (see Fig. 8). For example, the deviation on the first and second mirrors is δ, and on the third and fourth mirrors is $-\delta$, so there is no net change in angle. Although the PD does not change linearly with angle, if δ is small the change in PD will be small. With the three-mirror polarizer, this is not so. If the incoming ray deviates by δ, it exits by $-\delta$, so the change in PD is that associated with an angular change of 2δ.

Although Fig. 15 shows that for large angles of incidence the change in PD with index, that is, with wavelength, is slow, and the double rhomb does not approach achromatism. Some research has shown that rhombs can be achromatized in the visible region by applying a suitable thin dielectric film to the totally reflecting surfaces [48, 49]. Nagib et al. [50] have attempted to achromatize rhomb-type quarter-wave retarders in the visible by allowing the incident radiation to enter the rhomb at an oblique angle, as shown in Fig. 13d. Thus, as the wavelength changes, the angle of total internal reflection changes as does the PD. They have been able to achieve a total PD error of 0.02 deg over an index range of 1.625–1.65. However, on entry to, and exit from, the rhomb at an oblique angle, the s and p components are no longer equal. Nagib et al. claim that by changing the azimuth of the incident linearly polarized radiation with respect to the planes of incidence within the rhomb this inequality can be overcome. Achromatization of dielectric rhombs in the VUV has not been investigated.

Under certain conditions the proper PD for circular polarization can be obtained by a single reflection from a metal surface. Saito et al. [51] report using reflection of linearly polarized radiation from an aluminum mirror with the natural oxide at an angle of incidence of 58 deg in lieu of a quarter-wave plate at 1216 Å. Their paper shows that the PD on reflection from the aluminum is very close to $\pi/2$ from about 1000 to 1800 Å.

At shorter wavelengths where no transmitting materials are available, one must rely on circular polarization by reflection. If the radiation incident on a mirror is polarized with the azimuth of the incident beam at 45 deg to the plane of incidence, the polarization after reflection is elliptical, as mentioned earlier. Although the phase difference between the two components may be $\pi/2$ at the principal angle of incidence, the inequality of the reflected intensities prevents the polarization from being circular. One must adjust the azimuth of the incident radiation and the angle of incidence to achieve equality of the components and a PD of $\pi/2$. Azzam [52] and Kan'an and Azzam [53] have investigated the relation between principal angles and principal azimuths extensively.

McIlrath [54] built such a polarizer for the 1216-Å line of hydrogen. He used LiF plates for the linear polarizer and the analyzer and determined the proper azimuth by experimentation. Westerveld et al. [55] used Mueller matrices and a knowledge of the optical constants of gold to calculate the azimuthal and other angles involved. Nonpolarized radiation was polarized with a three-mirror polarizer and then reflected from a gold mirror. The azimuth of the linearly polarized radiation and the angle of incidence on the gold mirror were adjusted so that $R_s = R_p$ and the phase difference was $\pi/2$. This arrangement produced circular polarization in the wavelength region from about 550 to 1250 Å.

Kortright and Underwood [56] have calculated relative phase changes and R_s and R_p for Bragg reflection and transmission through multilayers and found that it is possible to adjust for equal R_s and R_p with phase shifts of $\pi/2$. Di Fonzo and Jark [57] have applied the same type of calculation to the region close to the carbon K edge. Kim et al. [58] have designed quarter-wave retarders using layers of $MgF_2/Al/MgF_2$ for use at 1216 Å. Their design called for an angle of incidence of 45 deg but they claim that by adjusting the thicknesses of the different layers the retarder can be made to work at any angle of incidence.

There have been a number of reports of circular polarization by means of reflection from three- and four-mirror polarizers. Schledermann and Skibowski [59] used two polarizers to measure the optical constants of a surface and to determine the ellipticity of radiation reflected from that surface. They demonstrated the technique by measuring the optical constants of gold at 700 Å. Johnson and Smith [60] have calculated the angles required for a three-mirror polarizer using either gold or platinum as the mirror coatings. Smith and Howells [61] studied the use of whispering gallery mirrors to produce circularly polarized radiation with incident synchrotron radiation. A whispering gallery mirror is, in essence, a curved mirror on which the radiation is incident at a very small glancing angle. In traversing the mirror, many reflections at the same angle of incidence occur. The glancing angle must be adjusted so that after the traverse the appropriate PD has occurred. The advantage, according to the authors, is that the many reflections at very small glancing angles do not cause as much intensity loss as, say, four reflections at larger glancing angles, for example, the four-mirror polarizer.

Höchst et al. [62, 63] have described and built a four-mirror circular polarizer for use with highly polarized synchrotron radiation to operate in the range from 8 to 100 eV. It uses gold-coated mirrors. The azimuth of the polarizer can be adjusted so that the PD can be made $\pi/2$, and the angles of incidence on the mirrors can also be adjusted simultaneously. The authors point out the importance of keeping the mirrors at the same angle of incidence for any azimuthal value. In their instrument the angles of incidence on the mirrors track within 1.0 arcsec.

12.3.2 Circular Polarization by Transmission

Quarter-wave plates can be made of any of the birefringent crystals transparent in the VUV. Crystalline quartz, sapphire, and MgF_2 are three possibilities. One must remember that the index is changing much more rapidly than in the visible region so that a proper quarter-wave plate design at a given VUV wavelength might give elliptical polarization at another wavelength close by.

Heinzmann [64] has developed an apparatus for polarizing VUV radiation circularly to study spin-polarized photoelectrons. Radiation from a hydrogen discharge lamp was dispersed by a Seya–Namioka monochromator, sent through a MgF_2 double Senarmont prism for linear polarizing, then through two MgF_2 quarter-wave plates to produce circular polarization. Analyzing was done with another double Senarmont prism. The apparatus was usable from the visible up to 9 eV. Winter and Ortjohann [65] developed an analyzer for elliptically polarized Lyman-α radiation that incorporated a MgF_2 quarter-wave plate followed by a pile-of-plates linear polarizer using MgF_2 plates. In operation the quarter-wave plate is rotated to perform the analysis.

The phenomenon of photoelasticity, or the piezo-optical effect, has been known and used for many years. It is commonly used to model strain by placing transparent models between crossed polarizers. The stressed transparent model acts as a phase retarder in which the retardation is proportional to the strain. Strain contours show in different colors while the contours of zero strain are dark. The implication is that any transparent material, under stress, acts as a retarder. Metcalf and Baird [66] have used this effect to achieve polarization ranging from linear to circular by compressing a crystal of LiF. They used a LiF plate in transmission at about a 60-deg angle of incidence to partially linearly polarize the radiation that impinged on the retardation plate. For a retardation plate they used LiF plates 2 to 4 mm thick and applied compression between opposite edges using either a screw or a pan holding weights, up to 10 kg. It was important to apply pressure evenly on the opposing edges to prevent nonuniform strain in the crystal, which would modulate the polarization and prevent accurate measurements. The analyzer was a single LiF plate, used in reflection and inclined at 60 deg to the retarded beam, and that could be rotated around the beam to analyze the elliptically polarized radiation. The measured intensities were analyzed using Mueller matrices. They found that the piezobirefringence varied from one sample to the next, which they attribute to mechanical properties of the crystals rather than optical properties.

Kemp [67] has discussed the angular limits within which the incident radiation will still have approximately the desired retardation for noncrystalline transparent materials. According to his analysis, lack of birefringence is an asset because the retardation is of order $(\Delta n/n)t/\lambda$, where t is the thickness and n refers to the index of refraction. In birefringent materials the retardation is proportional to path differences of order $\Delta nN\lambda$ because retardation plates cannot be made just one wavelength thick, but rather N wavelengths thick. Birefringence restricts the angular limits to a few degrees but a much wider angle is possible for nonbirefringent materials. For example, Pockel cells are limited to less than 10 deg for a 10% retardation spread while fused silica can have a half-angle of 25 deg. Although Kemp did not work in the UV or VUV, he suggested using CaF_2 plates as modulators for the VUV.

Gedanken and Levy [68] have devised an arrangement for measuring circular dichroism in the VUV using a biotite plate as a linear polarizer and CaF_2 as the stress plate. The shortest wavelength at which they can measure is about 1300 Å, limited by the stress plate transmittance.

The phenomenon of magnetic circular dichroism (MCD) has also been used to detect circular polarization in the soft x-ray region. Holldack et al. [69] used MCD to make a detector for circularly polarized radiation. Their detector is a 500-Å-thick layer of Fe on a film of 3-μm-thick Mylar foil mounted on a magnetic yoke, and inclined at an angle of 45 deg to the photon stream. A electromagnet, capable of supplying fields at the Fe foil of 200 G/A, was used to magnetize the Fe. As a dispersing element, they used a multilayer consisting of 100 periods of C/Cr at an angle of incidence of about 16 deg. The radiation source was a bending magnet of a storage ring. By raising the detector above and below the plane of the ring current they were able to calculate the degree of elliptical polarization of the synchrotron radiation.

Reference

1. J. M. Bennett and H. E. Bennett, Sec. 10 in "Handbook of Optics" (W. G. Driscoll and W. Vaughan, eds.), McGraw-Hill Book Co., New York (1978).
2. J. M. Bennett, Chapt. 5 in "Handbook of Optics," 2nd ed. (M. Bass, E. W. Van Stryland, D. R. Williams, and W. L. Wolfe, eds.), McGraw-Hill Book Co., New York (1995).
3. W. C. Johnson, Jr., *Rev. Sci. Instrum.* **35**, 1375–1376 (1964).
4. D. L. Steinmetz, W. G. Phillips, M. Wirick, and F. F. Forbes, *Appl. Opt.* **6**, 1001–1004 (1967).
5. W. R. S. Garton, private communication.
6. W. C. Walker, *Appl. Opt.* **3**, 1457–1459 (1964).
7. D. C. HInson, *J. Opt. Soc. Am.* **56**, 408 (1966).
8. J. Schellman, V. Chandrasekharan, and H. Danny, *C. R. Acad. Sci. Paris* **259**, 4560–4563 (1964).
9. W. R. Hunter, *Jpn. J. Appl. Phys.* **4**(suppl 1), 520–526 (1965).
10. J. Schellman, V. Chandrasekharan, H. Damany, and J. Romand, *C. R. Acad. Sci. Paris* **260**, 117–120 (1965).
11. W. König, in "Handbuch der Physik," Vol. 20 (H. Geiger and K. Scheel, eds.), J. Springer, Berlin, p. 242 (1928).
12. F. Abeles, *C. R. Acad. Sci. Paris* **230**, 1942–1943 (1950).
13. P. W. Berning, *Phys. Thin Films* **1**, 69–121 (1963).

14. S. P. F. Humphreys-Owen, *Proc. Phys. Soc.* **77**, 949–957 (1961).
15. H. Damany, *Optica Acta* **12**, 95–107 (1965).
16. W. R. Hunter, *Appl. Opt.* **17**, 1259–1270 (1978).
17. L. R. Canfield, G. Hass, and W. R. Hunter, *J. Phys. (Paris)* **25**, 124–129 (1964).
18. T. Sasaki and H. Fukutani, *Jpn. J. Appl. Phys.* **3**, 125–126 (1964).
19. W. R. Hanson, E. T. Aarakawa, and M. W. Williams, *J. Appl. Phys.* **43**, 1661–1665 (1972).
20. G. Hass, Night Vision Laboratories, Ft. Belvoir, VA, unpublished data.
21. C. G. Barkla, *Proc. Roy. Soc.* **77**, 247–255 (1906).
22. G. Stephan, J.-C. Lemonnier, Y. Le Calvez, and S. Robin, *C. R. Acad. Sci. Paris* **262**, 1272–1275 (1966).
23. J. Cazaux, *Nouv. Rev. Opt. Appl.* **2**, 397–403 (1971).
24. W. H. Hancock and J. A. R. Samson, *J. Electron Spectrosc. Relat. Phenom.* **9**, 211–216 (1976).
25. H. Winter, H. H. Bukow, and P. H. Heckman, *Opt. Commun.* **11**, 299–300 (1974).
26. M. B. Robin, N. A. Kuebler, and Yoh-Han Pao, *Rev. Sci. Instrum.* **37**, 922–924 (1966).
27. R. N. Hamm, R. A. MacRae, and E. T. Arakawa, *J. Opt. Soc. Am.* **55**, 1460–1463 (1965).
28. V. G. Horton, E. T. Aarakawa, R. N. Hamm, and M. W. Williams, *Appl. Opt.* **8**, 667–670 (1969).
29. A. Matsui and W. C. Walker, *J. Opt. Soc. Am.* **60**, 64–65 (1970).
30. G. Hass and W. R. Hunter, *Appl. Opt.* **17**, 76–82 (1978).
31. T. A. Remneva, A. V. Kozhevnikov, and M. M. Nikitin, *J. Appl. Spectrosc.* **25**, 1587–1590 (1977).
32. T. Saito, A. Ejiri, and H. Onuki, *Appl. Opt.* **29**, 4538–4540 (1990).
33. T. Koide, T. Shidara, M. Yuri, N. Kandaka, H. Fukutani, and K. Yamaguchi, *Rev. Sci. Instrum.* **63**, 1458–1461 (1992).
34. G. Rosenbaum, B. Feuerbacher, R. P. Godwin, and M. Skibowski, *Appl. Opt.* **7**, 1917–1920 (1968).
35. N. Watanabe, *J. Phys. E* 546 (1975).
36. W. R. Hunter and J. C. Rife, unpublished manuscript.
37. E. S. Gluskin, E. M. Trakhtenberg, and A. S. Vinogradov, *Nucl. Instrum. Methods* **152**, 133–134 (1978).
38. W. R. Hunter, *Appl. Opt.* **21**, 2103–2114 (1982).
39. J. Kim, M. Zukic, and D. G. Torr, *Proc. SPIE* **1742**, 413–422 (1992).
40. A Khandar, P. Dhez, and M. Berland, *Proc. SPIE* **688**, 176–183 (1986).
41. M. Yanagihara, T. Maehara, H. Nomura, T. Yamamoto, T. Namioka, and H. Kimura, *Rev. Sci. Instrum.* **63**, 1516–1518 (1992).
42. K. Rabinovitch, L. R. Canfield, and R. P. Madden, *Appl. Opt.* **4**, 1005–1010 (1965).
43. J. A. R. Samson, *Rev. Sci. Instrum.* **47**, 859–869 (1976).
44. J. A. R. Samson, *Nucl. Instrum. Methods* **152**, 225–230 (1978).
45. E. S. Gluskin, *Rev. Sci. Instrum.* **63**, 1523–1524 (1992).
46. F. A. Jenkins and H. E. White, "Fundamentals of Physical Optics," McGraw-Hill Book Co., New York (1937).
47. J. A. Stratton, "Electromagnetic Theory," McGraw-Hill Book Co., New York, p. 500 (1941).
48. V. A. Kizel, Yu. I. Krasilov, and V. N. Shamraev, *Opt. Spectrosc.* **17**, 248–249 (1964).

49. P. B. Clapham, M. J. Downs, and R. J. KIng, *Appl. Opt.* **8**, 1965–1974 (1969).
50. N. N. Nagib, *Appl. Opt.* **36**, 1547–1552 (1997); N. N. Nagib and S. A. Khodier, *Appl. Opt.* **34**, 2927–2930 (1995); N. N. Nagib and M. S. El-Bahrawy, *Appl. Opt.* **33**, 1218–1222 (1994).
51. T. Saito, A. Ejiri, and H. Onuki, *Appl. Opt.* **29**, 4538–4540 (1990).
52. R. M. A. Azzam, *J. Opt. Soc. Am.* **71**, 1523–1528 (1981).
53. A. M. Kan'an and R. M. A. Azzam, *Opt. Eng.* **34**, 1551–1556 (1995).
54. T. J. McIlrath, *J. Opt. Soc. Am.* **58**, 506–510 (1968).
55. W. B. Westerveld, K. Becker, P. W. Zetner, and J. W. McConkey, *Appl. Opt.* **24**, 2256–2262 (1985).
56. J. B. Kortright and J. H. Underwood, *Nucl. Instrum. Methods* **A291**, 272–277 (1990).
57. S. Di Fonzo and W. Jark, *Rev. Sci. Instrum.* **63**, 1375–1378 (1992).
58. J. Kim, M. Zukic, D. G. Torr, and M. M. Wilson, *Proc. SPIE* **1742**, 403–412 (1992).
59. M. Schledermann and M. Skibowski, *Appl. Opt.* **10**, 321–326 (1971).
60. P. D. Johnson and N. V. Smith, *Nucl. Instrum. Methods* **214**, 505–508 (1983).
61. N. V. Smith and M. R. Howells, *Nucl. Instrum. Methods* **A347**, 115–118 (1994).
62. H. Höchst, R. Patel, and F. Middleton, *Nucl. Instrum. Methods* *A*347, 107–114 (1994).
63. H. Höchst, P. Bulicke, T. Nelson, and F. Middleton, *Rev. Sci. Instrum.* **66**, 1598–1600 (1995).
64. U. Heinzmann, *J. Phys. E* **10**, 1001–1005 (1977).
65. H. Winter and H. W. Ortjohann, *Rev. Sci. Instrum.* **58**, 359–362 (1987).
66. H. Metcalf and J. C. Baird, *Appl. Opt.* **5**, 1407–1410 (1966).
67. J. C. Kemp, *J. Opt. Soc. Am.* **59**, 950–954 (1969).
68. A. Gedanken and M. Levy, *Rev. Sci. Instrum.* **48**, 1661–1664 (1977).
69. K. Holldack, T. Kachel, F. Schäfers, and I. Packe, *Rev. Sci. Instrum.* **67**, 2485–2489 (1996).

13. OPTICAL PROPERTIES OF MATERIALS

E. M. Gullikson

Center for X-Ray Optics
Materials Sciences Division
Lawrence Berkeley National Laboratory
Berkeley, CA

13.1 Introduction

Accurate knowledge of the optical properties of materials is required in order to describe anything involving the interaction of radiation with matter. This includes predicting or modeling the behavior of an optical element, such as mirror reflectivity, filter transmission, detector response, diffraction grating efficiency, and multilayer reflectivity.

This chapter gives a brief summary of the essential features of the optical properties of materials in the vacuum ultraviolet (VUV) through the x-ray regions. First the classical model of the electromagnetic response of free electrons is reviewed. The atomic scattering factor is defined, which may be used to calculate material optical constants. The propagation of a wave within an absorbing material is reviewed in Section 13.3. Finally, the reflection and transmission of radiation from a boundary are summarized.

13.2 Optical Constants

13.2.1 Classical Model of Dispersion

Many of the important features of the optical properties of materials can be seen from the classical response of an electron in an oscillating electric field **E**. This is the Drude model, which is covered in many texts [1, 2]. The equation of motion for a free electron in an electric field is

$$m\left(\frac{d^2\mathbf{x}}{dt^2} + \gamma \frac{d\mathbf{x}}{dt}\right) = -e\mathbf{E}, \tag{1}$$

where γ is a phenomenological damping constant. If the electric field varies in time [3, 4] as $e^{-i\omega t}$, the dipole moment **p** produced by the motion of the electron relative to a stationary ion will result in a polarizability

$$\alpha = \frac{\mathbf{p}}{\mathbf{E}} = \frac{-e\mathbf{x}}{\mathbf{E}} = \frac{-e^2/m}{\omega^2 + i\omega\gamma}. \tag{2}$$

For a solid with N_e "free" electrons per unit volume, the macroscopic polarization is given by

$$\mathbf{P} = N_e \alpha \mathbf{E}. \tag{3}$$

The dielectric constant ε is defined by $\mathbf{D} = \varepsilon \mathbf{E} = \mathbf{E} + 4\pi \mathbf{P}$. Then, using Eq. (3), the dielectric constant is

$$\varepsilon = 1 + 4\pi N_e \alpha. \tag{4}$$

For simplicity we neglect any difference between the applied field and the local field felt by the electron. The dielectric constant is then

$$\varepsilon = 1 - \frac{\omega_p^2}{\omega^2 + i\omega\gamma}, \tag{5}$$

where $\omega_p^2 = 4\pi N_e e^2/m = 4\pi N_e r_e c^2$ is the plasma frequency squared and $r_e = e^2/mc^2 = 2.818 \times 10^{-13}$ cm is the classical electron radius. The index of refraction is related to the dielectric constant by

$$n = \sqrt{\varepsilon} = 1 - \delta + \imath\beta, \tag{6}$$

where for energies above about 50 eV δ and β are small compared to one. If for the moment we neglect the damping constant, we be see from Eq. (5) that for frequencies above ω_p, $\varepsilon > 0$ and the index of refraction is real. For frequencies below ω_p, $\varepsilon < 0$, the index of refraction is imaginary, and an electromagnetic wave will be damped.

The plasma frequency is often one of the most dramatic features in the optical response of materials in the VUV. In particular, this is the case for free-electron metals where interband transitions do not contribute significantly to the absorption in the region of ω_p. This can be seen for Al in Fig. 1, which shows the absorption coefficient from [5, 6]. The plasma energy $\hbar\omega_p$ is at about 15 eV and is evidenced by a sharp drop in the absorption for energies above 15 eV. The dashed curve in Fig. 1 is calculated from Eq. (5) with $\hbar\omega_p = 14.8$ eV and $\hbar\gamma = 0.6$ eV. Below 15 eV the absorption coefficient saturates at a value of $\mu = 2\omega_p/c = 2\sqrt{4\pi r_e N_e} = 150$ μm^{-1}. At high energies, jumps in absorption are observed at the L$_{2,3}$ (2p) edge at 72.7 eV and the K (1s) edge at 1560 eV. There is fine structure above these edges that is not shown. Above the plasma energy the absorption coefficient drops more slowly than predicted by Eq. (5). At energies below the L edge the absorption coefficient is close to what is obtained using the theoretical photoabsorption cross section for atomic Al [7]. As discussed in the next section the optical response of materials at higher energies is "atomic-like."

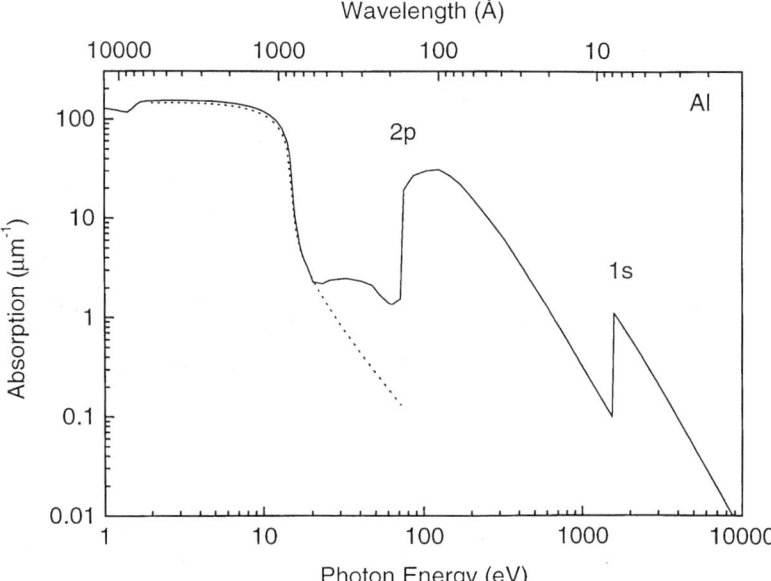

FIG. 1. The absorption coefficient of Al versus energy/wavelength (solid curve) from [5, 6]. The measured absorption is close to the classical Drude model (dashed curve) for energies near and below the plasma energy (15 eV). Steps in absorption are observed at the $L_{2,3}$ (2p) edge at 72.7 eV and the K (1s) edge at 1560 eV. The fine structure at energies above these edges is not shown.

13.2.2 Atomic Scattering Factor

The optical properties of a material in the visible region through much of the VUV region are strongly influenced by chemical state or band structure. This is common knowledge based on our everyday experience. For example, both graphite and diamond are made up completely of carbon atoms and yet their optical properties in the visible region are quite different. For these regions, the optical constants must be compiled for each material of interest. There have been many excellent compilations of the optical constants of materials in the VUV through the soft x-ray region [8, 9] with the most recent and complete being the *Handbook of Optical Constants of Solids* [10, 11]. In the EUV and soft x-ray regions, the photon energy is large compared to molecular binding energies and the optical response of a material is often determined by tightly bound core electrons that are insensitive to the chemical environment of an atom. In this case, the optical properties of a material are given to a good approximation by summing the responses of the individual atoms.

The *atomic scattering factor* $f = f_1 - if_2$ may be defined as the complex number which multiplies the polarizability of a single free electron to give the atomic polarizability

$$\alpha = -\frac{e^2}{m\omega^2}f = -r_e\frac{\lambda^2}{4\pi^2}f. \tag{7}$$

The polarizability of a single free electron may be seen from Eq. (2) with $\gamma = 0$. The atomic scattering factor describes the electromagnetic response of the atom. For photon energies that are high compared to the binding energies of the electrons, f approaches the atomic number Z. It is assumed here that the wavelength is large compared to atomic dimensions, or the scattering angle is small, so the atoms scatter as dipoles. The index of refraction of a material may then be obtained from

$$n = 1 - \delta + i\beta \approx 1 + 2\pi \sum_i N_i \alpha_i = 1 - \frac{r_e \lambda^2}{2\pi} \sum_i N_i f_i, \tag{8}$$

where the sum is over the different atomic species with atomic density N_i (atoms/cm^3) and atomic scattering factor f_i. The real and imaginary parts of the index of refraction are then

$$\delta \approx \frac{r_e \lambda^2}{2\pi} \sum_i N_i f_{i1} \quad \text{and} \quad \beta \approx \frac{r_e \lambda^2}{2\pi} \sum_i N_i f_{i2}. \tag{9}$$

The real and imaginary parts of the atomic scattering factor of Al are shown in Fig. 2. The real part may be thought of as the effective number of free electrons. From Eq. (9), we can see that the imaginary part f_2 is proportional to β, which, as shown in the next section, is proportional to the absorption coefficient.

This is an extremely useful and powerful approximation because it allows one to calculate the optical properties of any material given the atomic scattering factors of its constituent atoms. There are, of course, regions where the chemical state of a material is important, and in those cases this atomic approach breaks down. This is true in particular in the vicinity of absorption edges where the final states are influenced by the chemical environment of an atom, resulting in near-edge structure in absorption spectra. However, for most of the EUV and soft x-ray regions Eqs. (8) and (9) provide a good approach to calculating the optical constants of materials. The atomic scattering factors for the first 92 elements have been tabulated in the energy range from 30 eV to 30 keV in [5, 12].

13.3 Wave Propagation in a Solid

All practical materials become absorbing for photon energies above 11.9 eV (LiF). In this section the propagation of an electromagnetic wave in an

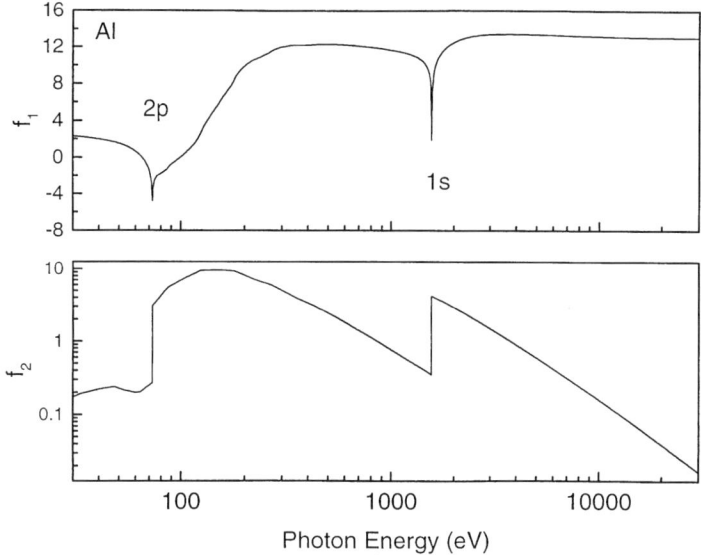

FIG. 2. Real and imaginary parts of the atomic scattering factor of Al. The real part f_1 approaches the atomic number Z at high energy. The imaginary part is proportional to the absorption coefficient.

absorbing material is reviewed [13, 14]. A very thorough treatment is given by Stern [15]. Consider a plane wave where the electric field is

$$\mathbf{E} = \mathbf{E}_0 e^{i(\mathbf{k}\cdot\mathbf{r} - \omega t)}. \tag{10}$$

In the presence of absorption the wave vector is complex

$$\mathbf{k} = \mathbf{k}' + i\mathbf{k}''; \tag{11}$$

thus, the electric field becomes

$$\mathbf{E} = \mathbf{E}_0 e^{-\mathbf{k}''\cdot\mathbf{r}} e^{i(\mathbf{k}'\cdot\mathbf{r} - \omega t)}. \tag{12}$$

The real part of the wave vector $\mathbf{k}'$ is perpendicular to the planes of constant phase and the imaginary part $\mathbf{k}''$ is perpendicular to the planes of constant amplitude. In general, $\mathbf{k}'$ and $\mathbf{k}''$ are not parallel and the wave is sometimes referred to as *inhomogeneous*.

For a plane wave of the form given in Eq. (10), Maxwell's equations can be written as

$$\mathbf{k}\cdot\mathbf{H} = 0, \quad \mathbf{k}\cdot\mathbf{E} = 0, \quad \mathbf{k}\times\mathbf{E} = \frac{\omega}{c}\mathbf{H}, \quad \mathbf{k}\times\mathbf{H} = -\frac{\omega}{c}\varepsilon\mathbf{E}, \tag{13}$$

where it is assumed that the magnetic permeability is unity. From the last two equations it follows that

$$\mathbf{k} \times (\mathbf{k} \times \mathbf{E}) = \frac{\omega}{c} \mathbf{k} \times \mathbf{H} = -\left(\frac{\omega}{c}\right)^2 \varepsilon \mathbf{E}$$
$$= (\mathbf{k} \cdot \mathbf{E})\mathbf{k} - (\mathbf{k} \cdot \mathbf{k})\mathbf{E} = -(\mathbf{k} \cdot \mathbf{k})\mathbf{E}, \quad (14)$$

$$\mathbf{k} \cdot \mathbf{k} = \left(\frac{\omega}{c}\right)^2 \varepsilon = \left(\frac{2\pi}{\lambda}\right)^2 n^2. \quad (15)$$

Now let us consider a wave incident from vacuum onto the surface of a material with dielectric constant ε with the geometry shown in Fig. 3. In order for the fields to be continuous everywhere on the surface, the in-plane components of all the wave vectors must be equal:

$$k_{ix} = k_{rx} = k_{tx} = \frac{2\pi}{\lambda} \cos \theta. \quad (16)$$

In the absence of absorption the condition of Eq. (16) leads to Snell's law of refraction. For an absorbing material, since the incident and reflected wave vectors are real, Eq. (16) requires that the x component of the transmitted wave vector also be real. Thus the imaginary part $\mathbf{k}''_t$ must be in the z direction; that is, the planes of constant amplitude must be parallel to the surface. From Eqs. (15) and (16) it follows that

$$k_{tz} = \sqrt{\left(\frac{2\pi}{\lambda}\right)^2 n^2 - k_{tx}^2} = \frac{2\pi}{\lambda} \sqrt{n^2 - \cos^2 \theta}. \quad (17)$$

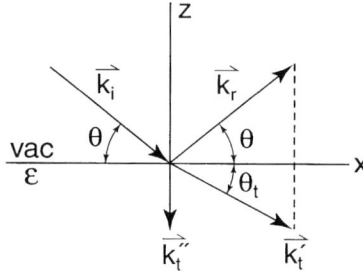

FIG. 3. Defining the geometry for a wave incident from vacuum onto the surface of a material with dielectric constant ε. The incident and reflected wave vectors are $\mathbf{k}_i$ and $\mathbf{k}_r$, respectively. Inside the material the wave is inhomogeneous and propagates along $\mathbf{k}'_t$. The imaginary part of the wave vector $\mathbf{k}''_t$ is normal to the surface, that is, the planes of constant amplitude are parallel to the surface.

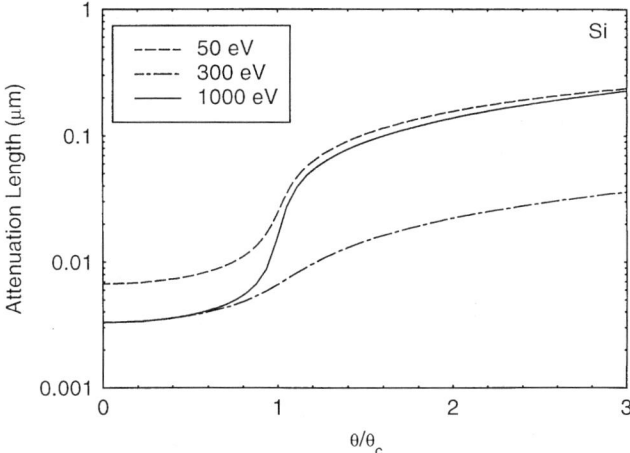

FIG. 4. The attenuation length along the surface normal versus angle of incidence for Si at three energies. The angle is normalized to the critical angle θ_c = 16.9, 5.5, and 1.7 deg for E = 50, 300, and 1000 eV, respectively.

In general, k_{tz} is complex. The real part k'_{tz} is related to the propagation direction of the transmitted wave by

$$\tan \theta_t = \frac{k'_{tz}}{k_{tx}} = \frac{\text{Re}\{\sqrt{n^2 - \cos^2 \theta}\}}{\cos \theta}, \qquad (18)$$

where θ_t as defined in Fig. 3 is real. Note that it is not necessary to introduce complex angles as is often done.

It is instructive to consider the case when the absorption is negligible and $n = 1 - \delta$. The quantity under the square root, $n^2 - \cos^2 \theta$, changes sign when $\cos \theta = n = 1 - \delta$. The angle where this occurs is the *critical angle*, which for small δ is given by $\theta_c = \sqrt{2\delta}$. For angles below the critical angle the quantity under the square root is negative and θ_t is small or zero in the absence of absorption. Thus, the transmitted wave propagates almost parallel to the surface. In this case there is total external reflection as is discussed in the next section.

The imaginary part k''_{tz} gives the absorption coefficient

$$\mu = 2k''_{tz} = 2 \times \text{Im}\{\sqrt{n^2 - \cos^2 \theta}\}, \qquad (19)$$

in the expression for the transmitted intensity at a depth z

$$I = I_0 e^{-\mu z}, \qquad (20)$$

where I_0 is the incident intensity.

The attenuation length in the z direction μ^{-1} is shown in Fig. 4 versus angle of incidence for Si at three energies. The attenuation length decreases with decreasing angle and drops sharply at the critical angle. The value below the critical angle is approximately $\mu^{-1} = (2k_p)^{-1}$ where $k_p = \omega_p/c = \sqrt{4\pi r_e N f_1}$. Except for the variation in f_1 with energy, this is just the low-frequency attenuation length at normal incidence in the visible region where $\omega < \omega_p$. The attenuation length at angles below the critical angle is generally in the range of 30–100 Å for all materials.

For normal incidence $\theta = 90$ deg, the wave is homogeneous and the familiar relationship applies:

$$k_{tz} = \frac{2\pi}{\lambda} n = \frac{2\pi}{\lambda}(1 - \delta + i\beta). \tag{21}$$

In this case the absorption coefficient is simply proportional to β:

$$\mu = \frac{4\pi\beta}{\lambda}. \tag{22}$$

In the soft and hard x-ray regions the *mass* absorption coefficient $\mu_m = \mu/\rho$ is often used since it is independent of the material density ρ. As long as δ and β are small, the absorption coefficient can be expressed as a simple average over the atomic photoabsorption cross sections, σ_i

$$\mu = \sum_i N_i \sigma_i = \frac{\rho N_A}{\text{MW}} \sum_i x_i \sigma_i, \tag{23}$$

where $N_A = 6.022 \times 10^{23}$ is Avogadro's number, $\text{MW} = \sum_i x_i \text{AW}_i$ is the molecular weight and x_i is the fraction of the element with atomic weight AW_i. Thus μ is proportional to ρ. In the VUV and at lower energies this is, of course, not necessarily the case.

13.4 Reflection and Transmission

In this section, the reflection and transmission of a wave incident on the boundary between two materials is reviewed and the Fresnel equations for the reflection and transmission coefficients are given. Consider a wave incident from medium 1 with a dielectric constant ε_1 onto a boundary with medium 2 having a dielectric constant ε_2. The reflection and transmission coefficients for s polarization ($\mathbf{E}_0 = E_0 \hat{\mathbf{y}}$ in Fig. 3) are defined by

$$\mathbf{E}_{r0} = r_s \mathbf{E}_{i0} \quad \text{and} \quad \mathbf{E}_{t0} = t_s \mathbf{E}_{i0}. \tag{24}$$

Since the component of $\mathbf{E}$ in the surface plane must be continuous,

$$E_{t0} = E_{i0} + E_{r0}, \quad \text{thus } t_s = 1 + r_s. \tag{25}$$

For the component of **H** in the surface plane to be continuous,

$$k_{tz} E_{t0} = k_{iz}(E_{i0} - E_{r0}), \quad \text{thus } k_{tz} t_s = k_{iz}(1 - r_s). \tag{26}$$

These two conditions yield for the reflection and transmission coefficients

$$r_s = \frac{k_{iz} - k_{tz}}{k_{iz} + k_{tz}}, \tag{27}$$

$$t_s = 1 + r_s = \frac{2k_{iz}}{k_{iz} + k_{tz}}. \tag{28}$$

For p polarization ($\mathbf{H}_0 = H_0 \hat{\mathbf{y}}$) it is more convenient to define the reflection and transmission coefficients in terms of **H**:

$$\mathbf{H}_{r0} = r_p \mathbf{H}_{i0} \quad \text{and} \quad \mathbf{H}_{t0} = t_p \mathbf{H}_{i0} \tag{29}$$

and

$$r_p = \frac{n_2^2 k_{iz} - n_1^2 k_{tz}}{n_2^2 k_{iz} + n_1^2 k_{tz}}, \tag{30}$$

$$t_p = 1 + r_p = \frac{2n_2^2 k_{iz}}{n_2^2 k_{iz} + n_1^2 k_{tz}}. \tag{31}$$

The reflectivity for an arbitrary linear polarization is given by

$$R = \frac{1+P}{2} |r_s|^2 + \frac{1-P}{2} |r_p|^2, \tag{32}$$

where P is the degree of polarization specified by $P = (I_s - I_p)/(I_s + I_p)$.

In Fig. 5 the reflectivity for s polarization is calculated as a function of angle relative to the critical angle. In the absence of absorption, $\beta = 0$, the reflectivity goes to one at the critical angle. In this case, the transmitted wave propagates parallel to the surface and the component of energy flow across the surface is zero. For increasing absorption the reflectivity below the critical angle is reduced and there is a corresponding increase in the transmission into the substrate. Above the critical angle the reflectivity falls off as $(\sin \theta)^{-4}$.

The transmission for either s or p polarization is given by

$$T = |t|^2 \frac{k'_{tz}}{k_{iz}} = |t|^2 \frac{\tan \theta_t}{\tan \theta}. \tag{33}$$

Because T is defined in terms of the energy flow across a plane that is parallel to the surface, the factor $\tan \theta_t / \tan \theta$ accounts for the change in the propagation direction of the transmitted wave. It can be verified that $R + T = 1$, that is, energy is conserved.

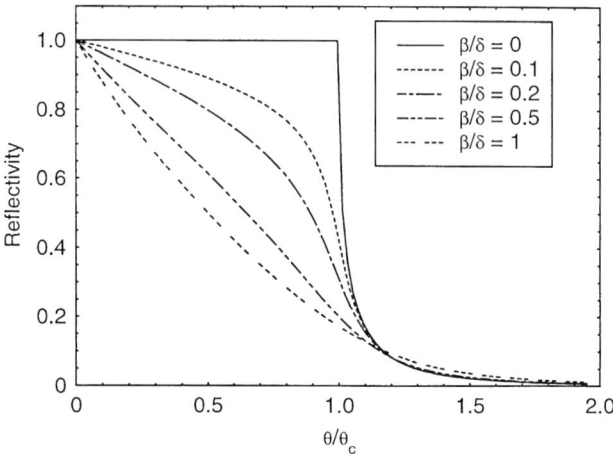

FIG. 5. Reflectivity versus angle for varying amounts of absorption. For $\beta = 0$ the reflectivity is unity below the critical angle and there is total external reflection. As the absorption increases the reflectivity below the critical angle is reduced and for $\beta/\delta \geq 1$ there is no well-defined critical angle.

We have assumed thus far that the interface between the two materials is infinitely sharp. In reality all interfaces have a finite width due to high-frequency roughness or interdiffusion between two materials. For roughness of sufficiently high spatial frequencies that the scattering is negligible, the reflectivity is commonly calculated by assuming a density that changes smoothly over a finite distance at the surface. For low-frequency roughness the situation can be more complicated [16]. The reflection and transmission from a graded interface may be calculated numerically by dividing the interface profile into discrete layers. Alternatively, one can use the approximate factors originally derived by Névot and Croce [17] and more recently by Sinha *et al.* [18] for an interface with an error function profile. The reflection coefficient is reduced from that of a sharp interface to

$$r = r_0 e^{-2k_{iz}k_{tz}\sigma^2} \tag{34}$$

and the transmission coefficient is increased to

$$t = t_0 e^{(k_{iz} - k_{tz})^2 \sigma^2 / 2}. \tag{35}$$

Here r_0 and t_0 are the reflection and transmission coefficients, respectively, of an ideal interface as given above for either s or p polarization.

In Fig. 6 the effect of roughness on reflectivity is illustrated for a Au surface at 150 eV. The reflectivity is reduced from that of an ideal interface with the largest reduction occurring at normal incidence. The points were calculated by

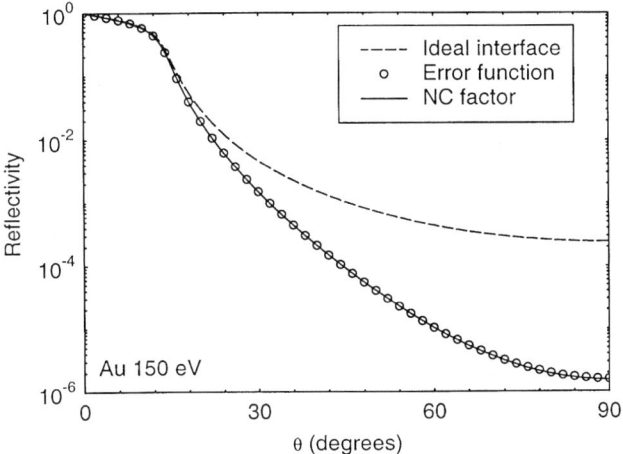

FIG. 6. The effect of roughness on the reflectivity of a Au surface at 150 eV. The points are calculated for an error function profile ($\sigma = 15$ Å), which is divided into discrete layers. The solid line is obtained using the approximate formula given in Eq. (34).

dividing the error function transition region into discrete layers. The solid curve, which was calculated using the Névot–Croce factor in Eq. (34), provides an excellent approximation.

13.4 Determination of Optical Constants

The material optical constants form the basis for the interpretation of measurements and the design of optical components and systems. The x-ray optical properties of materials have been studied since the early 1900s. Even so, for many of the elements there is little or no available data for large spectral regions of the VUV through the soft x-ray regions. In such cases, current tabulations [5] rely largely on theoretical atomic scattering factors. For some materials, high-quality measurements have been performed over a wide range of energies from the visible to the x-ray regions. A good example of a well-studied material is Al, where the optical constants based on numerous experimental studies have been determined over a wide energy range[19, 20]. However, even for Al new measurements in the EUV [7] have revealed previously unobserved structure and discrepancies of up to a factor of 2 with the tabulated absorption coefficient. Thus, there is still a need for high-quality measurements of the optical properties of materials particularly in the VUV through the soft x-ray regions.

In this section, a brief overview is given of the various methods for determining the optical constants. The Kramers–Kronig or dispersion relations relate the real and imaginary parts and provide an important and useful constraint on the optical constants.

The absorption coefficient μ or the imaginary part of the index of refraction β is directly determined by measuring the transmission of a thin film. The dispersive part δ of the refractive index may then be determined from the Kramers–Kronig relation

$$\delta(E) = \frac{hc}{2\pi^2} P \int_0^\infty \frac{\mu(E')}{E^2 - E'^2} dE'. \tag{36}$$

Once the values of both δ and β are known the atomic scattering factor may be obtained, for example, in [5] Eqs. (9) were used. Note that for energies above about 100 eV, the approximation used in Eq. (8) is good, whereas for lower energies δ and β may not always be small compared to unity. However, when calculating δ and β from the atomic scattering factors it is important to use the same relationship as was initially used to derive the scattering factors.

In the VUV where the absorption is high and extremely thin samples are required for transmission measurements, it is often preferable to measure the normal incidence reflectivity. In this case the phase shift dispersion relation [15] can be used:

$$\theta(E) = \frac{E}{\pi} \int_0^\infty \frac{\ln R(E') - \ln R(E)}{E^2 - E'^2} dE', \tag{37}$$

where $R(E)$ is the measured reflectivity and θ is the phase. Thus, the reflection coefficient is

$$r = \sqrt{R} e^{i\theta}. \tag{38}$$

Once the reflection coefficient is known, the complex index of refraction may be obtained by inverting the Fresnel equation, Eq. (27). Although this method is most often used in the VUV region for normal incidence reflectance measurements, it is also useful at higher energies and at more grazing angles of incidence.

By measuring the angle dependence of the reflectivity it is possible to determine experimentally both δ and β at each energy [21, 22]. This is extremely valuable in that it provides a independent check of the optical constants derived from the Kramers–Kronig relations. Since the Kramers–Kronig analysis involves an integral over a wide energy range, errors in one region may affect the derived values of the optical constant at other energies. Extremely smooth and clean surfaces are required in order to obtain accurate values with this method. Surface contamination is an issue of particular interest for the VUV and EUV regions where the absorption of the contaminant layer is often high. As can be seen from Fig. 5, when the absorption is small, the critical angle $\theta_c = \sqrt{2\delta}$ is clearly seen and β may be determined from the "rounding" of the curve in the region of the critical angle. However, when the absorption is high, $\beta/\delta > 1$, the reflectance has no distinguishing feature at the critical angle and it becomes difficult to determine uniquely [23] both δ and β.

Many more methods are available that are useful for deriving the optical properties of materials in the VUV through the soft x-ray range. These include electron energy loss spectroscopy [24, 25], ellipsometry [26], the angle dependence of the photoelectric yield [27], and interferometry [28]. Space prohibits going into the details of these methods here. Finally, it is important to mention the sum rules [29], which are extremely useful in evaluating the accuracy of measured optical constants over a wide range of energies.

References

1. J. D. Jackson, "Classical Electrodynamics," Wiley, New York (1975).
2. F. Wooten, "Optical Properties of Solids," Academic Press, New York (1972).
3. The complex conjugate can also be used. In that case, the imaginary part of all quantities would have the opposite sign. This choice of sign is arbitrary and both conventions are used.
4. R. T. Holm, in "Handbook of Optical Constants of Solids II" (E. D. Palik, ed.), Academic Press, New York, pp. 21–55 (1991).
5. B. L. Henke, E. M. Gullikson, and J. C. Davis, *Atom. Data Nucl. Data Tables* **54**, 181–342 (1993).
6. D. Y. Smith, E. Shiles, and M. Inokuti, in "Handbook of Optical Constants of Solids" (E. D. Palik, ed.), Academic Press, New York, pp. 369–406 (1985).
7. E. M. Gullikson, P. Denham, S. Mrowka, and J. H. Underwood, *Phys. Rev. B* **49**, 16283–16288 (1994).
8. J. H. Weaver, C. Krafka, D. W. Lynch, and E. E. Koch, *Phys. Data* 18–1, 2 (1981).
9. H.-J. Hagemann, W. Gudat, and C. Kunz, Technical Report No. SR-74/7, DESY (1974).
10. E. D. Palik, ed., "Handbook of Optical Constants of Solids," Academic Press, New York (1985).
11. E. D. Palik, ed., "Handbook of Optical Constants of Solids II," Academic Press, New York (1991) and "Handbook of Optical Constants of Solids III," Academic Press, New York (1998).
12. E. M. Gullikson, atomic scattering factor files and calculations of transmission and reflectivity of mirrors and multilayers; http://www-cxro.lbl.gov/optical_constants/.
13. M. Born and E. Wolf, "Principles of Optics," 5th ed., Pergamon Press, Oxford (1975).
14. L. D. Landau and E. M. Lifshitz, "Electrodynamics of Continuous Media," Pergamon Press, Oxford, Chaps. 9 and 10 (1960).
15. F. Stern, in "Solid State Physics," Vol. 15 (F. Seitz and D. Turnbull, eds.), Academic Press, New York, pp. 299–408 (1963).
16. D. K. G. de Boer, *Phys. Rev. B* **51**, 5297–5305 (1995).
17. L. Névot and P. Croce, *Rev. Phys. Appl.* **15**, 761–779 (1980) (in French).
18. S. K. Sinha, E. B. Sirota, S. Garoff, and H. B. Stanley, *Phys. Rev. B* **38**, 2297–2311 (1988).

19. H. R. Philip and H. Ehrenreich, *J. Appl. Phys.* **35**, 1416–1419 (1964).
20. E. Shiles, T. Sasaki, M. Inokuti, and D. Y. Smith, *Phys. Rev. B* **22**, 1612–1628 (1980).
21. W. R. Hunter, *J. Opt. Soc. Am.* **54**, 15–19 (1964).
22. D. L. Windt, J. W. C. Cash, M. Scott, P. Arendt, B. Newnam, R. F. Fisher, A. B. Swartzlander, P. Z. Takacs, and J. M. Pinneo, *Appl. Opt.* **27**, 249–295 (1988).
23. R. Soufli and E. M. Gullikson, *Appl. Opt.* **36**, 5499–5507 (1997).
24. H. Raether, in "Springer Tracts in Modern Physics," Vol. 88 (G. Hoehler, ed.), Springer-Verlag, Berlin (1980).
25. S. E. Schnatterly, in "Solid State Physics," Vol. 34 (H. Ehrenreich, F. Seitz, and D. Turnbull, eds.), Academic Press, New York, pp. 275–358 (1979).
26. J. Barth, R. L. Johnson, and M. Cardona, in "Handbook of Optical Constants of Solids II," (E. D. Palik, ed.), Academic Press, New York, pp. 213–246 (1991).
27. H.-G. Birken, C. Blessing, C. Kunz, and R. Wolf, *Rev. Sci. Instrum.* **60**, 2223–2226 (1989).
28. J. Svatos, D. Joyeux, D. Phalippou, and F. Polack, *Opt. Lett.* **18**, 1367–1369 (1993).
29. M. Altarelli, D. L. Dexter, H. M. Nussenzveig, and D. Y. Smith, *Phys. Rev. B* **6**, 4502–4509 (1972).

14. REFLECTING OPTICS: MULTILAYERS

Eberhard Spiller
Spiller X-Ray Optics
Mt. Kisco, New York

14.1 Introduction

Multilayer coatings of thin film are used to modify the optical properties of surfaces. Enhancement or reduction of reflectivity or transmission of mirrors and lenses are well-known examples in the visible region. Coatings are also used as spectral filters and as polarizers and phase retarders. The possible performance of a multilayer coating is limited by the optical constants of the materials available as thin films. The ideal situation, that absorption- and scatter-free films of different refractive index are available, permits practically unlimited optical performance for a coating. Coatings can be designed for nearly any specification; reflectivities can be 100% and one can design mirrors to produce any arbitrary reflectivity curve $R(\lambda)$.

The absorption of materials is the most severe limitation for the performance of coatings in the vacuum ultraviolet (VUV) region. Absorption-free materials of "high index" are only available for $\lambda > 150$ nm and multilayer mirrors with reflectivities $R > 95\%$ are still available down to this wavelength [1, 2]. Single films of Al and Be have good reflectivities close to 90% for photon energies lower than their plasma resonance (Fig. 1) [3, 4]. However, this reflectiviy can only be obtained when oxidation of the surface is prevented requiring evaporating and using the film in ultra high vacuum without ever exposing it to oxygen. Films of LiF and MgF_2 still have very low absorption for wavelengths $\lambda > 110$ nm and can be used to overcoat aluminum to prevent oxidation. The thickness of the films can be adjusted such that the amplitude reflected from the top of this film adds in phase to that reflected from Al, thus enhancing the reflectivity. Multilayers of Al and MgF_2 can enhance the reflectivity even further up to $R = 96\%$ [5].

There are no absorption-free thin-film materials for $\lambda < 110$ nm. In the wavelength region $\lambda = 80$–110 nm the absorption index β—the imaginary part of the complex index of refraction $\tilde{n} = 1 - \delta + i\beta$—of all stable materials is between 0.1 and 1. Overcoating an Al or Be mirror with any other material does not produce a substantial enhancement because the absorption in the overlayer attenuates the amplitude from the metal more than the additional boundary adds. However, it is possible to use such overlayers to suppress undesired wavelengths

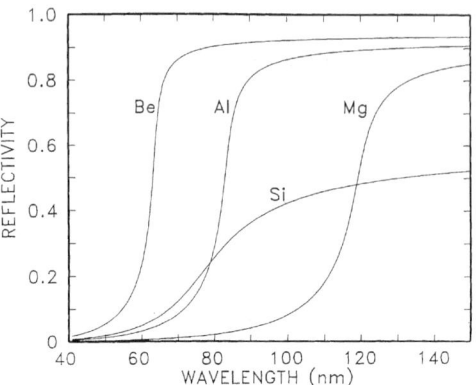

FIG. 1. Normal incidence reflectivity of some materials near their plasma frequency calculated with the Drude model with parameters from [3].

more than desired ones. This option has been used to image the $\lambda = 83.4$ nm emission from O^+ ions in the ionosphere while suppressing the much stronger hydrogen lines at $\lambda = 102.5$ and 121.6 nm [6, 7].

The alkali metals with plasma frequencies in the 3- to 5-eV range have low absorption [8, 9] in the $\lambda \simeq 100$ nm region and could be suitable for multilayer coatings. They have up to now not been used in multilayer structures because of their reactivity. Theoretically, a peak reflectivity $R \simeq 40\%$ with a bandwidth around 60 nm can be obtained with a K-C multilayer at $\lambda = 100$ nm, and a top layer of carbon might be sufficient to seal the alkali metal from the atmosphere.

The absorption of the lighter elements decreases to values close to $\beta = 0.01$ around $\lambda = 70$ nm. Radiation can propagate several wavelengths into materials allowing deeper boundaries to add their reflected amplitudes to that of the top surface, making it possible to enhance or modify the reflectivity of the best single-film materials with multilayer structures. Absorption of all materials decreases dramatically toward still shorter wavelengths, roughly proportional to λ^3 for $\lambda < 20$ nm and away from absorption edges. Simultaneously the refractive index of all materials approaches 1 with $\delta \propto \lambda^2$, while the reflectivity from a single boundary becomes very small with $R \propto \lambda^4$ and a value around $R = 10^{-5}$ for $\lambda = 6$ nm. However, it is now possible to enhance this small reflectivity by adding the reflected amplitudes from a large number of boundaries in a multilayer coating. Theoretically, one can obtain useful reflectivities above $R = 10\%$ for any wavelength below $\lambda = 20$ nm with values of $R > 80\%$ in some regions [10]. The decrease in the reflectivity of a single boundary with decreasing wavelengths is compensated by increasing the number of layers in a coating. Reflectivity enhancements over that of a single boundary can be higher than a factor of 10,000.

14.2 Multilayer Theory

To calculate the optical properties of a multilayer structure one needs Snell's law to obtain the direction of propagation in each of the layers and the Fresnel formulas for the amplitude reflection and transmission coefficients r_{ij} and t_{ij} at the boundaries between the layers. One assumes that there is no scattering in the volume of a film; only at the boundaries is an incoming wave split into a transmitted and reflected wave.

By introducing the parameters $q_i = (4\pi/\lambda)\tilde{n}_i \cos \phi_i$ for s polarization or $q_i = (4\pi/\lambda\tilde{n}_i) \cos \phi_i$ for p polarization, we can express the reflected and transmitted amplitudes at the boundary between two materials as

$$r_{12} = \frac{q_1 - q_2}{q_1 + q_2}, \qquad (1)$$

$$t_{12} = \frac{2q_1}{q_1 + q_2}. \qquad (2)$$

The q-values for s polarization are proportional to the change in the momentum of a photon perpendicular to the boundary, often called the momentum transfer in x-ray optics, while the name "effective index" is used in the literature on optical coatings (usually defined without the factor $4\pi/\lambda$). The angle ϕ_i is the propagation angle in each of the materials and is obtained from the angle ϕ_0 in the incident medium of index n_0 with Snell's law $\tilde{n}_i \sin \phi_i = n_0 \sin \phi_0$ as

$$\cos \phi_i = \sqrt{1 - (n_0/\tilde{n}_i)^2 \sin^2 \phi_0}. \qquad (3)$$

Figure 2 shows the field amplitudes a_i and b_i of the forward and backward running waves in each of the layers. The amplitude reflection in the incident

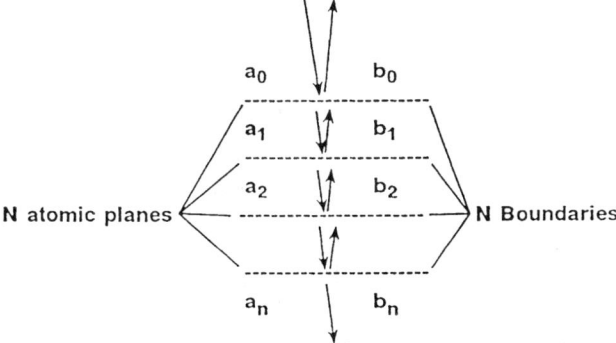

FIG. 2. Geometry of a multilayer strucure with forward running waves a_i and backward running waves b_i. The dashed lines represent the boundaries of the films in the optical theories or the atomic planes in x-ray diffraction.

medium is given by $r = b_0/a_0$ and the transmitted amplitude by $t = a_{n+1}/a_0$, if $n + 1$ represents the medium below the multilayer structure. For all calculations we assume that no radiation enters the structure from below: $b_{n+1} = 0$.

The field amplitudes in each layer in Fig. 2 are coupled to those of the adjacent layers by linear equations:

$$\begin{aligned} b_0 &= a_0 r_{01} + b_1 e^{i\varphi_1} t_{10}, \\ b_1 &= a_1 r_{12} + b_2 e^{i\varphi_2} t_{21}, \\ a_1 &= a_0 t_{01} e^{i\varphi_1} + b_1 e^{2i\varphi_1} r_{10}, \\ a_2 &= a_1 t_{12} e^{i\varphi_2} + b_2 e^{2i\varphi_2} r_{21}, \\ &\vdots \\ a_n &= a_{n-1} t_{n-1,n} e^{i\varphi_n} + b_n e^{2i\varphi_n} r_{n,n-1}, \\ b_n &= a_n r_{n,n+1} + b_{n+1} e^{i\varphi_{n+1}} t_{n+1,n}. \end{aligned} \quad (4)$$

The r_{ij} and t_{ij} are the Fresnel reflection and transmission coefficients, respectively, for the i, j boundary and ϕ_i is the phase delay due to the propagation through layer i:

$$\varphi_i = \frac{2\pi}{\lambda} \tilde{n}_i d_i \cos \phi_i. \quad (5)$$

Numerous methods are available for solving the system of linear equations in Eq. (1) that are described in many textbooks and papers [11–15].

14.2.1 Recursive Method [16]

By solving Eq. (1) first for a single film with two boundaries (layer 3 in Fig. 2 represents the substrate) and using the identity

$$t_{12} t_{21} + r_{12}^2 = 1 \quad (6)$$

for the Fresnel coefficients, one obtains for the reflected amplitude

$$r_f = \frac{r_t + r_b \exp(2i\varphi)}{1 + r_t r_b \exp(2i\varphi)}, \quad (7)$$

where $r_t = r_{01}$ and $r_b = r_{12}$ are the reflection coefficients of the top and bottom boundaries, respectively, of the film. Equation (7) remains valid if the thin film is deposited on top of a multilayer structure; for that case r_b represents the amplitude reflectivity of the multilayer into the film. Therefore, we can calculate the reflectivity of any multilayer structure by repeated application of Eq. (7) starting from the bottom layer on the substrate and continuing until the boundary between the top layer and the incident medium is reached.

14.2.2 Matrix Methods [11, 17]

Equations (4) can be rearranged into matrix form. One can describe the transfer of the field amplitudes over a boundary and the propagation through a layer with 2 × 2 matrices to obtain

$$\begin{pmatrix} a_i \\ b_i \end{pmatrix} = \frac{1}{t_{i,i+1}} \begin{pmatrix} e^{i\varphi_i} & r_{i,i+1}e^{-i\varphi_i} \\ r_{i,i+1}e^{i\varphi_i} & e^{-i\varphi_i} \end{pmatrix} \begin{pmatrix} a_{i+1} \\ b_{i+1} \end{pmatrix}. \tag{8}$$

It follows that the amplitudes in the incident medium and the substrate are connected by

$$\begin{pmatrix} a_0 \\ b_0 \end{pmatrix} = \frac{\prod_{i=1}^{i=n+1} M_i}{\prod_{i=1}^{i=n+1} t_{i,i+1}} \begin{pmatrix} a_{n+1} \\ b_{n+1} \end{pmatrix}, \tag{9}$$

with M_i the matrix of each layer as defined in Eq. (8). Using $\phi_{n+1} = 0$ in the matrix for the substrate gives the fields at the top of the substrate. The reflected and transmitted amplitudes from the multilayer structure can be obtained from the elements of the product matrix m_{ij} defined in Eq. (9):

$$r = b_0/a_0 = m_{21}/m_{11},$$
$$t = a_{n+1}/a_0 = \prod t_{i,i+1}/m_{11}. \tag{10}$$

The intensity reflectiviy becomes $R = rr^*$, while the transmitted intensity is $T = tt^*(q_{n+1}/q_0)$, where the values for q_0 and q_{n+1} are those for s polarization; they are proportional to the momentum of the photon perpendicular to the boundaries. For p polarization one defines $q_i = (4\pi/\lambda\tilde{n}_i)\cos\phi_i$. Polarization is included in the multilayer program through the proper Fresnel coefficients, most conveniently by expressing them as a function of q. In the optics literature the variable q (without the factor $4\pi/\lambda$) is often called the *effective refractive index*.

Another convenient matrix formalism due to Abelés [12, 17] introduces a matrix for each film that contains only parameters of this film. However, it is more convenient to include the influence of boundary roughness as a reduction in the amplitude reflection coefficients r_{ij} in the matrix form given in Eq. (8).

14.2.3 Boundary Imperfections and Reflectivity Reduction

The atomic structure of matter makes it impossible to produce sharp boundaries between two materials. All real boundaries have a certain width due either to diffusion of the two materials or to roughness of the interface. The reflectivity of each boundary, represented by a δ function at the interface for sharp interfaces, is spread over the width of the interface and is often represented as a Gaussian

$$r(z) = \frac{r_0}{\sigma\sqrt{2\pi}} \exp\left(-\frac{z^2}{2\sigma}\right), \tag{11}$$

where σ is the width of the boundary and r_0 is the amplitude reflectivity of the ideal sharp interface. The total reflectivity is reduced because the amplitudes reflected from different depth within the transition layer now add with different phases. For the case that $r_0 \ll 1$ and refraction can be neglected, one can obtain the reflectivity reduction as a function of wavelength or incidence angle by Fourier transform from coordinates z to momentum transfer q [18] of Eq. (11):

$$r(q) = r_0 \exp\left(-\frac{q^2\sigma^2}{2}\right) \simeq \exp\left(-\frac{2\pi^2 m^2 \sigma^2}{\Lambda^2}\right), \quad (12)$$

The second part of Eq. (12) is obtained by replacing q with the period Λ of a multilayer that has maximum reflectivity at order m using Eq. (21) below and neglecting refraction.

It has been shown [18–22] that one can describe the reduction of the reflectivity, even for the case of very small grazing angles of incidence ($\phi_0 \approx 90°$), where refraction and reflectivity are large, with a small modification of Eq. (12):

$$r(q_1, q_2) = r_0 e^{-0.5 q_1 q_2 \sigma^2}, \quad (13)$$

where q_1 and q_2 are defined in the two media far from the boundaries.

The reduction factor for the intensities—the square of Eqs. (12) or (13)—is called the Debye–Waller factor and was originally derived to describe the reduction of the x-ray diffraction peaks by thermal motion of the atoms [23–25]. While the q values are complex numbers for absorbing media, one sometimes uses only their real part to calculate the reflectivity reduction.

One can also calculate the influence of the transition layer on the reflectivity by dividing it into very thin homogenous films with an index distribution that describes the transition. Inserting the reduced reflectivity values from Eqs. (12) or (13) into Eqs. (7) or (8) reduces the computation time.

It is easy to write computer programs that calculate the performance of any multilayer structure, and personal computers are sufficiently fast to give results within seconds. Programming is especially convenient with modern high-level mathematics packages. Arithmetic with complex matrices is often included as a building block, making the matrix methods easy to program and fast to run. Programs using the recursive methods are slower because they require a loop from layer to layer. For good speed one should use a compiled program for the recursive method.

14.2.4 Boundary Roughness and Diffuse Scattering

A two-dimensional (2-D) Fourier transform of the deviation of the boundary heights $z(\vec{r})$ from its mean gives the 2-D power spectral density (PSD) of the boundary roughness as a function of spatial frequency $\vec{f} = (f_x, f_y)$ or spatial period $\Lambda_x = 1/f_x$, $\Lambda_y = 1/f_y$:

$$\text{PSD}_2(\vec{f}) = \frac{1}{A} \left| \int_A z(\vec{r}) e^{2\pi i \vec{f} \cdot \vec{r}} \, d\vec{r} \right|^2. \tag{14}$$

Each spatial frequency in the roughness spectrum scatters radiation in a direction determined by

$$\vec{k}_{\text{out}} = \vec{k}_{\text{in}} + m\vec{Q}_s, \tag{15}$$

where $\vec{k}$ is proportional to the photon momentum ($|k| = 2\pi n/\lambda$) and $\vec{Q}_s = 2\pi \vec{f}$ is the momentum parallel to the surface that is transferred to the photon from this Fourier component of the roughness. Equation (15) is equivalent to the grating equation

$$\sin \phi_{\text{out}} - \sin \phi_{\text{in}} = m\lambda/\Lambda_x. \tag{16}$$

For small roughness there is only scattering into the first order ($m = \pm 1$) and we have a one-to-one relationship between each spatial period and the direction of the scattered radiation. The total roughness of a surface that is isotropic is given by

$$\sigma^2 = 2\pi \int_0^\infty \text{PSD}_2(f) f \, df, \tag{17}$$

with $f^2 = f_x^2 + f_y^2$. All experiments measure roughness only over a finite bandwidth and the value obtained is that obtained from Eq. (17) over that finite band. The band is limited by the resolution and image size in microscopy and by the range of angles in a scattering experiment. Measurements of scattered light have a theoretical limit for the smallest period of $\lambda/2$ for a scattering angle of 180°; the experimental limit is in most cases determined by the lowest detectable intensity and is around 50 nm for the boundaries in good multilayer mirrors. All spatial periods, even those smaller than $\lambda/2$ up to a largest period, determined by the acceptance angle of the detector, reduce the reflectivity. Large spatial periods diffract only into very small angles from the scattered beam and are not easily distinguished from the specular reflectivity in an experiment.

The amount of light scattered from a surface with small roughness into a range of small angles from specular can be estimated from the bidirectional reflectivity distribution function (BRDF) [26–31]:

$$\frac{1}{I_0} \frac{dI}{d\phi_s} \simeq 16 \frac{\pi^2}{\lambda^4} \cos \phi_i \cos^2 \phi_s \sqrt{R(\phi_i) R(\phi_s)} \, \text{PSD}_2(f), \tag{18}$$

where ϕ_i and ϕ_s are the incident and scattering angles, respectively, in the plane of incidence and $R(\phi_i)$ and $R(\phi_s)$ are the specular reflectivities for these angles. The $\cos \phi$ factors describe the reduction in phase shift produced by a change in heights for off-normal incidence and the change in the widths of the scattered beam. Polarization effects have been omitted in Eq. (18); they can be neglected at grazing and near-normal incidence and only become important at intermediate angles.

Calculating the amount of scattering from a multilayer structure is considerably more difficult than the calculation of the specular reflectivity and most authors have used approximations in their theories [32–35]. The usual multilayer calculation including a value σ for the width of the boundaries is used to calculate all specular amplitudes a_i and b_i (see Fig. 2) within the multilayer structure. The amount of scattering of these amplitudes at the boundaries is obtained from the PSD of the boundaries and each scattered amplitude is propagated to the surface of the structure. For the addition of the scattered amplitudes from different layers it is important to know the degree of correlation between the roughness of different boundaries. For perfect correlation (i.e., all boundaries have the same shape, there is perfect replication of the roughness from layer to layer) the phase differences between the waves scattered from different boundaries are the same as those for a specular beam of the same direction. A scan through the scattered radiation (with a fixed input beam) shows similar interference structure as the specularly reflected beam at that angle. When there is no correlation between the contributions from different boundaries the scattered waves are added with random phases, which is equivalent to just adding the intensities. Practically no interference structure due to the multilayer is visible in the scattered field. The strength of the interference structure has been used to determine the degree of correlation between the roughness of different boundaries. Long spatial periods are usually replicated from layer to layer, producing strong correlation while small period roughness is uncorrelated. The degree of correlation between boundaries has been determined for many systems and the transition between uncorrelated and correlated roughness occurs at periods around 10 nm [36–41].

Theoreticians often prefer to use autocorrelation functions instead of power spectra to characterize rough surfaces [32, 42]. While that approach is in principle equivalent—the autocorrelation function and the power spectrum are Fourier transform pairs—the power spectrum is much more useful for experimental data [39, 43, 44]. Instrument resolution enters as a multiplication factor in the power spectrum and can easily be recognized and corrected for in a plot of the data; it is straightforward to combine partial power spectra obtained from different instruments. Scattering is directly given by the PSD at the spatial frequency that corresponds to the scattering angle. When using autocorrelation functions one usually has to extrapolate the experimental data with an analytical function to calculate the amount of scattering.

14.3 Multilayer Design

14.3.1 High-Reflectivity Mirrors

The standard design for high reflectivity is the quarter-wave stack. One deposits two materials of different refractive index (high = H, low = L) on top of each other. The thickness of the layers is selected such that all boundaries add in phase to the reflected wave. This requires that the phase delay in propagating each layer [Eq. (5)] is 90° or $\lambda/4$, producing a round-trip propagation delay of 180°; because $r_{12} = -r_{21}$, a 360° phase shift occurs between the reflected amplitudes from adjacent boundaries. For normal incidence the optical thickness of each layer is $nd = \lambda/4$. Selecting two materials with a large difference in the refractive index produces a large reflected amplitude at each boundary, reduces the number of layers required for good reflectivity, and increases the spectral or angular bandwidth. For the case in which all layers are absorption free one can obtain a reflectivity very close to 100% by using a sufficiently large number of layers, even if the refractive index difference between layers is small. A first estimate for the number of layers required is obtained from $Nr_{12} = 1$ and for the spectral resolution from $\lambda/\Delta\lambda = N$.

The quarter-wave stack is the design of choice for high-reflectivity mirrors in the visible region and is used in the UV region for wavelengths $\lambda > 150$ nm. At shorter wavelengths good-quality, absorption-free high-index materials are no longer available, and for $\lambda < 110$ nm no absorption-free material is available. Absorbing materials require a modified design. One can understand the basic design ideas by noting that the superposition of the forward and backward running waves (a_i and b_i in Fig. 2) produces a standing wave field inside the multilayer structure and that the absorption of a thin, strongly absorbing material can be very small if it is located at a node of this standing wave field. Alternating thin absorber films with spacer layers of low absorption located around the antinodes allow one to produce multilayer structures with minimized total absorption and maximum reflectivity. In these structures, all periods still add in phase, but the two boundaries within one period are not in phase.

The optimum thickness of the two materials can easily be found numerically. One usually defines the ratio* $\gamma = d_H/(d_H + d_L)$, where d_H and d_L represent the thickness of the reflector (H) and spacer (L) layer. For a large number of layers the optimum ratio can be estimated from [45]

$$\tan(\pi\gamma_{opt}) = \pi\left(\gamma_{opt} + \frac{n_L\beta_H}{n_H\beta_H - n_L\beta_L}\right), \qquad (19)$$

---

*The definition $\gamma = q_H d_H/(q_H d_H + q_L d_L)$ is more suitable for the discussion of the performance of a structure when one varies wavelength or incidence angle, and we will assume this definition when discussing multilayer performance.

which for the case $\beta_L \ll \beta_H$ can be approximated with

$$\gamma_{opt} \simeq \frac{1}{\pi}\left(\frac{3\pi n_L \beta_L}{n_H \beta_H}\right)^{1/3}. \qquad (20)$$

In general the optimum thickness ratio depends on the number of layers in the structure *and* on the location in the stack. The standing wave is most pronounced near the top of a good reflector where the a_i and b_i (see Fig. 2) are of the same magnitude. The absorber layers near the top will be thinner to take advantage of the low intensity near the nodes for reduced absorption; however, there is barely any modulation of the intensity by the standing wave near the bottom of the structure and film thickness is close to a quarter-wave there. Figure 3 shows an example for the change of γ in such an optimized design. From the bottom to the top of the stack the thickness ratio changes from the quarter-wave stack ($\gamma = 0.5$) to a value close to $\gamma = 0.31$, the value obtained from Eq. (19) for the optimum periodic design with a large number of layers. The reflectivity and the ratio γ saturate when the absorption in the spacer layer prevents radiation from reaching the deeper layers. Therefore, one has to select the material with the lowest absorption at a specific wavelength if one wants to reach high reflectivities. Table I is a list of such best materials at their respective absorption edges where they have the smallest absorption. The table also gives the number of periods N_{max} that radiation can travel through each of the materials at normal incidence (see later discussion).

In, practice the reflectivity curves $R(\lambda)$ for the design illustrated in Fig. 3 and the optimum periodic design are indistinguishable, because the bottom of the stack, where the two designs differ, contributes very little weight to the reflectivity. The differences between the two designs become significant at longer

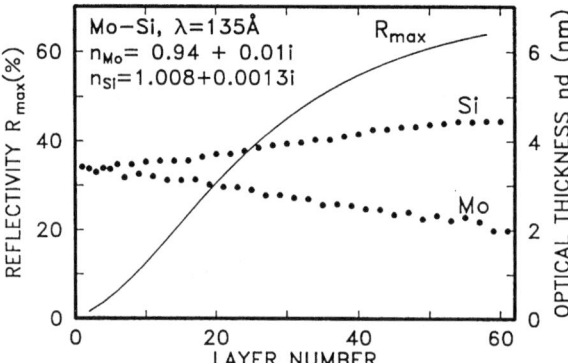

FIG. 3. Calculated increase of the reflectivity (full curve) versus the number of layers for a Mo-Si multilayer mirror, where the thickness ratio is optimized in each period to give the largest increase in reflectivity [18].

TABLE I. Absorption Index β, Linear Absorption Coefficient $\alpha = 4\pi\beta/\lambda$, and $N_{\max}$ at Normal Incidence for Good Spacer Materials at Their Best Wavelength Near Absorption Edges

	λ_{edge} (nm)	Density	β	α (µm^{-1})	$N_{\max}$
Mg-L	25.1	1.74	7.5×10^{-3}	3.8	21
Al-L	17.1	2.7	4.2×10^{-3}	3.0	38
Si-L	12.3	2.33	1.6×10^{-3}	1.6	99
Be-K	11.1	1.85	1.0×10^{-3}	1.16	155
Y-M	8.0	4.47	3.5×10^{-3}	5.5	45
B-K	6.6	2.34	4.1×10^{-4}	0.78	390
C-K	4.37	2.26	1.9×10^{-4}	0.53	850
Ti-L	3.14	4.51	4.9×10^{-4}	1.9	327
N-K	3.1	1.0	4.4×10^{-5}	0.18	3,580
Sc-L	3.19	3.0	2.9×10^{-4}	1.16	557
V-L	2.43	6.1	3.4×10^{-4}	1.75	469
O-K	2.33	1.0	2.2×10^{-5}	0.12	7,230
Mg-K	0.99	1.74	6.6×10^{-6}	0.084	24,000
Al-K	0.795	2.7	6.5×10^{-6}	0.10	24,500
Si-K	0.674	2.33	4.2×10^{-6}	0.078	38,000
SiC	0.674	3.2	6.2×10^{-6}	0.11	25,500
TiN	3.15	5.22	4.9×10^{-3}	1.96	324
Mg$_2$Si	25.1	1.94	7.4×10^{-3}	3.7	21
Mg$_2$Si	0.99	1.94	6.8×10^{-6}	0.09	23,200

wavelengths ($\lambda > 20$ nm), where fewer layers are needed and absorption is stronger.

The designs for high reflectivity differ in the thickness distribution within one period but fulfill the Bragg condition for the layer pair to ensure that all periods add in phase:

$$m\lambda = 2n_{\text{eff}}\Lambda \sin\theta_i, \quad (21)$$

where Λ is the period thickness, n_{eff} is the effective refractive index of the multilayer material, and θ_i is the propagation angle from grazing in the structure. We can replace the angle θ_i with the grazing angle of incidence in vacuum using Snell's law [Eq. (3)] and obtain

$$m\lambda = 2\Lambda \sin\theta_0 \sqrt{1 - \frac{2\delta}{\sin^2\theta_0}}, \quad \text{for } \delta \ll 1, \beta \ll \delta. \quad (22)$$

The value δ_{eff} in Eq. (22) is the weighted average index of the materials in the structure:

$$\delta_{\text{eff}} = \frac{d_H \delta_H + d_L \delta_L}{d_H + d_L}. \quad (23)$$

14.3.2 Multilayers as Filters

Multilayer mirrors reflect only within a wavelength band around the Bragg peaks, and the reflectivity curve can be tuned in λ by changing the angle of incidence. The width of the reflectivity curve is determined by the number of contributing boundaries and a large bandwidth can simply be obtained by using fewer layers. The smallest bandwidth or the largest possible number of contributing layers is determined either by absorption or by depletion of the incident beam due to reflection. Adding layers beyond that limit does not change the reflectivity curves because the bottom layers are not reached by radiation.

The limit by depletion of the incident beam can be changed by changing the reflectivity of each period with an adjustment of the thickness of the reflector layer. In analogy to x-ray diffraction, the reflectivity of the reflector layer in one period is called the *form factor* of the structure with the value

$$R_f = 4R_{12} \sin^2 m\pi\gamma. \quad (24)$$

This equation is obtained from Eq. (7) with the assumption $r_t, r_b \ll 1$, using $r_t = -r_b$, and selecting the period Λ to fulfill the Bragg condition [Eq. (22)] for order m. Reduction of γ to $\gamma < 0.5$ for $m = 1$ produces smaller reflectivity per period, deeper penetration toward the bottom of the stack, and a corresponding reduction in bandwidth because more layers contribute. The limit on the penetration depth for this case is due to absorption in the spacer layer and defines the smallest bandwidth or highest resolution:

$$N_{\max} = \frac{\sin^2 \theta}{2\pi\beta}; \qquad \frac{\lambda}{\Delta\lambda} = N_{\max}. \quad (25)$$

The largest bandwidth is obtained with a coating close to the quarter-wave stack ($\gamma = 0.5$) where all boundaries add in phase, and we need

$$N > N_{\min} = \frac{1}{2|r_{12}|} \quad (26)$$

layers to saturate the reflectivity. Selection of materials with the largest possible Fresnel coefficient produces the broadest reflectivity curves.

The Fresnel reflection coefficients depend on the angle of incidence and on the polarization. For s polarization the reflectivity of a single boundary increases monotonously with the angle of incidence (or decreases with increasing grazing angle) to a value close to 100% at the critical angle. Therefore, a mirror designed to operate at small grazing angles can have high reflectivity and a large bandwidth even at a very short wavelength. With the angle of incidence as a free parameter we can produce mirrors with a spectral or angular resolution between 1 near the critical angle and the value of $N_{\max}$ at normal incidence (see Table I). Very high resolution above 10^4 is theoretically possible at x-ray wavelengths and is realized in the Bragg reflection from crystals. Due to roughness of the interfaces the reflectivity of deposited, amorphous, or polycrystalline multilayers becomes too small for most applications for periods $\Lambda < 20$ Å. The resolution of multilayer mirrors is usually in the 50–100 range with a maximum around 250 [46, 47].

Substantially higher resolution than is possible with a multilayer mirror can be obtained with multilayer gratings. The blazed grating in Fig. 4 can be described in two ways: It can be seen as a multilayer that contains steps, making it possible to reach deeper into the structure by eliminating the depletion of the incident beam in the deeper layers. It can also be seen as a blazed grating that is overcoated with a multilayer in such a way that all periods of grating and multilayer add in phase [48]. The resolution is given by the maximum phase difference between interfering beams and is equal to the number of steps times the step heights expressed as a multiple of Λ; this value is $m = 5$ for the example of Fig. 4. In the picture where the grating is the primary element, the resolution is the number of grating lines times m, and the multilayer just enhances the reflectivity. Chapter 18 gives a detailed discussion of multilayer gratings.

14.3.3 Supermirrors

Mirrors with very large bandwidth beyond the limit given by $N_{\min}$ in Eq. (26) can be obtained by depositing multilayers with different periods, one for each

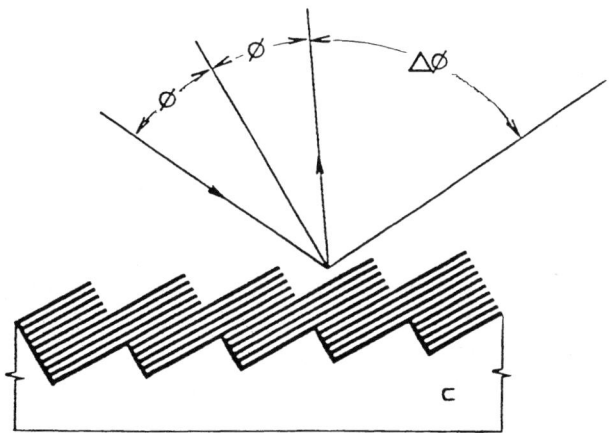

FIG. 4. A blazed grating overcoated with a matching multilayer structure can be described as a thick multilayer with steps in order to reach deeper layers and higher resolution. The drawing, where the step height is five multilayer periods, corresponds to a blazed grating used in the fifth order [48].

desired wavelength band, on top of each other. Absorption limits this procedure, and the ratio N_{max}/N_{min} or δ/β is a quality measure for the design freedom. The ratio is too small in the VUV or XUV region for practical use. Where absorption-free materials are available, for example for visible light or cold neutrons, such mirrors are common, and the term *supermirrors* has been coined for them. The ratio δ/β increases with higher photon energy and it becomes practical to produce such mirrors for x-rays for photon energies above 10 keV. The main application is the extension of the photon energy of grazing incidence optics beyond the critical angle. An example of such a mirror is a graded period W/Si mirror with 1200 layers and a reflectivity larger than 30% for photon energies between 20 and 70 keV at a grazing angle of 3 mrad [49, 50].

14.3.4 Multilayer Polarizers, Phase Retarders, and Beamsplitters

Any mirror used at nonnormal incidence has lower reflectivity for p polarization than for s polarization and can be used as polarizer. The minimum p polarization reflectivity at the Brewster angle occurs close to 45° in the VUV region, where the refractive index is close to one. The reflectivity for p polarization is zero at the Brewster angle for absorption-free materials, but increases with absorption. Therefore, the achievable degree of polarization is higher at the shorter wavelengths where absorption is smaller. Typical values for R_s/R_p are 10 around $\lambda = 30$ nm and over 1000 around $\lambda = 5$ nm. However, the reflectivity of single boundaries is much too low ($R \simeq 10^{-4}$ for s polarization and $\lambda = 5$ nm) to make such polarizers useful. Multilayer structures change the theoretically possible degree of polarization very little, but can enhance the reflectivity to useful values [48]. Therefore, all multilayer x-ray mirrors are efficient polarizers when designed for high reflectivity at the Brewster angle.

Phase retarders are more difficult to realize. The geometrical phase differences in a multilayer structure do not depend on polarization, and one cannot produce a reflector with high reflectivity for both polarizations and a 90° phase delay [51, 52]. The situation is different in transmission. The p polarization passes the structure without reflection near the Brewster angle and is just attenuated by absorption. Some of the s polarization bounces back at each boundary. If we use the multilayer structure off-resonance, for instance, at one of the side minima in the reflectivity curve, we also have high transmission for s polarization, but due to the internal reflection, s-polarized radiation is delayed [53]. By selecting the number of layers and the materials one can find designs that produce 90° phase delays. Using the transmission maximum of a Fabry–Perot interference filter is another possibility. It is, of course, necessary to fabricate these structures on thin transmitting substrates or use them in a self-supporting way. Multilayer phase retarders have been used as polarimeters in the 100-eV range [52, 54–57].

Roughness of the boundaries reduces the reflectivity at each boundary and the phase retardation and has to be included when the performance of a design is calculated. Roughness makes it more difficult to reach large delays at higher photon energies. At $\lambda = 50$ Å only a phase delay of 5° has been obtained up to now [58].

14.4 Multilayer Fabrication and Performance

Every thin-film deposition method can be used to fabricate multilayer x-ray mirrors. Thickness errors and boundary roughness are the most important parameters that have to be kept smaller than about $\Lambda/10$ for good performance. Thickness errors can be controlled as required in many systems. The deposition rate in sputtering systems [59] can be very well stabilized, and thickness control just by timing has produced multilayer structures with thickness errors below 0.1 Å per layer. Thermal deposition systems have larger variations; in these

TABLE II. Multilayer Systems and Their Performance in the VUV

$\lambda > 150$ nm	Absorption-free quarter-wave stack, $R > 95\%$.
	Al-MgF$_2$ mirrors and interference filters in transmission.
$\lambda = 110–150$ nm	Good absorption-free high index materials not available.
	MgF$_2$, LiF still transparent.
	Mirrors: Al with MgF$_2$ or LiF protection, $R \simeq 90\%$.
	Transmission filters: Al-MgF$_2$ Fabry–Perot.
$\lambda < 110$ nm	No absorption-free material available.
$\lambda = 80–110$ nm	Good reflection from unoxidized Al and Be; no protective overcoat available; $\beta = 0.1–1$ for all materials except alkali metals. Multilayers can not enhance reflectivity of clean Al or Be, but can be used to suppress an undesired λ more than a desired λ.
$\lambda < 700$ nm	$\beta < 0.01$, decreases with decreasing λ faster than δ. Multilayers can enhance reflectivity of best metals.
$\lambda < 20$ nm	Large enhancement of reflectivity with multilayers. For single surface $R \propto \lambda^4$, but multilayer with large N can compensate for the loss in reflectivity.
$\lambda = 12.5–20$ nm	$R \simeq 67\%$ with Mo-Si multilayers.
$\lambda = 11.4–12.5$ nm	$R = 70\%$ with Mo-Be multilayers.
$\lambda < 10$ nm	Boundary roughness affects the performance at normal incidence; high reflectivity can be maintained off-normal.
$\lambda = 6.7–10$ nm	$R \simeq 25\%$ at normal incidence with B$_4$C spacers.
$\lambda = 4.5–6.7$ nm	$R \simeq 10\%$ at normal incidence with C spacers.
$\lambda < 4.5$ nm	$R < 5\%$ at normal incidence, higher off normal.
$\lambda < 1$ nm	$\delta/\beta \gg 1$. High-quality multilayers with little absorption possible, but roughness requires $\Lambda > 2.5$ nm and use at grazing incidence. $R > 80\%$ for $\lambda < 0.2$ nm. Used as collimators and filters for synchrotron radiation and x-ray tubes.

systems one can prevent the accumulation of thickness errors with *in situ* monitoring of the soft x-ray reflectivity during deposition [60]. A thickness error in one layer can be compensated for in the next layer and the accumulated error can be kept under 2 Å, even after the deposition of hundreds of layers.

Reduction of boundary roughness is a bigger challenge. The random arrival of atoms produces large roughness at high spatial frequencies and it is important to reduce that roughness as much as possible to get good reflectivity. Increased substrate temperatures, high kinetic energies of the evaporant, bombardment of the growing film with the sputter gas, or polishing the films with an ion beam are methods that have been used to move atoms away from the peaks of a surface and reduce roughness. Good multilayer mirrors have roughness values around $\sigma = 0.3$ nm. We want to note that roughness values reported in the literature cannot be compared easily, because different authors or the same authors at different times use different methods to analyze their structures.

A summary of the types of multilayer coatings used in the VUV region and their performance is given in Table II, and Fig. 5 is a plot of the peak normal incidence reflectivity achieved up to now. The drop in peak reflectivity toward shorter wavelengths for $\lambda < 10$ nm is due to the increased influence of boundary roughness, whereas the drop at long wavelength ($\lambda > 13$ nm) is due to the

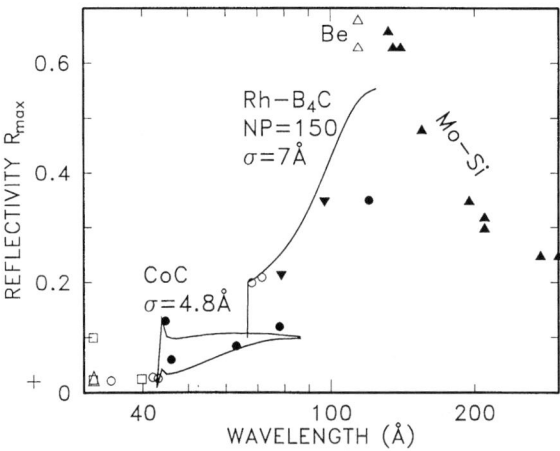

FIG. 5. Achieved peak reflectivity near normal incidence of multilayer structures with spacer layers of C, ●●; B$_4$C, ○○; Y, ▼▼; Si, ▲▲; Sc, □□; TiN, △△. (From a database at www-cxro.lbl.gov/multilayer/survey.html and [18].) The full curves are calculated; the two curves for Co-C are for 70 and 150 periods. The reflectivity of 10% at $\lambda = 3.14$ nm (□) represents a 370-layer mirror of Cr-Sc with a period $\Lambda = 16.3$ Å that was tuned to the region of anomalous dispersion of Sc by tilting it to 18° from normal [61].

increased absorption of the Si spacer layer. There is also a drop at the short-wavelength side of absorption edges (see Table I) where the best spacer layer material changes. One can maintain high reflectivities for shorter wavelengths than those in Fig. 5 by keeping the multilayer period fixed (around $\Lambda > 3$ nm) and tuning the Bragg peak toward shorter wavelengths by increasing the angle of incidence. At short wavelengths very high reflectivities are routinely being obtained at grazing incidence, for example, $R > 80\%$ for $\lambda = 1.54$ Å.

References

1. B. K. Flint, *Adv. Space Res.* **2**, 135–142 (1983).
2. M. Zukic and D. G. Torr, *Appl. Opt.* **31**(10), 1588–1596 (1992).
3. W. R. Hunter, *J. Phys.* **25**, 154–160 (1960).
4. G. Hass and W. R. Hunter, in "Physics of Thin Films," Vol. 10 (G. Hass and M. H. Francombe, eds.), Academic Press, New York, pp. 72–166 (1978).
5. E. Spiller, in "Space Optics" (B. J. Thompson and R. R. Shannon, eds.), National Academy of Sciences, Washington, DC, pp. 581–597 (1974).
6. J. F. Seely and W. R. Hunter, *Appl. Opt.* **30**(19), 2788–2794 (1991).
7. S. Chakrabarti, J. Edelstein, R. A. M. Keski-Kuha et al., *Opt. Eng.* **33**(2), 409–413 (1994).
8. S. E. Schnatterly, in "Solid State Physics," Vol. 34 (H. Ehrenreich, F. Seitz, and D. Turnbull, eds.), Academic Press, New York, pp. 275–358 (1979).
9. H. Raether, in "Springer Tracts in Modern Physics," Vol. 88 (G. Hoehler, ed.), Springer-Verlag, Berlin (1980).
10. A. E. Rosenbluth, *Rev. Phys. Appl.* **23**, 1599–1621 (1988).
11. O. S. Heavens, "Optical Properties of Thin Solid Films," Dover, New York (1966).
12. M. Born and E. Wolf, "Principles of Optics," 5th ed., Pergamon Press, Oxford (1975).
13. H. A. Macleod, "Thin-Film Optical Filters," Adam Hilgers, Bristol (1989).
14. L. G. Parratt, *Phys. Rev.* **95**, 359–368 (1954).
15. J. H. Underwood and J. T.W. Barbee, *Appl. Opt.* **20**, 3027–3034 (1981).
16. P. Rouard, *Ann. Phys.* **7**, 291–384 (1937).
17. F. Abelés, *Ann. Phys.* **5**, 596–639 (1950).
18. E. Spiller, "Soft X-Ray Optics," SPIE Optical Engineering Press, Bellingham, WA (1994).
19. P. Croce, L. Névot, and B. Pardo, *Nouv. Rev. Opt. Appl.* **3**, 37–50 (1972).
20. P. Croce and L. Névot, *J. Phys. Appl.* **11**, 113–125 (1976).
21. L. Névot and P. Croce, *Rev. Phys. Appl.* **15**, 761–779 (1980).
22. F. Stanglmeier, B. Lengeler, and W. Weber, *Acta Cryst.* **A48**, 626–639 (1992).
23. P. Debye,*Verh. Deutsch. Phys. Ges.* **15**, 738 (1913).
24. I. Waller, *Z. Physik* **17**, 398 (1923).
25. J. Laval, *Rev. Mod. Phys.* **30**, 222–227 (1958).
26. E. L. Church, H. A. Jenkinson, and J. M. Zavada, *Opt. Eng.* **18**, 125–136 (1979).
27. E. L. Church and P. Z. Takacs, *Proc. SPIE* **645**, 107–115 (1986).
28. E. L. Church and P. Z. Takacs, *Proc. SPIE* **640**, 126–133 (1986).
29. E. L. Church and P. Z. Takacs, *Proc. SPIE* **1165**, 136–150 (1989).
30. J. C. Stover, "Optical Scattering," McGraw-Hill, New York (1990).

31. J. M. Bennett and L. Mattson, "Introduction to Surface Roughness and Scattering," Optical Society of America, Washington, DC (1990).
32. S. K. Sinha, E. B. Sirota, S. Garoff et al., *Phys. Rev. B* **38**, 2297–2311 (1988).
33. D. G. Stearns, *J. Appl. Phys.* **65**, 491–506 (1989).
34. D. G. Stearns, *J. Appl. Phys.* **71**, 4286–4296 (1992).
35. M. Kopecky, *J. Appl. Phys.* **77**(6), 2380–2387 (1995).
36. D. E. Savage, J. Kleiner, N. Schimke et al., *J. Appl. Phys.* **69**, 1411–1424 (1991).
37. D. E. Savage, Y. H. Phang, J. J. Rownd et al., *J. Appl. Phys.* **74**, 6158–6164 (1993).
38. J. B. Kortright, *J. Appl. Phys.* **70**, 3620–3625 (1991).
39. E. Spiller, D. G. Stearns, and M. Krumrey, *J. Appl. Phys.* **74**, 107–118 (1993).
40. J. Slaughter and C. Falco, *Proc. SPIE* **1742**, 365–372 (1992).
41. X. M. Jiang, T. H. Metzger, and J. Peisl, *Appl. Phys. Lett.* **61**, 904–906 (1992).
42. A. L. Barabasi and H. E. Stanley, "Fractal Concepts in Surface Growth," Cambridge University Press, New York (1995).
43. E. Church, *Appl. Opt.* **27**(8), 1518–1526 (1988).
44. J. A. Ogilvy, "Theory of Wave Scattering from Random Rough Surfaces," IOP Publishing, Bristol, England (1991).
45. A. V. Vinogradov and B. Y. Zel'dovich, *Appl. Optics* **16**, 89–93 (1977).
46. M. Bruijn, J. Verhoeven, M. v. d. Wiel et al., *Opt. Eng.* **25**, 679–684 (1987).
47. E. J. Puik, M. J. v. d. Wiel, H. Zeijlemaker et al., *Vacuum* **38**, 707–709 (1988).
48. E. Spiller, in "Low Energy X-Ray Diagnostics," Vol. 75 (D. T. Attwood and B. L. Henke, eds.), American Institute of Physics, Monterey CA, pp. 124–130 (1981).
49. K. D. Joensen, P. Høghøj, F. E. Christensen et al., *Proc. SPIE* **2011**, 360–372 (1993).
50. K. D. Joensen, P. Voutov, A. Szentgyorgyi et al., *Appl. Opt.* **34**(34), 7935–7944 (1995).
51. E. Spiller, in "New Techniques in X-ray and XUV Optics" (B. Y. Kent and B. E. Patchett, eds.), Rutherford Appleton Laboratory, Chilton, UK, pp. 50–69 (1982).
52. J. B. Kortright and J. H. Underwood, *Nucl. Instrum. Methods* **A291**, 272–277 (1990).
53. N. B. Baranova and B. Y. Zel'dovich, *Sov. Phys. JETP* **52**(5), 900–904 (1980).
54. E. S. Gluskin, *Rev. Sci. Instrum.* **63**, 1523–1524 (1992).
55. M. Yamamoto, M. Yanagihara, H. Nomura et al., *Rev. Sci. Instrum.* **63**, 1510-1512 (1992).
56. H. Kimura, T. Miyahara, Y. Goto et al., *Rev. Sci. Instrum.* **66**(2), 1920–1922 (1995).
57. J. B. Kortright, M. Rice, and K. D. Franck, *Rev. Sci. Instrum.* **66**(2), 1567–1569 (1995).
58. S. Di Fonzo, B. R. Muller, W. Jark et al., *Rev. Sci. Instrum.* **66**(2), 1513–1516 (1995).
59. T. W. Barbee, Jr., *Opt. Eng.* **25**, 893–915 (1986).
60. E. Spiller, *Proc. SPIE* **563**, 367–375 (1985).
61. N. N. Salashchenko and E. A. Shamov, *Opt. Commun.* **134**, 7–10 (1997).

15. ZONE PLATES

Yuli Vladimrsky

Center for X-ray Lithography
University of Wisconsin-Madison
Stoughton, Wisconsin

15.1 Coherent Imaging

15.1.1 Fresnel Lenses

Imaging in a projection or a probe (scanning) mode requires a focusing element. A lens is used for this purpose in visible and near-ultraviolet (UV) light. The major obstacle for constructing a deep-UV or x-ray lens is a low refractive $(1 - \delta)$ index and a high extinction coefficient (β) $(n = 1 - \delta - i\beta)$. The optical quality of a material can be defined as a ratio [1] $\delta/2\pi\beta$, and material is sufficient for an acceptable lens when this quantity is $\delta/2\pi\beta > 10$. It is theoretically possible to produce lenses with good efficiency in the deep-UV and x-ray regions if the thickness of the lens is kept small [1, 2]. In this case the curvature of the surface and the effective diameter of the lens will be of the order of several microns. Due to the fact that only a few Fresnel zones are utilized, the resolution is of the order of the lens diameter, which makes this lens hardly more useful than a pinhole [2]. To reduce the thickness of the lens (and improve transmission), one can remove material while maintaining proper optical path differences, thus creating a *Fresnel lens* [1, 3], such as is used in lighthouses, searchlights, and overhead projectors. In a final state, the thickness variation within every Fresnel zone of the lens provides a phase shift from 0 to 360 deg. This *coherent Fresnel lens* [2] or *phase Fresnel lens* [4] acts as a diffractive element and can be considered [1] a fully *blazed zone plate*.

15.1.2 Ideal Zone Plate Geometry

Now, let's consider focusing [2] from one focal S point into another S'. By tracking the optical path we can construct a set of confocal prolate ellipsoids (Fig. 1) with the distance D between foci S and S', origin located in the middle, and path difference of $P_{n+1} - P_n = \lambda/2$ between two neighboring ellipsoids with indices $n + 1$ and n:

$$\frac{16x^2}{(2D + n\lambda)} + \frac{16(y^2 + z^2)}{n\lambda(4D + n\lambda)} = 1. \tag{1}$$

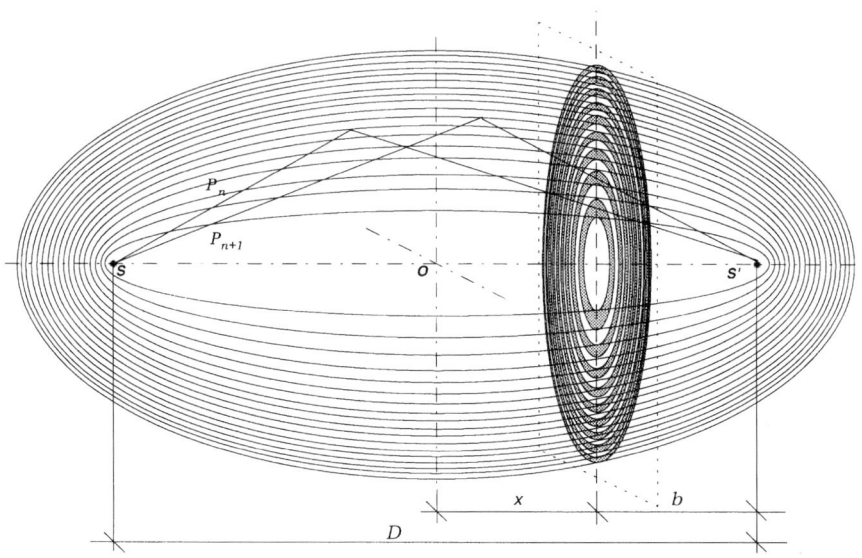

FIG. 1. A set of confocal "diffraction" ellipsoids representing imaging of a point S into a point S'. An optical path difference of $\lambda/2$ is formed by any two neighboring ellipsoids.

An intersection of a two-dimensional surface with this set of ellipsoids will produce a focusing element [2].

An expression for the transmission zone plate, positioned at a distance b from the image point S' can be obtained by introducing a plane $x = D/2 - b$ perpendicular to the x-axis. Using conventional notation for the radii of boundaries between zones $R_n^2 = y^2 + z^2$ and number of zones $n = 1, 2, ..., N$, we find

$$R_n^2 = \frac{n\lambda(4D + n\lambda)[16(D - b)b + n^2\lambda^2 + 4Dn\lambda]}{16(2D + n\lambda)^2}. \tag{2}$$

When the proper phase relations are fulfilled within each zone, the zone pate performs like a thin lens (Fig. 2) with a focal length f and magnification $M = b/(D - b)$ in accordance with the lens formula $(D - b)^{-1} + b^{-1} = f^{-1}$. A simplified zone plate, a Fresnel zone plate (Fig. 3), is formed when alternative zones are blocked [5, 6], thus achieving constructive interference at a focal point. Introducing the magnification parameter $\mu = b(D - b)/D^2 = M/(M + 1)^2$ that changes in an interval $0 \leq \mu \leq \frac{1}{4}$, when the magnification changes in the range of $0 \leq M \leq \infty$, and using the relation $f = DM/(M + 1)^2 = D\mu$, we can present the zone plate equation, Eq. (2), in a form [7] that has familiar terms:

$$R_n^2 = nf\lambda + \frac{n^2\lambda^2}{4}(1 - 3\mu) - \varnothing(f, n\lambda, \mu). \tag{3}$$

It can be seen that the term $\varnothing = \mu n^3 \lambda^3 (1 - 4\mu)(8f + 3\mu n\lambda)/16(2f + \mu n\lambda)^2$ is overall very small compared with the second term ($\sim \mu n\lambda/2f$ times less) and has a maximum in the vicinity of $\mu \approx 1/8$. When using the magnification M explicitly, Eq. (3) can be presented [2, 8] as

$$R_n^2 = nf\lambda + \frac{n^2\lambda^2}{4}\frac{M^3 + 1}{(M + 1)^3}\cdots. \quad (4)$$

From a geometrical point of view the zone plate equation simply states that the projected area of each zone in the direction of the focus is the same and each zone makes the same contribution to the amplitude at the focal point [2]. The first and dominating term is $nf\lambda$. The higher order terms are small for zone plates with numerical apertures up to 0.5 and reflect the dependence of the zone area on the obliquity factor. The zone area is proportional to the zone width $d_n = R_n - R_{n-1}$:

$$d_n = \frac{f\lambda}{2R_n} + \frac{n\lambda^2}{4R_n}(1 - 3\mu) \approx \frac{\sqrt{f\lambda}}{\sqrt{n}} + \frac{3n\lambda^2}{16\sqrt{nf\lambda}}(1 - 3\mu) \approx \frac{\sqrt{f\lambda}}{2\sqrt{n}}. \quad (5)$$

Clearly, the radius of the zone plate and the outermost zone width are $R_N \approx \sqrt{Nf\lambda}$ and $d_N \approx \sqrt{f\lambda}/2\sqrt{N}$, respectively.

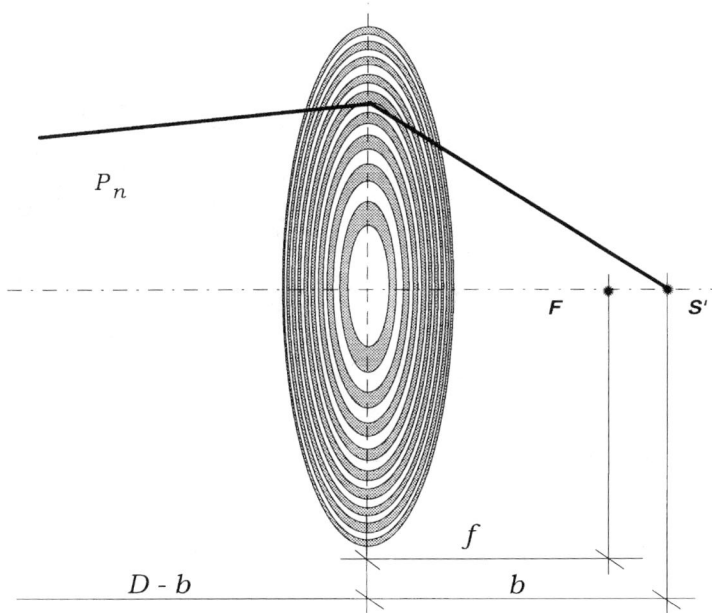

FIG. 2. Zone plate as a thin lens.

When the third term in Eq. (3) vanishes ($\emptyset \to 0$), two special cases are obtained:

1. $\mu = \frac{1}{4}$ corresponds to the formation of an image with magnification $M = 1$ and the zone plate located midway between foci:

$$R_n^2 = nf\lambda + \frac{n^2\lambda^2}{16}. \tag{6}$$

2. $\mu = 0$ is the most commonly referred and discussed case. When $M = \infty$, plane waves are formed; and when $M = 0$, Fresnel lens focuses a plane wave into a point

$$R_n^2 = nf\lambda + \frac{n^2\lambda^2}{4}. \tag{7}$$

FIG. 3. Fresnel zone plate with transparent and opaque zones.

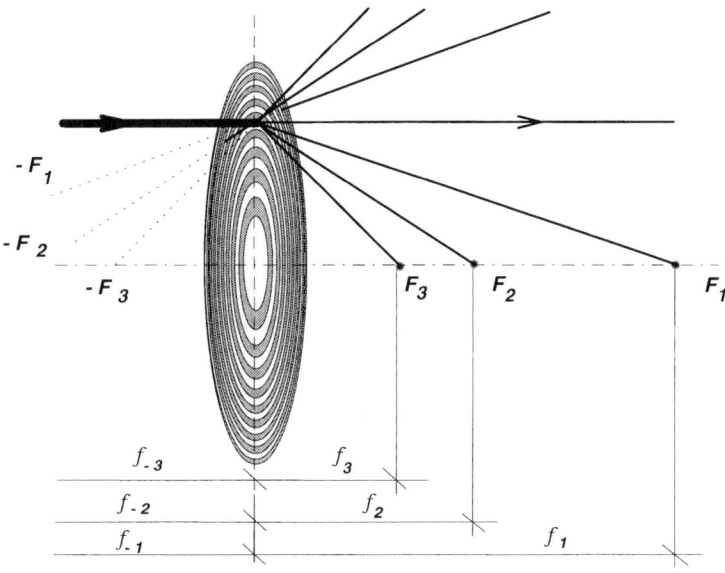

FIG. 4. Zone plate foci.

15.2 Diffractive Focusing with a Zone Plate

The focusing ability of a zone plate as a diffractive element is characterized by multiple foci and strong chromaticity or chromatic aberration.

15.2.1 Multiple Order Foci

In the general case, the phase relations in the diffracted wave provide constructive interference not only for an optical path difference of $\pm\lambda$, but also for $\pm 2\lambda, \pm 3\lambda, \ldots, \pm m\lambda, \ldots$. These images are focused as high orders, similar to the principal maxima distribution of a grating [9]. The corresponding focal lengths are

$$f_{\pm 1} = \pm f, \quad f_{\pm 2} = \pm \frac{f}{2}, \quad f_{\pm 3} = \pm \frac{f}{3}, \quad \ldots, \quad f_{\pm m} = \pm \frac{f}{m}, \quad \ldots, \quad (8)$$

where m is the diffraction order and a negative sign indicates virtual foci (see Fig. 4).

The zeroth-order undiffracted beam is also passing through the zone plate. The beams of all orders, or spatial components, are reaching the focal planes in different stages of convergence and divergence. For effective performance of a zone plate lens in an imaging or scanning microscope it is necessary to block the undesirable spatial components, especially the zeroth-order beam. This can be achieved by forming a central stop [2, 10–12], which blocks a number of central zones but still lets the desirable order pass through an aperture positioned coaxially in the shadow of the central stop. Another way is to focus off-axis rays using only a portion of the outer zones [2, 13].

15.2.2 Chromaticity

The zone plate is inherently chromatic and the strong spectral dispersion of a zone plate is reflected in the dependence of the focal length on the wavelength. For a zone plate with a small numerical aperture this relation is reasonably simple [14]:

$$f \approx \frac{R_N^2}{N\lambda}. \qquad (9)$$

The dependence of the focal length on the wavelength allows us to use a zone plate as a monochromator for wavelength tuning [2]. A very important requirement, associated with spectral dispersion of a zone plate, is a restricted spectral bandwidth:

$$\frac{\Delta\lambda}{\lambda} < \frac{1}{2N}. \qquad (10)$$

An appropriate degree of monochromatization is usually provided by a condenser zone plate [15–17] or multilayer double-crystal monochromator [18] positioned upstream of a zone plate.

15.2.3 Resolution

The similarity of a zone plate and a lens is not limited by the lens formula. The radial resolution of a zone plate is very close to that of a lens [14]. For a zone plate with the radius $R_N = \sqrt{Nf\lambda}$ the resolution is proportional and approximately equal to the width of the outermost zone:

$$\omega \approx 1.22 \frac{f\lambda}{2R_N} \approx 1.22 \frac{\sqrt{f\lambda}}{2\sqrt{N}} = 1.22\, d_N. \qquad (11)$$

A rigorous analysis [19] shows that for a zone plate with transparent positive zones the resolution is slightly worse than the resolution of a lens of the same diameter, while for a zone plate with negative transparent zones the resolution is slightly better than for the lens. For a zone plate with just few (3–5) zones the resolution difference is within ±15%, and for a zone plate with 100 zones this difference is less than ±1%. The focal length of a zone plate is inversely proportional to the order of diffraction, Eq. (8). Accordingly, the resolution of a zone plate in higher orders is $\omega_m \approx \omega_1/m \approx 1.22\, d_N/m$.

15.2.4 Efficiency

The intensity of light focused by a zone plate is calculated by integrating the amplitude of a diffracted wave over a pair of adjacent zones and summation of the contributions of all zone pairs. The resulting intensity normalized to the incident flux is the efficiency of the zone plate. For equal width of obstructed and clear zones the efficiency in different orders can be presented [7] as:

$$\eta_m = \frac{1}{m^2\pi^2}(1 + e^{-2\kappa\phi} - e^{-\kappa\phi}\cos\phi), \quad \text{for } m = \pm 1, \pm 3, \pm 5, \ldots, \quad (12)$$

where $\kappa = \beta/\delta$, $\phi = 2\pi t \delta/\lambda$, and t is the thickness of the zone plate material. For opaque zones ($\kappa = \infty$) the diffraction efficiency in the first order is $\eta_1 = 1/\pi^2 \approx 10.1\%$, and in the third order it is $\eta_3 = 1/9\pi^2 \approx 1.1\%$. When absorption is very weak ($\kappa \to 0$) and the phase reversal is introduced ($\phi = \pi$), the diffraction efficiency can be as high as $\eta_1 = 4/\pi^2 \approx 40.5\%$. Recently an efficiency of 33% for a Fresnel phase plate was obtained [20]. Optimization of the zone plate performance requires a blazed profile (Fig. 5), which will depend on maximization of a system transfer function [7]. Fabrication of an arbitrary blaze shape could be very difficult. The blaze shape can be approximated by "staircase" profile [22]. Each pair of zones is sectioned into p bands with annular radii expressed by

$$r_n(v) = \sqrt{(n + 2v - 2)f\lambda}, \quad (13)$$

where $v = 1, 2, \ldots, p$. A cross-section of a three-band ($p = 3$) blazed zone plate is presented in Fig. 6. The efficiency of a phase zone plate with a staircase profile, fabricated of a nonabsorbing material is [22]:

$$\eta_m(p) = \frac{1}{m^2}\left(\frac{\sin(\pi/p)}{\pi/p}\right)^2. \quad (14)$$

In case of $p = 2$ we have a familiar binary zone plate (see Fig. 3), and efficiency is $\eta_1(2) = 40.5\%$. When $p = 3$ (see Fig. 6), $p = 4$ and $p = \infty$ the efficiency will be $\eta_1(3) = 68.5.5\%$ and $\eta_1(4) = 81.2\%$, and $\eta_1 = 100\%$, respectively.

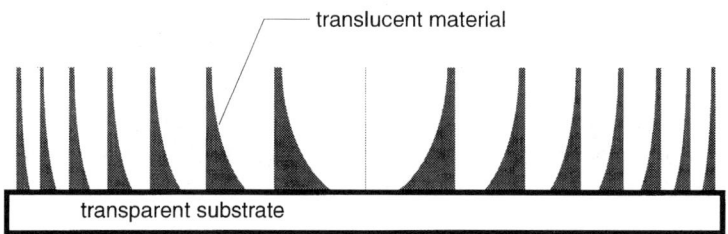

FIG. 5. Cross section of a zone plate with a blaze profile.

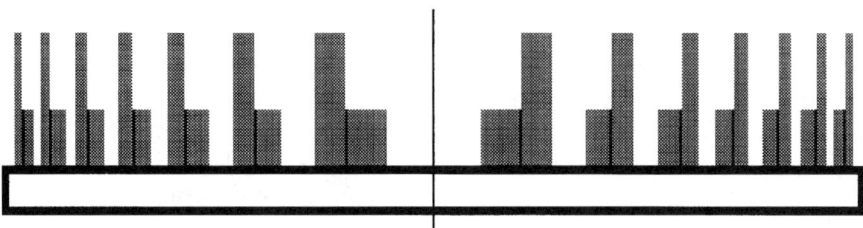

FIG. 6. Cross section of a blazed zone plate with three-band "staircase" profile.

15.3 Multilayer Zone Plates

Simultaneous monochromatization and focusing is possible by combining the multilayer or a crystal with a zone plate [24–27]. This combination is realized by modulation or profiling the multilayer or crystal surface with a zone plate pattern, producing *reflection* zone plates. The multilayer zone plates (also called Bragg–Fresnel lenses) offer the convenience of a solid substrate. As a consequence of the off-axis geometry, these zone plates suffer from strong coma and astigmatism that seriously limits their field of diffraction-limited performance [27]. Another problem of the Bragg–Fresnel zone plates is the very narrow energy band of focused radiation and, consequently, low flux. The multilayer zone plates in the soft x-ray region are used at near-normal incidence [25] and modulated crystal zone plates, linear and circular, are finding application in the hard x-ray region from 6 to 30 keV [24, 26].

15.4 Zone Plate Aberrations

The performance of a zone plate in terms of resolution and efficiency depends on the optical scheme that is used and on the accuracy of the shape and the actual position of the zones relative to each other. In this respect the zone plate aberration can be caused by the off-axis imaging and by the inaccuracies of the zone plate pattern.

15.4.1 Off-Axis Aberrations

In addition to chromatic aberration, imaging with a zone plate is restricted by the same off-axis aberrations as a lens [28]. The wavefront deformation caused by the marginal ray propagating at an angle α compared with the axial ray can be expressed by

$$\partial W = \frac{\alpha R_N^3}{2f^2} \cos \phi - \frac{\alpha^2 R_N^2}{2f} \cos^2 \phi - \frac{\alpha^2 R_N^2}{4f}. \tag{15}$$

The first term in Eq. (15) represents coma, the second astigmatism, and the third field curvature. All are characteristic of off-axis imaging. But, unlike a lens, the zone plate is completely free of linear distortion [28]. Spherical aberration is also absent for on- and off-axis imaging when an ideal zone plate is used for a design wavelength. However, the zone plates with large numerical aperture will reveal a certain degree of spherical aberration, when it is used at a wavelength other than that for which it was designed [2], as can be seen from the Eqs. (4) and (5). When the number of zones N is large, the diffraction-limited field is small and the coma term dominates. Setting this term to $\lambda/4$ (the Rayleigh condition), we find that the half-field within which the aberrations are tolerable, is $\alpha = 1/2(f/\lambda)^{1/2}N^{-3/2}$. For a zone plate with a few zones the astigmatism and field curvature dominate, but the diffraction-limited field can be large $[\alpha = (3N)^{-1/2}]$ [28].

15.4.2 Aberrations Caused by Zone Plate Pattern Errors

The performance of a zone plate in terms of resolution and efficiency depends on the accuracy of the shape of the zone plate bands and their actual positioning during fabrication processes [29–32]. The computations of the effect of circular and noncircular zone boundary errors, performed for a zone plate with 100 zones [29, 30], show that those errors have to be within ~40% of the outermost zone width $\Delta R < 0.4\delta_N$. The likely types of imperfections are [32] zone ellipticity, due to slightly different x and y magnifications or deflection axis nonorthogonality of the electron-beam writer, radial zone error, and nonconcentricity of the zone bands due to a stage or beam drift during writing. These defects and associated aberrations for on-axis imaging (astigmatism, spherical aberration and coma respectively) are presented in Table I. Defining ε as the eccentricity of the zones,

TABLE I. Zone Plate Pattern Inaccuracies, Associated Aberrations and Tolerances

Pattern inaccuracy	Ellipticity ε^2	Nonconcentricity γ	Radial error σ
Aberration	Astigmatism	Coma	Spherical
Foci distance	$\varepsilon^2 f$	$\gamma N\sqrt{\lambda f}$	$2\sigma f N$
Best focus position	$\frac{1}{2}\varepsilon^2 f$	$\frac{1}{3}\gamma N\sqrt{\lambda f}$	$\sigma f N$
Tolerance	$\varepsilon^2 \leq \dfrac{0.7}{N}$	$\gamma \leq \dfrac{1.2}{N\sqrt{N}}$	$\sigma \leq \dfrac{1.9}{N^2}$
Outer zone displacement	$\Delta R \leq 0.7 d_N$	$\Delta R \leq 1.2 d_N$	$\Delta R \leq 2 d_N$
Moiré method sensitivity	$\varepsilon^2 \geq \dfrac{0.5}{N}$ any N	$\gamma \geq \dfrac{0.25}{N\sqrt{N}}$ any N	$\sigma \geq \dfrac{0.05}{2N}$ $N \leq 76$

γ as the zone nonconcentricity parameter in the Φ direction, and σ as a coefficient of zone radial error, Eq. (7) can be generalized [32] as

$$R_n^2(1 - \varepsilon^2 \cos^2 \Phi) + \frac{1}{\sqrt{f\lambda}} \gamma R_n^3 \cos(\Theta + \Phi) - \frac{1}{f\lambda}\left(\sigma + \frac{\lambda}{4\varphi}\right)R_n^4 = nf\lambda. \quad (16)$$

The primary aberrations can be balanced to a certain degree by an appropriate defocusing [33] and tilting. The tolerances for the zone plate inaccuracies were obtained by [32] using a Strehl ratio (normalized intensity at the diffraction focus) of $I_s \leq 0.8$ or the $\lambda/4$ condition. The aberration balancing occurs naturally when the "best" focus is obtained during alignment. The astigmatism of an elliptical zone plate can also be compensated by tilting, but this will introduce coma into the zone plate performance. Table I shows the best focus positions, tolerances for zone plate defect parameters and corresponding outermost zone displacements [32].

15.4.3 Detection of Zone Plate Pattern Inaccuracies

Moiré patterns, resulting from two images of the same zone plate superimposed with rotation and/or a shift, can be used to determine the geometrical quality of a zone plate [32]. The fringe patterns ($|K|$ is the fringe number) and the equations in polar coordinates ($\rho_K^2 = r_K^2/NF\lambda$) are shown in Fig. 7. The estimates of moiré technique sensitivity when $|K| = 0.5$ and $\rho_K = 1$ are also included in Table I.

- *Zone plate pattern ellipticity* can be detected by superposition of two images rotated $\Omega = \pi/2$ relative to each other revealing a characteristic moiré pattern with fourfold symmetry (Fig. 7a). From the Kth fringe position ($\Theta = 0$, π, $1 \pm \pi/2$) the value of eccentricity can be calculated $\varepsilon_{mr}^2 = |K|/N\rho_K^2$.
- *Zone plate pattern nonconcentricity* can be observed when the zone plate images are rotated at $\Omega = \pi$ (see Fig. 7b). The resulting moiré picture has a twofold symmetry. The Kth fringe position (ρ_K) determines the value of nonconcentricity: $\gamma_{mr} = |K|/2N\sqrt{N\rho_K^3}$.
- *Zone plate pattern radial error* can be estimated from a moiré pattern generated by introducing a shift τ between images and representing a set of curved lines (see Fig. 7c). One can calculate the radial error parameter $\sigma_{mr} = (K\rho_L - L\rho_K)/2N(L\rho_K^3 - K\rho_L^3)$ by measuring positions of any two fringes ρ_L and ρ_K in the horizontal direction ($\Theta = 0$).

15.5 Zone Plate Fabrication

The fabrication of zone plates has a history extending more than a century [2] The first zone plates were produced by photographically demagnifying drawings

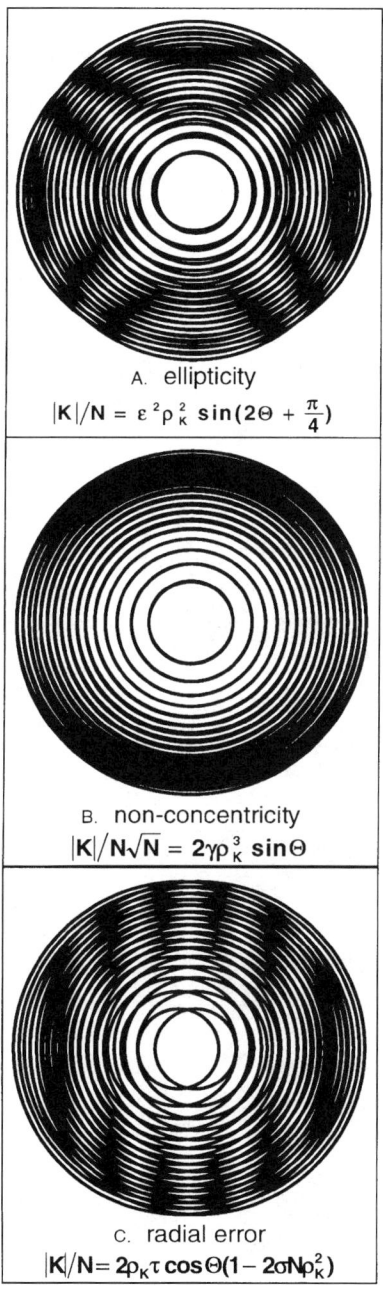

FIG. 7. Moiré fringe patterns for zone plates with defects.

of zone plates. Another method was based on photographing Newton's rings, produced by a double convex lens [14]. Zone plates were also fabricated by ruling the alternate zones on a transparent plate coated with an opaque substance [14]. The construction of the first self-supported gold Fresnel zone plate [34] was performed in 1961 by etching and gold microelectrodeposition. This zone plate was ~10 μm thick and had 19 open zones with an outermost zone width of ~17 μm. Since then remarkable progress in zone plate fabrication has been made by the development of advanced lithographic techniques. Holographic [8, 16, 17] and interferometric [13] lithography produced zone plate patterns having many hundreds of zones with an outermost zone width as small as 60 nm. The latest method is based on the interference of two spherical waves generated by splitting a UV laser beam. A certain degree of spherical aberration had to be introduced in the optical system to fulfill conditions required for proper performance of the zone plate in the soft x-ray region. The interferogram

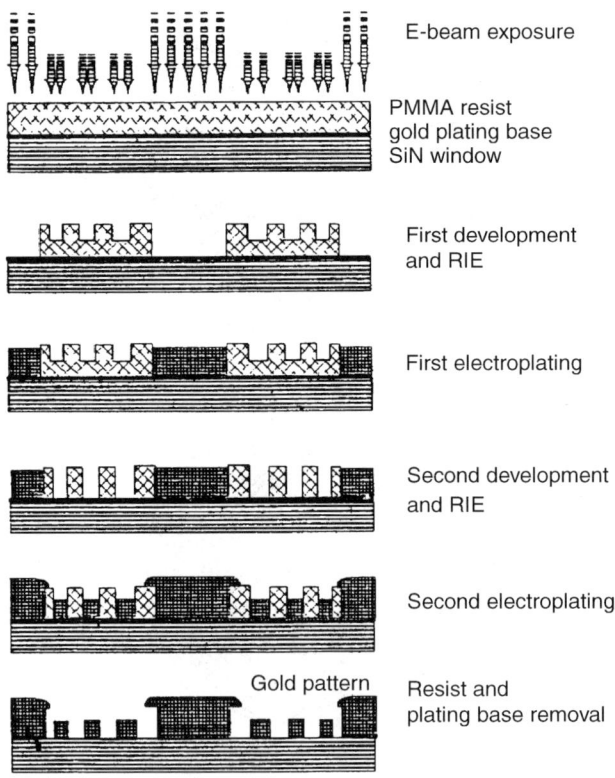

FIG. 8. Single-exposure double-development/double-plating zone plate fabrication process.

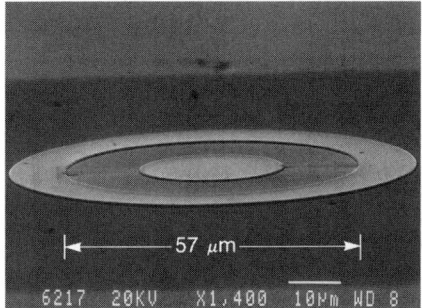 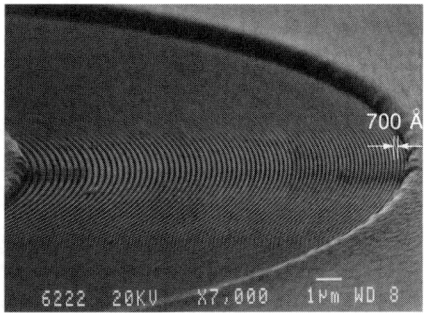

FIG. 9. Gold Fresnel zone plate with 70-nm nominal zone width. $f = 1.3$ mm at $\lambda = 31$ Å, 156 working zones 130 nm thick. Central stop: 28.5 μm diamter, ~1 μm thick. 100 nm thick silicon nitride substrate.

was recorded in the photoresist. The direct write electron-beam lithographic approach allowed the fabrication of zone plates with outer zone widths as small as 40–30 nm [10, 15, 35–37]. The processing sequence of an electron-beam lithography technique for fabricating zone plates with the central stop is presented in Fig. 8. The pattern generator is programmed to produce the desired patterns. After development the zone plate is formed by electroplating gold [35–37], Ni [15], or another suitable metal. A gold zone plate fabricated using a single-exposure double-development/double-plating process is shown in Fig. 9. Advanced reactive ion etching (RIE) processes can be used to transfer the zone plate patterns with an outermost zone width as small as 19 to 55 nm in Ge and Ta [15, 38]. High-resolution zone plates require the use of a relatively thin resist, ~250 nm thick, which limits the thickness of the absorber formation. Thin polyimide and silicon nitride films (50–500 nm thick) are used to support the zone plate. To increase absorber thickness the gold zone plate patterns are used as masks in x-ray lithographic replication processes [13, 20, 22, 23, 39, 40].

The most recent progress in zone plate fabrication was demonstrated by producing a high-performance phase [20] and blazed [22, 23] zone plates for hard x-rays. This fabrication technique involves a three-level lithographic process and requires three masks and high-precision alignment. The parabolic or optimized blaze within the zones [7, 21] was approximated by a four-step staircase thickness profile [23] of the gold phase shifter (Fig. 10). Another approach is based on depositing many alternate layers of x-ray "opaque" and "transparent" thin films on a thin rotating cylinder (wire) [9, 17, 18]. The thicknesses of these layers correspond to the zone plate equation. A sufficiently thin slice of that cylindrical structure will act as a zone plate. Ideally, this technique can produce zone plates with resolution on the order of 10–15 Å. But at this point, this method is still in the development stage [41].

FIG. 10. Blazed Fresnel zone plate for hard x-rays with 0.5 μm nominal zone width. (Courtesy of Zheng Chen, CXrL UW-Madson.) f = 400 mm at λ = 1.54 Å, 61 zones, diameter 122 μm. Four-band zones, 2.25 μm total thickness.

References

1. B. X. Yang, *Nucl. Instrum. Methods A* **328**, 578–587 (1993).
2. E. Spiller, "Soft X-Ray Optics," SPIE Engineering Press, Bellingham, WA, pp. 41–57, 81–97, 250–251 (1994).
3. "The Handbook of Plastic Optics," U.S. Precision Lens, Cincinnati, OH, p. 35 (1988).
4. K. Miyamoto, *J. Opt. Soc. Am.* **51**(1), 17–20 (1961).
5. F. A. Jenkins and H. E. White, "Fundamentals of Optics," McGraw-Hill Book Company, New York, pp. 355–356 (1950).
6. A. Walther, Chap. 7, in "Applied Optics and Optical Engineering," Vol. 1, Academic Press, New York, pp. 254–255 (1965).
7. J. Kirz, *J. Opt. Soc. Am.* **64**, 301–309 (1974).
8. B. Niemann, D. Rudolph, and G. Schmahl, *Opt. Commun.* **12**, 160–163 (1974).
9. M. Sussman, *Am. J. Phys.* **28**, 394–398 (1960).
10. Y. Vladimirsky, D. P. Kern, T. H. P. Chang, D. T. Attwood, N. Iskander, S. Rothman, K. McQuaide, J. Kirz, H. Ade, I. McNulty, H. Rarback, and D. Shu, in "Synchrotron Radiation Instrumentation Proc.," Madison, WI, June 1987, pp. 324–328, North-Holland (1988).
11. H. Rarback, D. Shu, S. C. Feng, H. Ade, J. Kirz, I. McNulty, D. P. Kern, T. H. P. Chang, Y. Vladimirsky, N. Iskander, D. T. Attwood, K. McQuaid, and S. Rothman, *Rev. Sci. Instrum.* **59**(1), 52–59 (1988).
12. H. Rarback, D. Shu, S. C. Feng, H. Ade, C. Jacobsen, J. Kirz, I. McNulty, Y. Vladimirsky, D. Kern, and T. H. P. Chang, in "X-Ray Microscopy II" (D. Sayre, M. Howels, J. Kirz, and H. Rarback, eds.), Springer-Verlag, pp. 194–200 (1988).
13. Y. Vladimirsky, E. Källne, and E. Spiller, *Proc. SPIE* **448**, 25-37 (1984).
14. O. E. Myers, *Am. J. Phys.* **19**, 339–365 (1951).

15. G. Schneider, T. Wilhein, B. Nieman, P. Guttmann, T. Schliebe, J. Lehr, H. Aschoff, J. Thieme, D. Rudolph, and G. Schmahl, *Proc. SPIE* **2516**, 90–101 (1995).
16. D. Rudolph and G. Schmahl, *Ann. NY Acad. Sci.* **342**, 94–104 (1980).
17. G. Schmahl, D. Rudolph, and B. Niemann, in "Scanned Image Microscopy" (E. A. Ash, ed.), Academic Press, London, pp. 393–411 (1980).
18. N. Koyama, H. Tsuyuzaki, K. Kuroda, and A. Shribayama, *Jpn. J. Appl. Phys.* **34**, 6748–6753 (1995).
19. D. J. Stigliani, R. Mitra, and R. G. Semonin, *J. Opt. Soc. Am.* **57**(5), 610–613 (1967).
20. A. A. Krasnoperova, J. Xiao, F. Cerrina, E. Di Fabrizio, L. Luciani, M. Figliomeni, M. Gentili, W. Yun, B. Lai, and E. Gluskin, *J. Vac. Sci. Technol.* **B11**(6), 2588–2591 (1993).
21. R. O. Tatchyn, in "X-Ray Microscopy," Springer Series in Optical Sciences, Vol. 43 (G. Schnahl and D. Rudolph, eds.), Springer Verlag, Berlin, pp. 90–50 (1984).
22. E. Di Fabrizio, M. Gentili, L. Grella, M. Baciocchi, A. A. Krasnoperova, F. Cerrina,Yun, B. Lai, and E. Gluskin, *J. Vac. Sci. Technol.* **B12**(6), 3979–3985 (1994).
23. A. A. Krasnoperova, Z. Chen, F. Cerrina, E. Di Fabrizio, M. Gentili, W. Yun, B. Lai, and E. Gluskin, *Proc. SPIE* **2516**, 15–26 (1995).
24. A. Snigirev, *Rev. Sci. Instrum.* **66**(2), 2053–2058 (1995).
25. K. Holldack, A. Erco, and W. B. Peatman, *Proc. SPIE* **2516**, 210–216 (1995).
26. Ya. Hartman, E. Tarazona, P. Elleaume, I. Snigireva, and A. Snigirev, *Rev. Sci. Instrum.* **66**(2) 1978–1980 (1995).
27. A. I. Erko, V. V. Aristov, and B. Vidal, "Diffraction X-ray Optics," Institute of Physics Publishing, Bristol and Philadelphia, pp. 73–93 (1996).
28. M. Young, *J. Opt. Soc. Am.* **62**(8), 972–976 (1972).
29. M. J. Simpson and A. G. Michette, *Optica Acta* **30**(10), 1455–1462 (1983).
30. A. G. Michette, "Optical Systems for Soft X-rays," Plenum Press, New York, pp. 217–224 (1986).
31. A. G. Michette and C. J. Buckley, eds., "X-Ray Science and Technology," Institute of Physics Publishing, Bristol and Philadelphia p. 334 (1993).
32. Y. Vladimirsky, H. W. P. Koops, *J. Vac. Sci. Technol.* **B6**(6), 2142–2146 (1988).
33. M. Born and E.Wolf, "Principles of Optics," Pergamon Press, New York, pp. 462, 469–473 (1980).
34. A. V. Baez, *J. Opt. Soc. Am.* **51**(4), 405–412 (1961).
35. Y. Vladimirsky, D. P. Kern, T. H. P. Chang, D. T. Attwood, H. Ade, J. Kirz, I. McNulty, H. Rarback, and D. Shu, *J. Vac. Sci. Technol.* **B6**(1), 311–315 (1988).
36. Y. Vladimirsky, D. Kern, T. H. P.Chang, W. Meyer-Ilse, P. Guttmann, B. Greinke, and D. Attwood, "OSA Proc. Short Wavelength Coherent Radiation Meeting" (R. W. Falcone and J. Kirz, eds.), p. 325 (1988).
37. E. H. Anderson and D. Kern, in "X-Ray Microscopy III" (A. G. Michette, G. R. Morisson, and C. J. Buckley, eds.), Springer, Berlin, p. 75 (1992).
38. C. David, B. Kaulich, R. Medenwaldt, M. Hettwer, N. Fay, M. Diehl, J. Thieme, and G. Schmahl, *J. Vac. Sci. Technol.* **B13**(6), 2762–2766 (1988).
39. E. Spiller, in "Scanned Image Microscopy" (Erick A. Ash, ed.), Academic Press, London, pp. 365–391 (1980).
40. D. C Shaver, D. C. Flanders, N. M. Seglio, and H.I. Smith, *J. Vac. Sci. Technol.* **16**(1), 1626–1630 (1979).
41. N. Kamijo, S. Tamura, Y. Suzuki, and H. Kihara., *Rev. Sci. Instrum.* **66**(2), 2132–2134 (1995).

16. WINDOWS AND FILTERS

W. R. Hunter
SFA Inc.
Largo, Maryland

16.1 Introduction

For vacuum ultraviolet (VUV) applications the differences between windows and transmission filters is more one of semantics than fact. The term *filter* implies separation, or isolation, of spectral regions, and *window* implies separation of spatial regions, for example, volumes with different gas pressures. In fact, windows also separate spectral regions and filters often separate volumes at different pressures. Both have the common characteristic of transmitting radiation, therefore, in what follows we will not distinguish between the two. In the final part of this discussion, we include a section on a type of filter that neither transmits radiation nor isolates spatial regions: reflecting filters.

Transmitting filters take three different forms: dielectrics, thin films, and gases. The solid filters are used mostly at normal incidence because reflection losses, and possible interference and polarization effects, are at a minimum in that orientation. Unless specified otherwise, the following descriptions of solid filters involve radiation impinging at normal incidence.

16.2 Dielectric Materials as Transmitting Filters

A number of dielectrics with large bandgaps (>6 eV) can be used as transmitting filters. Their transmission windows extend from the infrared, or near infrared, to the intrinsic absorption in the VUV. Table I, taken from Hunter and Malo [1], lists some of these materials, their short wavelength transmittance limits, or "cutoff wavelengths," at room temperature, the state of the sample (polished or cleaved), the thickness, the supplier (if known), and the solubility in water. The solubility data are included because some of the crystals are slightly soluble in water and are adversely affected by water vapor in the air. One should remember that the term *cutoff wavelength* is not a well-defined physical characteristic but depends somewhat on the whim of the investigator. The author has adopted an arbitrary definition: that wavelength at which the transmittance has dropped to 0.5%.

The dielectric with the shortest cutoff wavelength, at 1050 Å, is LiF. This material can be obtained either as cleaved plates or with ground and polished

TABLE I. Short Wavelength Cutoff, or Transmittance Limit, at 20°C of Dielectric Filter Materials for the VUV [1]

Material	λ_c	Polished	Cleaved	Supplier	Thickness (mm)	Solubility[a] (g/100 cm^3 water)
LiF	1045	×		Harshaw	3.0	0.27
MgF$_2$	1130	×		Harshaw	3.0	0.0076
					1.0	
CaF$_2$	1220	×		Harshar	1.0	0.0016
LaF$_3$	1245	×		?	1.1	?
BaF$_2$	1340	×		Harshaw	0.8	0.12
			×	Koch-Light		
Synthetic sapphire	14.25	×		Linde	2.5	Insoluble
Cultured crystal SiO$_2$	1455	×		Sawyer	1.1	Insoluble
Fused SiO$_2$	1525	×		?	1.3	Insoluble

[a] Data on solubility was taken from the *Handbook of Chemistry and Physics*, 49th ed. (R. C. Weast, ed.), Chemical Rubber Co., Cleveland, OH (1968). Reproduced by courtesy of Elsevier Science Ltd.

surfaces. The transmittance values at the longer wavelengths may be somewhat larger for cleaved plates than for ground and polished surfaces because of damage to the surface due to the polishing procedures. Prior to about 1960 cleaving LiF was a rather simple task because the crystals were fairly hard due to the presence of small amounts of divalent impurities [2]. Subsequently, when LiF of higher purity was produced, the crystals were rather soft. When cleaved, they deformed plastically, many dislocations were introduced, and the crystals were often visibly bent. Furthermore, fracture tended to occur along (110) planes rather than the normal (100) cleavage planes. Nadeau and Johnston [3] found that it is possible to harden the soft LiF crystals by exposure to gamma irradiation from Co-60 using an exposure of 1.6×10^6 rad at room temperature. This treatment permitted successful cleaving operations without the deleterious effects described. Annealing for about 2 h at 450°C apparently removes all optical and mechanical effects of the irradiation and returns the crystals to their original condition (however, see Section 16.1.1 later).

Although cleaved LiF may have a somewhat better transmittance than ground and polished samples it is not suitable for imaging systems because of the unevenness of its surface. Figure 1, taken from Patterson and Vaughan [4], is a contact photograph, using 1216-Å radiation, showing tear marks and cleavage planes that occur when LiF is cleaved. LiF is attacked by water vapor in the air. A chemical reaction takes place rather than a simple adsorption of water, hence the transmittance loss is not reversible with cleaning [4].

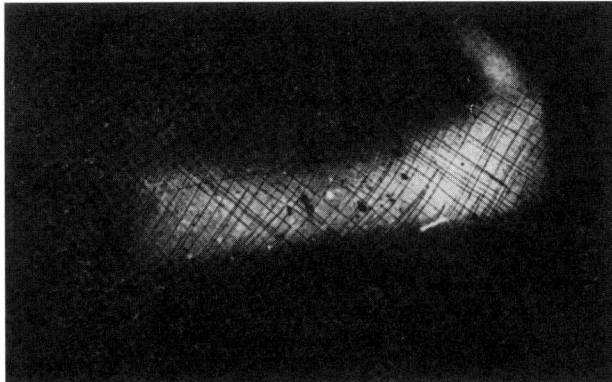

FIG. 1. A transmission photograph of a cleaved LiF crystal made at 1216 Å. The tear markings on both crystal surfaces are clearly visible [4].

MgF$_2$ has its cutoff at 1130 Å. Its water solubility is much less than that of LiF and it is not affected to any extent by atmospheric water vapor. CaF$_2$ has its cutoff at 1220 Å, LaF$_3$ at 1245 Å, and BaF$_2$ at 1350 Å. Synthetic sapphire and cultured, or synthetic, crystal quartz have cutoff wavelengths about 30 Å apart, at 1425 and 1455 Å, respectively. Finally, "good" fused silica has the longest cutoff wavelength, at 1525 Å, of all the samples measured. It should also be mentioned that fused silica samples have the widest range of cutoff wavelengths of any of the materials measured—from 1525 to about 2000 Å.

The transmittance spectra of these materials behave in the same general manner. The transmittance rises abruptly from the cutoff value to a large value over about a 50-Å range. In principle, at full transmittance the only losses are reflectance losses. As the wavelength increases the transmittance rises slightly because the index of refraction decreases, reducing the reflectance losses. Figure 2 [5] is a textbook example of the transmittance of a dielectric, in this case a cultured crystal quartz sample cut in three different orientations. This material is discussed later.

Needless to say, the condition vis-à-vis cleanliness of the surfaces of windows affects the transmission in the VUV. Before use the windows should be cleaned as thoroughly as possible. Materials such as sapphire and quartz can be washed with water and a detergent; however, slightly water-soluble materials such as LiF should not be cleaned with water. Rinsing with organic solvents will clean most of the gross contamination. The material should be warmed beforehand to prevent excessive water condensation due to cooling of the sample by evaporation of the solvents. Vangonen *et al.* [6] have reported a technique for cleaning polished MgF$_2$ using pulsed VUV radiation which might also be useful for other materials subject to water damage. They identified the contaminant via multiple frustrated total internal reflections as hydrocarbons adsorbed on the surfaces. They assert that photodissociation of the oxygen molecules by the VUV radiation promotes a reaction between the oxygen atoms and the contaminant that oxidizes the contaminant producing volatile compounds that leave the surface. Evidence supporting this assertion is obtained by the retardation of contaminant removal when helium is substituted for air to reduce the amount of oxygen present. By inserting filters with cutoff wavelengths of 1250, 1350, and 1800 Å, one at a time, the cleaning efficiency was reduced by 1.4, 2.2, and 9, respectively, thus demonstrating that the wavelength range from 1150 to 1800 Å causes most of the cleaning. This appears to be another manifestation of cleaning carbon contamination by excited oxygen, similar to rf cleaning [7–9].

The cutoff wavelengths in Table I are those to be expected from "good" samples. The cutoff wavelength depends strongly on the purity and the method of preparation of the sample and the listed values are as short as can be expected at room temperature. If the cutoff wavelength of a clean sample is more than about 10 Å longer than the values given here, the sample may have been contaminated during preparation. An exception to this rule is fused silica which

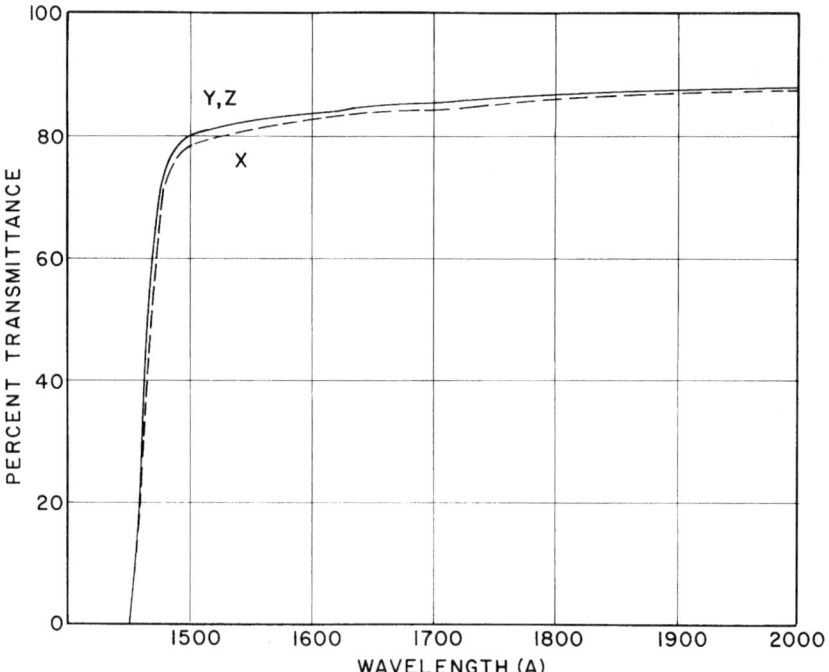

FIG. 2. Transmittance of x, y, and z cuts of cultured crystalline quartz. The samples were cut from the same crystal and had a thickness of 1 mm [5].

is prepared in a variety of different ways depending on its ultimate use. Hereaus Quarzglas [10] uses four methods for the preparation of fused silica for use as a transparent medium: (1) fusing of crystalline quartz in an electrically heated vacuum furnace, (2) fusing of crystalline quartz by a flame, (3) fabrication of fused silica from $SiCl_4$ in a flame, and (4) fabrication of fused silica from $SiCl_4$ using an electric plasma torch. The first two methods produce what is sometimes called *fused quartz*. When fused quartz produced by these methods is viewed between crossed polarizers one can see indistinct outlines of the quartz pieces that have been fused together, thus it should be used only for simple transmission windows and not for high-quality optical components. The best VUV transmission is found in samples made by methods (3) and (4), which are referred to as SUPRASIL (SUPRASIL I and SUPRASIL II) and SUPRASIL-W. Figure 3 shows transmittance curves for these materials adapted from information supplied by Hereaus Quarzglas GmbH [10]. SUPRASIL has its cutoff wavelength at a somewhat longer wavelength ($\approx$1570 Å) than SUPRASIL-W ($\approx$1530 Å), but SUPRASIL has a more rapid change in transmission with wavelength than SUPRASIL-W. SUPRASIL reaches its full transmittance at about 1800 Å but SUPRASIL-W does not become fully transmitting until about 1900 Å. SUPRASIL has some strong absorptances in the infrared caused by OH groups arising from water in the flame. These OH groups are present in about 1200 ppm by weight. In SUPRASIL-W these features are much reduced as a result of using the electric plasma torch. The OH content of SUPRASIL-W is about 5 ppm by weight.

The OH content may also influence the absorption coefficient of SUPRASIL and SUPRASIL-W in the region of the cutoff wavelength. Kaminow *et al.* [11] have measured the absorption coefficient of SUPRASIL 2, SUPRASIL-W2, SPECTROSIL WF, and SPECTROSIL A. Presumably SUPRASIL 2 and SUPRASIL II are equivalent as are SUPRASIL-W and SUPRASIL W2. The SPECTROSIL samples were obtained from Thermal Syndicate Ltd., Wallsend, England. Kaminow *et al.* found evidence of structure at about 7.6 eV (1630 Å) with the so-called dry (no water content) samples—SUPRASIL W, etc.—but no such indications for the wet (slight water content) samples. The features they found were on the order of a few tens of inverse centimeters.

Cultured crystal quartz, sometimes referred to as *synthetic crystal quartz*, is grown under controlled conditions with constituents of known purity. It was originally grown for use in electronic circuits but, with improvements in the method of preparation, it has become useful in optics as well. It has a rather short cutoff wavelength of about 1450 Å and the transmission of a good sample rises abruptly to almost full transmission by 1500 Å (see Fig. 2).

With the advent of cultured crystal quartz and the control over its constituents, one would expect little variability in transmittance from sample to sample or over the same sample, unlike natural crystalline quartz [12–14]. However, Hass and

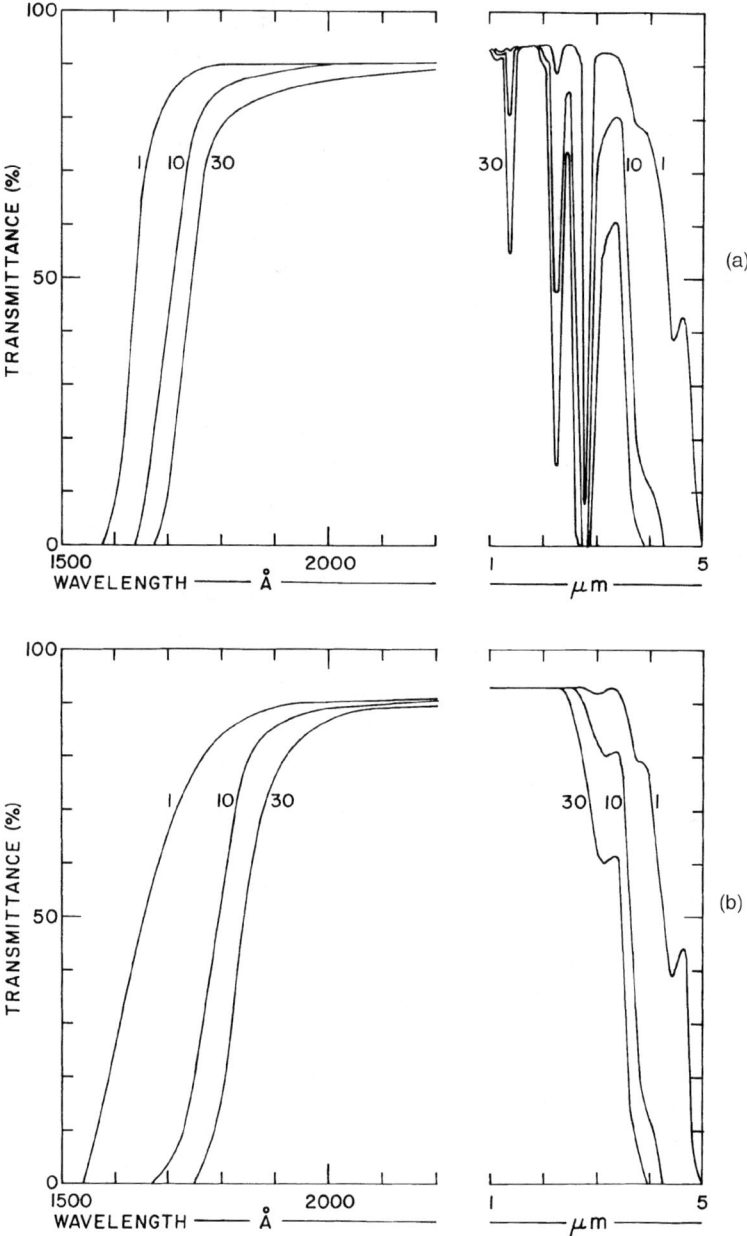

FIG. 3. Transmittance curves of (a) SUPRASIL and (b) SUPRASIL-W showing the cutoff region in the VUV and the response in the infrared. (With permission of Heraeus Quarzglas, GmbH).

Hunter [5] have investigated the transmittance of cultured quartz and found that the transmittance can vary considerably from sample to sample and even over the same sample. Figure 2 illustrates the transmittance of x, y, and z cuts of cultured crystalline quartz. The samples were cut from the same crystal and had a thickness of 1 mm. These measurements represent the transmittance at a position on the crystal where maximum transmittance occurs for a given wavelength. They do not, therefore, represent the transmittance at a fixed position on the sample. The cutoff wavelength is at 1455 Å and the transmittances of the samples increase rapidly toward longer wavelengths. For all samples measured, the x cut had a slightly smaller transmittance than the other two cuts for reasons unknown. Gerasimova *et al.* [15] have measured the transmittance as a function of wavelength for x, y, and z cuts from the same crystal and found consistently smaller transmittance values for the z cut.

Figure 4 [5] shows transmittance measurements at four wavelengths on a 3-mm-thick x cut sample from a crystal other than that shown in Fig. 2. The abscissas represent the distances across the crystal, 38 mm, and the solid and dashed lines show the transmittances for two orientations of the crystal 90 deg apart with respect to the incident radiation. Measurements were made at two orientations to try and obtain a better idea of the uniformity, or lack thereof, of transmittance over the sample surface. The two short solid lines at the upper left represent the width of the VUV beam at the sample. At the four wavelengths measured, the nonuniformities appear to be largest at 1502 Å and decrease toward the longer wavelengths. The sample showing the largest nonuniformities was a z cut taken from a third crystal. The measurements are shown in Fig. 5 [5], which includes photographs in white light with the samples between crossed polarizers. This sample was 24.5 mm across. Samples from this crystal were the only samples of optical quality in which non-uniformities could be seen between crossed polarizers.

Figure 6 [5] shows the transmittance vs wavelength of x, y, and z cut crystal quartz samples of electronic grade taken from the same crystal. The distance across the sample was 32 mm. The x, and z cuts have essentially no transmittance at wavelengths less than 1600 Å. The peculiar transmittance of the y cut is an example of extreme non-uniformity. Figure 7 [5] shows transmittance vs position on the crystal surface of the y cut sample of Fig. 6. There is a strip approximately in the center of the sample that has a transmittance almost as good as any optical quality sample. The transmittance values on either side of this strip are, however, rather small. The reason for the anomolous transmittance spectrum of Fig. 6 for wavelengths less than 1800 Å is that the radiation beam from the monochromator straddled the highly transmitting strip, thus producing the peculiar curve shown in the figure.

It is evident from the results shown in Figs. 4 and 5 that cultured quartz of optical quality can be quite nonuniform in its transmission characteristics. It is

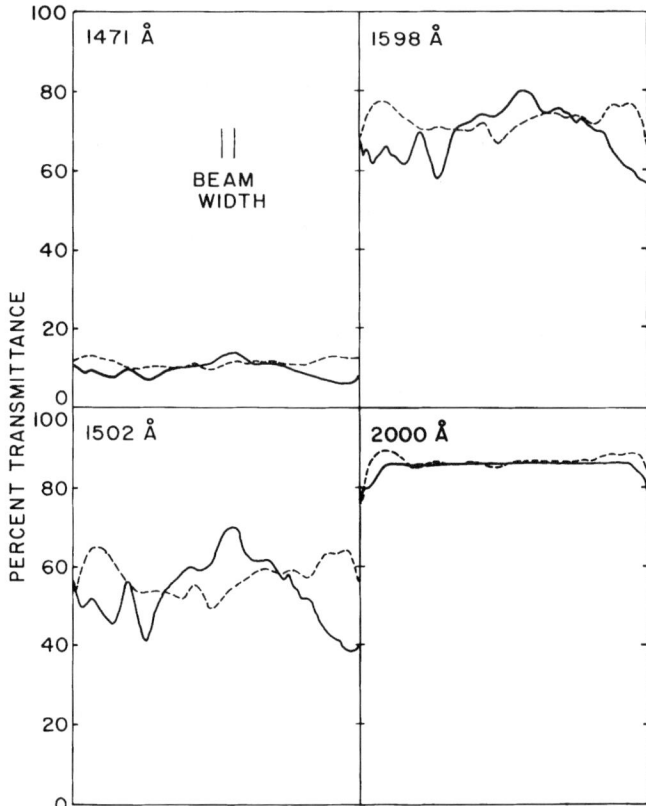

FIG. 4. Transmittance of optical quality, x cut, cultured crystalline quartz 3 mm thick as a function of position on the surface. The solid and dashed lines indicate measurements at orientations 90 deg apart. The abscissas represent the distance across the crystal, which is 38 mm [5].

also evident that electronic-grade cultured quartz is not suitable for optical purposes. Although uniformity measurements of other crystals (LiF, MgF_2, etc.) have not been made, there is no reason to believe *a priori* that they have a uniform spatial transmission characteristic.

Figure 8 [16] shows measured transmittance values for polished LiF and MgF_2 crystals about 3 mm thick. Note that the LiF transmittance spectrum does not have the shape of the cultured crystal quartz spectrum shown in Fig. 2. The transmittance begins to decrease at about 2000 Å and eventually reaches the cutoff wavelength at 1050 Å. Apparently, polishing LiF spoils the transmission as the cutoff wavelength is approached. The transmittance of the MgF_2 sample begins to decrease at about 1800 Å, appears to recover at about 1400 Å, then decreases rapidly to the cutoff at 1130 Å. These values should not be taken as typical of these two materials because there may be some variability in the quality of the crystals, especially MgF_2.

An unusual VUV filter has been described by Falcone and Bokor [17] who have used quartz to separate UV from VUV radiation, in effect a dichroic filter. They made a beamsplitter that transmits the second order (2800 Å) of a pulsed dye laser while reflecting the third harmonic (933 Å) subsequently generated in a Hg vapor cell. The quartz plate, wedge shaped to avoid reflections from the back surface, was given an antireflection coating on its front surface to reduce the reflection of the 2800-Å radiation. This treatment was not sufficient to reduce the 2800-Å component to the desired level, so an indium filter was used in the reflected beam. Radiation at 933 Å is transmitted by a thin indium film (see discussion later) while the longer wavelength is not. The authors also point

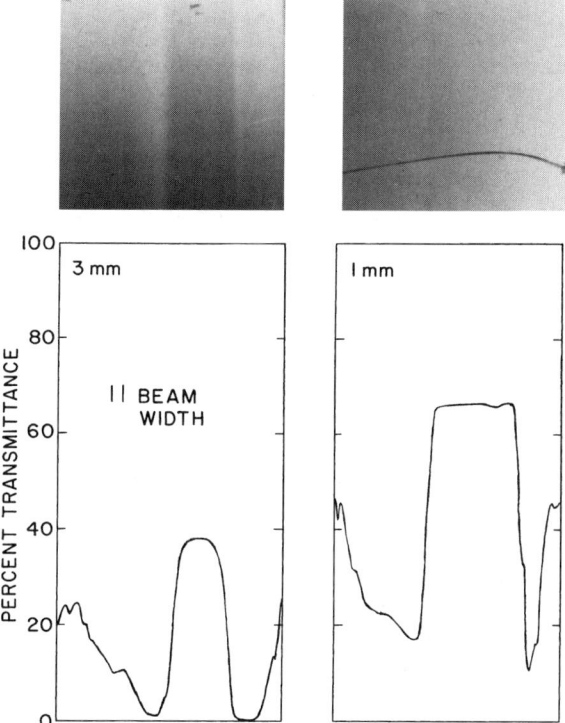

FIG. 5. Photographs of z cut quartz between crossed polaroids using white light and transmittance of the quartz at 1490 Å. The sizes of the photographs were adjusted to match the recording shown underneath them. The abscissas represent the distance across the crystal, which is 24.5 mm. The curved line on the 1 mm thick sample is a crack [5].

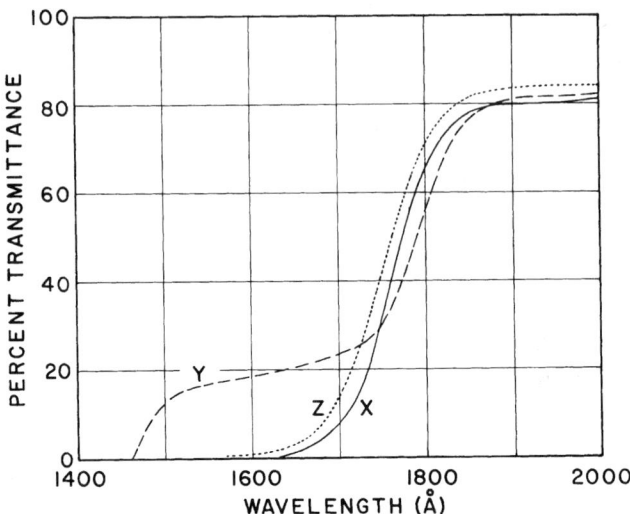

FIG. 6. Transmittance of x, y, and z cuts of cultured crystalline quartz from differentcrystals. The quartz was of electronic quality [5].

out that if the longer wavelength is plane polarized it may be possible to use the dielectric plate at the Brewster angle, possibly avoiding the need for an antireflection coating.

16.2.1 Environmental Effects: Radiation Damage

Researchers have long known that dielectric materials can be damaged by radiation. Kato *et al.* [18] have irradiated LiF with β and γ radiation and observed formation of the F band at about 5 eV and other extra absorptions closer to the fundamental absorption, at 11.75, 11.1, and 10.7 eV.

Figure 8 also shows that LiF and MgF_2 are damaged by electron bombardment. The samples were exposed to 1-MeV electrons, a dose of 2×10^{14} electrons/cm^2 for the LiF and 10^{15} electrons/cm^2 for the MgF_2. Irradiation alters the transmission characteristics of the LiF drastically and makes it practically useless as a transmitting material. By contrast, MgF_2 suffers only a moderate reduction in transmittance. Heath and Sacher [19] have investigated a number of VUV transmitting dielectrics, including LiF and MgF_2, to determine their suitability for use in satellites exposed to the energetic electron environment of space. Of the fluorides listed in Table I, crystals of high-purity BaF_2 were found to have the highest radiation resistance. Synthetic sapphire was least affected by electron bombardment. It showed a slight color change, to beige, after exposure but no change in the transmission from 3000 Å to the cutoff wavelength. It is considered a good material for shielding the more sensitive dielectrics from electron bombardment. By contrast, fused silica remained clear after exposure but the transmittance at wavelengths shorter than 3000 Å was decreased by about 10%.

Electron-induced absorption in fused silica and other quartz glasses (and MgF_2) can be reduced, or bleached, by irradiation with near UV from a quartz-mercury lamp. Complete restoration cannot be achieved even with prolonged bleaching. Annealing or heat treatments can also be used to restore transmission in electron-irradiated samples. It is possible to have memory effects, however, as illustrated in Fig. 9 [16]. Two high-purity fused silica glasses were irradiated with 1 and 5×10^{15} electrons/cm^2 of 1 MeV. The smaller dose produced only a slight transmission loss but the larger dose reduced the transmittance considerably. Heating at 500°C for 18 h removed most of the induced absorptance, however, when the annealed samples were exposed to an additional 1×10^{15} electrons/cm^2 of 1 MeV their transmittances dropped to much lower values than

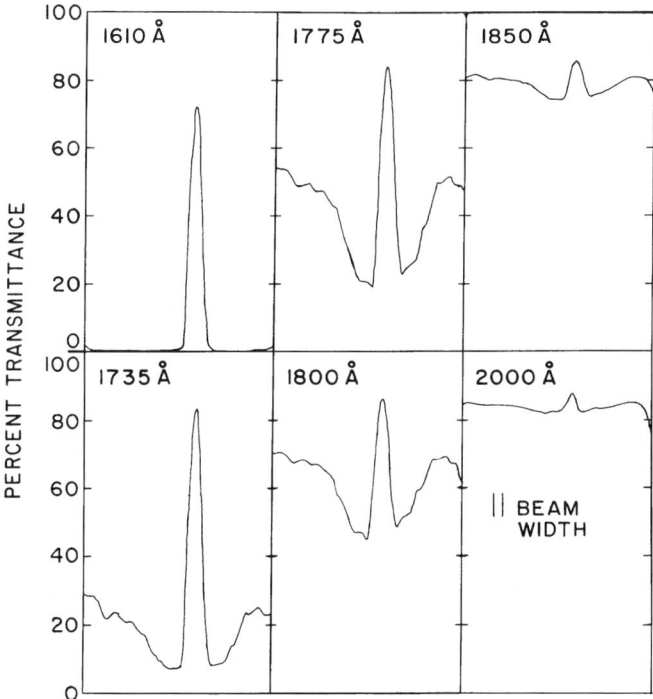

FIG. 7. Transmittance vs position on the crystal surface of electronic quality, y cut cultured crystalline quartz 2 mm thick from Fig. 6. The abscissas of each block represent a distance of 32 mm [5].

those measured after the initial 1×10^{15} electrons/cm^2 exposure. The transmittance of one of the annealed samples dropped during the postannealing irradiation to an even lower value than that induced by the preannealing dose of 5×10^{15} electrons/cm^2. Such memory effects may be present in other VUV window materials.

Cultured crystalline quartz is quite resistant to radiation damage. Good quality samples, such as those shown in Figs. 2 and 4, are only slightly affected by exposures of up to 10^{15} electrons/cm^2 of 1-MeV electrons. At 1417 Å the effect is pronounced, reducing the transmittance from 13 to 1%; however, at 1502 Å the effect is minor and becomes even less to longer wavelengths. Quartz of doubtful quality, such as that shown in Fig. 5, can be quite strongly affected. Figure 10 shows the sample of Fig. 5 after irradiation with 10^{15} electrons/cm^2 of 1-MeV electrons. The photographs were made with unpolarized white light. Two regions of high transmittance slightly to the right of center were only

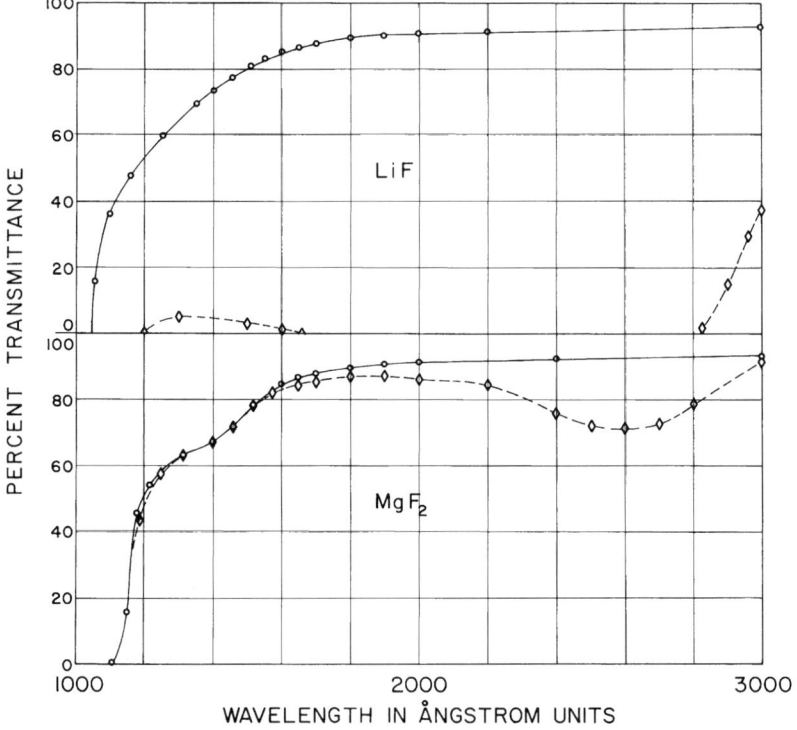

FIG. 8. Measured transmittance of 3-mm-thick LiF and MgF$_2$ plates before (solid lines) and after (dashed lines) irradiation with 1-MeV electrons as a function of wavelength from 1000 to 3000 Å. Dose for LiF = 2×10^{14} electrons/cm, and for MgF^2 = 1×10^{15} electrons/cm^2. (From [16], adapted from [19].)

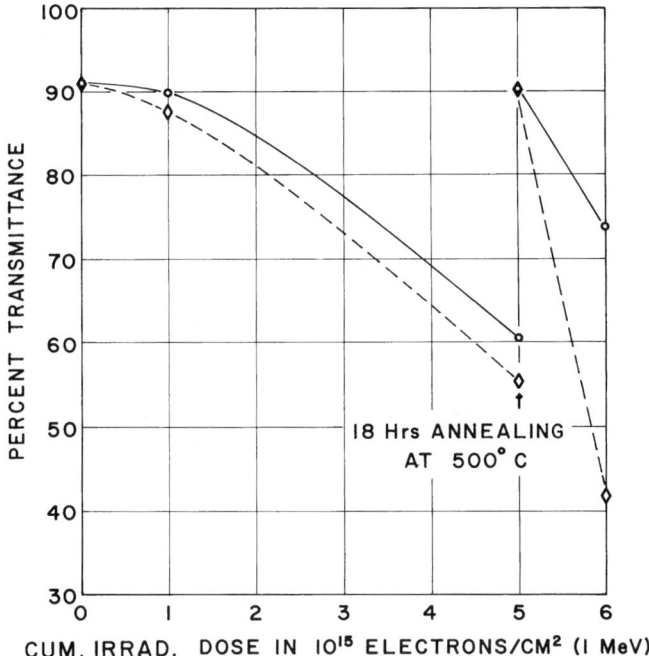

FIG. 9. Measured effect of electron irradiation, followed by annealing at 500°C, followed by a second electron irradiation on the transmittance of fused silica at 2200 Å [16].

slightly affected. At the far right of each sample are dark lines, some as narrow as 0.1 mm. On the left there is a large region where considerable blackening has occured and the transmittance at 1490 Å has been reduced practically to zero. Electron paramagnetic resonance analysis of this sample showed large concentrations of aluminum and alkali impurities. Why these regions should be so sharply delineated is unknown.

16.2.2 Temperature Effects

The cut-off wavelength also depends on the temperature of the dielectric. Hunter and Malo [1] have measured the change in cutoff wavelength with temperature of a number of dielectrics used as filters in the VUV. Figure 11 shows their results. The materials correspond to those in Table I, and all behave in the same general manner with the exception of LaF_3 whose cutoff wavelength appears to show no dependence on temperature. The transmittance of this sample of LaF_3 had rather small values (30%) at room temperature as compared with materials of VUV grade for which the transmittance would be about 70–80%. These small values suggest that it may not have been VUV grade, and it is possible that the grade affects the dependence of cutoff wavelength on temperature. Small transmittance values have also been obtained with samples of synthetic sapphire, MgF_2, and LiF that were not of VUV grade.

The dependence of cutoff wavelength on temperature suggests the use of temperature changes to make special filters. For example, CaF_2 has its room temperature cutoff wavelength practically at the Ly-α line of hydrogen, 1216 Å. By cycling the temperature of the CaF_2 above and below room temperature, the Ly-α component of a spectrum could be modulated, thus permitting its subtraction from the accompanying spectrum.

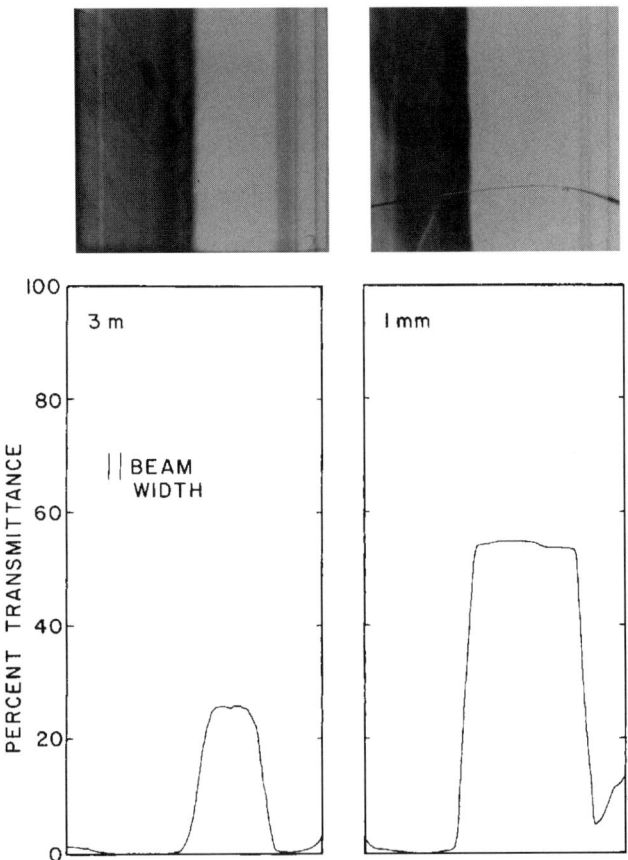

FIG. 10. The z cut quartz shown in Fig. 5 after irradiation with 10^{15} electrons/cm^2 of 1-MeV energy at 1490 Å. The photographs were made with white light and were adjusted in size to match the recordings shown underneath. The abscissas represent the distance across the crystals, which is 24.5 mm [5].

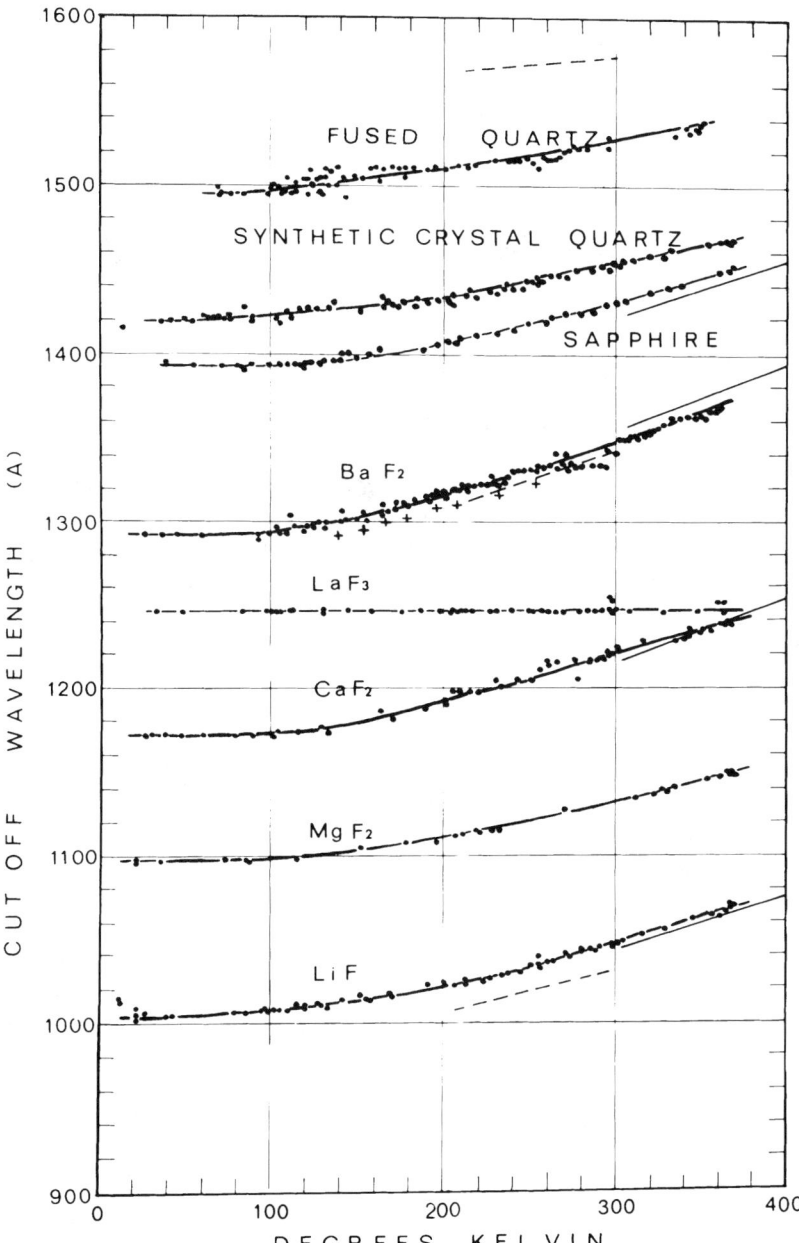

FIG. 11. The temperature dependence of the short-wavelength transmittance limit of various VUV materials. (From [1], with permission of Elsevier Science Ltd.)

Carruthers [20] has used the temperature dependence of LiF to construct a filter that isolates a range of wavelengths about the Ly-β line of hydrogen at 1026 Å. He coated a LiF crystal with In, which has a transmittance window extending from 736 to about 1100 Å. By cooling the coated LiF to liquid nitrogen temperature a passband filter was created that transmitted approximately from 1000 to 1080 Å.

16.3 Thin Solid Films as Transmitting Filters

16.3.1 Metals and Semiconductor Unbacked Films

16.3.1.1 Transmission Characterstics.

In 1918, R. W. Wood [21] discovered that thin films of sodium and potassium were transparent in the ultraviolet. With further research [22, 23] he showed that all the alkali metals have the same general characteristics, high transparency in the ultraviolet and nonmeasurable transmittance in the visible, and that the onset of transmission occurred at longer wavelengths as the atomic weight increased. Zener [24] attempted to explain this effect using classical electromagnetic theory. He was able to show that the onset of transmission depended on the valence electron density; the larger the density, the shorter the wavelength at which the onset of transmission took place.

As with the cutoff wavelength, the onset of transmission is a term that can only be defined by the user. Zener's theory results in a critical wavelength, however, which is a characteristic of the medium:

$$L_c = \sqrt{[(2\pi c/e)(m/N)]},$$

where N is the density of the valence electrons, e and m are the electronic charge and mass, respectively, and c is the velocity of light. If the wavelength of the incident radiation is longer than the critical wavelength, the medium is totally reflecting, and if the wavelength is less than L_c the radiation is transmitted.

Kronig [25] modified Zener's theory by including the damping of the free-electron motion by collisions with the lattice so that the reflectance at the longer wavelengths was not total. This modification gave a more accurate description of the optical behavior of the alkali metals and is equivalent to the free-electron theory of Drude [26].

A number of experimental studies of the transmission characteristics of thin metal films have been made. Tomboulian and Bedo [27] have measured the properties of Be, Mg, and Al from 50 to 500 Å, and both Si and Ge [28] in the soft x-ray region, Hunter et al. [29] have studied Al, In, Bi, Si, Mg, and Ge, Codling et al. [30] have studied Sn, Rustgi [31] has studied Ti, Te, Sb, and Be, and Samson [32] has studied carbon films. Much of the recent research has been carried out by Powell [33, 34] and his associates who have studied both metals and organics. Choi and Favre [35] have used silver as a soft x-ray filter for plasma diagnostics. Huizenga et al. [36] have investigated lexan as a window for a proportional counter for the wavelength region from 10 to 250 Å, and Barstow et al. [37] have made measurements of the transmittance of carbon and beryllium in combination with organic films from 160 to 2000 Å preparatory to using them on rockets to look at hot stars. Keski-Kuha [38] has studied the performance of aluminum films and aluminum-lexan combinations at temperatures down to 77 K for use with low heat capacity quantum calorimeters that operate at 0.1 K. Spiller et al. [39] have conducted a systematic search for thin-film filter materials suitable for x-ray solar telescopes. Figure 12 shows typical transmittances of a number of metals in the form of thin unbacked films, and Fig. 13 [34] shows the linear absorption coefficient of boron and carbon.

Although the metal films are often referred to as unbacked, or unsupported, they must be supported by a frame and are frequently supported across the aperture by a fine metal mesh, usually nickel mesh.

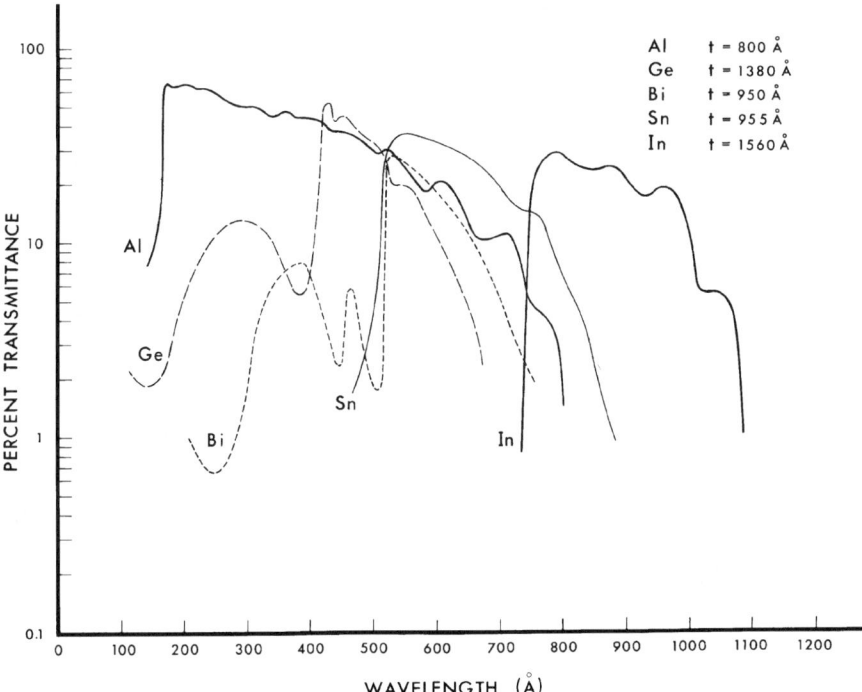

FIG. 12. Measured transmittances of a number of unbacked thin metal films as a function of wavelength.

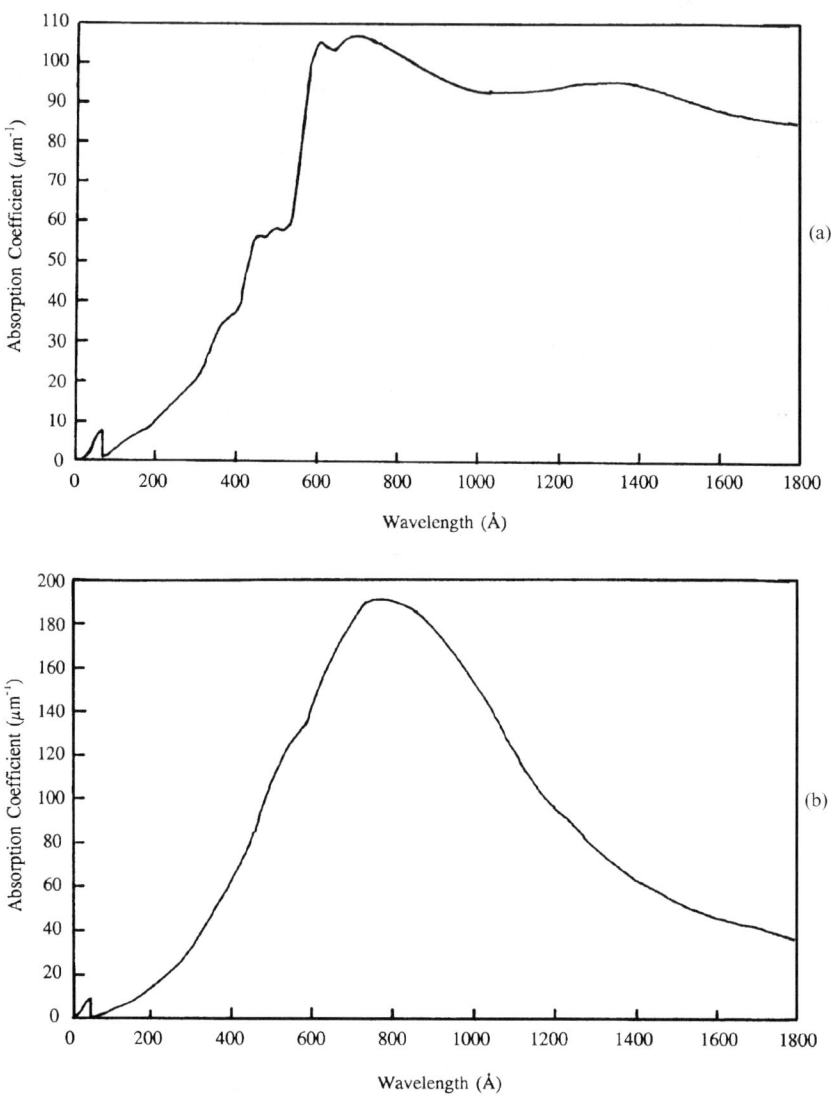

FIG. 13. Linear absorption coefficient of (a) boron and (b) carbon as a function of wavelength. (From [34], with permission from SPIE.)

Different metals require different techniques in preparation and removal from the substrates. Hunter [40] has published a detailed review of the methods for making thin metal film filters using thermal evaporation *in vacuo*. A common procedure is to use a massive, nonsoluble substrate on which a soluble parting agent is placed and to evaporate the metal onto the parting agent. Steele [41] evaporated aluminum onto a glass substrate previously coated with fluorescein, glued a fine nickel mesh to the aluminum surface, then immersed the substrate–film–mesh completely in acetone until the fluorescein dissolved. The resulting mesh–film combination was then dried and stored for future use. An alternate method is to use a soluble substrate (no parting agent) and coat it directly with the metal. NaCl is often used although it may react with the metal film when in solution. Diamond-like thin films can be made in this manner by forming the film on a silicon wafer and then dissolving the silicon wafer.

16.3.1.2 Environmental Effects. One of the plagues of thin metal film filters is pinholes. They permit radiation outside the desired spectral region to reach the detector and they are mechanically weak spots that may become enlarged when subjected to vibration. Steele [41] has made an exhaustive study of the conditions under which pinholes form in Al films and how they may be prevented. Pinholes can form during any of the three stages of production of filters: during deposition, during separation from the substrate, and during mounting and storage. Generally pinholes that appear large to the unaided eye are quite small and can be seen only because the eye is sensitive over a very large range of light levels. Steele has devised a technique for locating pinholes and measuring their transmission. He uses an extended light source a short distance in front of the filter and puts a piece of photographic film just behind the filter. A suitable exposure gives a "pinhole map," which shows the spatial distribution of pinholes and can be used, in a semiquantitative way, to judge their size. Figure 14 [42] shows a pinhole image map, a pinhole diagram of the map, and a pinhole fogging map for a large (4.1- $\times$ 24.4-cm) aluminum filter used in the extreme ultraviolet spectroheliograph [43] flown in the Apollo Telescope Mount (ATM) of SKYLAB. The pinhole image map was obtained as described earlier, the diagram shows the character of the larger pinholes as determined with the aid of the microscope, and the pinhole fogging maps were obtained by mounting the filter in a mock-up of the ATM slitless spectrograph, which was then illuminated by sunlight via a ceolostat mirror, as it would be during flight. Two exposures were used to produce the fogging maps; 8 min on 104 UV film for the topmost exposure, and 32 min on 101 UV film for the bottom exposure.

For a quantitative measurement of pinhole transmission per square centimeter, Steele [41] replaces the photographic film with a photoelectric detector. He has found that for small areas, ≈ 1 cm^2, the transmission per square centimeter can be as low as 10^{-9} although for larger areas the transmission is usually about 2 orders of magnitude larger. Powell and Fox [44] have also devised a photometer for quantitative measurements of pinhole transmission.

Whether or not the pinhole transmittance of the filter will compromise the purpose of the filter depends on the relative intensities of the total spectrum of the source and that part of it to be transmitted by the film. For laboratory light

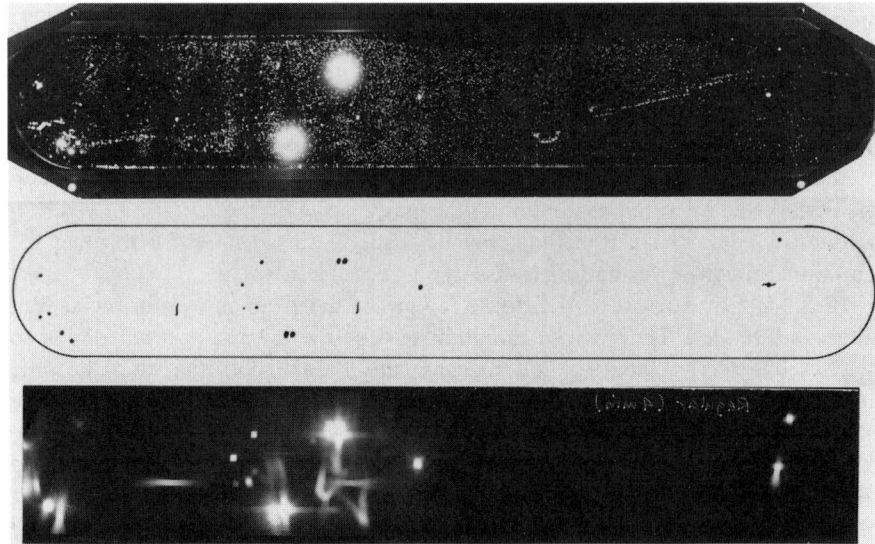

FIG. 14. Pinhole image map (top), pinhole diagram, and pinhole fogging maps (bottom). The fogging maps were obtained with a spectroheliograph pointed at the sun [42].

sources, such as dc glow discharges in gases, a transmission of 10^{-2} is usually acceptable. In contrast, for solar spectroscopy from spacecraft, the integrated visible and near-UV spectrum of the sun is about 6 orders of magnitude larger than the integrated XUV spectrum transmitted by an aluminum filter of about 1000-Å thickness; consequently, a pinhole transmittance of 10^{-7} is necessary.

In laboratory instruments the vibration amplitudes are usually small enough to have no effect on the lifetimes of the filters. In rocket launches, however, the vibration and acoustic fields may be quite intense and a film with excessive pinholes may not survive a launch. Pinholes tend to form around the boundary where they are attached to the frame because that is where most of the flexing occurs.

16.3.2 Thin Organic Film and Organic/Metal Film Filters

Parlodion has been measured by Hunter et al. [29]. Tomboulian and Bedo [27] have studied Zapon (cellulose acetate), Formvar (polyvinyl formal), Mylar (polyethylene tetraphthalate), Teflon (tetrafluoroethylene), Collodion (cellulose polynitrate) and polystyrene. These materials are often used as thin-film substrates onto which metals are deposited. When the metal to be evaporated requires high temperatures, the organic film can be laid over a glass backing to prevent damage during heating. It is wise to put a parting agent on the glass backing before putting the organic film on it to ensure subsequent removal of the filter. Very dilute solutions of the organic materials can also be sprayed onto a glass-parting agent surface. One can also dip a glass plate into a dilute solution of the organic material. Tomboulian and Bedo give the following organic materials and their solvents: Zapon in iso-amyl acetate, polystyrene in benzene, and Mylar in ethylene bromide.

Powell *et al.* [34] prefer Lexan as a substrate for metal films because of its strength and absorption in the VUV. Its linear absorption coefficient is shown in Fig. 15, and is quite similar to that of carbon, in Fig. 13, because of the carbon content of the material. Most of these organic materials have absorption spectra similar to those of carbon and Lexan until the wavelength drops below about 100 Å, where soft x-ray structure becomes dominant.

Polypropylene and parylene C can also be used as filters or thin-film substrates. Caruso [45] has measured the mass absorption coefficients of both these materials between 8.34 and 452 Å, and Stern and Paresce [46] have measured those of parylene N from 44 to 2500 Å. Many of these materials are also used as windows for counters in the soft x-ray region.

Seely [47] has developed a "water-window" for the wavelength region 23–44 Å by using thin layers of Sn and Al on Formvar, and Seely *et al.* [48] have reported a transmission filter for the 61- to 120-Å region made of Saran. The Saran filter is 4000 Å thick and is made by a spinning technique.

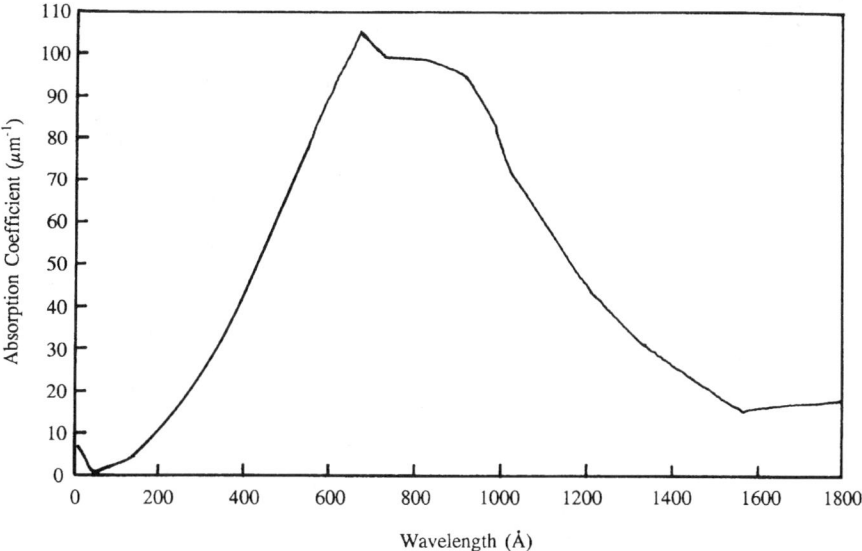

FIG. 15. Linear absorption coefficient of Lexan as a function of wavelength. Note the similarity to the absorption spectrum of carbon (Fig. 13b). (From [34], with permission from SPIE.)

16.3.2.1 Environmental Effects.
No systematic study of enviromental effects of plastics and metal/plastic combinations has been made. On continued exposure to VUV, and ions often associated with VUV experiments, for example, ion pumps, many plastics deteriorate and have to be replaced.

16.3.3 Dieletrics and Metals on Dielectric Substrates

Bates and Bradley [49] have succeeded in making transmission filters using alternate layers of Al and MgF_2 on suitable substrates, such as SPECTROSIL (fused silica) and sapphire. They have achieved maximum transmittance values of as much as 29% at 1780 Å with a halfwidth of about 400 Å. This type of filter is asymmetrical, with a transmitting tail toward longer wavelengths due to the transmitting characteristics of the substrate.

Malherbe and Guillard [50] and Malherbe [51] have also developed interference filters consisting of alternating layers of Al and MgF_2. Using a LiF substrate and three periods of Al/MgF_2, they were able to construct a filter centered on 1216 Å that had about 15% transmittance and a halfwidth of 90 Å. Because the Al layer thickness had to be controlled, they could not use fast evaporation in a conventional system but used a UHV system with a base pressure of 10^{-9} Torr and produced their coatings at pressures of about 10^{-8} Torr.

Malherbe [52, 53] has also worked on all-dielectric filters but has not succeeded in making any with passbands at wavelengths shorter than 2000 Å. His first filters [52] were made using LaF_3 or NdF_3 as the high index layers and chiolithe as the low index layer. He achieved a peak transmittance of about 60% and a halfwidth of about 20 Å. Subsequently [53], he used the same high index materials but MgF_2 as the low index material. With LaF_3/MgF_2 he made a filter centered at 2025 Å with more than 80% transmittance and a halfwidth of 25 Å. Filters of this type, as with the filters of Bates and Bradley, have a transmitting tail toward longer wavelengths. To eliminate this tail, the filters require "blocking," that is, a secondary filter to eliminate the long wavelength tail. Malherbe used two schemes for blocking, one was a thin coating of metal, probably Al, that reduced the transmittance to about 20% but reduced the adjacent wavelengths to 10^{-3}. The second scheme made use of four dichroic mirrors in series with the filter that effectively suppressed the adjacent wavelengths.

16.4 Gases as Transmitting Filters

The visual experience shows that gases are excellent transmitting media. There are, however, isolated wavelengths at which a gas can absorb rather strongly. As one approaches the VUV region these wavelengths occur more and more often until, at the ionization limit, the gas becomes strongly absorbing at all wavelengths shorter than the ionization limit. Table II lists the elemental

TABLE II. Elemental Gases that Might be Useful as Filters, Their First Ionization Potential, and the Equivalent Wavelength in Å

Element	Ionization potential[a]	Wavelength
H	13.598	912
He	24.587	504
N	14.534	853
O	13.618	910
Cl	12.968	956
F	17.423	712
Ne	21.564	575
Ar	15.760	787
Kr	14.000	886
Xe	12.130	1022

[a] Ionization potentials taken from *Handbook of Chemistry and Physics*, 76th ed. (D. R. Lide, ed.), Chemical Rubber Co. Press, New York (1995). Conversion factor from eV to Å is 12398.52.

gases, their first ionization limits, and the equivalent wavelength in angstoms. In principle, all of these gases can be used in spectrographs to remove wavelengths shorter than their ionization limits; however, some precautions must be taken. In practice the fact that hydrogen, nitrogen, oxygen, chlorine, and fluorine exist as molecules under ordinary conditions means that their absorption spectra are complicated and not simply related to the ionization limits of the atoms. Furthermore, two of these gases, Cl_2 and F_2, are quite active chemically, can damage delicate optical components, and are dangerous to handle.

16.4.1 Noble Gases

The absorption spectra of the noble, or rare, gases are shown in Fig. 16 [54] from 1000 to 280 Å. The absorption coefficients shown may be superseded by now but the general shapes of the curves remain the same. The vertical lines shown on the Ar, Kr, and Xe spectrum indicate the regions in which autoionization takes place. In these regions there are fairly narrow areas (windows) in which the absorption coefficient dips to small values compared to the average value in that region, for example, 200 cm^{-1}. If these gases are to be used to absorb all the shorter wavelengths in a spectrum, they must be used at pressures large enough to absorb completely radiation in the region of the autoionization windows.

He and Ne also have regions of autoionization but at wavelengths shorter than 280 Å. Economic concerns may restrict the use of Ne, Kr, and Xe for absorption filters because recovery is usually not feasible.

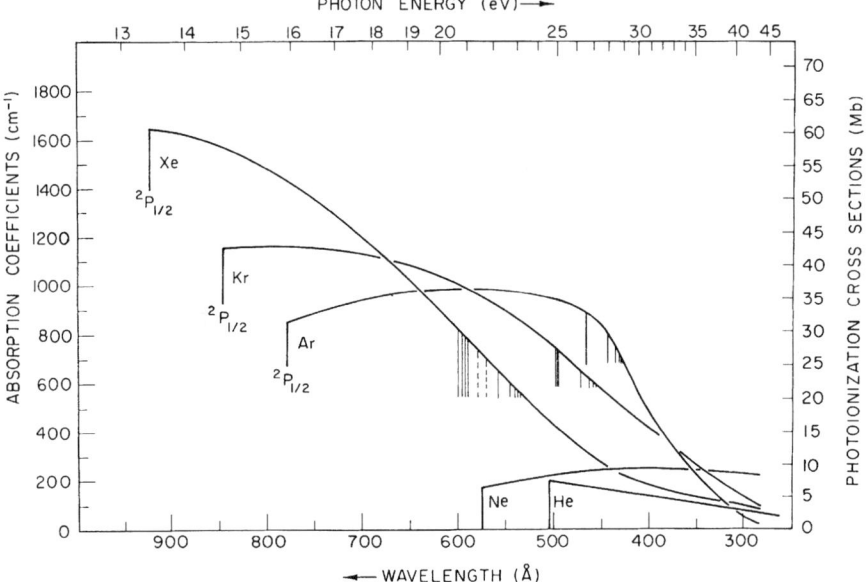

Fig. 16. Absorption spectrum of the rare gases from 1000 to 280 Å. The vertical lines in the Ar, Kr, and Xe spectra represent regions of autoionization. (From [54], with permission of J. A. R. Samson.)

16.4.2 Gases in Atomic Form

Hydrogen forms a natural barrier to astronomical research in the extreme ultraviolet because it absorbs strongly at its ionization limit. Although the density of hydrogen in interstellar space is small, the distances are immense so that the column absorption is strong enough to prevent radiation shortward of 912 Å from reaching the vicinity of earth. For some time astronomers despaired of obtaining stellar spectra shortward of 912 Å, however, to ever shorter wavelengths the hydrogen absorption weakens and rocket and satellite flights have seen wavelengths as long as 304 Å from stars. Bowyer has reviewed this subject [55, 56].

Morton and Purcell [57] developed a filter for the Ly-α line of hydrogen by sealing molecular hydrogen in a container with LiF windows and partially dissociating the H_2 with a heated tungsten filament so that the Ly-α radiation could be absorbed. The filter was intended to be used to determine the emission line widths of the Ly-α line emitted in the night sky. Winter and Chubb [58] have also used the same filter for night sky research.

16.4.3 Molecular Gases

The absorption spectra of molecular gases are complicated and are not considered in detail here. Those interested may consult the many texts on molecular spectra. What follows has been taken from Sullivan and Holland [59] and Hudson [60].

Air exhibits absorption structure by virtue of its component gases. For example, air transmits radiation without absorption from the visible and near-IR wavelengths down to about 2600 Å where a weak continuum absorption in oxygen commences ($\alpha \approx 10^{-4}$ cm^{-1}) and continues to about 1600 Å. The Schumann–Runge bands of molecular oxygen begin absorbing at about 1950 Å but the absorption does not become very strong until 1850 Å ($\alpha \approx 0.02$ cm^{-1}); to shorter wavelengths α increases very rapidly. Thus, the onset of strong absorption bands at 1850 Å can act as a "cutoff" if the pressure is high enough (atmospheric). This characteristic of oxygen absorption has been used to evaluate stray light of diffraction gratings longward of 1850 Å [61] at atmospheric pressure.

Molecular nitrogen is practically transparent to wavelengths longer than 1450 Å. Between 1450 and 910 Å there are small regions of very low absorption coefficient that would be atmospheric windows in the absence of absorption by other gases.

There are a number of wavelengths in the vicinity of 1216 Å at which both molecular nitrogen and oxygen have very small absorption coefficients; $\alpha \approx 1.0$ cm^{-1} or less. One of these windows has been used to make reflection measurements at ≈ 1216 Å at atmospheric pressure [62].

16.5 Thin Films on Substrates as Reflecting Filters

One of the earliest broadband reflecting filters consists of a layer of MgF$_2$ on an opaque Al layer. The MgF$_2$ prevents oxidation of the Al thus preserving the intrinsic high reflectance of Al from the visible to about 1200 Å where the MgF$_2$ transmittance begins to decrease. Some enhancement is also provided through interference. The same effect can be achieved using a coating of LiF over Al, which produces a highly reflecting coating to wavelengths as short as 1025 Å. Figures 17 and 18 [63] show the measured reflectance values to be expected from both types of coatings at normal incidence, 35 deg corresponding to a Seya–Namioka monochromator, and 85 deg.

The success of the Al/MgF$_2$ broadband filters prompted research on selective filters for the 1216-Å line of hydrogen. Measurements of two early attempts are shown in Fig. 19 [64]. To the left is a quadruple layer; the first Al layer is opaque, the two MgF$_2$ layers are 250 Å thick, and the intermediate Al layer is thin. The maximum reflectance is 55%, the minimum of 2% occurs at 1350 Å,

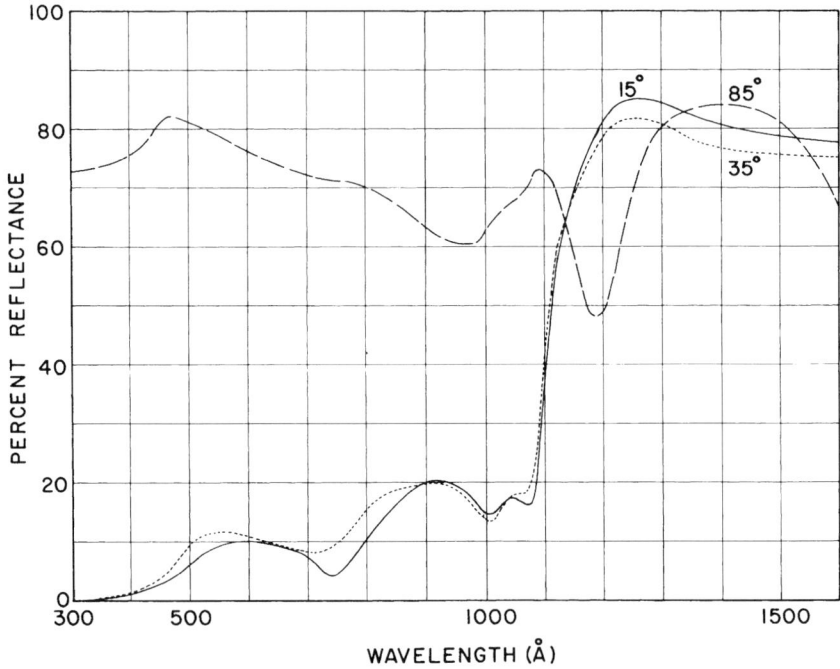

FIG. 17. Measured reflectance of an Al + MgF$_2$ mirror from 300 to 1600 Å. The MgF$_2$ thickness is 250 Å for greatest reflectance at 1216 Å [63].

and the halfwidth is about 200 Å. A blocking filter would be required to suppress the longer wavelengths. To the right is a double layer; the first layer is ZnS, thick enough to be opaque to 1216 Å, overcoated with a layer of MgF$_2$ 250 Å thick. The maximum reflectance is 46%, the minimum is about 10%, and the halfwidth is fairly large.

16.5.1 Multilayers

Currently perhaps the best known examples of VUV reflecting filters on substrates are multilayer coatings. Spiller has discussed the theory of such coatings in Chapter 14, so only some typical examples are given here.

Many different materials can be paired to make multilayer coatings. Rosenbluth [65] has compiled a list of possible pairs. His criterion was the difference in soft x-ray scattering factors of the different materials; the larger the differences the better the choice. However, other factors must also be considered, one of the most important being the tendency of the two materials to interdiffuse, thus adding extra layers and possibly rendering sharp boundaries

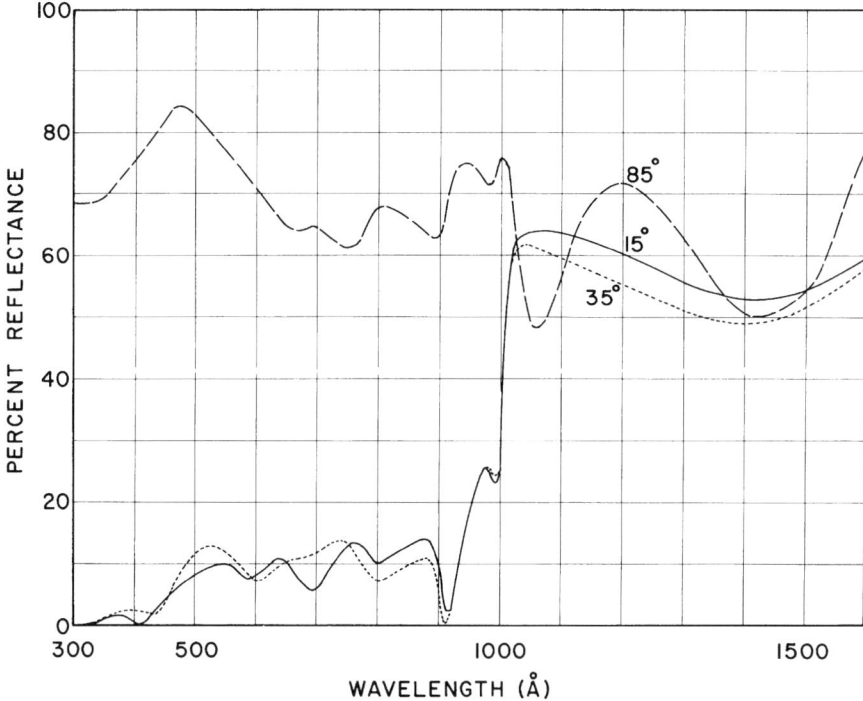

FIG. 18. Measured reflectance of an Al + LiF mirror from 300 to 1600 Å. The LiF thickness is 140 Å for greatest reflectance at 1025 Å [63].

diffuse. Long-term stability under varying environmental conditions and the difficulty of the evaporation, sputtering, or whatever process is used to produce a layer of the material can also be important. Finally, the location of the absorption edges can affect the spectral range over which the multilayer is useful if it is used at different angles of incidence. Chapter 11 shows the location of soft x-ray absorption edges of some materials used in making multilayers.

A good example of the wavelength dependence of the bandpass of a multilayer coating is shown in Figs. 20a and b, taken from Yanagihara et al. [66]. The multilayer materials were Ru and C and the multilayer consisted of 21 layers on a polished Si wafer. The peak reflectance of 34% for s polarized radiation occured at 99 eV at an angle of incidence of 45 deg. After fabrication the mirror was cut into two parts, which were mounted in an arrangement resembling a double crystal x-ray monochromator. Thus the two mirrors could be rotated and translated so that the angles of incidence on each remained constant. Figure 20a shows the two-mirror reflectance at different energies for mostly s polarized radiation emerging from a 2-m grasshopper monochromator attached to beamline

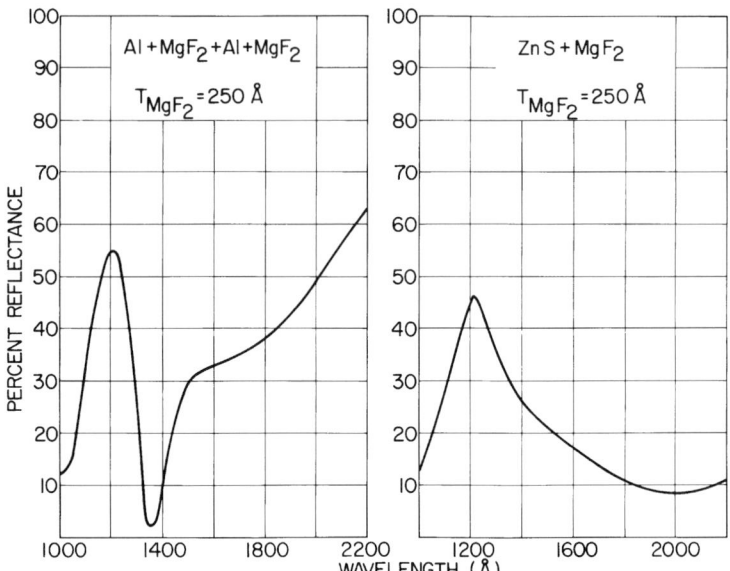

Fig. 19. Reflecting interference filters designed for maximum reflectance at 1216 Å. (From [64], with permission from Taylor & Francis.)

11A at the Photon Factory. Above each black circle is the angle of incidence at which the measurement at that wavelength was made. The dotted curves associated with each black circle shows reflectance vs wavelength at that angle of incidence. As the energy decreases from about 150 Å, the peak reflectance increases to a maximum of 11.5% at about 47 deg and 100 eV, then decreases rapidly to the lowest energy measured, about 40 eV. Figure 20b shows the same sort of measurement but with mostly *p*-polarized radiation. Here the maximum reflectance of slightly more than 0.8% occurs at about 140 eV. The upper curve in Fig. 20a is the polarizance of the coating, that is, $(R_s - R_p)R_s + R_p)$.

Two pairs of materials that are commonly used are Si/Mo and B_4C/W. Silicon and Mo interdiffuse and the thickness of the diffusion layer depends on which material is being deposited on which. If Si is deposited on Mo the interdiffusion layer thickness is less than when Mo is deposited on Si. Stearns *et al.* [67] report interdiffusion layer thicknesses of 10–12 Å for Mo sputtered onto Si and 5–6 Å for Si sputtered onto Mo. Boher *et al.* [68] found interdiffusion layer thicknesses of 15 Å for Mo on Si and 8 Å for Si on Mo. Thus, there is some latitude in choosing the interdiffusion layer thickness for modeling. The interdiffusion layers are usually assumed to have the composition $MoSi_2$. When sputtered, B_4C and W apparently do not interdiffuse, or if they do the interdiffusion layer is very small.

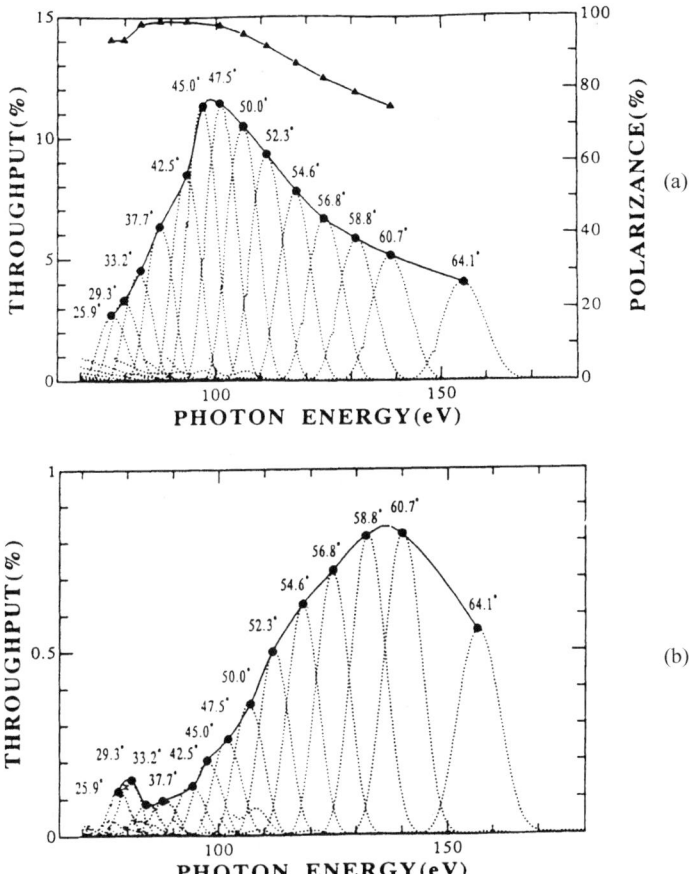

FIG. 20. (a) Measured transmittance of the C/Ru double multilayer for s-polarized radiation as a function of photon energy. (b) Measured transmittance of the C/Ru double multilayer for p-polarized radiation as a function of photon energy. The uppermost curve in part a is the polarizance of the double multilayer. (From [66], with permission from the American Institute of Physics.)

The top part of Figure 21 [69] shows measured reflectances of an Si/Mo multilayer consisting of 40 periods, that is, 40 layer pairs, overcoated with a 10-Å-thick layer of carbon to retard oxidation of the final Mo layer. The multilayer design required that each period be 72 Å thick with Si and Mo thicknesses of 57.6 and 14.4 Å, respectively. Measurements were made at two angles of incidence, 25 and 10 deg, and with s and p polarization. The polarization, about 95% complete, was due to synchrotron radiation filtered through a monochromator. At 10 deg the s and p components are almost equal, the s

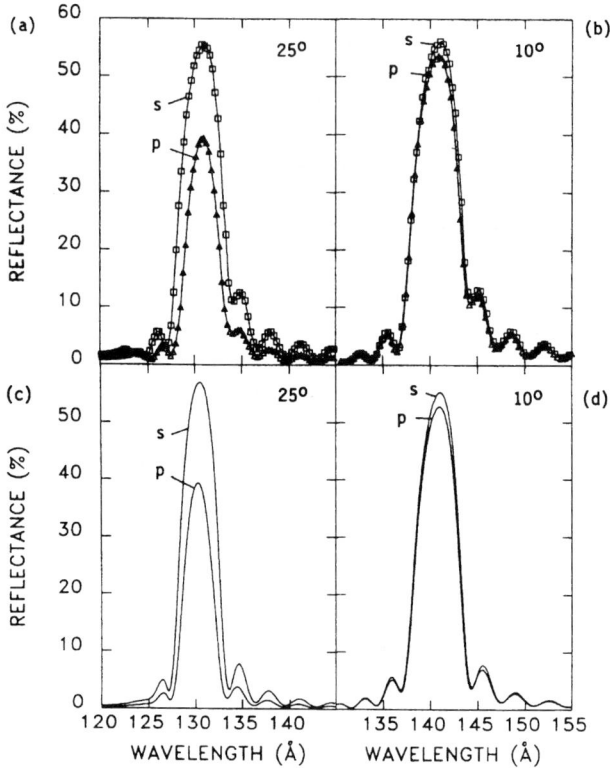

FIG. 21. (a, b) Measured reflectances of a multilayer coating illuminated at angles of incidence of 25 and 10 deg. The measurements were made with 90% s-polarized radiation or 90% p-polarized radiation. (c, d) Corresponding calculated reflectances [69].

component being about 56% and the p component being about 53%. The halfwidths of the s components are about 5 Å. The undulations at the bottom are caused by interference effects involving the substrate. In changing the angle of incidence from 10 to 25 deg, the reflectance maximum has shifted from about 142 to 131 Å. Hence, the position of the reflectance maximum depends on the angle of incidence and shifts the wavelength in accordance with the corrected Bragg law. Note, also, that the p component for the 25-deg measurement has decreased to about 39%. In the spectral regions where multilayers are most useful, the indices of refraction of most materials are very close to unity and the extinction coefficients are small; consequently, p-component reflectance values are strongly dependent on angle of incidence and are very small at 45 deg.

Modeling of the preceding multilayer using the design values did not agree with the measured values because of interdiffusion layers and roughness. Best agreement with measurements was obtained using thicknesses of 42.6 and 8.6 Å for the Si and Mo layers, respectively, and assuming two different interdiffusion layer thicknesses of $MoSi_2$, 15 Å where Mo is deposited onto Si and 7 Å where Si is deposited onto Mo. Thus, for purposes of calculation the multilayer consists of the substrate overcoated with three layers followed by a repeating four-component system as shown:

substrate/Si(42.6)/MoSi2(15)/Mo(8.6)/MoSi2(7)/Si(42.6)/
MoSi2(15)/Mo(8.6)/...../C(10).

Periodicity begins after the first Mo layer and the four-component system has 39 periods. For each layer an rms roughness of 5 Å was assumed. The results of the calculations are shown in the bottom part of Fig. 21 and show excellent agreement with the measured values.

Underwood *et al.* [70, 71] found that the reflectance of Si/Mo multilayers deteriorated with time due to oxidation of the uppermost Mo layer to MoO_3 and MoO_2, as determined by photoelectron spectroscopy. This oxidation appears to consume the entire uppermost Mo layer down to the $MoSi_2$ interdiffusion, which is inert. Oxidation can be prevented by storing the mirrors in vacuum or an inert atmosphere. Alternatively, the oxide layer can be removed by sputtering or various chemical means, including an etch in distilled water at 100°C for 15 min. To avoid the Mo oxide entirely, the final layer can be Si, which forms a stable oxide of from 10 to 20 Å and has very little effect on the reflectance of the multilayer. Carbon contamination was also present in the uppermost oxidized Mo layer.

Figure 22 [72] shows measured reflectances of two B_4C/W coatings, at normal incidence each consisting of 100 periods but with different period thicknesses, 17.54 Å at the top and 25.74 Å at the bottom. Modeling of these results showed that no interdiffusion layers were required, only rms roughnesses of 3.1 and 3.7 Å for the 17.54- and 25.74-Å periods, respectively. The calculations were quite sensitive to the roughness values; changing the roughness by 0.1 Å changed the calculated value significantly.

Figure 23 [73] shows the measured reflectances of two B_4C/W multilayers at 1.543-Å CuKα radiation as a function of glancing angle. This coating was designed to have a broad band, rather than the usual narrow band, and is intended for use on a Kirkpatrick–Baez telescope configuration. It has 200 periods of B_4C/W with the periods graded so that at the top of the structure the period is 18 Å while at the bottom, next to the substrate, the period is 22 Å. The ratio of W to period thickness, Γ, is maintained throughout the structure. Because two mirrors are used in the Kirkpatrick–Baez configuration, both mirrors were measured to ensure that they had the same reflectance, as shown in Figs. 23a and b. Figure 23c shows the calculated reflectance assuming an rms roughness at each layer of 2.6 Å. Figure 24 [73] shows a calculation at a 1-deg glancing angle showing how wide in wavelength the graded period coating is compared to a coating with constant period. The solid line represents the graded period coating, the dashed line represents a B_4C/W multilayer with the same number of periods but with a constant period thickness of 19 Å, and the dashed-dotted curve is the reflectance of tungsten; all calculations assume a 2.6-Å rms roughness. Although graded period multilayers have a reduced maximum reflectance compared to a coating with constant period thickness, the wavelength-integrated reflectance is greatly increased. The constant period multilayer has a halfwidth of 0.01 Å, while the graded period coating halfwidth is 0.13 Å.

Using a graded period also increases the angular halfwidth of the coating. The coatings of Fig. 23 have an angular halfwidth of 0.4 deg in contrast to the halfwidth of a constant period coating, which is calculated to be 0.03 deg.

Multilayers using Be as a spacer have been investigated by a number of workers [74–77]. Most of these studies were concerned with Be multilayers for use at grazing incidence but Skulina *et al.* [77] have conducted a systematic study of Be/Mo multilayers for use at normal incidence. Be has no L edges, only a K edge at 112 Å so Skulina *et al.* concentrated their efforts on the spectral

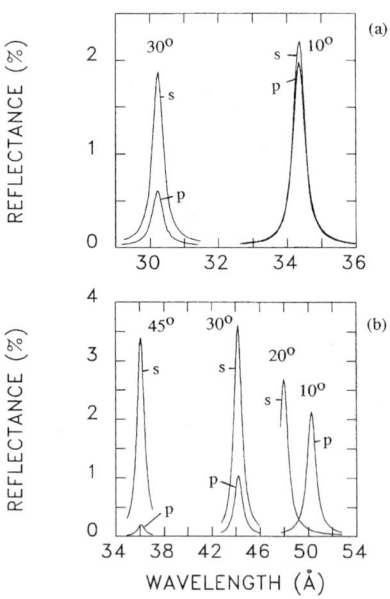

FIG. 22. Measured reflectances of two B_4C/W multilayer mirrors at the indicated angles of incidence and for *s*- or *p*-polarized radiation. The period thicknesses of the two mirrors are (a) $d = 17.54$ Å and (b) $d = 25.74$ Å [72].

region just longward of the K edge. To summarize their findings, approximately 70 Be/Mo periods are required to reach maximum reflectance of about 68%. Rutherford backscattering from a Be layer 5000 Å thick showed the existence of 0.1 at. % level in the sputtered Be, and a stable oxide layer of 29 Å on the surface. When incorporated into multilayers with Mo, the interface roughness value, obtained from modeling, is 6.8 Å. Multilayers ending in Be showed that a BeO layer 30 Å thick formed on the surface. Repeated measuring of some of

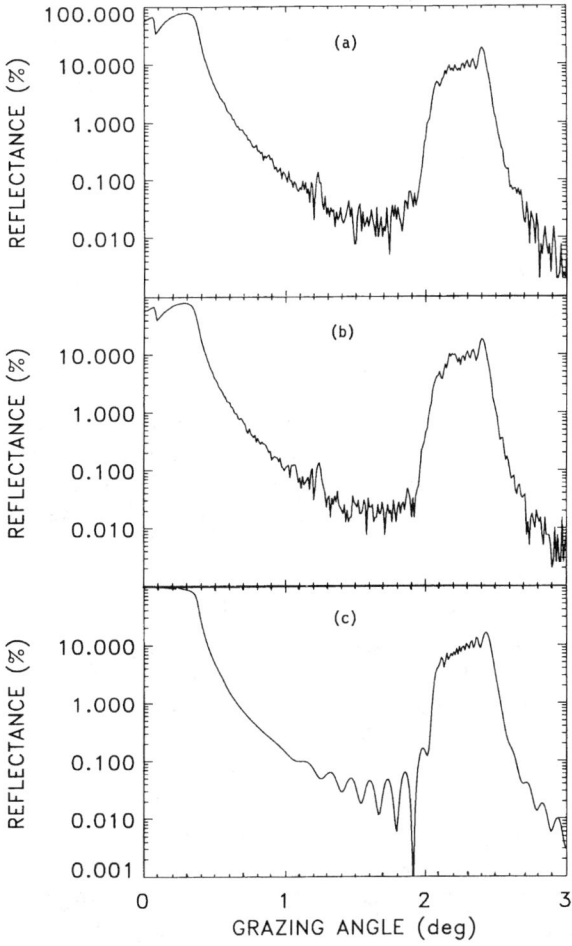

FIG. 23. (a, b) Reflectances of two B_4C/W depth-graded multilayer mirrors measured using CuKα radiation with a wavelength of 1.543 Å. The detector partially occulted the radiation beam at grazing angles of less than 0.3 deg. (c) Calculated reflectance [73].

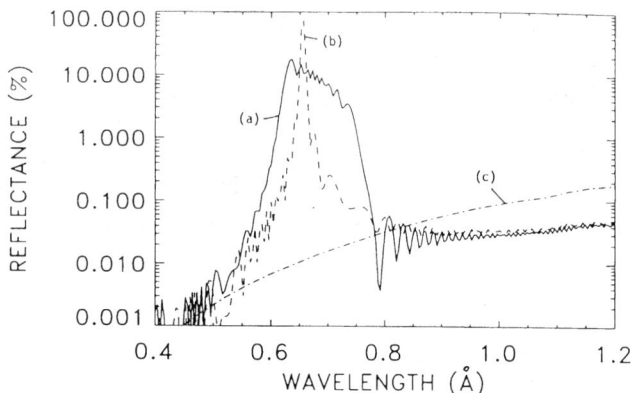

FIG. 24. Reflectances calculated at a 1 deg glancing angle of (a) the B_4C/W depth-graded multilayer (solid curve), (b) the B_4C/W multilayer with a constant period thickness of 19 Å (dashed curve) and (c) a thick W coating (dashed-dotted curve) [73].

these multilayers after 4 months of storage in air showed no loss in reflectance within experimental error.

16.5.2 Critical Angle and Interference Filters

Some materials have a natural filtering characteristic in reflection, for example, silver which has large reflectance values at normal incidence in the visible and to about 3200 Å at which point, the critical wavelength, the reflectance decreases rapidly because of the plasmon oscillation in the metal. A number of quasi-free electron materials have this property: a critical wavelength where a rapid normal incidence reflectance decrease occurs in the VUV. A calculation showing a number of reflectance spectra that illustrate this property is shown in Fig. 1 of Chapter 14. Thus, at wavelengths longer than the critical wavelength, the material behaves as a metal with large reflectance values, while at the shorter wavelengths the material behaves as a slightly lossy dielectric. In order for a material to behave in this manner it must have an index of refraction less than unity and an extinction coefficent that is small, <0.1.

At wavelengths shorter than the critical wavelength, the material has a critical angle of incidence. At larger angles the reflectance values are large (frustrated total external reflection, FTER) and at smaller angles the reflectance values are very small. By exploiting this characteristic it is possible to make tunable filters that reflect longer wavelengths while suppressing shorter wavelengths simply by changing the angle of incidence on the reflector, a critical angle mirror (CAM). Terminello et al.. [78] have used this effect to suppress higher order radiation in monochromators by using a Si crystal from which the oxide layer has been removed, and Sainctavit *et al.* [79] have used glass mirrors at higher energies (2–12 keV) to suppress higher orders.

Hunter and Long [80] found that suppression could also be obtained by combining the critical angle technique and interference. The principle is illustrated in Fig. 25, which shows the calculated reflectances, at 620 and 310 Å, of an oxidized Si film 600 Å thick on a fused silica substrate. The oxide film is 30 Å thick. Both wavelengths are shorter than the critical wavelength of 760 Å. The abrupt drops in reflectance, at ≈30 deg for 620 Å and 70 deg for 310 Å, locate the critical angles for these wavelengths. The maxima and minima in the curves are due to interference of the radiation reflected from the vacuum–layer and layer–substrate interfaces. The persistence of large reflectance values at angles of incidence between the critical angle and grazing incidence are examples of FTER.

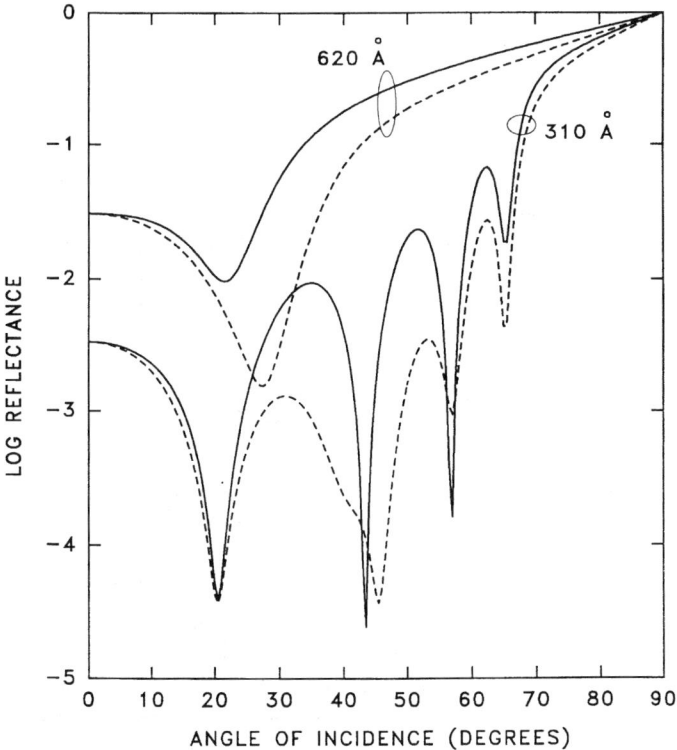

FIG. 25. Calculated reflectance vs angle of incidence at two wavelengths, 620 and 310 Å, of a Si film 600 Å thick on a SiO_2 substrate. The film is overcoated with 30 Å of oxide. The solid curves are the *s* component, and the dashed lines are the *p* component [80].

For their reflecting interference filter (RIF), Hunter and Long chose to work with the minimum in the 310-Å curve, the *second harmonic*, at 57 deg. At this angle the calculated reflectance for the perpendicular component is $\approx 1.5 \times 10^{-4}$, while the corresponding value at 57 deg for 620 Å, the *fundamental wavelength*, is 0.39, a factor of 2600 greater. This factor is sensitive to the oxide thickness and calculations showed that increasing that thickness to 32 Å increased the factor by some orders of magnitude although it also reduced the reflectance at 620 Å by a few percentage points. As the angle of incidence is changed, the minima shift in wavelength, which permits the RIF to be tuned.

Hunter and Long [80] also prepared a Si wafer for use as a CAM by removing the oxide layer with HF, and they compared both CAM and RIF. Figure 26 shows the results. The two upper curves show the fundamental wavelength of the CAM (topmost curve) and the RIF (next curve down). The bottom two curves show the reflectance at the second harmonic, that is, at half of the plotted wavelength, for the CAM (third curve down) and the RIF (bottom curve). Note that the measured reflectance of the RIF at 620 Å is ≈ 0.39, in agreement with the calculation, but the measured reflectance at 310 Å is $\approx 3.5 \times 10^{-3}$, more than an order of magnitude larger than the calculated value. This discrepancy may be caused by the sensitivity of the depth of the interference minimum to the oxide thickness. It appears that the oxide thickness on the RIF is different from the 30 Å used in the calculations. Even though the reflectance of the second harmonic is larger than the calculated values, useful rejection ratios are still obtained. Whether or not to choose a CAM or a RIF depends, in part, on the stability of the Si with respect to oxidation. An oxide layer is likely to grow on the CAM surface, albeit slowly, but the oxide layer on the RIF is already present and should be stable with time.

16.6 Multiplate Resonant Reflectors

Multiplate resonant reflectors (MRRs) were introduced as mode selectors for giant pulsed ruby lasers [81]. Such a resonator consists of two or more plates of a transparent material. The surfaces of each plate are plane-parallel, and the individual plates must be coaligned so that all surfaces are parallel. In the first designs of MRRs [82], the thicknesses of the plates and spacers were chosen by approximate methods. However, Watts [83] showed that the true properties of an MRR can only be obtained using a thin-film calculation, that is, assuming that the components of the electromagnetic waves are coherent over the length of the resonator.

Because the plates of the resonator are not coated, the chance of damage to the resonator is reduced. Furthermore, when the beams reflected from the plate surfaces are in phase, the reflectance of the resonator can be quite large. These properties seem almost ideal for lasers operating in the VUV. Experiments in tuning an Ar excimer laser to 1216 Å, in which the author participated, showed that the output mirror coating, semitransparent Al+MgF$_2$ on an MgF$_2$ substrate, was easily damaged. In an attempt to produce a VUV laser output mirror, an MRR was designed and manufactured [84]. MgF$_2$ plates were used because of the difficulty in polishing LiF plates and their propensity to damage by chemisorption of water vapor.

The maximum reflectance of an MRR can be calculated using the formula

$$R = [(n^{2N} - 1)/(n^{2N} + 1)]^2,$$

where n is the index of refraction of the plates and N the number of plates. Using

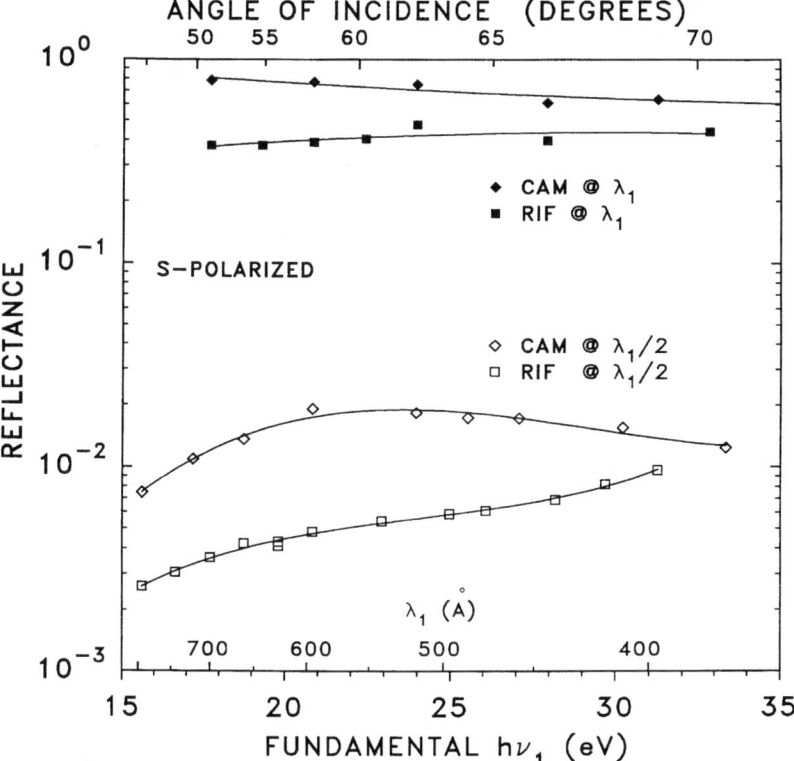

FIG. 26. Measured performance of the RIF and the CAM. At each photon energy $h\nu_1$ the angle of incidence was adjusted, as given in the top axis, to the value it would have if the suppressor were used as a replacement for one of the pair of gratings in the GCM. The 3/2 order interference minimum of the RIF closely tracked $h\nu_2$ until the upturn in reflectance for $h\nu_2$ seen for $h\nu_1$ was larger than $\approx$28 eV. The solid curves are meant to aid the eye [80].

index of refraction values obtained from Hass [85], the calculated reflectance of a three-plate resonant reflector at 1216 Å ($n = 1.73$) is 86.1%. Figure 27 shows the calculated reflectance of a three-plate resonator from 1215.9 to1216 Å. The plates are MgF_2, and the optical thicknesses of spacings and plates are multiples of 1216 Å; the spacings are 1.216 mm and the plate thicknesses are 0.70289 mm (1.216/1.73). Over this small spectral range the index is assumed to be constant. The calculation shows large maxima located about 0.06 Å apart, with reflectance values of 86% and halfwidths slightly larger than 0.01 Å. Note that none of the large maxima occurs at 1216 Å despite the fact that the thicknesses are multiples of that wavelength, probably because of round-off errors in the calculation.

Although MgF_2 is usually considered to be a transparent material in the VUV, it does have a slight extinction coefficient, k, when the wavelength is less than 1300 Å. Since the laser radiation traverses the plate many times, an appreciable extinction coefficient would decrease the reflectance of the MRR. Figure 28 shows how the maximum reflectance of the previous figure decreases at the extinction coefficient is increased. The extinction coefficient is increased from 10^{-7} ($\alpha = 0.1$ cm^{-1}) to 10^{-4} ($\alpha = 100$ cm^{-1}). For $k = 10^{-7}$ the reflectance is unaffected but at $k = 10^{-4}$ the reflectance has been effectively reduced to that of a single MgF_2 surface. The k value of the plates used in making the resonator was estimated to be 10^{-6} [86].

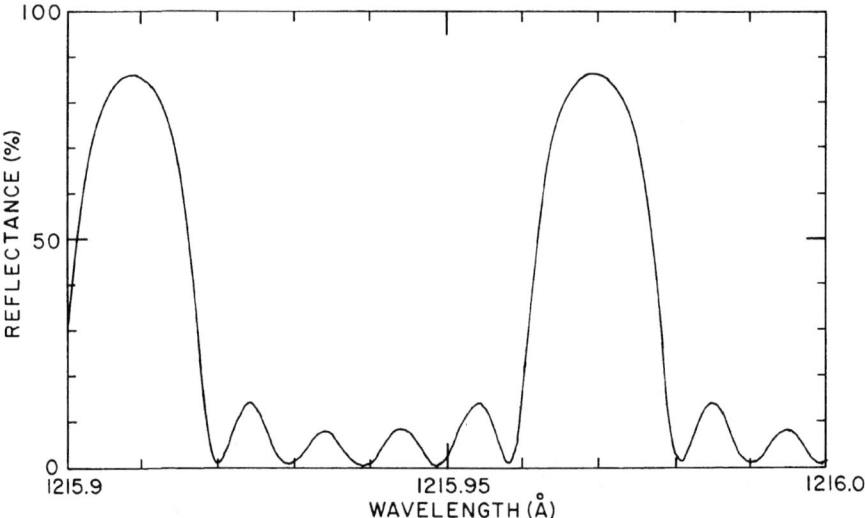

FIG. 27. Calculated reflectance as a function of wavelength of a three-plate resonant reflector from 1215.9 to 1216 Å. Plates are of MgF_2, 0.70289 mm thick, and the spacers are 1.216 mm thick. In this wavelength region the index of MgF_2 is approximately 1.73 [84].

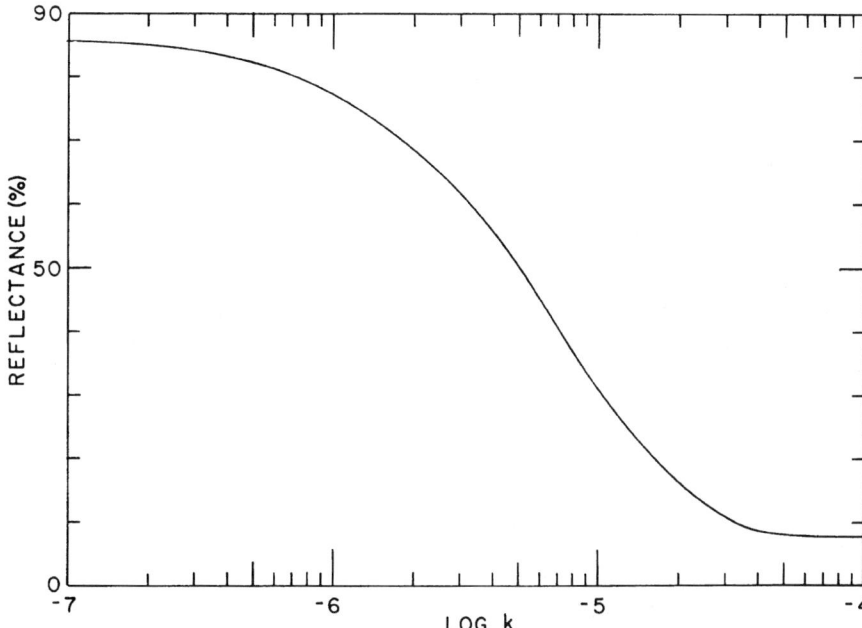

FIG. 28. The effect of absorption on the reflectance maxima of Fig. 26. The curve shows how the maxima of Fig. 26 decrease as the extinction coefficient k increases [84].

A direct measurement of the reflectance of the three-plate MgF_2 resonator was not possible but a crude estimate was obtained by comparing laser damage thresholds when the MRR and a conventional Al + MgF_2 mirror of known reflectance were used as output couplers. A value consistent with that calculated was obtained.

References

1. W. R. Hunter and S. A. Malo, *J. Phys. Chem. Solids* **30**, 2739–2745 (1969).
2. W. G. Johnston, *J. Appl. Phys.* **33**, 2050–2058 (1962).
3. J. S. Nadeau and W. G. Johnston, *J. Appl. Phys.* **32**, 2563–2565 (1961).
4. D. A. Patterson and W. H. Vaughan, *J. Opt. Soc. Am.* **53**, 851–855 (1963).
5. G. Hass and W. R. Hunter, *Appl. Optics* **17**, 2310–2315 (1978).
6. A. I. Vangonen, V. M. Zolotarev, T. S. Lyubarskaya, and A. M. Pukhov, *Sov. J. Opt. Technol.* **54**, 273–275 (1987).
7. R. B. Gillette, J. R. Hollohan, and G. L. Carlson, *J. Vac. Sci. Technol.* **7**, 534–537 (1970).
8. E. D. Johnson, S. L. Hulbert, R. F. Garrett, G. P. Williams, and M. L. Knotek, *Rev. Sci. Instrum.* **58**, 1042–1045 (1987).

9. T. Koide, T. Shidara, K. Tanaka, A. Yagishita, and S. Sato, *Rev. Sci. Instrum.* **60**, 2034–2037 (1989).
10. Heraeus Quarzglas, Postfach 15 54, 63405 Hanau, Germany, private communication.
11. I. P. Kaminow, B. G. Bagley, and C. G. Olson, *Appl. Phys. Lett.* **32**, 98–99 (1978).
12. A. Pfluger, *Phys. Z.* **5**, 215–216 (1904).
13. F. Zernicke, *Physica* **8**, 81–87 (1928).
14. W. M. Powel Jr., *Phys. Rev.* **46**, 43–46 (1934).
15. N. G. Gerasimova, V. H. Kreyskop, I. G. Panova, and V. Shin, *Sov. J. Opt. Technol.* **40**, 4–7 (1973).
16. G. Hass and W. R. Hunter, *Appl. Optics* **9**, 2101–2110 (1970).
17. R. W. Falcone and J. Bokor, *Opt. Lett.* **8**, 21–23 (1983).
18. R. Kato, S. Nakashima, K, Nakamura, and Y. Uchida, *J. Phys. Soc. Jpn.* **15**, 2111–2112 (1960).
19. D. F. Heath and P. A. Sacher, *Appl. Opt.* **5**, 937–943 (1966).
20. G. R. Carruthers, *Appl. Opt.* **10**, 1461–1463 (1971).
21. R. W. Wood, *Phil. Mag.*(Series 6) **38**, 98–112 (1933).
22. R. W. Wood, *Phys. Rev.* **44**, 353–360 (1933).
23. R. W. Wood, *Nature (London)* **131**, 582 (1933).
24. C. Zener, *Nature (London)* **131**, 968 (1933).
25. R. Kronig, *Nature (London)* **132**, 211–212 (1934).
26. P. Drude, "Theory of Optics," Dover, New York (1959).
27. D. H. Tomboulian and D. E. Bedo, *Rev. Sci. Instrum.* **26**, 747–750 (1955).
28. D. H. Tomboulian and D. E. Bedo, *Phys. Rev.* **104**, 590–597 (1956).
29. W. R. Hunter, D. W. Angel, and R. Tousey, *Appl. Opt.* **4**, 891–898 (1965).
30. K. Codling, R. P. Madden, W. R. Hunter, and D. W. Angel, *J. Opt. Soc. Am.* **56**, 189–192 (1966).
31. O. P. Rustgi, *J. Opt. Soc. Am.* **55**, 630–634 (1965).
32. J. A. R. Samson, *J. Opt. Soc. Am.* **54**, 1491 (1964).
33. F. R. Powell, *Proc SPIE* **1160**, 37–48 (1989).
34. F. R. Powell, P. W. Vedder, J. F. Lindblom, and S. T. Powell, *Opt. Eng.* **29**, 614–624 (1990).
35. P. Choi and M. B. Favre, *J. Phys. D* **20**, 169–173 (1987).
36. H. Huizenga, J. A. M. Bleeker, W. H. Diemer, and A. P. Huben, *Rev. Sci. Instrum.* **52**, 673–677 (1981).
37. M. A. Barstow, B. J. Kent, M. J. Whiteley, and P. H. Spurrett, *J. Mod. Opt.* **34**, 1491–1500 (1987).
38. R. A. M. Keski-Kuha, *Appl. Opt.* **28**, 2965–2968 (1989).
39. E. Spiller, K. Grebe, and L. Golub, *Opt. Eng.* **29**, 625–631 (1990).
40. W. R. Hunter, *Phys. Thin Films* **7**, 43–114 (1973).
41. G. N. Steele, in "Space Optics, Proc. IX Int. Congress of the ICO" (B. J. Thompson and R. R. Shannon, eds.), National Academy of Sciences, Washington, DC, p. 367–389 (1974).
42. W. R. Hunter, J. D. Purcell, and G. N. Steele, *Appl. Opt.* **12**, 1874–1879 (1973).
43. R. Tousey, J-D. F. Bartoe, G. E. Brueckner, and J. D. Purcell, *Appl. Opt.* **16**, 870–878 (1977).
44. F. R. Powell and J. R. Fox, *Proc SPIE* **2011**, 428–437 (1993).
45. A. J. Caruso, *Appl. Opt.* **13**, 1744–1745 (1974).
46. R. Stern and F. Paresce, *J. Opt. Soc. Am.* **65**, 1515–1516 (1975).

47. J. F. Seely, *Opt. Commun.* **70**, 207–212 (1989).
48. J. F. Seely, L. Shirey, and A. Kingman, *Appl. Opt.* **28**, 1818–1821 (1989).
49. B. Bates and D. J. Bradley, *Appl. Opt.* **5**, 971–975 (1966).
50. A. Malherbe and M. Guillard, *Nouv. Rev. Opt. Appl.* **1**, 401–404 (1970).
51. A. Malherbe, *Appl. Opt.* **13**, 1275–1276 (1974).
52. A. Malherbe, *Nouv. Rev. Opt. Appl.* **2**, 337–344 (1971).
53. A. Malherbe, *Appl. Opt.* **13**, 1276 (1974).
54. J. A. R. Samson, "Techniques of Vacuum Ultraviolet Spectroscopy," Pied Publications, Lincoln, Nebraska (1967).
55. S. Bowyer, *Proc. SPIE* **2517**, 90–96 (1995).
56. S. Bowyer, *Proc. SPIE* **2517**, 97–106 (1995).
57. D. C. Morton and J. D. Purcell, *Planet. Space Sci.* **9**, 455–458 (1962).
58. T. C. Winter Jr., and T. A. Chubb, *J. Geophys. Res.* **72**, 4405–4414 (1967).
59. J. O. Sullivan and A. C. Holland, NASA Report CR-371, Washington, DC (1966).
60. R. D. Hudson, *Rev. Geophys. Space Phys.* **9**, 306–408 (1971); also available as NBS Report NSRDS-NBS 38, Washington, DC (1971).
61. F. S. Johnson and R. Tousey, *J. Opt. Soc. Am.* **38**, A1103 (1948).
62. J. D. Purcell, *J. Opt. Soc. Am.* **43**, 1166–1169 (1953).
63. W. R. Hunter, J. F. Osantowski, and G. Hass, *Appl. Opt.* **10**, 540–544 (1971).
64. W. R. Hunter, *Opt. Acta* **9**, 255–268 (1962).
65. A. E. Rosenbluth, *Rev. Phys. Appl.* **23**, 1599–1621 (1988).
66. M. Yanagihara, T. Maehara, H. Nomura, M. Yamamoto, T. Namioka, and H. Kimura, *Rev. Sci. Instrum.* **63**, 1516–1518 (1992).
67. D. G. Stearns, R. S. Rosen, and S. P. Vernon, *Proc. SPIE* **1547**, 2–13 (1991); and *J. Vac. Sci. Technol.* **A9**, 2662–2669 (1991).
68. P. Boher, Ph. Houdy, L. Hennet, M. Kühne, P. Müller, J. P. Frontier, P. Trouslard, C. Senillou, J. C. Joud, and P. Ruterana, *Proc. SPIE* **1547**, 21–38 (1991).
69. J. F. Seely, M. P. Kowalski, W. R. Hunter, J. C. Rife, T. W. Barbee, Jr., G. E. Holland, C. N. Boyer, and C. M. Brown, *Appl. Opt.* **32**, 4890–4897 (1993).
70. J. H. Underwood, E. M. Gullikson, and K. Nguyen, *Appl. Opt.* **32**, 6985–6990 (1993).
71. J. H. Underwood, E. M. Gullikson, W. Ng, A. Ray-Chaudhuri, and F. Cerrina, in "OSA Proceedings on Extreme Ultraviolet Lithography" (F. Zernike and D. T. Attwood, eds.), Optical Society of America, Washington, DC (1995).
72. J. F. Seely, G. Gutman, J. Wood, G. S. Herman, M. P. Kowalski, J. C. Rife, and W. R. Hunter, *Appl. Opt.* **32**, 3541–3543 (1993).
73. J. F. Seely, M. P. Kowalski, W. R. Hunter, and G. Gutman, *Appl. Opt.* **35**, 4408–4412 (1996).
74. Y. Utsumi, H. Kyuragi, T. Urisu, and H. Maezawa, *Appl. Opt.* **27**, 3933–3936 (1988).
75. O. Renner, M. Kopecky, E. Krousky, F. Schafers, B. R. Muller, and N. I. Chkhalo, *Rev. Sci. Instrum.* **63**, 1478–1481 (1992).
76. Y. Utsumi, H. Kyuragi, and T. Urisu, *J. Vac. Sci. Technol.* **B8**, 436–438 (1990).
77. K. M. Skulina, C. S. Alford, R. M. Bionta, D. M. Makowiecki, E. M. Gullikson, R. Soufli, J. B. Kortright, and J. H. Underwood, *Appl. Opt.* **34**, 3727–3730 (1995).
78. L. J. Terminello, A. B. McLean, A. Santoni, E. Spiller, and F. J. Himpsel, *Rev. Sci. Instrum.* **61**, 1626–1628) (1990).
79. Ph. Sainctavit, J. Petiau, A. Manceau, R. Rivallant, M. Belakhovsky, and G. Renaud, *Nucl. Instrum. Methods* **A273**, 423–428 (1988).
80. W. R. Hunter and J. P. Long, *Appl. Opt.* **33**, 1264–1269 (1994).

81. J. M. Burch, in "Quantum Electronics, Proc. Third Int. Congress" (P. Grivet and N. Bloembergen, eds), Columbia University Press, New York, pp. 1187–1202 (1964); and M. Hercher, *Appl. Phys. Lett.* **7**, 39–41 (1965).
82. G. Magyar, *Rev. Sci. Instrum.* **38**, 517–519 (1967).
83. J. K. Watts, *Appl. Opt.* **7**, 1621–1623 (1968).
84. W. R. Hunter, M. H. R. Hutchinson, and M. R. O. Jones, *Appl. Opt.* **20**, 770–772 (1981).
85. G. Hass, Night Vision Laboratory, Fort Belvoir, VA, private communication.
86. G. H. C. Freeman, National Physical Laboratory, Teddington, Middlesex, England, private communication.

17. DIFFRACTION GRATINGS

Takeshi Namioka
Tohoku University
Sendai, Japan

17.1 Introduction

The history of diffraction gratings [1, 2] can be traced back to 1785 when Rittenhouse made the first known transmission grating by winding human hair between two parallel fine screws. However, we had to wait until the late nineteenth century to see the beginning of the modern era of the ruled grating. In the 1880s Rowland invented the concave grating [3] and produced both plane and concave gratings of high quality by using his sophisticated ruling engine. In the late 1940s Harrison applied interferometry to ruling-engine control [4, 5]. This novel method of engine control and the development of high-fidelity replication processes made a variety of high-quality gratings commercially available in the 1950s. In these developments, every effort had been directed to rule perfectly equispaced parallel grooves.

Ruling of gratings, however, had various problems such as diamond wear, adverse environmental effects, limitation in groove positioning accuracy, and so on. To avoid these difficulties inherent to mechanical ruling, Michelson [6] suggested in 1927 a nonmechanical method of grating production by photographing interference fringes. Following his suggestion many researchers tried in the 1950s to make gratings, but they failed to produce high-quality gratings for lack of an intense coherent light source and a suitable recording material. The advent of argon ion lasers and photoresist changed the situation. In 1967–68, Rudolph and Schmahl, and Labeyrie and Flamand succeeded independently in producing plane gratings suitable for spectroscopic work by recording interference fringes, which were formed between two plane wavefronts, in photoresist coated on a glass blank [7, 8]. Subsequently, various types of concave spherical gratings having reduced aberrations and specific focal curves, including stigmatic gratings, were produced by recording interference fringes formed between two spherical wavefronts [7–10]. This new breed of gratings is called a *holographic grating* and is now available in a variety of forms. The capability of the holographic grating to correct aberrations was expanded by Koike *et al.* [11] in 1987 using aspheric wavefronts to produce interference fringes. Holographic gratings of this kind are called *holographic gratings recorded with aspheric wavefronts* to distinguish them from those recorded with spherical (or plane) wavefronts. The grooves of the holographic grating are in general neither straight nor equally spaced when projected onto the plane tangent to the grating surface at its vertex. One of the most attractive features of the holographic grating is that the aberrations and focal property of the grating can be adjusted to meet specific requirements by altering the fringe pattern for changing the curvature and spacing of the grooves. In contrast, a grating with equally spaced and straight grooves does not have such a property. Gratings having constant spacing and straight grooves are often called *conventional gratings*, regardless of whether they are ruled or holographically produced.

Despite its novel imaging property, the holographic grating suffered from lack of freedom in choosing a recording laser wavelength best suited for the production of aberration-reduced vacuum ultraviolet (VUV) gratings: Lasers currently available for fringe recording can provide several visible and ultraviolet lines only. This drawback in producing highly aberration-corrected VUV gratings was remedied by means of mechanical rulings. In the mid-1970s Harada and his collaborators [12] succeeded in ruling varied-line-spacing (VLS) gratings with curved grooves, which were equivalent to stigmatic holographic gratings, by incorporating a computer control system into his interferometrically controlled ruling engine. With this ruling engine, the amount of spacing variation and the curvature of the grooves can be chosen without much restriction, providing more freedom to choose conditions for correcting aberrations as compared with interference fringe recordings. Therefore, VLS ruled gratings can be tailored to meet specific requirements and are used widely in various research fields [12–17].

17.2 VLS Ruled Grating with Curved (or Straight) Grooves

The scheme devised by Harada *et al.* [12, 13] for ruling a concave grating with varied spacing and curved grooves is schematically shown in Fig. 1. The vertex O of the grating blank is taken as the origin, the blank normal at O as the

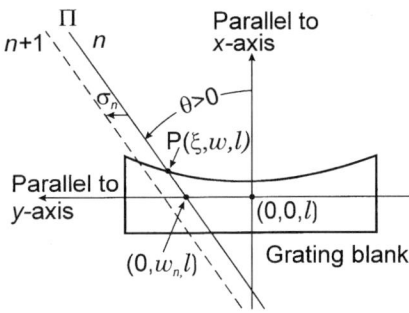

FIG. 1. Scheme for ruling a VLS grating with curved grooves (a cross-sectional view at $l = $ constant).

x axis, and the symmetry plane of the blank as the x-y plane. Here, the zeroth groove is assumed to pass through O, and a point P on the nth groove is denoted by coordinates (ξ, w, l). The figure is a cross-sectional view at l ($\neq 0$).

Curved grooves are ruled by constraining the reciprocating movement of the diamond tool in a reference plane Π, which makes angles θ and 90 deg with the x-z and the x-y planes, respectively. After the nth groove is ruled, the reference plane is advanced in the y direction to the next ruling position. For practical convenience we assume that the amount of this advance σ_n is given by

$$\sigma_n = w_{n+1} - w_n = \sigma_0 + 2an + 6bn^2 + 4cn^3 \quad (1)$$
$$\sigma_0 \gg a \gg b \gg c,$$

where σ_0, a, b, and c are constants. These constants and θ are called the *ruling parameters*. The amount of advance w_n of the reference plane from the zeroth to the nth groove is then

$$w_n = w - \xi \tan\theta = \sum_{n=0}^{n-1} \sigma_n$$
$$= n(\sigma_0 - a + b) + n^2(a - 3b + c) + 2n^3(b - c) + cn^4. \quad (2)$$

As is shown in Section 17.6.2, the effective grating constant σ is defined by $1/[\partial n/\partial w]_{w=l=0}$. Using this definition with $(\partial \xi/\partial w)_{w=l=0} = 0$ and $(n)_{w=l=0} = 0$, we obtain from Eq. (2)

$$\sigma \equiv 1/[\partial n/\partial w]_{w=l=0} = \sigma_0 - a + b. \quad (3)$$

Equation (2) also represents the equation of the groove pattern projected onto the y-z plane. We consider a concave spherical grating whose surface figure is expressed by

$$\xi = R - \sqrt{R^2 - (w^2 + l^2)}. \quad (4)$$

Substitution of Eq. (4) into Eq. (2) yields the equation of the groove pattern projected onto the y-z plane:

$$\frac{[w - (R\tan\theta + w_n)\cos^2\theta]^2}{[\sqrt{R^2 - 2Rw_n\tan\theta - w_n^2\sin\theta\cos\theta}]^2} + \frac{l^2}{[\sqrt{R^2 - 2Rw_n\tan\theta - w_n^2\cos\theta}]^2} = 1. \quad (5)$$

When $\theta \neq 0$, Eq. (5) represents a family of ellipses with the parameter n. When $\theta = 0$, Eq. (2) becomes $w = n\sigma + n^2(a - 3b + c) + 2n^3(b - c) + cn^4$. The grooves are a family of straight lines whose spacing varies with n. Furthermore, if $a, b, c, \theta = 0$, we have $w = n\sigma_0$, that is, a conventional grating having straight and equispaced grooves.

With the aid of power series expansions we obtain, from Eq. (2), the groove function $n = h(w, l)$ in a power series [18]

$$n\sigma = w + \Gamma[\tfrac{1}{2}(n_{20}w^2 + n_{02}l^2 + n_{30}w^3 + n_{12}wl^2) + \tfrac{1}{8}(n_{40}w^4 + 2n_{22}w^2l^2 + n_{04}l^4) + \cdots], \quad (6)$$

where Γ is the parameter introduced for unifying the expressions of the groove functions for the ruled and the holographic grating. The expansion coefficients n_{ij} are referred to as the groove parameters. The parameters Γ and n_{ij} are expressed as

$$\Gamma = 1, \qquad n_{20} = -(\tan\theta/R) - J, \qquad n_{02} = -\tan\theta/R$$
$$n_{30} = J(\tan\theta/R) + J^2 - K, \qquad n_{12} = J(\tan\theta/R),$$
$$n_{40} = -J(\tan\theta/R)^2$$
$$\qquad + [6K - 6J^2 - (1/R^2)](\tan\theta/R) + 5J(2K - J^2) - (8c/\sigma^4),$$
$$n_{22} = -J(\tan\theta/R)^2 + (\tan\theta/R)[3K - 3J^2 - (1/R^2)], \qquad (7)$$
$$n_{04} = -J(\tan\theta/R)^2 - (\tan\theta/R^3),$$
$$J = 2(a - 3b + c)/\sigma^2, \qquad K = 4(b - c)/\sigma^3.$$

For conventional ruled gratings ($a = b = c = \theta = 0$), we have $\sigma_0 = \sigma$, $n_{ij} = 0$, and $n\sigma = w$. For the groove parameters of an ellipsoidal and a toroidal grating, refer to [18] and [19], respectively.

17.3 Holographic Grating Recorded with Spherical Wavefronts

When a grating blank coated with a suitable photoresist is exposed to monochromatic beams originating from two coherent point sources C and D (Fig. 2), interference fringes are recorded in the photoresist, in accordance with the irradiance of the fringes. The fringe pattern formed on the grating blank is

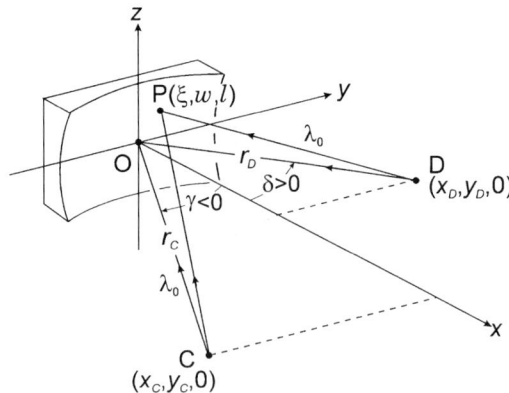

FIG. 2. Spherical wavefront recording system for producing a holographic grating.

transformed, on development, to the grating grooves of a sinusoidal form. The fringes formed on the blank are the intersections of the blank surface and a family of confocal hyperboloids of revolution about the line CD with the foci at C and D, when C and D are both real or both virtual. By a real (or a virtual) point source we mean the source that generates a spherical wave diverging from (or converging to) it. When C is real and D is virtual (or vice versa), the resulting interference fringes are a family of confocal ellipsoids of revolution about the line CD with the foci at C and D. Only when C and D are placed at infinity and symmetrically with respect to the grating blank normal at the vertex O, do they produce a family of equally spaced planes parallel to the x-z plane, resulting in a conventional grating. The grooves of a holographic grating are generally neither straight nor equally spaced when projected onto the plane tangent to the grating surface at its vertex O. The curvature and spacing and, consequently, the imaging property of the grating can be adjusted by altering the recording geometry (the positions of C and D).

To investigate the relation between the groove pattern and the positions of C and D, we refer to Fig. 2 to define a rectangular coordinate system: Let the origin be at the vertex O of the grating blank; let the x axis be the normal to the blank at O; and let the x-y plane be the symmetry plane of the blank. We assume that C and D lie in the x-y plane and the difference between the distances of C and D from O is an integer multiple of λ_0, the wavelength of the recording laser light, and that the zeroth groove passes through O. Then, the nth groove is formed according to

$$n\lambda_0 = (CP - DP) - (CO - DO), \quad (8)$$

where $P(\xi, w, l)$ is a point on the nth groove, CP is the distance between C and P, and so on. If C is a virtual source, the corresponding distance CP takes negative sign, and the same applies to DP.

For a concave spherical holographic grating recorded with spherical waves of wavelength λ_0 originating from two coherent point sources $C(x_C = r_C \cos \gamma, y_C = r_C \sin \gamma, 0)$ and $D(x_D = r_D \cos \delta, y_D = r_D \sin \delta, 0)$, the groove function is obtained from Eq. (8) by expanding CP and DP in power series:

$$\frac{n\lambda_0}{\sin \delta - \sin \gamma} = w + \frac{1}{\sin \delta - \sin \gamma} \left[\frac{1}{2}(n_{20} w^2 + n_{02} l^2 + n_{30} w^3 + n_{12} wl^2) \right.$$
$$\left. + \frac{1}{8}(n_{40} w^4 + 2n_{22} w^2 l^2 + n_{04} l^4) + \cdots \right]. \quad (9)$$

Since the effective grating constant σ is by definition

$$\sigma \equiv \frac{1}{[\partial n/\partial w]_{w=l=0}} = \frac{\lambda_0}{\sin \delta - \sin \gamma}, \quad (10)$$

Eq. (9) is rewritten in the same form as Eq. (6) with $\Gamma = 1/(\sin\delta - \sin\gamma)$. The groove parameters are expressed as follows [18].

$$\Gamma = (\sin\delta - \sin\gamma)^{-1} = \sigma/\lambda_0, \quad n_{20} = T_C - T_D, \quad n_{02} = S_C - S_D,$$

$$n_{30} = \frac{T_C \sin\gamma}{r_C} - \frac{T_D \sin\delta}{r_D}, \quad n_{12} = \frac{S_C \sin\gamma}{r_C} - \frac{S_D \sin\delta}{r_D},$$

$$n_{40} = \frac{4T_C \sin^2\gamma}{r_C^2} - \frac{4T_D \sin^2\delta}{r_D^2} - \frac{T_C^2}{r_C} + \frac{T_D^2}{r_D} + \frac{S_C - S_D}{R^2}, \quad (11a)$$

$$n_{22} = \frac{2S_C \sin^2\gamma}{r_C^2} - \frac{2S_D \sin^2\delta}{r_D^2} - \frac{T_C S_C}{r_C} + \frac{T_D S_D}{r_D} + \frac{S_C - S_D}{R^2},$$

$$n_{04} = -\frac{S_C^2}{r_C} + \frac{S_D^2}{r_D} + \frac{S_C - S_D}{R^2},$$

where

$$T_C = \frac{\cos^2\gamma}{r_C} - \frac{\cos\gamma}{R}, \quad T_D = \frac{\cos^2\delta}{r_D} - \frac{\cos\delta}{R},$$

$$S_C = \frac{1}{r_C} - \frac{\cos\gamma}{R}, \quad S_D = \frac{1}{r_D} - \frac{\cos\delta}{R}. \quad (11b)$$

The parameters r_C, γ, r_D, and δ are called the *recording parameters* and are used to adjust the imaging property of the holographic grating. For the groove parameters of ellipsoidal and toroidal gratings, refer to [18] and [19], respectively.

17.4 Holographic Grating Recorded with Aspheric Wavefronts

Design flexibility of holographic gratings can be increased by introducing a spherical mirror into one or each one of the optical branches of the point-source recording system. Figure 3 shows a recording system that has an auxiliary spherical mirror M_1 in between the point source C and the grating blank G. The recording parameters in this case are the radius of curvature of M_1, the angle of incidence η_C at the vertex O_1 of M_1, the distances $CO_1 = p_C$ and $O_1O = q_C$, γ, r_D, and δ. This increased freedom will facilitate better designs.

This freedom will be further increased when another spherical mirror M_2 is placed in between the point source D and G. The auxiliary spherical mirrors modify the incident spherical wavefronts and generate aspheric wavefronts for recordings. The resulting grating is called a *holographic grating recorded with aspheric wavefronts*. The groove function for this type of grating has the same

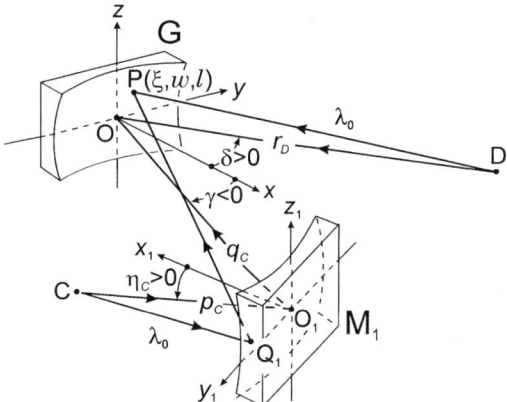

FIG. 3. Aspheric wavefront recording system for producing a holographic grating.

form as Eq. (6) with $\Gamma = 1/(\sin \delta - \sin \gamma)$. The expressions of the groove parameters n_{ij} are not given here because of their complexity. For explicit expressions and further details, refer to [20].

17.5 Transmission Grating

In recent years, transmission gratings have become increasingly important in x-ray astronomy because they allow spectroscopy with imaging telescopes without major changes of the overall design. These gratings are required to have a resolution of ~100–1000 and a dispersion as high as possible. This leads to the necessity of a high groove density because transmission gratings are used at normal or very near normal incidence. Furthermore, gratings should either be free standing or be supported on a thin transmissive membrane such as polyimide because no material is perfectly transparent to soft x-rays (~0.2–30 nm). This presents a difficulty when producing large transmission gratings. The technological advances in microelectronics have made it possible to fabricate high-groove-density free-standing or thin-film-backed transmission gratings. Fabrication methods are roughly classified into two categories. One is application of lithography [21–25] and the other is the stripping of a holographic grating relief pattern with metallic bars and spaces off its photoresist-coated substrate [26]. The former has been developed to produce gratings for x-ray astronomy and the latter for synchrotron radiation research.

As an example of the lithographic method, the procedure used by Canizares et al. [23] is described next. It involves three processes: (1) fabrication of a mask, (2) replication of the mask, and (3) electroplating. In step 1, a holographic grating with a desired grating constant is formed in photoresist on a 1-μm-thick polyimide film coated on a silicon wafer using a UV laser. A high aspect ratio laminar resist grating is made using a procedure involving oblique shadowing followed by oxygen reactive ion etching. Then, about 100 nm of gold is evaporated over the grating, and the photoresist is dissolved. A mask is finished by back etching of the wafer. Step 2 uses x-ray lithography to transfer the mask pattern into a 1- to 1.5-μm-thick film of polymethylmethacrylate (PMMA) coated on a plating base. The plating base is prepared by spin coating 1-μm-thick polyimide on a silicon wafer, followed by the evaporation of 10 nm of chromium and 20 nm of gold. In step 3 the PMMA is developed down to the plating base. The grating structure thus prepared is further electroplated with gold, and the PMMA is etched away in an oxygen plasma. Finally, the gold transmission grating of $\sim 25 \times 25$ mm^2 on a silicon frame is made by back etching of the wafer. With this procedure, gold gratings with 5000 and 3333 lines/mm and silver gratings with 1666 lines/mm were fabricated. There are many other lithographic procedures: contamination lithography [24], a combination of mechanical ruling and photolithography [21], and a combination of holographic recording and electron-beam lithography [25] to mention a few.

The stripping-off method is a simple method of producing a free-standing metallic transmission grating [26]. A holographic grating with a line density of ~ 2100 lines/mm is made by recording interference fringes in photoresist coated on a glass microscope slide. The sinusoidal grating profile formed in the photoresist is used as a mold for a metallic transmission grating. A uniform aluminum film of ~ 100–140 nm, which is fairly transparent to soft x-rays of long wavelengths, is deposited on the holographic grating to form a continuous support. Absorbing grating bars are produced by depositing silver to a thickness of 10–70 nm at an angle between 60 and 70 deg from the normal such that only one side of the sinusoidal profile is coated. A sample holder with a hole 9 mm in diameter is glued to the grating surface, and then the grating is removed from the surface of the photoresist by dissolving the photoresist in acetone. With this method a slight distortion always occurs that leads to $\sim 1\%$ changes in the grating constant.

The bars and spaces of a transmission grating for soft x-rays of longer wavelengths are considered in practice to be completely absorbing and totally transmitting, respectively. The grating can be treated as the amplitude grating. For soft x-rays of shorter wavelengths the bars are partially transmitting with a phase shift. The grating therefore becomes a phase grating. The grating efficiency of the first order can be maximized by choosing a proper material for the bars so that the bars cause a phase shift of 180 deg relative to radiation passing through the spaces, creating destructive interference in the zeroth order and enhancement in the first order. The grating efficiency of the transmission grating is discussed in Section 17.6.9.2.

An effort has also been directed toward the realization of aberration-corrected transmission gratings for use with a space telescope: use of a VLS transmission grating for coma elimination in a telescope-grating system [27, 28] and a significant reduction of third-order aberrations by arranging ~1000 individual equally spaced plane transmission gratings on a toric surface [29, 30]. The optimum support structures for free-standing transmission gratings were studied to keep the obstruction as low as possible while guaranteeing the mechanical strength to withstand acoustic loads during the launch of a space vehicle [31].

17.6 Theory and Basic Properties of Diffraction Gratings

17.6.1 Fermat's Principle

The geometric theory of the diffraction grating can be developed on the basis of a single hypothesis, known as Fermat's principle [32]. Fermat's principle states that the pathlength of an actual ray traveling from a point A to a point B takes an extremal or stationary value, and this distinguishes the actual ray from other curves connecting these two points. Mathematically, this is expressed by $\delta F = 0$, where F is the pathlength from A to B. We call this F the *light path function*.

To apply Fermat's principle to the grating, we refer the reader to Fig. 4. Introducing a Cartesian coordinate system with the origin O located at the center of the grating rulings, let the x axis be the grating normal at O with the x-y plane in the plane of symmetry of the grating. Gratings are used almost exclusively in the in-plane mounting in which the center of the entrance slit is set in the

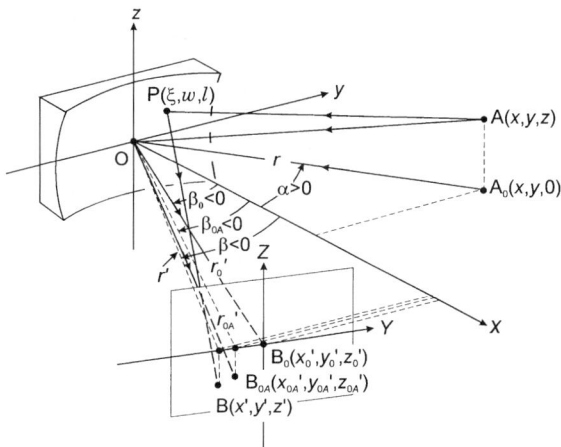

FIG. 4. Spectral image formation by the grating.

356 DIFFRACTION GRATINGS

symmetry plane of the grating. We therefore consider a ray originating from a point $A(x, y, z)$ on the entrance slit whose center $A_0(x, y, z)$ lies in the x-y plane. This ray meets the grating at a point $P(\xi, w, l)$ on the nth groove, the zeroth groove being assumed to pass through the origin. The ray AP is diffracted at P in the direction PB, $B(x', y', z')$ being a point on the diffracted ray of wavelength λ in mth order ($m = 0, \pm 1, \pm 2, \cdots$). Two rays diffracted from the zeroth and the nth grooves are reinforced when their path difference is equal to $nm\lambda$. Thus, for the diffraction grating the light path function for an arbitrarily chosen ray APB can be expressed by

$$F = AP + PB + nm\lambda. \tag{12}$$

Since the surface of the grating blank is expressed by $\xi = f(w, l)$, F is a function of w and l only. Therefore, application of Fermat's principle to the light path function yields

$$\delta F = \frac{\partial F}{\partial w} \delta w + \frac{\partial F}{\partial l} \delta l = 0. \tag{13}$$

Fermat's principle as applied to the diffraction grating is now stated as

$$\frac{\partial F}{\partial w} = \frac{\partial AP}{\partial w} + \frac{\partial PB}{\partial w} + m\lambda \frac{\partial n}{\partial w} = 0,$$

$$\frac{\partial F}{\partial l} = \frac{\partial AP}{\partial l} + \frac{\partial PB}{\partial l} + m\lambda \frac{\partial n}{\partial l} = 0. \tag{14}$$

To express Eq. (14) explicitly, we consider a concave spherical grating having a radius of curvature of R, and then expand the light path AP in a power series of w and l:

$$AP = \sqrt{(\xi - x)^2 + (w - y)^2 + (l - z)^2} \tag{15a}$$

$$= r\left(1 + \frac{z^2}{r^2}\right)^{1/2} - w \sin \alpha \left(1 + \frac{z^2}{r^2}\right)^{-1/2} - \frac{lz}{r}\left(1 + \frac{z^2}{r^2}\right)^{-1/2}$$

$$+ \frac{w^2}{2}\left(\frac{\cos^2 \alpha}{r} - \frac{\cos \alpha}{R}\right) + \frac{l^2}{2}\left(\frac{1}{r} - \frac{\cos \alpha}{R}\right) + O\left(\frac{w^3}{R^2}\right). \tag{15b}$$

Similarly we obtain for the light path PB

$$PB = r'\left(1 + \frac{z'^2}{r'^2}\right)^{1/2} - w \sin \beta \left(1 + \frac{z'^2}{r'^2}\right)^{-1/2} - \frac{lz'}{r'}\left(1 + \frac{z'^2}{r'^2}\right)^{-1/2}$$

$$+ \frac{w^2}{2}\left(\frac{\cos^2 \beta}{r'} - \frac{\cos \beta}{R}\right) + \frac{l^2}{2}\left(\frac{1}{r'} - \frac{\cos \beta}{R}\right) + O\left(\frac{w^3}{R^2}\right). \tag{16}$$

THEORY AND BASIC PROPERTIES OF DIFFRACTION GRATINGS 357

In deriving Eqs. (15b) and (16) we used the expanded form of Eq. (4) and the cylindrical coordinates for x, y, x', and y':

$$\xi = \frac{w^2}{2R} + \frac{l^2}{2R} + \frac{w^4}{8R^3} + \frac{w^2 l^2}{4R^3} + \frac{l^4}{8R^3} + O\left(\frac{w^6}{R^5}\right). \qquad (17)$$

$$x = r \cos \alpha, \qquad y = r \sin \alpha, \qquad x' = r' \cos \beta, \qquad y' = r' \sin \beta. \qquad (18)$$

Here, α and β are the angles of incidence and diffraction of the ray APB, both measured in the x-y plane, respectively, and r and r' are the distances between O and the projections of A and B onto the x-y plane, respectively (see Fig. 4). Substituting Eqs. (15b) and (16) into Eq. (14) we obtain

$$\begin{aligned}
\frac{\partial F}{\partial w} &= -\left(1 + \frac{z^2}{r^2}\right)^{-1/2} \sin \alpha - \left(1 + \frac{z'^2}{r'^2}\right)^{-1/2} \sin \beta + m\lambda \frac{\partial n}{\partial w} \\
&\quad + w\left(\frac{\cos^2 \alpha}{r} + \frac{\cos^2 \beta}{r'} - \frac{\cos \alpha + \cos \beta}{R}\right) + O\left(\frac{w^2}{R^2}\right) = 0, \\
\frac{\partial F}{\partial l} &= -\frac{z}{r}\left(1 + \frac{z^2}{r^2}\right)^{-1/2} - \frac{z'}{r'}\left(1 + \frac{z'^2}{r'^2}\right)^{-1/2} + m\lambda \frac{\partial n}{\partial l} \\
&\quad + l\left(\frac{1}{r} + \frac{1}{r'} - \frac{\cos \alpha + \cos \beta}{R}\right) + O\left(\frac{w^2}{R^2}\right) = 0.
\end{aligned} \qquad (19)$$

These two simultaneous equations provide all the information on the geometrical properties of the diffraction grating.

17.6.2 Grating Equation and Lateral Magnification

For the central ray AOB_{0A} that goes through O (see Fig. 4), Eqs. (19) are reduced to

$$\begin{aligned}
\left[\frac{\partial F}{\partial w}\right]_{w=l=0} &= -\left(1 + \frac{z^2}{r^2}\right)^{-1/2} \sin \alpha - \left(1 + \frac{z'^2_{0A}}{r'^2_{0A}}\right)^{-1/2} \\
&\quad \times \sin \beta_{0A} + m\lambda \left[\frac{\partial n}{\partial w}\right]_{w=l=0} = 0, \qquad (20) \\
\left[\frac{\partial F}{\partial l}\right]_{w=l=0} &= -\frac{z}{r}\left(1 + \frac{z^2}{r^2}\right)^{-1/2} - \frac{z'_{0A}}{r'_{0A}}\left(1 + \frac{z'^2_{0A}}{r'^2_{0A}}\right)^{-1/2} = 0.
\end{aligned}$$

Note here that $[\partial n/\partial l]_{w=l=0} = 0$ from Eq. (6). This is true also for aspheric gratings as far as their grooves are symmetric with respect to the x-y plane because the groove function $n = h(w, l)$ that defines such a symmetric groove pattern should not have terms containing odd powers of l. From these two equations

[Eqs. (20)] we have for the central ray AOB_{0A} the lateral magnification M

$$M = -\frac{z'_{0A}}{z} = \frac{r'_{0A}}{r} \qquad (21)$$

and the grating equation

$$\left(1 + \frac{z^2}{r^2}\right)^{-1/2} (\sin\alpha + \sin\beta_{0A}) = m\lambda \left[\frac{\partial n}{\partial w}\right]_{w=l=0}. \qquad (22)$$

For the principal ray $A_0 OB_0$, $z = 0$ and $\beta_{0A} = \beta_0$. Therefore, we have

$$(\sin\alpha + \sin\beta_0) = m\lambda[\partial n/\partial w]_{w=l=0}. \qquad (23)$$

Equation (23) states that the effective grating constant of a grating having varied spacing and curved grooves, both ruled and holographic, should be defined as

$$\frac{1}{\sigma} \equiv \left[\frac{\partial n}{\partial w}\right]_{w=l=0}. \qquad (24)$$

This definition yields $\sigma = \sigma_0$ for a conventional grating that has straight grooves with a constant spacing σ_0 when projected onto the y-z plane, because $n = w/\sigma_0$. Therefore, the grating equation for the in-plane mounting is written as

$$\sigma(\sin\alpha + \sin\beta_0) = m\lambda. \qquad (25)$$

The grating equation shows that the zeroth order ($m = 0$) spectrum appears in the same direction as mirror reflection, $\beta_0 = -\alpha$. When the spectrum lies in a region $-\alpha < \beta_0 < 90$ deg (or -90 deg $< \beta_0 < -\alpha$) the spectral order m must be positive (or negative), and the spectrum is referred to as the positive order or sometimes the inside order (or the negative order or the outside order), respectively.

On the other hand, the grating equation

$$\left(1 + \frac{z_0^2}{r^2}\right)^{-1/2} \sigma(\sin\alpha + \sin\beta_0) = (\sigma\cos\varepsilon)(\sin\alpha + \sin\beta_0) = m\lambda \qquad (26)$$

must be used for the off-plane mounting [33] and the conical-diffraction mounting [34] (also see Section 17.6.9.1), where z_0 is the vertical displacement of A_0, the center of the entrance slit, from the x-y plane and ε is the elevation angle of the principal ray $A_0 O$ from the x-y plane. Note here that the apparent grating constant $\sigma\cos\varepsilon$ becomes smaller as z_0 increases.

17.6.3 Ray Tracing through a Grating

The spectral imaging property of the diffraction grating can be determined, in principle, by solving the simultaneous equations [Eqs. (14)] with respect to the unknown quantities r', β, and z', that define the spectral image point of the ray

APB. One practical approach to do so is to derive exact ray-tracing formulas, which are applicable to all types of modern gratings. This, in turn, allows us to develop the third-order aberration theory of the grating, as is shown in Section 17.6.4.

Referring to Fig. 4 we consider the incident ray AP with the direction cosines (L, M, N) and its diffracted ray PB of wavelength λ in mth order with the direction cosines (L', M', N'). We define the direction cosines of the rays AP and PB as

$$L \equiv (\xi - x)/AP, \qquad M \equiv (w - y)/AP, \qquad N \equiv (l - z)/AP,$$
$$L' \equiv (x' - \xi)/PB, \qquad M' \equiv (y' - w)/PB, \qquad N' \equiv (z' - l)/PB. \tag{27}$$

Substituting Eq. (15a) and a similar expression for PB into Eq. (14) and performing partial differentiations, we obtain with the aid of Eq. (27)

$$\frac{\partial F}{\partial w} = (L - L')\frac{\partial \xi}{\partial w} + (M - M') + m\lambda \frac{\partial n}{\partial w} = 0,$$
$$\frac{\partial F}{\partial l} = (L - L')\frac{\partial \xi}{\partial l} + (N - N') + m\lambda \frac{\partial n}{\partial l} = 0. \tag{28}$$

Equation (28) is rewritten in a form

$$L' - L = \frac{M - M' + m\lambda(\partial n/\partial w)}{\partial \xi/\partial w} = \frac{N - N' + m\lambda(\partial n/\partial l)}{\partial \xi/\partial l} = T \tag{29}$$

or

$$L' = L + T, \qquad M' = M + m\lambda \frac{\partial n}{\partial w} - T\frac{\partial \xi}{\partial w},$$
$$N' = N + m\lambda \frac{\partial n}{\partial l} - T\frac{\partial \xi}{\partial l}. \tag{30}$$

Here, T is an unknown constant, and it can be determined by solving a quadratic equation resulting from substitution of Eq. (30) into $L'^2 + M'^2 + N'^2 = 1$:

$$eT^2 - 2pT + q = 0, \tag{31}$$

where

$$e = 1 + \left(\frac{\partial \xi}{\partial w}\right)^2 + \left(\frac{\partial \xi}{\partial l}\right)^2,$$
$$p = -L + \left(M + m\lambda \frac{\partial n}{\partial w}\right)\frac{\partial \xi}{\partial w} + \left(N + m\lambda \frac{\partial n}{\partial l}\right)\frac{\partial \xi}{\partial l}, \tag{32}$$
$$q = 2m\lambda\left(M\frac{\partial n}{\partial w} + N\frac{\partial n}{\partial l}\right) + (m\lambda)^2\left[\left(\frac{\partial n}{\partial w}\right)^2 + \left(\frac{\partial n}{\partial l}\right)^2\right].$$

The proper root of Eq. (31) is found to be

$$T = \frac{1}{e}\left(p + \sqrt{p^2 - eq}\right) \quad (33)$$

from the condition that the ray AOB_{0A} in zeroth order should satisfy the law of reflection at O, that is, $L' = -L$, $M' = M$, and $N' = N$. Note here that $(\partial \xi/\partial w)_{w=l=0} = (\partial \xi/\partial l)_{w=l=0} = 0$. Therefore, Eq. (30) with Eq. (33) yields the exact direction cosines (L', M', N') of the diffracted ray PB.

The diffracted ray PB of wavelength λ in mth order is now expressed as

$$\frac{x' - \xi}{L'} = \frac{y' - w}{M'} = \frac{z' - l}{N'} = d, \quad (34)$$

where d is an unknown constant. When an image plane Σ is specified, the intersection of the ray PB with the plane Σ forms a ray-traced spot.

We now assume $z/r \ll 1$, which is true in most practical cases, and take the ray $A_0 OB_0$ as the principal ray. We consider two image planes Σ and Σ_0, both of which are perpendicular to the x-y plane and pass through a point $B_0(x'_0, y'_0, 0)$ on the diffracted principal ray of wavelength λ in mth order (Fig. 5). The plane Σ_0 is perpendicular to the diffracted principal ray OB_0. The plane Σ makes an angle ϕ with the plane Σ_0. The angle ϕ is positive when measured counterclockwise from Σ_0 toward Σ, and $|\phi| \leq \pi/2$. Denoting $OB_0 = r'_0$, $x'_0 = r'_0 \cos \beta_0$, and $y'_0 = r'_0 \sin \beta_0$, we express the plane Σ as

$$x' \cos(\beta_0 + \phi) + y' \sin(\beta_0 + \phi) = r'_0 \cos \phi. \quad (35)$$

For a monochromator the image plane is the plane of the exit slit and thus $\phi = 0$. For a Rowland-circle spectrograph (see Section 17.6.6), the image plane

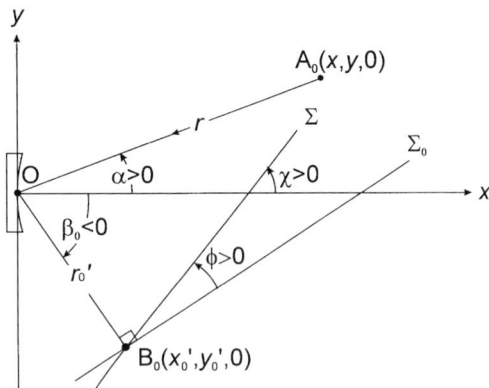

FIG. 5. Schematic diagram showing two image planes Σ_0 and Σ in relation to the diffracted principal ray OB_0.

is the Rowland cylinder $x'^2 + y'^2 - Rx' = 0$, and it can also be approximated by a series of image planes Σ with $\phi = \beta_0$, β_0 being a variable in a range from $\pi/2$ to $-\pi/2$. For a flat-field spectrograph we have $\phi = \chi - \beta_0 - \pi/2$ because the image plane is common to all the wavelengths of interest and makes a constant angle χ with the x-z plane. The angle χ is positive when measured counterclockwise from the positive direction of the x axis.

The unknown constant d in Eq. (34) is determined by solving Eqs. (34) and (35) simultaneously. This leads to the determination of the intersection $B(x', y', z')$ of the ray PB with the plane Σ, that is, the coordinates of the ray-traced spot:

$$d = \frac{r'_0 \cos\phi - \xi \cos(\beta_0 + \phi) - w \sin(\beta_0 + \phi)}{L' \cos(\beta_0 + \phi) + M' \sin(\beta_0 + \phi)}, \qquad (36)$$

$$x' = \xi + L'd, \qquad y' = w + M'd, \qquad z' = l + N'd. \qquad (37)$$

For convenience we introduce in the plane Σ a new rectangular coordinate system whose Y axis lies in the x-y plane and Z axis is parallel to the z axis (see Fig. 4). The ray-traced image point $B(Y, Z)$ is then expressed as

$$Y = (y' - r'_0 \sin\beta_0) \sec(\beta_0 + \phi), \qquad Z = z'. \qquad (38)$$

Equations (30), (33), and (35)–(38) with the grating equation, Eq. (25), give a complete set of ray-tracing formulas, provided that the surface figure $\xi = f(w, l)$ and the groove function $n = h(w, l)$ of the grating are known.

17.6.4 Third-Order Aberration Theory

The third-order aberration theory of the grating can be developed by analytically following the exact ray-tracing formalism described in Section 17.6.3 with the aid of power series expansions. For other aberration theories, refer to [35]. For simplicity, we consider a spherical concave grating whose surface figure is expressed by Eq. (4). As to the theory of the ellipsoidal and the toroidal grating, refer to [18] and [19], respectively.

The coordinates of the spectral image point B formed in the image plane Σ for the ray APB of wavelength λ in mth order are expressed as [18]

$$Y = r'_0 \sec(\beta_0 + \phi)(1 - \tan\beta_0 \tan\phi)[wf_{100} + w^2 f_{200} + l^2 f_{020} + lz f_{011}$$
$$+ z^2 f_{002} + w^3 f_{300} + wl^2 f_{120} + wlz f_{111} + wz^2 f_{102} + O(w^4/R^4)], \qquad (39)$$

$$Z = r'_0[-z/r + lg_{010} + wlg_{110} + wzg_{101} + w^2 lg_{210} + w^2 zg_{201}$$
$$+ l^3 g_{030} + l^2 zg_{021} + lz^2 g_{012} + z^3 g_{003} + O(w^4/R^4)]. \qquad (40)$$

In Eqs. (39) and (40), the aberration coefficients f_{ijk} and g_{ijk} are

$$f_{100} = F_{200}, \tag{41a}$$

$$f_{200} = \tfrac{3}{2} F_{300} + \tfrac{1}{2}(F_{200})^2 \sec \beta_0 (\tan \beta_0 - 2 \tan \phi)$$
$$+ F_{200}\left[\frac{\tan \beta_0}{R} - \frac{\cos \beta_0}{r_0'}(2 \tan \beta_0 - \tan \phi)\right], \tag{41b}$$

$$f_{020} = \tfrac{1}{2} F_{120} + \tfrac{1}{2}(F_{020})^2 \sin \beta_0 - F_{020}(\sin \beta_0/r_0'), \tag{41c}$$

$$f_{011} = F_{111} - F_{020}(\sin \beta_0/r), \tag{41d}$$

$$f_{002} = \tfrac{1}{2} F_{102}, \tag{41e}$$

$$g_{010} = F_{020}, \tag{42a}$$

$$g_{110} = F_{120} - F_{200} F_{020} \sec \beta_0 \tan \phi$$
$$+ F_{200} \sec \beta_0 \left(\frac{\sin \beta_0}{R} + \frac{\tan \phi}{r_0'}\right) - F_{020}\frac{\sin \beta_0}{r_0'}, \tag{42b}$$

$$g_{101} = F_{111} + \frac{1}{r} F_{200} \sec \beta_0 \tan \phi. \tag{42c}$$

Owing to limited space, the aberration coefficients f_{ijk} and g_{ijk} with $i+j+k=3$ are not given here. Their explicit expressions are found in [18] and [19]. The F_{ijk} in Eqs. (41) and (42) are

$$F_{200} = T_A + T_B + n_{20} \Lambda, \tag{43a}$$

$$F_{020} = S_A + S_B + n_{02} \Lambda, \tag{43b}$$

$$F_{300} = (T_A \sin \alpha/r) + (T_B \sin \beta_0/r_0') + n_{30} \Lambda, \tag{43c}$$

$$F_{120} = (S_A \sin \alpha/r) + (S_B \sin \beta_0/r_0') + n_{12} \Lambda, \tag{43d}$$

$$F_{111} = -(\sin \alpha/r^2) + (\sin \beta_0/rr_0'), \tag{43e}$$

$$F_{102} = (\sin \alpha + \sin \beta_0)/r^2, \tag{43f}$$

where

$$T_A = (\cos^2 \alpha/r) - (\cos \alpha/R), \quad T_B = (\cos^2 \beta_0/r_0') - (\cos \beta_0/R),$$
$$S_A = (1/r) - (\cos \alpha/R), \quad S_B = (1/r_0') - (\cos \beta_0/R), \tag{44}$$
$$\Lambda = \Gamma(\sin \alpha + \sin \beta_0).$$

Equations (41)–(44) suggest that the F_{ijk} and, in turn, the aberration coefficients f_{ij0} and g_{ij0} can be minimized by choosing proper groove parameters n_{ij}. This is the basis of the design of modern gratings. For actual design procedures refer to [19] and [36–38]. Furthermore, Eqs. (39) and (40) express the aberrations (up to third order) in spectral images formed in the plane Σ (see Fig. 5).

They represent ray-traced spot diagrams also, and they are called *spot diagram (SD) formulas*. The accuracy of Eqs. (39) and (40) is limited only by the neglect of the fifth and higher order terms in performing power series expansions of Eqs. (6) and (15)–(17).

It is of practical interest to examine, with reference to exact ray tracing, the validity of the SD formulas and also that of similar formulas based on the light path function (LPF). The LPF-based formulas, which are widely in use for the design of grating instruments, are given by [39]

$$Y = r'_0 \sec \beta_0 (\partial F/\partial w), \quad Z = -(zr'_0/r) + r'_0(\partial F/\partial l), \quad (45)$$

where the image plane is assumed to be perpendicular to the principal diffracted ray [Σ_0 in Fig. 5 and $\phi = 0$ in Eq. (35)]. The expansion coefficients of the light path function F contain the coordinates (r', β, z') of point B, which are unknown functions of the coordinates (ξ, w, l) of point P [see Fig. 4 and Eqs. (12), (15b), and (16)]. Partial differentiation of F therefore is not possible in principle. To circumvent this problem, it is customarily assumed that in the image plane, deviations of the spot B formed by the ray APB from the spot B_0 due to the principal ray $A_0 O B_0$ are negligible so that the unknown quantities r', β, and z' in the expansion coefficients can be approximated by the corresponding known quantities r'_0, β_0, and $-zr'_0/r$, which are all associated with the principal ray $A_0 O B_0$:

$$F \approx [F]_{r'=r'_0, \beta=\beta_0, z'=-zr'_0/r}. \quad (46)$$

With this approximation, Eqs. (45) are expressed as

$$Y = r'_0 \sec \beta_0 \left[wF_{200} + \frac{3}{2} w^2 F_{300} + \frac{1}{2} l^2 F_{120} \right.$$
$$\left. + lzF_{111} + \frac{1}{2} z^2 F_{102} + O\left(\frac{w^3}{R^3}\right) \right], \quad (47)$$

$$Z = r'_0 \left[-\frac{z}{r} + lF_{020} + wlF_{120} + wzF_{111} + O\left(\frac{w^3}{R^3}\right) \right]. \quad (48)$$

It can be shown analytically that the LPF-based formulas are correct to third order only when $\beta_0 = 0$ and the horizontal (or tangential) and vertical (or sagittal) focal curves cross on the grating normal [18]. The same results are obtained [18] for other ray-deviation formulas that are based on wavefront aberration (WFA) theory [40, 41].

In practice, these LPF- and WFA-based formulas can be applied safely to cases in which the defocus (F_{200}) and astigmatism (F_{020}) are small over the design wavelength range and the corresponding angles of diffraction β_0 are not too large. The validity of individual formulas is depicted clearly in Fig. 6, which compares the spot diagrams generated for a 142-deg constant-deviation grazing

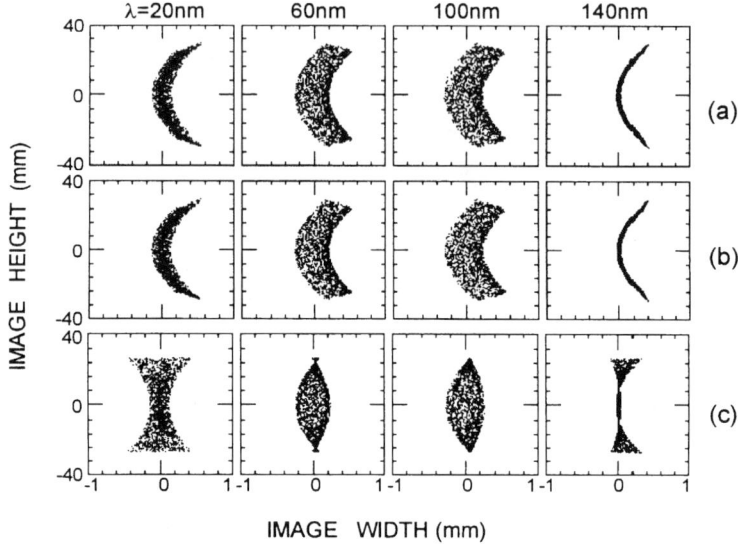

FIG. 6. Spot diagrams constructed for a 142-deg constant-deviation monochromator using (a) exact ray tracing, (b) SD formulas, and (c) LPF formulas.

incidence monochromator by the SD- and LPF-based formulas with those constructed by exact ray tracing. Specifications of the monochromator are wavelength range, 15–150 nm; $m = +1$; $r = 319.9$ mm; $r'_0 = 317.8$ mm; $R = 1000$ mm; $1/\sigma = 550$ grooves/mm; $2a = 3.877791 \times 10^{-9}$ mm; $6b = -6.844844 \times 10^{-14}$ mm; $4c = 3.654780 \times 10^{-20}$ mm; and ruled area, 60 (W) $\times$ 30 (L) mm^2. The monochromator has a rather large astigmatism as can be seen in Fig. 6. The spot diagrams in Fig. 6b (SD-based formulas) are very similar to the corresponding ones in Fig. 6a (ray tracing), proving the validity of the SD-based formulas. On the other hand, the spot diagrams generated from the LPF-based formulas (Fig. 6c) are quite different in shape and magnitude from those constructed by exact ray tracing and the SD-based formulas. Spot diagrams constructed by the WFA-based formulas also deviate considerably from Figs. 6a and b. These results show the limitation in the applicability of the LPF- and WFA-based formulas to a system in which astigmatism is not corrected to a sufficient degree.

17.6.5 Dispersion

When light is incident on a grating, the different wavelengths in the light are diffracted into different directions in accordance with the grating equation, Eq. (25). To express the angular separation of any two wavelengths, the rate of change of diffraction angle β_0 with change of wavelength λ is frequently used.

This quantity $d\beta_0/d\lambda$ is called the *angular dispersion*. Differentiating Eq. (25) with respect to λ, we obtain for a fixed angle of incidence α

$$\frac{d\beta_0}{d\lambda} = \frac{m}{\sigma \cos \beta_0}. \tag{49}$$

This shows that the angular dispersion is higher when a high order ($m \geq 2$), a smaller grating constant σ, and/or a larger angle of diffraction β_0 are used. For given values of m and σ, the angular dispersion will be minimum on the grating normal, where $\beta_0 = 0$. Around the grating normal, $\cos \beta_0 \cong 1$, and the dispersion of spectrum is very nearly constant. Such a spectrum is called the *normal spectrum*, and this provides a simple linear scale for wavelengths in the spectrum.

In specifying the dispersion of a grating spectrograph or a monochromator, it has been common practice to use the reciprocal linear dispersion in its focal plane or in the plane of the exit slit. This quantity is also known as the plate factor and is expressed in angstroms per millimeter or nanometers per millimeter. The reciprocal linear dispersion (or the plate factor) $d\lambda/ds$ in the image plane Σ of Fig. 5 [or Eq. (35)] is expressed by the reciprocal of the linear dispersion that is equal to $r_0' \sec \phi$ times the angular dispersion:

$$\frac{d\lambda}{ds} \equiv \frac{1}{r_0' \sec \phi} \cdot \frac{d\lambda}{d\beta_0} = \frac{\sigma \cos \beta_0}{mr_0' \sec \phi} \quad \text{(Å/mm or nm/mm)}, \tag{50}$$

where ds is measured along the Y axis of the image plane Σ. For a Rowland-circle mounting, which is explained in the next subsection, $r_0' = R \cos \beta_0$ and $\phi = \beta_0$, and thus the reciprocal linear dispersion is expressed by $\sigma \cos \beta_0/(mR)$, where σ and R are measured in units of angstroms (or nanometers) and millimeters, respectively. The reciprocal linear dispersion has smaller numerical values for larger spectral dispersions.

17.6.6 Focal Curves

Referring to the SD formulas, Eqs. (39) and (40), we immediately find that a greater part of the aberration can be removed by arranging the entrance slit and the image plane so as to make the lowest order aberration coefficients f_{100} and g_{010} zero or minimum over a wavelength range of interest. The mounting parameters can be determined from $f_{100} = F_{200} = 0$ and $g_{010} = F_{020} = 0$:

$$\frac{\cos^2 \alpha}{r} + \frac{\cos^2 \beta_0}{r_0'} - \frac{\cos \alpha + \cos \beta_0}{R} + n_{20}\Lambda = 0, \tag{51}$$

$$\frac{1}{r} + \frac{1}{r_0'} - \frac{\cos \alpha + \cos \beta_0}{R} + n_{02}\Lambda = 0. \tag{52}$$

For a given source point $(r, \alpha, 0)$, Eqs. (51) and (52) give the horizontal (or tangential) and the vertical (or sagittal) focal curve of the grating, respectively. In spectrographs and monochromators, the horizontal focal curve plays an extremely important role in obtaining a high-quality spectrum, because the formation of sharp spectral images comes before the control of elongation of images.

For conventional gratings, $n_{20} = n_{02} = 0$ and Eq. (51) has two well-known solutions:

$$r = R \cos \alpha, \quad r'_0 = R \cos \beta_0, \tag{53}$$

and

$$r = \infty, \quad r'_0 = \frac{R \cos^2 \beta_0}{\cos \alpha + \cos \beta_0}. \tag{54}$$

Equation (53) is the equation of a circle of diameter R equal to the radius of curvature of the grating, the vertex of the grating being the point of contact with the circle. This circle was discovered by Rowland [3b] in 1883, and it is known as the *Rowland circle*. The Rowland circle is fixed to the grating, so that it moves together with the grating whenever the grating is rotated or translated. When a point light source is placed on the Rowland circle, the spectrum of the source produced by the grating is focused horizontally on the circumference of this circle. Mountings that utilize the Rowland circle are called Rowland-circle mountings. Do not confuse this with the Rowland mounting [3b, 42], which is a specific type of Rowland-circle mounting. The merit of the Rowland circle is not only its novel focal property but also elimination and reduction of many of the higher order aberration coefficients in Eqs. (39) and (40). Because of this, the Rowland-circle mounting has been used most frequently in spectrographs and monochromators.

The second solution, Eq. (54), utilizes parallel incident light and is known as the Wadsworth mounting [43]. At $\beta_0 \sim 0$, both the horizontal and vertical focal curves are reduced to $r'_0 \simeq R/(1 + \cos \alpha)$. Therefore, a nearly stigmatic spectrum is obtained at around the grating normal. Furthermore, when a small angle of incidence is used, this has a small focal length $r'_0 \approx R/2$. These properties are needed to realize a fast spectrograph. The focal curve, however, is not a simple curve, and this offsets the advantages of the mounting.

The general solution of Eq. (51) has also been used in the design of non-Rowland-circle mountings such as Seya-Namioka monochromators [10, 13, 44], flat-field spectrographs [13, 14], normal incidence monochromators [16, 19], a zero-dispersion double-grating predisperser [45, 46], and VLS plane grating monochromators [15, 36–38]. It should be mentioned that the introduction of proper groove parameters contributed much to the successful development of non-Rowland-circle mountings.

Equation (52) gives the vertical focal curve. The spectrum of a point source placed on the Rowland circle ($r = R \cos \alpha$) is focused horizontally on the Rowland circle and vertically on a curve given by $r_0' = R/(\cos \beta_0 - \sin \alpha \tan \alpha)$. Thus, spectrum lines are focused as vertical and horizontal lines on the horizontal and the vertical focal curve, respectively. In the case of near-normal incidence, r_0' is positive, that is, the diffracted light is focused vertically on the vertical focal curve located in front of the grating, whereas in the case of grazing incidence it is negative, that is, the diffracted light is vertically diverging from the virtual vertical focal curve behind the grating.

17.6.7 Astigmatism

Astigmatism is the major aberration of a concave spherical grating and is characterized by the elongation and curvature of spectral images formed on the horizontal focal curve. The elongation and curvature of spectral lines arise from the finite illuminated length of the grooves and that of the entrance slit, respectively. Therefore, astigmatism on the Rowland circle ($\phi = \beta_0$) is represented by the terms of Eqs. (39) and (40) that depend only on l and z. Neglecting $O(w^4/R^4)$ we have

$$Y_{\text{ast}} = R \sec \beta_0 [l^2 f_{020} + lz f_{011} + z^2 f_{002}], \tag{55}$$

$$Z_{\text{ast}} = r_0'[-z/r + l g_{010}]. \tag{56}$$

Here, the subscript "ast" signifies the contributions from astigmatism. The length of the astigmatic image L_{ast} in the Rowland-circle mounting with a conventional grating is obtained from Eq. (56) as

$$L_{\text{ast}} = H(\cos \beta_0 / \cos \alpha) + L(\sin \alpha \tan \alpha \cos \beta_0 + \sin^2 \beta_0), \tag{57}$$

where H and L are the total illuminated length of the slit and that of the grooves, respectively. It can be seen from Eq. (57) that astigmatism is very large for grazing incidence, though it is small for near-normal incidence.

A spectral image of a point source on the entrance slit appears as curved. This curvature is called the *astigmatic curvature*. The equation of the astigmatic curvature can be obtained from Eqs. (55) and (56) simply by eliminating l. The image shape is a parabola. For a point source on the Rowland circle, the equation of its astigmatic image is obtained by eliminating l from Eqs. (55) and (56), both with $z = 0$ and Eq. (53):

$$Y_{\text{ast}} = \frac{Z_{\text{ast}}^2}{2\Psi^2 R \cos \beta_0} \left[\frac{\sin^3 \alpha}{\cos^2 \alpha} - \sin \beta_0 + \frac{\sin \beta_0}{\cos^2 \beta_0}(1 - \Psi)^2 \right], \tag{58a}$$

$$\Psi = \sin \alpha \tan \alpha \cos \beta_0 + \sin^2 \beta_0.$$

There appears another type of curvature in the spectral lines. This curvature, known as the *enveloping curvature*, was defined by Beutler [47] as the curve joining the image points formed by the central rays from every point on the infinitely narrow entrance slit (a line source). For the central ray AOB_{04} (see Fig. 4) in the Rowland-circle mounting with a conventional grating, we have from Eqs. (55) and (56), both with $l = 0$, on elimination of z

$$Y_{ast} = \frac{Z_{ast}^2(\sin\alpha + \sin\beta_0)}{2R\cos^3\beta_0} = \frac{m\lambda Z_{ast}^2}{2\sigma R\cos^3\beta_0}. \qquad (58b)$$

This is the equation of the enveloping curvature. As seen from Eq. (58b), the curvature of spectral lines rapidly changes with the angle of diffraction and is opposite on both sides of the zeroth-order image. Spectral lines are convex toward the zeroth-order image, which is always straight.

Astigmatism is unavoidable in conventional gratings. However, it can be reduced to a certain extent even with spherical concave gratings when ruled or recorded to introduce the right amount of varied line spacing and/or curved grooves [10, 12].

17.6.8 Resolving Power

According to the Rayleigh criterion, two similar spectrum lines of λ and $\lambda + \Delta\lambda$ are resolved when the principal maximum of one falls exactly on the first minimum of the other (Fig. 7a). In this case, the intensity and the full width of either contour at the central crossing point are $4/\pi^2$ of the maximum intensity and $\Delta\lambda$, respectively. This definition yields $\Re = mN$ for a conventional plane grating used in mth order, N being the total number of grooves illuminated. A modified Rayleigh criterion was used by Mack et al. [48] and Namioka [39, 49] in treating concave gratings.

A realistic definition of the resolving power can be developed by introducing an effective Gaussian line profile whose standard deviation is given by the standard deviation σ_Y of the spots computed from the SD formulas, Eqs. (39)

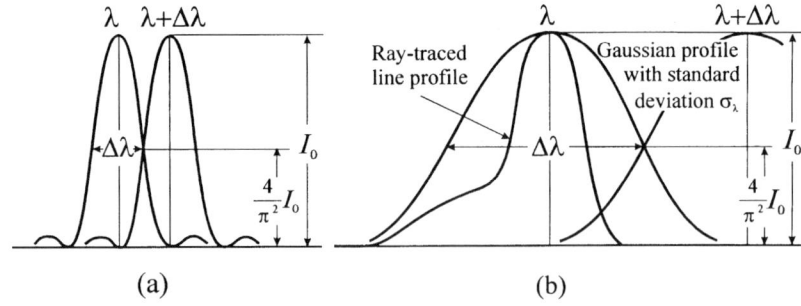

FIG. 7. Definitions of the resolving power based (a) on the Rayleigh criterion and (b) on the effective Gaussian line profiles.

and (40), for the ray of wavelength λ (Fig. 7b). We assume that two similar spectral lines of λ and $\lambda + \Delta\lambda$ are resolved when their effective Gaussian lines are separated by $\Delta\lambda = 2.643\sigma_Y(d\lambda/ds)$ to make the minimum resultant intensity between the lines $8/\pi^2$ as great as the resultant intensity at the central maximum of either of the lines. Thus the resolving power $\Re$ is defined by

$$\Re \equiv \frac{\lambda}{\Delta\lambda} = \frac{m\lambda r_0' \sec\phi}{2.643\sigma_Y \sigma \cos\beta_0}. \tag{59}$$

This definition has been applied to various cases involving varied line spacing ruled gratings and holographic gratings. The results show its validity and consistency with the spectral purity of the beam emerging through the exit slit [50].

With the definition of Eq. (59), the maximum usable ruled width $W_{\max}$ can be determined for a grating in a given mounting. For simplicity, we treat here a conventional spherical concave grating in the Rowland-circle mounting. The maximum attainable resolving power $\Re_{\max}$ of the grating could be represented by the resolving power obtainable at the limit where the groove length tends to zero and the entrance slit is replaced by a point source.

Spot diagrams constructed for this system are represented by Eqs. (39) and (40) with $l = z = f_{100} = f_{200} = 0$. The variance of the spots formed when all the possible rays from the source point are traced is approximated by

$$\sigma_Y^2 = \frac{1}{W}\int_{-W/2}^{W/2}(Y - \bar{Y})^2\, dw = \frac{1}{W}\int_{-W/2}^{W/2} Y^2\, dw, \tag{60}$$

where $\bar{Y}$ is the mean of Y and is zero in this case. Carrying out the integration we obtain

$$\sigma_Y^2 = \frac{W^6 R^2}{448} f_{300}^2 \sec^2\beta_0 = \frac{W^6 \sec^2\beta_0}{1792 R^4}(\sin\alpha\tan\alpha + \sin\beta_0\tan\beta_0)^2. \tag{61}$$

Substituting Eq. (61) and $\phi = \beta_0$ into Eq. (59), we obtain

$$\Re = \frac{16\sqrt{7}\, m\lambda R^3}{2.643\sigma W^3(\sin\alpha\tan\alpha + \sin\beta_0\tan\beta_0)}. \tag{62}$$

Approximating this $\Re$ with the resolving power of the plane grating, that is, mW/σ, and denoting W by $W_{\max}$, we obtain

$$W_{\max} = 2.0\left[\frac{R^3\lambda\cos\alpha\cos\beta_0}{\sin^2\alpha\cos\beta_0 + \cos\alpha\sin^2\beta_0}\right]^{1/4}. \tag{63}$$

This agrees fairly well with the result obtained by Mack et al. [48] using physical optics, and the only difference lies in their numerical factors: 2.0 versus 2.5. The maximum usable width of a modern grating in a non-Rowland-circle mounting can be calculated in a similar manner.

17.6.9 Grating Efficiency

The efficiency $\mathscr{E}(\lambda, m)$ of a grating at a given wavelength λ in a given order m is defined as the ratio, in percent, of the intensity of the diffracted radiation of λ in the mth order to that of the incident monochromatic radiation of wavelength λ. In a strict sense, it should be defined separately for the p and the s polarization of the incident radiation. It is, however, very difficult in practice to measure the grating efficiency for the two components of polarization because of lack of suitable polarizers in the vacuum ultraviolet below ~ 12 nm (note that synchrotron radiation has a degree of polarization less than $\sim 90\%$). In specifying the grating efficiency, the angle of incidence should be stated clearly.

17.6.9.1 Reflection Grating.
It is extremely desirable in spectroscopy to have a grating that concentrates the diffracted light into a single order. A practical solution to this end was found by Wood by following Lord Rayleigh who suggested in 1888 the possibility of concentrating all of the diffracted light into a single order by making the laminae of prismatic form on a plane surface [51]. Wood succeeded in ruling, in gold-plated copper plates, reflection gratings having triangular grooves with an apex angle of 120 deg by using the natural edge of a selected carborundum crystal as the ruling tool and observed a concentration of energy in one or two orders. With normal incidence he observed a concentration of energy at an angle of 40 deg with the grating, which had a groove shape with one side sloping at an angle of 20 deg with the original surface. Wood states in his book [51]: "These gratings may thus be regarded as reflecting echelons, of comparatively small retardation, and may be termed echelette, to distinguish them from the ordinary grating and the Michelson echelon." The word "echelette" faded away in the late 1950s because sawtooth-like grooves became standard in all commercial gratings. They are now called blazed gratings. Note that the grooves of holographic gratings can also be made triangular by means of ion-beam etching and such gratings are commercially available.

To quantify the observation of Wood, we refer to Fig. 8, which shows schematically the cross section of a blazed plane grating. One side of the groove makes an angle θ with the grating surface. The normals to the grating blank surface and the facet at a point P are denoted as N and N', respectively. Parallel light incident at P at an angle α is dispersed into spectrum in various orders. To concentrate the diffracted energy into a certain wavelength λ_b in mth order, the direction of this diffracted light, specified by the angle of diffraction β_0, should coincide with the direction of the specularly reflected light from P on the facet

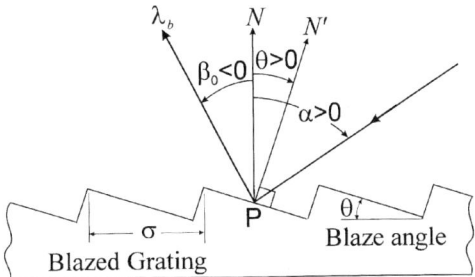

FIG. 8. Cross section of a blazed plane grating and the blaze condition.

surface. This condition is expressed by

$$\alpha - \theta = \theta - \beta_0 \quad \text{or} \quad \theta = (\alpha + \beta_0)/2. \tag{64}$$

Eliminating β_0 from Eqs. (25) and (64) we obtain

$$m\lambda_b = 2\sigma \sin\theta \cos(\alpha - \theta). \tag{65}$$

The grating is blazed at this wavelength λ_b, and λ_b and θ are called the *blaze wavelength* and *blaze angle*, respectively.

Although the blaze wavelength is a function of the angle of incidence and blaze angle, commercial brochures give one particular blaze wavelength together with the blaze angle. This particular blaze wavelength is defined as the one obtainable under the conditions $\alpha = \beta_0 = \theta$ and $m = 1$. These conditions mean that the incident light is made parallel to $N'P$, the normal to the large facet, and the retrodiffracted first-order light is observed (see Fig. 8). The blaze wavelength λ_{b0} listed in commercial brochures is expressed by

$$\lambda_{b0} = 2\sigma \sin\theta. \tag{66}$$

The expected blaze wavelength in a mounting should be calculated from Eq. (65) using the catalog values of λ_{b0} and θ, that is, from $m\lambda_b = \lambda_{b0} \cos(\alpha - \theta)$. A rule of thumb for the wavelength range effectively blazed is between $2\lambda_b$ and $\lambda_b/2$.

The grating can be illuminated from the direction opposite to that shown in Fig. 8. In this case a portion of the diffracted light will hit the steep side of the grooves, resulting in an increase in the scattered light level. It is therefore recommended that grating illumination such as that illustrated in Fig. 8 be used. To facilitate this, some manufacturers print an arrow on the back of their gratings that points the direction of the entrance slit to be placed.

The fact described above is true only for the plane grating. For the concave grating the blaze angle is not constant over the ruled area and varies from position to position on the grating surface. This naturally causes changes in the blaze angle across the grating surface that can be large especially for a small blaze angle. To ease this angle variation to a certain degree, concave gratings are often ruled in multiple sections by adjusting the diamond tool at every section, that is, by changing the blaze angle to concentrate the diffracted energy more efficiently into a single order. Gratings thus ruled are called multipartite gratings, for example, tripartite when ruled in three sections with slightly different blaze angles. The effects of the curvature of the blank on measured efficiencies of gratings are discussed by Neviere and Hunter [52].

High grating efficiencies can also be obtained by using multilayer coated gratings and extreme off-plane mountings called *conical diffraction mountings* [34, 53–56]. The former are described in detail in Chapter 18. The latter uses an incident beam nearly parallel to the grooves, and it is of particular interest in achieving very high efficiencies in the soft x-ray region. In an extreme off-plane mounting, the grating equation is given by Eq. (26), ε being usually greater than ~ 80 deg for soft x-rays. This mounting is called a conical diffraction mounting because the diffracted orders lie on a cone whose axis is tangent to the zeroth groove at the vertex of the grating and whose generator is the zeroth order with the half-angle (90 deg $-\varepsilon$). High first-order efficiencies greater than 40% at several wavelengths in a range of 0.83–6.70 nm were observed by Werner [54] for a 3600 groove/mm cylindrical grating with a blaze angle of 5 deg placed on an extreme off-plane mounting with $\varepsilon = \sim 84$–89 deg. This phenomenon was explained by Neviere *et al.* by means of the electromagnetic theory [55] and also by experiment [34]. It was shown that the efficiency for the $+1$st order is very high when the plane of the incidence is perpendicular to the face of the large facet and the incident and the diffracted $+1$st order light are symmetrical with respect to the normal to the large facet. This extreme off-plane mounting is sometimes called *generalized Marechal–Stroke (GMS) mounting* [34]. In this case the blaze wavelength is given by Eq. (65), with $m = 1$, on the substitution of $\sigma \cos \varepsilon$ and θ for σ and α, respectively:

$$\lambda_b = 2\sigma \sin \theta \cos \varepsilon. \tag{67}$$

It was also shown theoretically that even a laminar grating and a sinusoidal grating would have high efficiencies in the GMS mounting, though not as high as that of a blazed grating, if the groove depth could be optimized properly [56]. Comprehensive articles on the electromagnetic theory of diffraction gratings are found in [57].

Mention should be made in passing that the grating efficiency is very sensitive to the condition of the grating surface: carbon contamination of a grating on a synchrotron radiation beamline, for example. Various methods of cleaning contaminated gratings were reported. They are classified into two categories: oxygen-discharge cleaning [58] and a photo-assisted ashing in the presence of a proper amount of ozone [59]. Both the methods were found to be very effective in recovering the efficiency of contaminated gratings. The former method has the advantage of the *in situ* applicability and the latter is free from possible damages, such as roughening of the surface, due to discharge, and has the cleaning capability well proved in semiconductor productions.

17.6.9.2 Transmission Grating. We consider a grating consisting of a series of equispaced slits in an opaque screen. Denoting the slit width and the spacing (or the grating constant) by a and σ, we obtain the efficiency of this grating at a wavelength λ in mth order as [60]

$$\mathcal{E}(\lambda, m) = \left[\frac{\sin(m\pi a/\sigma)}{m\pi}\right]^2, \quad (68)$$

where $\alpha = 0$ and $m\lambda = \sigma \sin \beta_0$, with α and β_0 being the angles of incidence and diffraction, respectively. When $a/\sigma = 0.5$, it is seen from Eq. (68) that the efficiency reaches maxima for odd orders and vanishes for even orders: 10.13% for each first order and 25% for zeroth order.

The grating bars are not fully opaque to soft x-rays of short wavelengths, however. Constructive or destructive interference occurs between the radiation coming through the slit openings and the attenuated and phase-shifted radiation coming from the grating bars. The efficiency depends not only on the a/σ ratio but also on the thickness of the bars. We refer to the efficiency formulas derived by Tatchyn et al. [61], among others [62], to consider this effect. For the rectangular-profile transmission grating shown in Fig. 9a, the optimum thickness H_{opt} of the grating bar and the optimum width a_{opt} of the space that maximize the mth-order efficiency are given by

$$H_{opt} = \frac{\lambda}{2\pi\delta}\left\{\cos^{-1}\left[\frac{k}{\sqrt{\delta^2 + k^2}}\right] + \cos^{-1}\left[\left(\frac{k}{\sqrt{\delta^2 + k^2}}\right)\exp\left(-\frac{2\pi k H_{opt}}{\lambda}\right)\right]\right\}, \quad (69)$$

$$a_{opt} = \frac{\sigma}{2m},$$

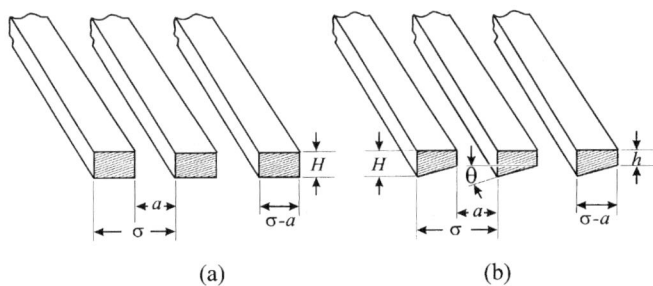

FIG. 9. Schematic diagrams of (a) a rectangular-profile transmission grating and (b) a blazed transmission grating.

where δ and k are the optical constants of the bar material, defined by $\tilde{n} = n + ik$ and $\delta = 1 - n$, $\tilde{n}$ being the complex index of refraction. The maximum attainable efficiency in mth order is

$$\mathcal{E}(\lambda, m)_{\max} = \left(\frac{1}{m\pi}\right)^2 \sin^2\left(\frac{2\pi}{\lambda} H_{\mathrm{opt}} \delta\right)\left(\frac{\delta^2 + k^2}{k^2}\right). \tag{70}$$

It can be seen from Eq. (69) that the ratio a_{opt}/σ is 0.5 for the first order.

Figure 9b shows a blazed transmission grating and its parameters. According to Tatchyn et al. [61], the optimum slope of the linearly blazed profile is

$$\tan \theta_{\mathrm{opt}} = -m\lambda/(2\sigma\delta). \tag{71}$$

The optimum a/σ, a_{opt}/σ, and the optimum h, h_{opt}, can be obtained by iterating the equation

$$\frac{a}{\sigma} = \frac{1}{m\pi}\left|\cos^{-1}\left[-\frac{\delta \sinh \eta}{\sqrt{\delta^2 + k^2} \sin(m\pi a/\sigma)} \exp\left(-\eta - \frac{2\pi k h}{\lambda}\right)\right]\right|$$
$$+ \frac{(2\pi/\lambda)h\delta - \gamma}{m\pi}, \tag{72}$$

$$\eta = \frac{m\pi k}{\delta}\left(1 - \frac{a}{\sigma}\right), \quad \gamma = \cos^{-1}\left(\frac{k}{\sqrt{\delta^2 + k^2}}\right).$$

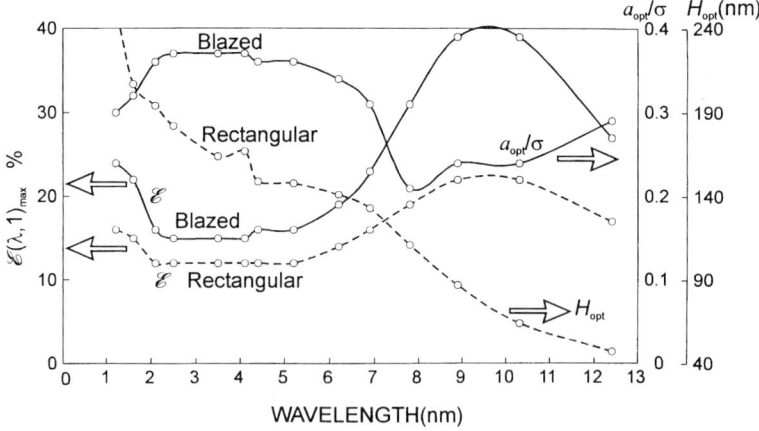

FIG. 10. Values of $\mathcal{E}(\lambda, 1)_{\max}$ and H_{opt} calculated for a rectangular-profile gold transmission grating with $a_{\mathrm{opt}}/\sigma = 0.5$, and $\mathcal{E}(\lambda, 1)_{\max}$ and a_{opt}/σ obtained for a blazed gold transmission grating with $h_{\mathrm{opt}} = 0$.

The maximum attainable efficiency $\mathscr{E}(\lambda, m)_{\max}$ is expressed in terms of h_{opt} and a_{opt}/σ:

$$\mathscr{E}(\lambda, m)_{\max} = \left(\frac{a_{\mathrm{opt}}}{\sigma}\right)^2 \frac{\delta^2 + k^2}{k^2} \mathrm{sinc}^2\left(\frac{m\pi a_{\mathrm{opt}}}{\sigma}\right) \sin^2\left(\frac{2\pi h_{\mathrm{opt}}\delta}{\lambda} - \frac{m\pi a_{\mathrm{opt}}}{\sigma}\right). \quad (73)$$

As an example, Fig. 10 shows H_{opt} and $\mathscr{E}(\lambda, 1)_{\max}$ calculated for a rectangular-profile gold transmission grating with $a_{\mathrm{opt}}/\sigma = 0.5$ and a_{opt}/σ and $\mathscr{E}(\lambda, 1)_{\max}$ obtained for a blazed gold transmission grating with $h_{\mathrm{opt}} = 0$ [61].

References

1. G. W. Stroke, in "Handbuch der Physik," Vol. XXIX (S. Flügge, ed.), Springer, Berlin, pp. 426–754 (1967).
2. E. G. Loewen, "Diffraction Grating Handbook," Bausch & Lomb, Rochester, NY, p. 4 (1970).
3. (a) H. A. Rowland, *Phil. Mag.* Ser. 5, **13**, 469–474 (1882); (b) H. A. Rowland, *Phil. Mag.* Ser. 5, **16**, 197–210 (1883).
4. G. R. Harrison and G. W. Stroke, *J. Opt. Soc. Am.* **45**, 112–121 (1955).
5. G. R. Stroke, in "Progress in Optics," Vol. II (E. Wolf, ed.), North-Holland, Amsterdam, pp. 1–72 (1963); G. R. Harrison, *Appl. Opt.* **12**, 2039–2049 (1973).
6. A. A. Michelson, "Studies in Optics," 2nd ed., University of Chicago Press, Chicago, pp. 99–103 (1962).
7. G. Schmahl and D. Rudolph, in "Progress in Optics," Vol. XIV (E. Wolf, ed.), North-Holland, Amsterdam, pp. 195–244 (1976).
8. M. C. Hutley, *Sci. Prog. Oxf.* **61**, 302 (1974); M. C. Hutley, "Diffraction Gratings," (N. H. March and H. N. Daglish, eds.), Techniques of Physics Series 6, Academic Press, London (1982).
9. J. Cordelle, J. Flamand, G. Pieuchard, and A. Labeyrie, in "Optical Instruments and Techniques," (J. Home Dickson, ed.), Oriel Press, Newcastle-upon-Tyne, pp. 117–124 (1970).
10. T. Namioka, M. Seya, and H. Noda, *Jpn. J. Appl. Phys.* **15**, 1181–1197 (1976).
11. M. Koike, Y. Harada, and H. Noda, *Proc. SPIE* **815**, 96–101 (1987); H. Noda, Y. Harada, and M. Koike, *Appl. Opt.* **28**, 4375–4380 (1989).
12. T. Harada, S. Moriyama, and T. Kita, *Jpn. J. Appl. Phys.* **14** (Suppl. 14-1), 175–179 (1975).
13. T. Harada and T. Kita, *Appl. Opt.* **19**, 3987–3993 (1980).
14. T. Kita, T. Harada, N. Nakano, and H. Kuroda, *Appl. Opt.* **22**, 512–513 (1983); N. Nakano, H. Kuroda, T. Kita, and T. Harada, *Appl. Opt.* **23**, 2386–2392 (1984).
15. T. Harada, T. Kita, M. Itou, H. Taira, and A. Mikuni, *Nucl. Instrum. Methods A* **246**, 272–277 (1986).
16. M. Itou, T. Harada, T. Kita, K. Hasumi, I. Koyano, and K. Tanaka, *Appl. Opt.* **25**, 2240–2242 (1986); M. Itou, T. Harada, and T. Kita, *Appl. Opt.* **28**, 146-153 (1989).
17. T. Harada, T. Kita, S. Bowyer, and M. Hurwitz, *Proc. SPIE* **1545**, 2–10 (1991); S. Bowyer, *Proc. SPIE* **2517**, 97–106 (1995); T. Harada, H. Sakuma, Y. Ikawa, T. Watanabe, and T. Kita, *Proc. SPIE* **2517**, 107–115 (1995).

18. T. Namioka, M. Koike, and D. Content, *Appl. Opt.* **33**, 7261-7274 (1994). Misprints found in Eqs. (4), (21), (23), (24), (35), and (36) should be corrected as follows. The numerator at the first line in Eq. (4) should read $M - M' + m\lambda(\partial n/\partial w)$. λ in the last line of Eq. (21) should read λ_0. The last terms in Eqs. (23) and (24) should read $O(w^4/R^4)$. A bracket] is missing at the end of Eq. (35). $O(w^3/R)$ in the last term of Eq. (36) should read $O(w^3/R^2)$.
19. T. Namioka and M. Koike, *Nucl. Instrum. Methods A* **319**, 219–227 (1992). Misprints found in Eqs. (22) and (24) should be corrected as follows. The integrand Y in Eq. (22) should read $[Y]_{l=z=0}$. On the right hand side of Eq. (24), w in the first term should read W.
20. T. Namioka and M. Koike, *Appl. Opt.* **34**, 2180–2186 (1995). Misprints found in Eqs. (23)–(25) should be corrected as follows. V_C in the last term of Eq. (23e) should read V_D. K_C in the fifth term of Eq. (23f) should read K_D. Both a_1 and a_2 in the first and second equations in Eq. (24) should read a. The minus signs at the third and fourth lines in Eq. (25b) should be plus signs. The third term of Eq. (25c) should be multiplied by $\cos \eta_C$.
21. H. Bräuninger, H. Kraus, H. Dangschat, K. P. Beuermann, P. Predehl, and J. Trümper, *Appl. Opt.* **18**, 3502–3505 (1979).
22. A. M. Hawryluk, N. M. Ceglio, R. H. Price, J. Melngailis, and H. I. Smith, *J. Vac. Sci. Technol.* **19**, 897–900 (1981).
23. C. R. Canizares, M. L. Schattenburg, and H. I. Smith, *Proc. SPIE* **597**, 253–260 (1985).
24. A. G. Michette, "Optical Systems for Soft X Rays," Plenum Press, London, p. 160 and pp. 224–233 (1986) and references therein.
25. H. Aritome, S. Matsui, K. Moriwaki, H. Aoki, S. Namba, S. Suga, A. Mikuni, M. Seki, and M. Taniguchi, *Nucl. Instrum. Methods* **208**, 233–236 (1983).
26. E. T. Arakawa and P. J. Caldwell, *Nucl. Instrum. Methods* **172**, 293–296 (1980).
27. K. P. Beuermann, R. Lenzen, and Bräuninger, *Appl. Opt.* **16**, 1425–1431 (1977).
28. J. H. Dijkstra, L. J. Lantwaad, C. Timmerman, *Proc. XX COSPAR Sympo.*, Tel Aviv, 1977, A.3.6 (1977).
29. K. P. Beuermann, H. Bräuninger, and J. Trümper, *Appl. Opt.* **17**, 2304–2309 (1978).
30. H. Lochbihler and P. Predehl, *Appl. Opt.* **31**, 964–971 (1992).
31. J. F. Meekins, *Appl. Opt.* **28**, 1221–1227 (1989).
32. R. J. Pegis, in "Progress in Optics," Vol. I (E. Wolf, ed.), North-Holland, Amsterdam, pp. 1–29 (1961).
33. T. Namioka, *J. Opt. Soc. Am.* **49**, 460–465 (1959).
34. M. Neviere, D. Maystre, and W. R. Hunter, *J. Opt. Soc. Am.* **68**, 1106–1113 (1978).
35. F. Zernike, in "Pieter Zeeman," Martinus Nijhoff, Hague, pp. 323–335 (1935); W. T. Werford, in "Progress in Optics," Vol. IV (E. Wolf, ed.), North-Holland, Amsterdam, pp. 241–280 (1965); W. Werner, *Appl. Opt.* **6**, 1691–1699 (1967); E. Ishiguro, R. Iwanaga, and T. Oshio, *J. Opt. Soc. Am.* **69**, 1530–1538 (1979); K. Goto and T. Kurosaki, *J. Opt. Soc. Am. A* **10**, 452–465 (1993).
36. M. Koike and T. Namioka, *Appl. Opt.* **33**, 2048–2056 (1994). In this paper the last term $2f_{120}f_{120}$ of Eq. (6) should read $2f_{120}f_{102}$. In line 10 on the right of p. 2051, line 12 on the left of p. 2052, and line 10 on the right of p. 2052, 178° should read 174°.
37. M. Koike and T. Namioka, *J. Electron Spectrosc. Relat. Phenom.* **80**, 303–308 (1996).

38. M. Koike, R. Beguiristain, H. Underwood, and T. Namioka, *Nucl. Instrum. Methods A* **347**, 273–277 (1994): M. Koike and T. Namioka, *Rev. Sci. Instrum.* **66**, 2144–2146 (1995).
39. T. Namioka, *J. Opt. Soc. Am.* **49**, 446–460 (1959).
40. W. R. McKinney and C. Palmer, *Appl. Opt.* **26**, 3108–3118 (1987); R. Grange, *Appl. Opt.* **31**, 3744–3749 (1992).
41. M. P. Chrisp, *Appl. Opt.* **22**, 1508–1518 (1983).
42. F. A. Jenkins and H. E. White, "Fundamentals of Optics," 3rd ed., McGraw-Hill, New York, pp. 339–340 (1950).
43. F. L. O. Wadsworth, *Astrophys. J.* **3**, 47–62 (1896); M. Seya and T. Namioka, *Sci. Light (Tokyo)* **16**, 158–168 (1967).
44. H. Noda, T. Namioka, and M. Seya, *J. Opt. Soc. Am.* **64**, 1043–1048 (1974); T. Kita and T. Harada, *J. Spectrosc. Soc. Jpn.* **29**, 256–262 (1980).
45. T. Namioka, H. Noda, K. Goto, and T. Katayama, *Nucl. Instrum. Methods* **208**, 215–222 (1983); K. Ito, T. Namioka, Y. Morioka, T. Sasaki, H. Noda, K. Goto, T. Katayama, and M. Koike, *Appl. Opt.* **25**, 837–847 (1986).
46. M. Koike, Y. Ueno, and T. Namioka, *Proc. SPIE* **3150**, 31–39 (1997).
47. H. G. Beutler, *J. Opt. Soc. Am.* **35**, 311–350 (1945).
48. J. E. Mack, J. R. Stehn, and B. Edlen, *J. Opt. Soc. Am.* **22**, 245–261 (1932).
49. T. Namioka, *J. Opt. Soc. Am.* **49**, 951–961 (1959).
50. M. Koike and T. Namioka, *Appl. Opt.* **36**, 6308–6318 (1997).
51. R. W. Wood, "Physical Optics," 2nd ed., Macmillan, New York, pp. 226–231 (1929).
52. M. Neviere and W. R. Hunter, *Appl. Opt.* **19**, 2059–2065 (1980).
53. G. H. Spencer and M. V. R. K. Murty, *J. Opt. Soc. Am.* **52**, 672–678 (1962); P. Lemaire, *Appl. Opt.* **30**, 1294–1302 (1991).
54. W. Werner, *Appl. Opt.* **16**, 2078–2080 (1977).
55. M. Nevière, P. Vincent, and D. Maystre, *Appl. Opt.* **17**, 843–845 (1978).
56. P. Vincent, M. Nevière, and D. Maystre, *Appl. Opt.* **18**, 1780–1783 (1979).
57. R. Petit, ed., "Electromagnetic Theory of Gratings," Topics in Current Physics, Vol. 22, Springer-Verlag, Berlin (1980).
58. W. R. McKinney and P. Z. Takacs, *Nucl. Instrum. Methods* **195**, 371–374 (1982); E. D. Johnson and R. F. Garrett, *Nucl. Instrum. Methods A* **266**, 381–385 (1988); T. Koide, T. Shidara, M. Yanagihara, and S. Sato, *Appl. Opt.* **27**, 4305–4313 (1988).
59. T. Harada, S. Yamaguchi, M. Itou, S. Mitani, H. Maezawa, A. Mikuni, W. Okamoto, and H. Yamaoka, *Appl. Opt.* **30**, 1165–1168 (1991).
60. M. Born and E. Wolf, "Principles of Optics," 5th ed., Pergamon Press, New York, p. 405 (1975).
61. R. Tatchyn, P. L. Csonka, and I. Lindau, *J. Opt. Soc. Am.* **72**, 1630–1638 (1982).
62. H. W. Schnopper, L. P. Van Speybroeck, J. P. Delvaille, A. Epstein, E. Källne, R. Z. Bachrach, J. Dijkstra, and L. Lantwaad, *Appl. Opt.* **16**, 1088–1091 (1977); H. Bräuninger, P. Predehl, and K. P. Beuermann, *Appl. Opt.* **18**, 368–373 (1979).

18. MULTILAYER GRATINGS

W. R. Hunter
SFA Inc.
Largo, Maryland

18.1 Introduction

The usefulness of diffraction gratings in the vacuum ultraviolet (VUV) is limited, in part, by the reflecting power of the grating coating. Heretofore, at near-normal incidence, the coatings were either a double layer of Al/dielectric or a single metal for wavelengths less than 1000 Å. The double layers have been extremely useful in increasing the efficiency of gratings used at normal incidence, sometimes by a factor of 3 or 4 [1] because they have large reflectance values from the visible to wavelengths as short as the intrinsic absorption of the dielectric; about 1200 Å for MgF_2 or 1000 Å for LiF (see Chapter 16). At the shorter wavelengths one uses single metals, for example, Au, Pt, Ir, Os, Re, Ru, or W, which have large reflectance values at normal incidence compared to Al/dielectric layers and cover a wide range of wavelengths (see Chapter 11). At grazing incidence there is no advantage in using the double Al/dielectric layers; one makes use of quasi-free-electron metals if the angle of incidence is greater than the critical angle (see Chapter 16), or one uses the heavier metals just listed. Grazing incidence, with its many aberrations, is disadvantageous if the instrument is intended for imaging, so for imaging there is an incentive to remain close to normal incidence.

An alternative to the single metal coating or double layer is the multilayer coating. Spiller has described the principles of these coatings in Chapter 14. A multilayer coating is designed for a specific wavelength, and has a bandwidth of about 5-10% of the design wavelength. This restricted bandwidth is a limitation for spectroscopes intended for wide wavelength coverage. In other applications, however, such as scanning monochromators, the necessary broadband operation can be achieved, despite the narrowband characteristic of the coating, because the coating essentially follows the Bragg law. Therefore, as the scanning angle (angle of incidence) changes the reflectance maximum shifts toward shorter wavelengths, almost in synchronism with the scanning angle. There are also many x-ray and VUV astronomy experiments concerned only with specific emission lines, not requiring broad wavelength ranges, to be studied to aid in modeling stellar atmospheres and coronas.

18.2 Designing Multilayers for Gratings

Before designing the multilayer, one must consider what part of the spectral region is of interest, what grating to choose, and how the grating is to be used, that is, at near-normal or grazing incidence. These are the conditions imposed by the experiment.

Because the grating response follows the simple Bragg law while the multilayer response is more closely described by the corrected Bragg law, the grating and multilayer will be accurately matched, that is, the grating will have maximum efficiency, at only one wavelength and angle of incidence; at other wavelengths and angles the efficiency will be less. The corrected Bragg law requires that both indices of refraction and extinction coefficients of the layers must be taken into account when calculating the multilayer reflectance. The multilayer must be optimized at the angle of incidence of the radiation on the grating *facet* for a blazed grating.

To design a multilayer for a grating such that the mismatch between multilayer and grating is minimized, one must solve the grating equation and the Bragg equation simultaneously, subject to experimental conditions. The grating equation is

$$N_g \lambda = d_g(\sin \alpha + \sin \beta),$$

where N_g is the grating order, d_g is the groove spacing in centimeters per groove, α is the angle of incidence, and β is the angle of diffraction. The equation can also be written as

$$N_g \lambda = 2d_g \sin \Theta_B \sin \sigma$$

where Θ_B is the blaze angle and σ is the glancing angle with respect to the facet surface. A schematic diagram showing the blazed grating and significant angles is shown at the top of Fig. 1 [2]. The bottom of the figure illustrates Bragg diffraction; σ is the glancing angle and d is the layer spacing. The Bragg equation is

$$N_b \lambda = 2d_b \sin \sigma$$

and the simultaneous solution is

$$d_b/N_b = (d_g \sin \Theta_B)/N_g.$$

A number of authors [3–7] have discussed the simultaneous solution, which matches the multilayer to the grating, primarily for laminar gratings. The reader is referred to their papers for more information.

There is an effective long wavelength limit to the usefulness of multilayers, imposed by the absorptance of the layer materials. A general rule used by the author is that multilayers are not useful at wavelengths longer than 400 Å. From this wavelength down to about 100 Å and less, the reflectances that can be obtained with multilayers permit the use of gratings near normal incidence. This long-wavelength limit restricts the choice of gratings in that their normal incidence region must be less than 400 Å. The simplest way to determine the long-wavelength region of a grating is to calculate the Littrow wavelength and angle of incidence. The Littrow wavelength is that wavelength for which the incident and diffracted rays coincide, that is, $\alpha = \beta$, and the grating equation becomes (without the subscripts)

$$N\lambda = d(2 \sin \alpha).$$

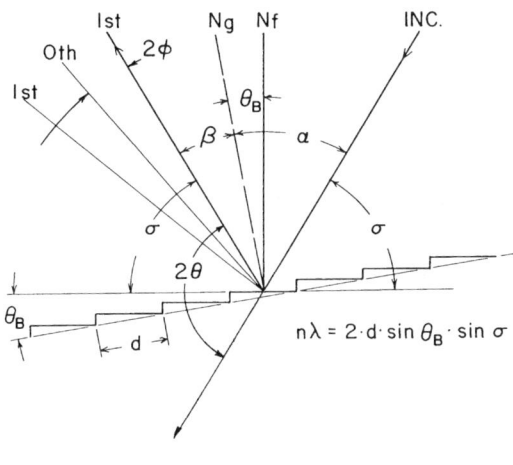

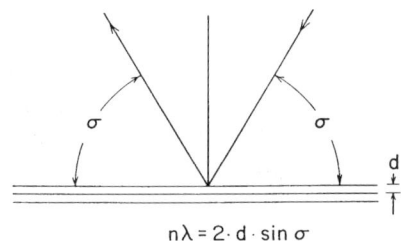

FIG. 1. Illustration of the blaze condition for a diffraction grating and how it is analogous to Bragg diffraction for a crystal, where N_g and N_f are the grating and facet normals, and α and β are the angles of incidence and diffraction, respectively. INC represents the incident beam. The incident beam and zeroth order make equal angles with N_g. The incident beam and inside first order, that is, that closest to N_g make equal angles with respect to N_f. 2Θ is the deviation angle of the desired order and 2ϕ is the deviation of the desired order from the zeroth order. If $\phi = \Theta_B$, then the desired order is on blaze. (Adapted from [2], with permission from Elsevier Science, Netherlands.)

TABLE I. Littrow Wavelengths (Å) vs Groove Density

Order	α	\multicolumn{5}{c}{Groove density (g/mm)}
		1000	2000	3000	4000	5000
1st	1	349	174	116	87	70
	2	698	349	233	174	140
	5	1743	872	581	436	349
	10	3473	1736	1158	868	694
2nd	1	174	87	58	44	35
	2	349	174	116	87	70
	5	872	436	291	218	175
	10	1736	868	579	434	347
3rd	1	116	58	39	29	23
	2	233	117	78	58	46
	5	581	291	194	145	116
	10	1158	579	386	289	232

Note: Littrow angles of incidence (α°) are show to the left.

Table I shows calculated values of the Littrow wavelength and angle of incidence for different line densities and orders. The dashed lines within each order separate the spectral regions wherein multilayers are useful (upper) from those where they are not useful (lower), according to the previously mentioned general rule. If the grating is blazed, then the Littrow angle of incidence is also the blaze angle. As the line densities and order numbers increase, the Littrow wavelengths decrease and permit the use of larger blaze angles if the grating is blazed.

Although the preceding equations are for blazed gratings they are general if one takes 2Θ as the deviation angle of the given grating order and 2ϕ as the deviation angle of the given order from the zeroth order. Then, if $\phi = \Theta_B$ the given order has maximum efficiency, that is, it is on-blaze.

The author and his colleagues at the Naval Research Laboratory use a numerical procedure for matching multilayer to grating. One decides on the design wavelength, compatible with the grating constants and angle of incidence, chooses nodal and spacer layer materials with no, or few, absorption edges, chooses a period thickness (approximately the design wavelength), and calculates the layer thicknesses using tabulated optical constants to give the desired results. The author uses an optimization program that calculates the layer thicknesses to give the maximum reflectance at any angle of incidence and for either the s or p component.

To illustrate the numerical design procedure, choose a grating with 3000 g/mm and a 5-deg blaze angle; the Littrow wavelength in first order is 581.038 Å. Because the requirement is for a wavelength of about 80 Å at normal incidence it is necessary to use the grating in a high order. In the seventh order the Littrow wavelength is 83.005 Å so that order is used. The *first*-order interference of the multilayer was chosen, rather than a higher order, because calculations showed that the first-order reflectance maximum was larger than that of the higher orders. The matching of grating and a multilayer of 50 periods of Be/Pt was done at 10 Å, which required an angle of incidence on the grating facet of 83.081 deg for the inside order. Optimizing the multilayer for this wavelength and angle ensures that the multilayer reflectance maximum occurs at 83.081 deg as shown in the upper part of Fig. 2. The reflectance for p polarization is slightly less than that for s polarization at this angle of incidence. The bottom part of the figure shows the calculated reflectance maximum of the multilayer at 50 Å and its displacement from the grating angle of 52.96 deg. Note that the reflectance of the p component for the 50-Å match drops to a small value as the Brewster angle ($\approx$45 deg) is approached. For this sort of application the p component is not useful.

A number of such calculations at different wavelengths are made to build a picture of the loss in efficiency of the coated grating as the wavelength changes. Some results are shown in Fig. 3. The solid lines represent the peak reflectances of the multilayers as a function of wavelength. The reflectances of the multilayers at the angle of diffraction of the seventh order are shown as the dashed line. To the left is the calculation for matching the Be/Pt multilayer and grating at 10 Å. As the wavelength increases the mismatch increases, reaching a maximum at about 45 deg. At the longer wavelengths ($\approx$80 Å) the mismatch is less. If the match had been made at a longer wavelength, say, 50 Å, then the efficiency of the grating would be reduced at wavelengths both longer and shorter than 50 Å. The right part of the figure shows a similar calculation for a C/W multilayer and grating, matched at 10 Å, for which the wavelength range straddles the K absorption of carbon. At wavelengths less than 44 Å the efficiency loss is small, but at 44 Å and to longer wavelengths the loss in efficiency increases abruptly to about 60%. Such abrupt efficiency changes may cause complications in interpreting experimental results, so coating materials without absorption edges should be chosen for the coating materials.

Two cautionary notes should be mentioned at this point. The first concerns the reflectance of the multilayer at longer wavelengths. Figure 4 shows the calculated reflectance of the Be/Pt multilayer, optimized at 10 Å, from about 10–150 Å at an angle of incidence of 83 deg. Note that the reflectance begins to increase appreciably between 40 and 50 Å and has practically the same reflectance at 130 Å as the peak at 10 Å. Long-wavelength stray light could be reflected from the multilayer to contaminate the spectrum.

The second concerns the width of the multilayer bandpass. The preceding procedure peaks the bandpass on the desired wavelength, grating order, and glancing angle on the grating facet. If the bandpass is wide enough, adjacent orders may be included in the bandpass and cause contamination of the spectrum. As an example, consider a multilayer of Si/Mo with 50 periods applied to a grating with 2000 g/mm and a blaze angle of 5 deg. At 291 Å the third order of the grating is used (see Table I) and the first order of the multilayer. Taking

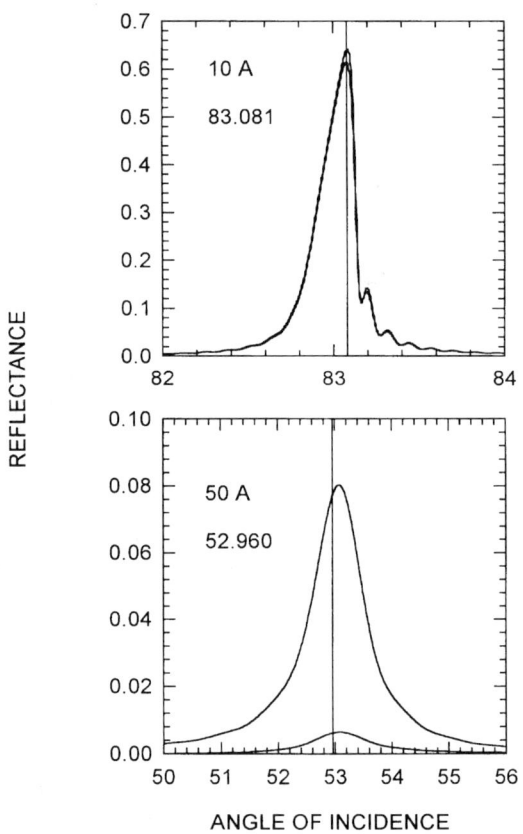

FIG. 2. Calculation showing the displacement from the angle of diffraction of the reflectance maximum of a Be/Pt multilayer reflecting coating. At the top, the multilayer has been optimized for 10 Å, and 83.083-deg angle of incidence on the facets, and its reflectance maximum coincides with the angle of diffraction. At the bottom is shown the reflectance calculation for a grating diffraction angle of 52.960 deg and a wavelength of 50 Å. The reflectance maximum is displaced from the angle of diffraction because actual values of the indices of refraction and extinction coefficients of the coating layers were used in the calculation. The reflectance at the intersection of the vertical line with the reflectance curve is the efficiency to be expected from the grating at 50 Å. Note the small reflectance value of the p component.

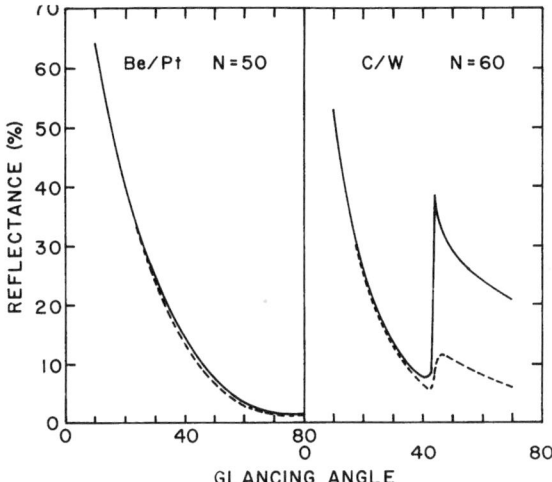

FIG. 3. Calculation showing the loss in efficiency as a function of wavelength caused by mismatch of the multilayer coating to the grating. The values were obtained by doing a number of calculations, at different wavelengths, of the type shown in Fig. 2. To the left a Be/Pt has been matched at 10 Å ($\approx$83-deg angle of incidence). To the right a C/W multilayer has been matched at 10 Å ($\approx$83-deg angle of incidence). Note the abrupt change in efficiency across the carbon K edge.

the angle of incidence as 5 deg, the outside (negative) second, third, and fourth orders have diffraction angles of 11.74, 15.17, and 18.66 deg, respectively. Since our interest is in the third order, the multilayer must be designed for that order, which means it must be optimized in its first order for the angle of incidence on the facet; (15.17 + 5)/2 = 10.085 deg. Doing so results in a multilayer with a maximum R_s = 26.15% and spacer and nodal layer thicknesses of 120.7 and 38.2 Å, respectively.

Figures 5a, b, and c show the angular response of the above multilayer at three wavelengths, 291, 208, and 130 Å, respectively, and the location of the second, third, and fourth grating orders. At normal incidence the second and fourth orders are well within the response curve of the multilayer and have efficiencies comparable to the third order. As the wavelength decreases, the multilayer bandpass shrinks while the angular distance between grating diffraction angles increases. Thus at 208 Å the second and fourth orders are less efficient than the third order, and at 130 Å, they are down by a factor of between 5 and 6. One should bear in mind that this calculation assumes a perfect facet. In practice, facets may be curved, rough, or have undulations, all of which can spread the angular response curve of the multilayer.

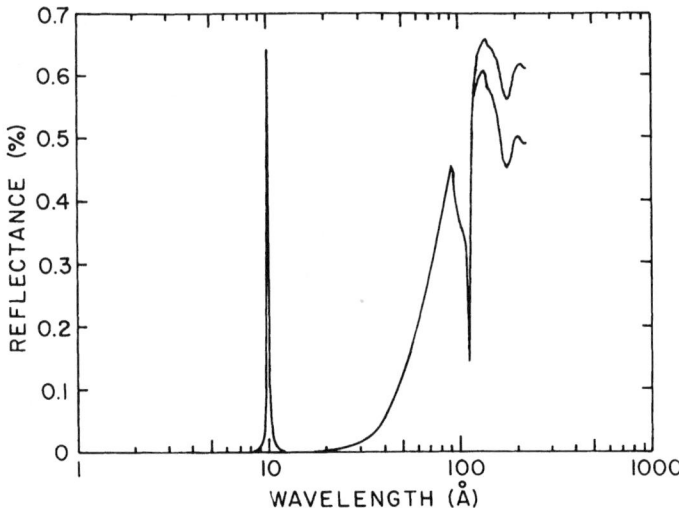

FIG. 4. Calculation showing the reflectance of the Be/Pt multilayer coating of Figs. 2 and 3 at an angle of incidence of ≈83 deg as a function of wavelength. As the wavelength increases the reflectance of the coating becomes as large as the peak reflectance at 10 Å. The curve splits into two parts just longward of 100 Å because of polarization. The upper branch is *s* polarization and the lower *p* polarization.

18.3 Efficiency Measurements of Multilayer-Coated Gratings

18.3.1 Blazed Gratings

An early attempt at applying a multilayer coating to a blazed grating was made by Jark [8]. He used a mismatched, three-period coating of C/Au with a 182-Å period to a 1200 g/mm grating blazed at 1.5 deg. His efficiency measurements off-blaze, at 50 and 100 Å, showed an improvement of five times over that of a gold-coated grating.

The first attempt to improve the efficiency of a blazed grating with a multilayer, in which the author was involved [9], used a 2000 g/mm holographic grating, ion etched into silica, with a nominal blaze angle of 4.1 deg. It was coated with a carbon/tungsten multilayer of 21 periods optimized for ninth order in normal incidence. This grating was a test grating 20 mm in diameter. A 25-mm-diameter flat was coated at the same time for reflectance measurements. Measurements with CuK_α radiation after coating showed a 2-D spacing of 81.72 Å and a Γ of 0.65. [Γ is defined as the absorbing layer (W) thickness divided by the period thickness.] The maximum efficiency of the grating, measured at 60.57 Å and a glancing angle of 43.8 deg, occurred in the eighth order and was 0.38% while the reflectance of the flat was 3.8%. Although a value of 0.38% does not seem very large, it is larger than the specular reflection of gold at this wavelength and angle of incidence. The flat and grating were installed in a grating/crystal monochromator [2] and a scan was made, on blaze, from 800 to 1700 eV using the eighth order and with an aluminum filter 0.00025 in. thick. The results are shown in Fig. 6a [9]. Clearly visible at the extremes of the scan are the Ni $L_{2,3}$ and the Al K edges. The Ni edge is probably the result of diffusion of Ni into the Pt coating placed on the collimating and focusing mirrors. In the center is the $L_{2,3}$ edge of Ga; the detector was a Schottky diode of GaAsP type. This extremely wide range of wavelengths illustrates the usefulness of multilayer coatings as applied to gratings.

A reflectance scan of a polished sample of lanthanum glass of unknown origin is shown in Fig. 6b [9], in which can be seen the La $M_{4,5}$ core levels. The La M_4 natural line width is 0.73 eV [10] but the line width in lanthanum glass is not

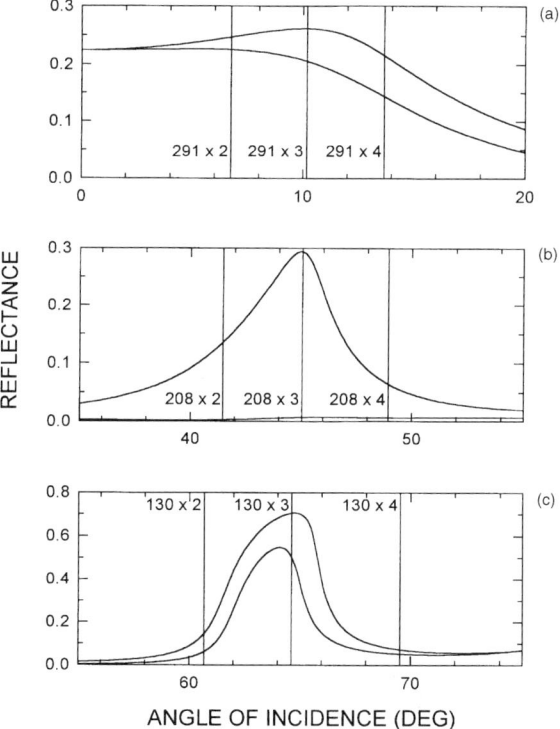

FIG. 5. This figure shows the relation between the calculated angular response of an Si/Mo multilayer designed for the first interference order of 291 Å at 10.085 deg, and the second, third, and fourth diffraction orders of a grating with 2000 g/mm and blaze angle of 5 deg at (a) 291 Å from 0 to 20 deg, (b) 208 Å from 35 to 55 deg, and (c) 130 Å from 55 to 75 deg.

known. The line width shown in the figure is 2.6 eV at 848 eV and that of the M_5 is 1.7 eV at 833 eV. Judging from the line widths, the resolving power of the instrument was greater than 490.

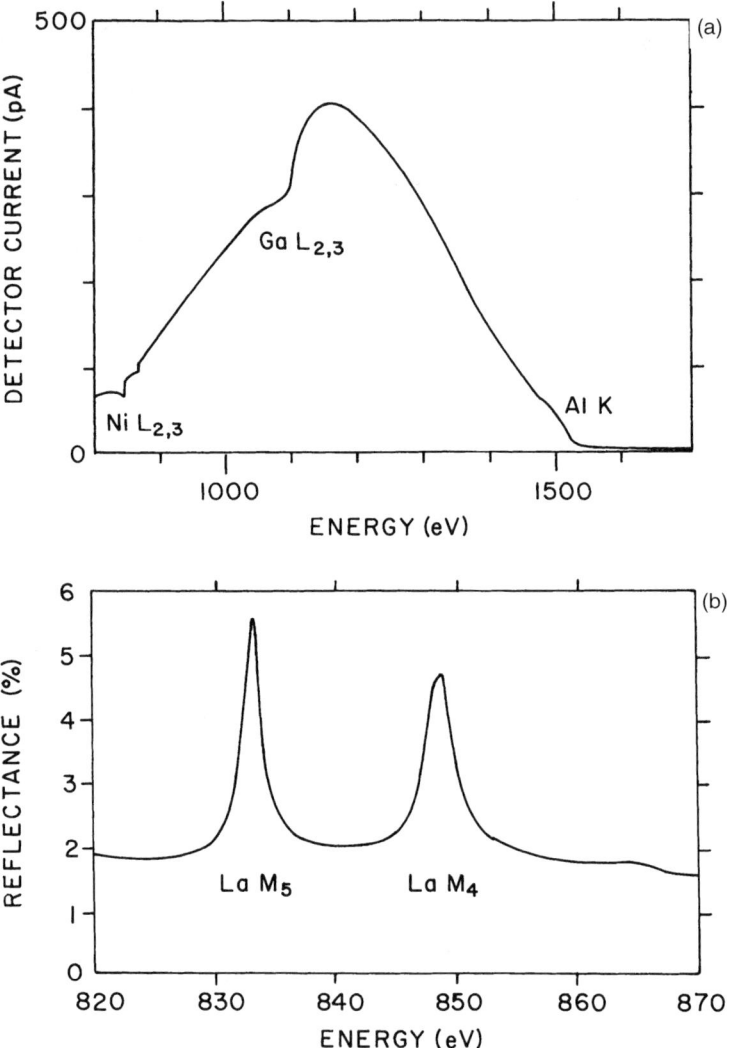

FIG. 6. (a) Monochromator scan from 800 to 1700 eV using a blazed grating and a flat mirror, both coated with a multilayer of C/W. (b) Monochromator scan from 820 to 870 eV showing the reflectance of lanthanum glass and the effect of the La $M_{4,5}$ absorption edges on the reflectance [9].

To evaluate any effects due to the large reflectance values of the multilayer at longer wavelengths, two spectral scans, shown in Fig. 7 [11] were made using the coated mirror and grating used to obtain Fig. 6. No filters were used in making these scans. The scanning mode was one in which the mirror angle of incidence

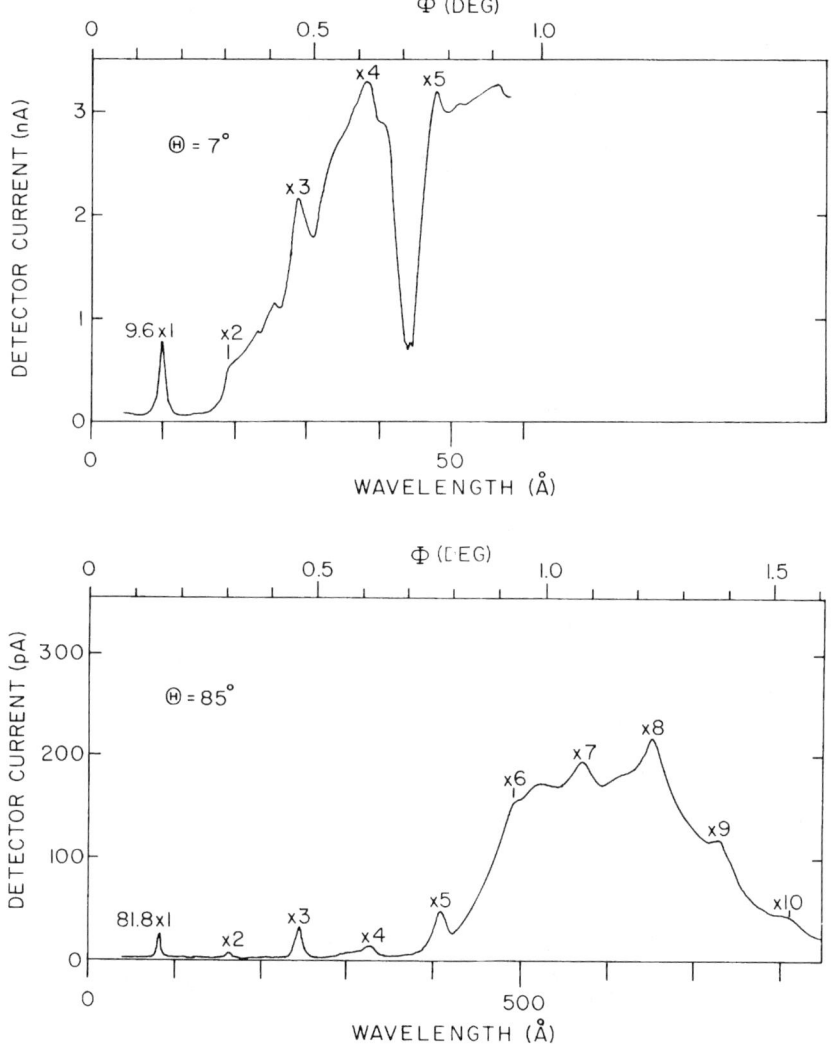

FIG. 7. Monochromator scans using the grating and mirror of Fig. 6 to show the presence of radiation at the longer wavelengths reflected by the multilayer. This measured result is similar to the calculated results of Fig. 4.

was fixed and the grating rotated, an off-blaze scan. The scales at the top show the angles through which the grating was rotated to obtain the scan and the scales at the bottom are the corresponding wavelengths. In the upper part of the figure the two components are close to grazing incidence ($\lambda = 9.6$ Å) at a nominal glancing angle of 7 deg. A rotation of the grating of approximately 1 deg scanned from the first order of 9.6 Å, which is isolated, through the higher orders which overlie a broad continuum consisting of lower order radiation diffracted by the grating, to the fifth order. The sharp dip between the fourth and fifth orders is caused by the carbon edge. With the components at near-normal incidence the same type of scan is shown in the lower part of the figure. A rotation of the grating through about 1.5 deg shows the first order of 81.8 Å through the tenth order and the long-wavelength background continuum. A proper choice of filters would eliminate some of this longer wavelength stray light.

For instruments with a fixed angle of incidence, the small bandpass of a multilayer restricts the wavelength range over which the grating is effective. However, higher order interferences of the multilayer can sometimes be used, thus providing more than one bandpass. Second-order interference has been used to create a dual-waveband grating using a concave, 2.2-m radius of curvature, replica in normal incidence [12]. The grating had a nominal 2-deg blaze angle and a ruling density of 2400 groove/mm. The coating was Si/Mo and consisted of 30 periods with a D spacing of 162.5 Å, thus optimizing for first interference order at about 308 and the second at 156 Å at normal incidence. Measurements of the witness blank, coated with the grating, at an angle of incidence of 20 deg are shown in Fig. 8a [12]. Because of the shift in angle of incidence, from 0 to 20 deg, the first and second interference orders have shifted to 290 and 148 Å and the match between grating and multilayer is not optimum. Although the coating was optimized for the first order, the second order has a larger reflectance maximum because the absorptance of the layers is less at that wavelength. Figure 8b [12] shows the calculated reflectance of the coating, assuming interdiffusion and surface roughness typical of sputtered Si/Mo coatings [13] (solid line) at 20 deg and compares it with the calculated reflectance at 4 deg (dashed line) and the calculated reflectance of gold at 4 deg (dashed-dotted line) assuming a gold surface roughness of 8-Å rms. Figure 8c [12] shows the measured efficiencies as a function of wavelength of the multilayer-coated grating (ML) and a sister gold-coated grating (Au) at a 14-deg angle of incidence in the first outside order. The efficiencies of the multilayer grating exceed those of the Au grating by a factor of 4 in the first order and 200 in the second order.

Figure 9a [12] shows the measured efficiency of the multilayer and Au gratings of Fig. 8 as a function of diffraction angle at a 14-deg angle of incidence and at a wavelength of 290 Å. Figure 9b [12] is the same type of measurement of the same two gratings at 151 Å. The different orders are indicated. Note that for the gold grating at 290 Å the strongest order is the −1 order while the multilayer grating has almost equal distribution of efficiency on the ±1 and 0 orders. At 151 Å the −2 order of the gold grating is strongest while the −1 order of the multilayer grating is strongest. This change in flux distribution from the gold-coated grating to the multilayer grating indicates that the groove profile was modified by the coating. Such a modification might be expected since for a nominal blaze angle of 2 deg and a grating spacing of 4167 Å the height of the step is only 145 Å, and the thickness of the coating is 4885 Å, almost 34 times larger than the groove depth.

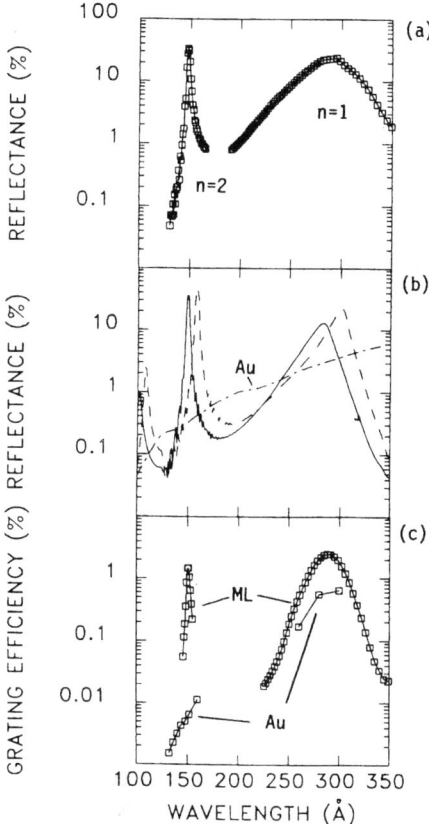

FIG. 8. (a) Measured reflectance of the multilayer witness flat at an angle of incidence of 20 deg. (b) Calculated reflectance of the witness flat for angles of incidence of 20 deg (solid curve) and 4 deg (dashed curve). The dashed-dotted curve is the calculated reflectance of gold at a 4-deg angle of incidence and assuming 8-Å roughness rms. (c) Measured efficiencies of the multilayer (ML) and gold (Au) gratings at a 14-deg nominal angle of incidence and in the first outside grating order [12].

Not only has the groove profile been modified, it appears that the multilayer has caused the stray light background to increase. The performance of the two gratings in a spectrograph showed that they had many ghosts in the spectral region shown. The most reasonable explanation is that the multilayer amplified the ghosts and brought up the stray light and that the reflectance of the gold coating is not strong enough for the ghosts to be seen during efficiency measurements. Similar increases in stray light were obtained in applying a multilayer coating to a grating for operation in the 136- to 142-Å wavelength region [14].

Figure 10 [12] shows the spectrum obtained on mounting the grating of Figs. 8 [10] and 9 [12] in a spectrograph and illuminating it with radiation from a Cu/W electrode that was sparked at 8 kV 800 times to provide a quasi-continuum light source. The first and second multilayer orders are indicated.

Figure 11 [12] shows the spectrum from a vanadium electrode (2200 sparks at 2 kV) recorded with the multilayer grating of Figs. 8 and 9 through a Be filter. The lines near 156 Å (second multilayer order) are shown in the first, second, and third grating orders. The resolution of the 156.608-Å vanadium line in the

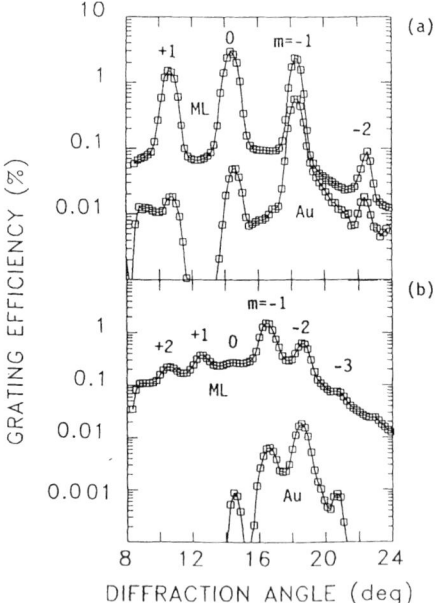

FIG. 9. Measured efficiencies of the multilayer (ML) and gold (Au) gratings as a function of diffraction angle at a 14-deg angle of incidence and at wavelengths of (a) 290 Å near the first multilayer order and (b) 151 Å near the second multilayer order. The inside ($m > 0$) and outside ($m < 0$) gratings orders are indicated [12].

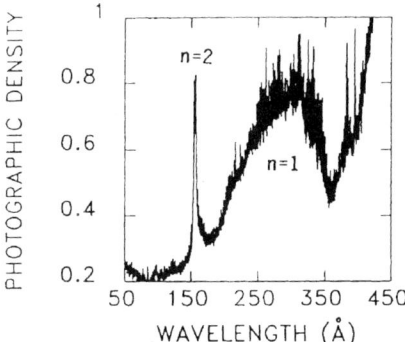

FIG. 10. Quasi-continuum spectrum from a copper/tungsten electrode (800 sparks at 8 kV) recorded with the multilayer grating of Figs. 8 and 9. The broad first-order, and narrow second-order, interference of the multilayer can be seen [12].

third order is 14,000 after removal of the instrumental broadening. Apparently the groove profile modification did not affect the resolving power of the grating. Others [14–16] have also used concave gratings with a multilayer coating, optimized for 304 Å, and found no deterioration of resolution caused by application of the multilayer.

Although amplification of ghosts by the multilayer coating is a cogent argument for the increase in stray light shown in Figs. 9a and b, it is not conclusive. However, subsequent efficiency measurements on a blazed, ion-etched, holographic grating showed a significant gain in the signal to background. This gain is attributed to the smoother grooves of the holographic

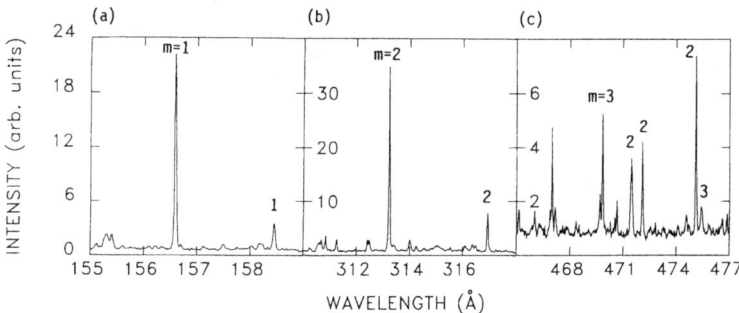

FIG. 11. Spectrum from a vanadium electrode (2200 sparks at 2 kV) recorded with the multilayer grating of Figs. 8 and 9 and a Be filter. The lines near 156 Å in the (a) first, (b) second, and (c) third orders are shown. The first, second, and third order lines are identified by 1, 2, and 3, respectively. The resolution of the 156.608 Å line in third order, after the instrumental broadening is removed, is 14,000 [12].

grating and consequent lack of ghosts. The grating has 2400 g/mm and a radius of curvature of 2.2 m. It was coated with 25 periods of Si/Mo with a period thickness of 154 Å. This grating was intended for use in a Seya–Namioka monochromator so that the first Bragg order occurred at 230 Å at an angle of incidence of about 35 deg and the second at 145 Å when the grating was rotated to about 10 deg. The grating was measured as a function of angle of incidence as shown in Figs. 12 and 13 [17]. To the left is shown the measurements, close to normal incidence, at three wavelengths close to the second Bragg order. The highest efficiency of 7.5% occurred in the -2 order at 146.7 Å. To the right, measurements in the region around a 35-deg angle of incidence are shown for the first Bragg order and for s and p polarization where the polarization was $\approx 80\%$. It is clear that the peak to background ratio is considerably larger than that obtained with the replica grating of Figs. 8 and 9, which can be attributed to the smoother grooves found in holographic gratings.

18.2.2 Laminar Gratings

Laminar gratings can also be matched to multilayer coatings using the same technique described for blazed gratings. However, the concept of the multilayer-coated laminar grating is slightly different than for a blazed grating. The lands and grooves of a laminar grating are surfaces parallel to the average grating surface. Thus, a multilayer on a laminar grating is, for conceptual purposes, optimized only for the zeroth grating order. It is not possible to optimize the coating to any other order, as is the case with a blazed grating, because the Bragg law requires that the incident and diffracted radiation make equal angles with respect to the surface. One must also consider the bandpass of the multilayer, which, because it is centered on the grating zeroth order, must be wide enough to cover the other desired orders. Of course, the multilayer also has higher order interferences so that higher grating orders can be included in these interference envelopes if conditions are right.

Although, in concept, the multilayer is optimized to the zeroth order of the grating, the zeroth order is not useful in spectroscopy. However, the ratio of zero order of the grating to the two first, or other, orders can be controlled by controlling the depth of the groove. Hellwege [18] has analyzed laminar gratings and gives the path difference between radiation reflected from the lands and groove as $(m + p)\lambda/2$ where

$$m\lambda/2 = d(\sin \alpha - \sin \beta),$$
$$p\lambda/2 = h(\cos \alpha + \cos \beta) \approx 2h.$$

The symbols have their usual meaning and h is the groove depth. To suppress the zeroth order, p should be odd and $h = p\lambda/4$. For a laminar groove depth of 220 Å and $p = 5$, the corresponding wavelength is 176 Å (for $d = 5000$ Å).

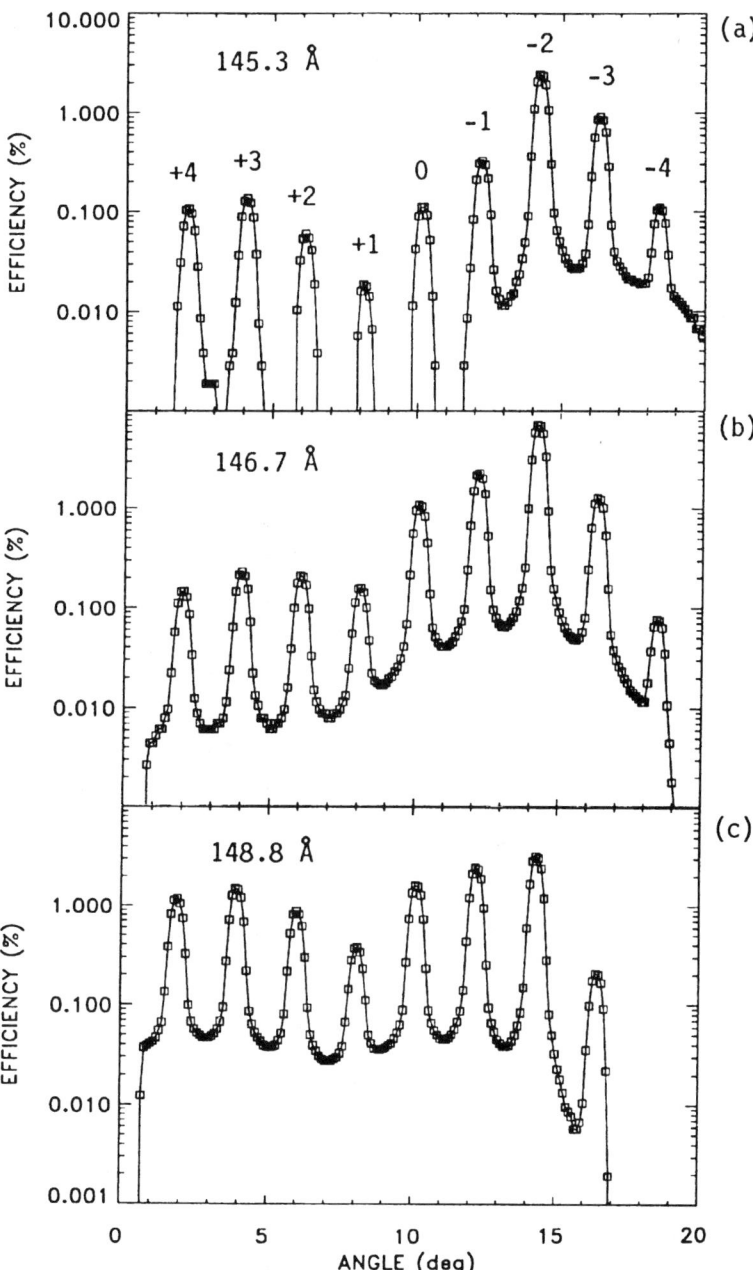

FIG. 12. Grating efficiency measurements at an angle of incidence of 10 deg (normal) and at (a–c) 145.3 through 148.8 Å. The inside ($m > 0$) and outside ($m < 0$) grating order are identified in (a) [17].

A laminar grating designed for the requirements just given was ion etched into silica. A talystep measurement indicated a groove depth of 220 Å. The multilayer coating was of Si/Mo using 25 periods, each of 92.5-Å thickness, deposited after etching. A problem occurred during the sputter deposition of the 20th silicon layer, which broadened the halfwidth of the reflectance spectrum somewhat but did not impair the performance of the grating significantly. Figure 14 [7] shows the grating efficiency at 9.8 deg angle of incidence. At 174 Å the zero order was not suppressed but at 164 Å it was, which indicates an error in the talystep groove depth measurement. The two first-order efficiencies are about 3.8%, which is much greater than could be achieved with a simple metal coating.

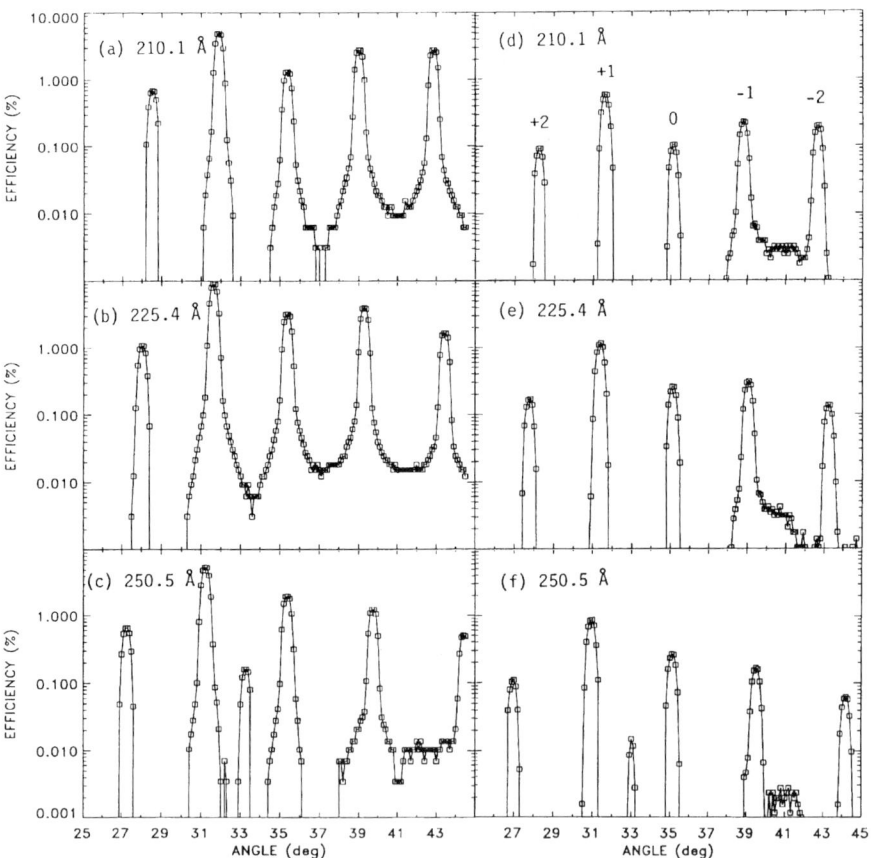

FIG. 13. Grating efficiency measurements at an angle of incidence of 35 deg at wavelengths 210.1 through 250.5 Å using (a–c) primarily s-polarized radiation and (d–f) primarily p-polarized radiation. The different grating orders are identified in (d) [17].

Laminar multilayer gratings are also made by depositing the multilayer on a mirror surface, then forming a holographic grating in an overcoating of photoresist. Etching through the exposed photoresist leaves a laminar grating whose lands are coated with the original multilayer but whose grooves are coated with a truncated, or no, multilayer. Such gratings are sometimes referred to as amplitude gratings, in contrast to phase gratings in which both lands and grooves have the same coating. In principle, amplitude gratings have the advantage in that the lands have the same rms roughness as the substrate, modified by the original coating. Such roughness may be less than that of the coated facet surfaces of a ruled grating but perhaps no smoother than the coated facet of a holographically produced blazed grating.

Finally, the characteristics of laminar gratings are also affected by the ratio of land width to groove width. Hellwege's [18] analysis assumes a ratio of unity

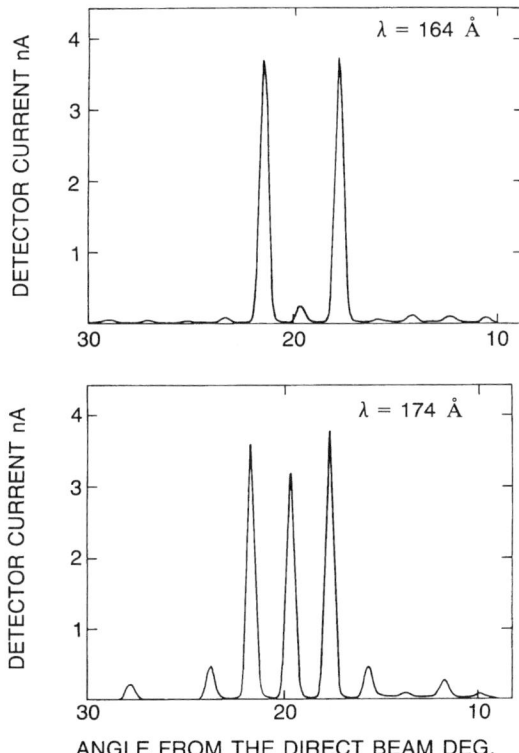

FIG. 14. The signal from the reflectometer detector as it scanned through the orders diffracted from a multilayer laminar grating. The results obtained at 164 and 174 Å are shown. In both cases the angle of incidence was 9.8 deg. The maximum efficiency was about 6.5% [7].

but others [5, 6] have investigated this ratio and the reader is referred to their work for more information.

In conclusion, multilayer coatings applied to gratings increase the grating efficiency within the bandpass of the multilayer, usually by a large factor. The coatings do nothing toward increasing the resolving power of the grating. In fact, if the coating, which is usually much thicker than the groove depth, changes the groove profile or spacing, the resolving power may become less. In the experience at the Naval Research Laboratory no degradation of the resolving power has been found. A degradation in efficiency has been found over a number of years [19], which is attributed to carbon contamination plus additional oxidation of the uppermost Mo layer of the multilayer coating.

The author would like to acknowledge the scientific and editorial advice of J. C. Rife, J. F. Seely, M. P. Kowalski, and R. G. Cruddace, all of the Naval Research Laboratory.

References

1. P. G. Wilkinson and D. W. Angel, *J. Opt. Soc. Am.* **52**, 1120–1122 (1962).
2. W. R. Hunter, R. T. Williams, J. C. Rife, J. P. Kirkland, and M. N. Kabler, *Nucl. Instrum. Methods* **195**, 141–153 (1982).
3. W. K. Warburton, *Nucl. Instrum. Methods* **A291**, 278–285 (1990).
4. E. J. Puik, M. J. Van der Wiel, P. Lambooy, J. Verhoeven, F. E. Christensen, and H. A. Padmore, *J. X-Ray Sci. Technol.* **3**, 19–34 (1991).
5. P. Troussel, D. Schirmann, J. M. Dalmasso, C. K. Malek, H. Berrouane, and R. Barchewitz, *Rev. Sci. Instrum.* **63**, 2125–2131 (1992).
6. H. Berrouane, C. K. Malek, J-M. Andre, F-R. Ladan, J. R. Rivoira, and R. Barchewitz, *Opt. Eng.* **31**, 213–217 (1992).
7. R. G. Cruddace, T. W. Barbee, Jr., J. C. Rife, and W. R. Hunter, *Phys. Scripta* **41**, 396–399 (1990).
8. W. Jark, *Opt. Commun.* **65**, 201–205 (1986).
9. J. C. Rife, T. W. Barbee, Jr., W. R. Hunter, and R. G. Cruddace,*Phys. Scripta* **41**, 418–421 (990).
10. E. J. McGuire, *Phys. Rev.* **A5**, 1043–1047 (1972).
11. J. C. Rife and W. R. Hunter, Naval Research Laboratory, Washington, DC, unpublished data.
12. M. P. Kowalski, J. F. Seely, W. R. Hunter, J. C. Rife, T. W. Barbee, Jr., G. E. Holland, C. N. Boyer, C. M. Brown, and R. G. Cruddace, *Appl. Opt.* **32**, 2422–2425 (1993).
13. D. G. Stearns, R. S. Rosen, and S. P. Vernon, *Proc. SPIE* **1547**, 2–13 (1991).
14. J. F. Seely, M. P. Kowalski, W. R. Hunter, J. C. Rife, T. W. Barbee, Jr., G. E. Holland, C. N. Boyer, and C. M. Brown, *Appl. Opt.* **32**, 4890–4897 (1993).
15. R. A. M. Keski-Kuha, R. J. Thomas, J. S. Gum, and C. E. Condor,*Appl. Opt.* **29**, 4529–4531 (1990).
16. R. J. Thomas, R. A. M. Keski-Kuha, W. M. Neupert, C. E. Condor, and J. S. Gum, *Appl. Opt.* **30**, 2245–2251 (1991).

17. J. F. Seely, R. G. Cruddace, M. P. Kowalski, W. R. Hunter, T. W. Barbee, Jr., J. C. Rife, R, Eby, and K. G. Stolt,*Appl. Opt.* **34**, 7347–7354 (1995).
18. K.-H. Hellwege, *Z. Phys.* **106**, 588–596 (1937).
19. M. P. Kolwalski, T. W. Barbee, Jr., R. G. Cruddace, J. F. Seely, J. C. Rife, and W. R. Hunter, *Appl. Opt.* **34**, 7338–7346 (1995).

19. CRYSTAL OPTICS

Eckhart Förster

X-Ray Optics Group
Institute of Optics and Quantumelectronics
Friedrich-Schiller University Jena
Jena, Germany

19.1 Introduction

The hundred years after the discovery of x-rays by Röntgen in 1896 have seen great progress as described in many reviews and books. Crystal optics is based on the discovery of x-ray diffraction in crystals in the well-known experiment of Max von Laue and his colleagues in 1912. The basic law for x-ray diffraction was formulated by Bragg [1]. The early history of x-ray diffraction has recently been reviewed by Hildebrandt [2]. Applications for optics followed surprisingly quickly; Goby [3] demonstrated that x-rays could be used for microradiography, Rutherford and Andrade [4] used a crystal in transmission mode to measure extinction coefficients, and de Broglie and Lindemann [5] and Rohmann [6] described broadband x-ray spectrometers with thin convexly bent mica crystals.

The next major advance took place when DuMond and Kirkpatrick [7] pointed out that concave crystals could be used to build focusing instruments. One-dimensional focusing spectrographs were built in the pioneering work of Johann [8], Cauchois [9], and Johansson [10].

The third major step took place in the early 1950s, when crystals could be bent in two dimensions. It thus became possible in principle to exploit the short wavelength of x-rays to build microscopes with much better spatial resolution than is obtainable with visible light. However, at that time crystals could not be bent to the accuracy required. Early attempts to build an x-ray microscope have been reviewed by Hildenbrand [11]. The development of imaging crystal optics then ceased because electron microscopy could provide better resolution. However, that technique has some disadvantages for imaging biological material because specimens have to be thinned, and heavy metal stains are needed to enhance the low natural contrast (e.g., Michette [12]).

The current renaissance of interest in crystal spectroscopy started in the 1970s due to the following new developments:

1. New intense x-ray sources
2. Advances in the manufacture of one- and two-dimensionally bent crystals
3. New types of detectors

The great advances in source brightness are shown in Figs. 1a and b. Figure 1a shows how the available peak brilliance has been increased by about 6 orders of magnitude by the advent of second-generation synchrotron light sources and how for instance a further 6 orders will become available from third-generation synchrotrons, e.g., the European Synchrotron Radiation Facility with its upgrades. The peak brilliance of the proposed x-ray free-electron lasers is expected to be 10 orders of magnitude higher still. Plasmas produced by ultrashort laser pulses are also efficient x-ray sources of high peak brilliance and have the advantage that they are much easier to use than synchrotrons or free-electron lasers.

Figure 1b shows how the peak brilliance depends on energy for various sources and how they peak in or close to the x-ray region. As an example of the peak brilliance of an ultrashort laser plasma, asterisk 1 in Fig. 1b shows values calculated from a recent measurement of the absolute photon number [13] and x-ray pulse duration [14] for a small laser system. Asterisk 2 shows the peak brilliance presently obtainable if the experimental conditions were to be optimized [15]. The curve LPP shows the incoherent x-ray emission from a

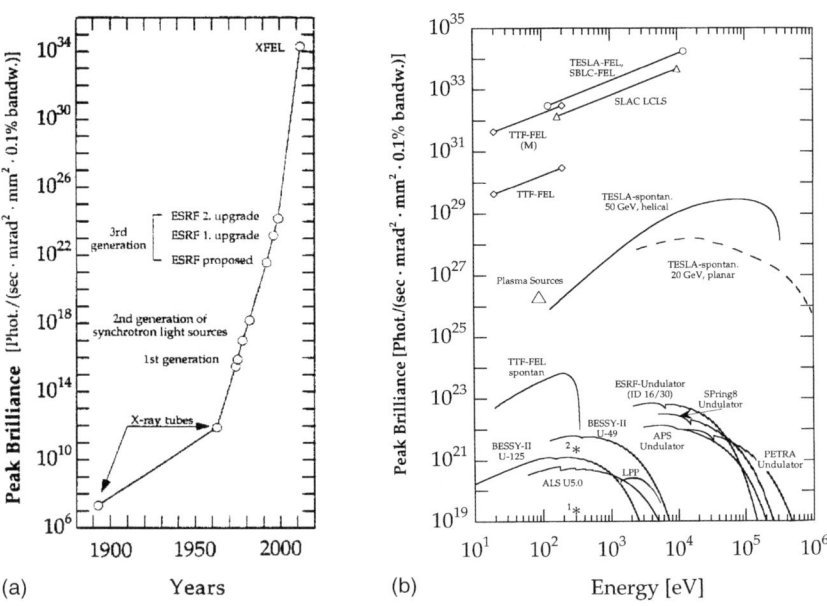

FIG. 1. Brilliance of x-ray sources. (a) Increase of peak brilliance of existing and planned sources since 1900. (b) Dependence of peak brilliance of illustrated x-ray sources with energy. The abbreviations refer to synchrotron and free-electron laser sources. The asterisks 1 and 2, the curve LPP, and the triangle plasma sources refer to laser-produced plasmas as explained in the text. (Courtesy P. Gürtler, U. Teubner.)

laser-produced plasma [16]; the position of the triangle plasma sources shows the brilliance of the amplified stimulated emission from the neon-like selenium laser developed at Lawrence Livermore National Laboratory. New types of sensitive area detectors include CCD cameras, streak and framing cameras, image plates, and multiwire detectors.

The diagnostics of physical parameters such as temperature, density and ionic composition in the newly developed high-temperature plasma sources require z-ray spectroscopy with both spatial and temporal resolution. Typical requirements are spatial resolution of 1 to 10 μm, temporal resolution of 1 to 10 ps, and a spectral resolution of 1000. For less intense sources a trade-off between recorded intensity and resolution may be necessary. Although other focusing methods in the x-ray region are available (namely, grazing incidence techniques, zone plates and capillary optics), they do not fulfill all these requirements. For example, the competing technique of x-ray zone plates (e.g., [12]) can routinely provide a spatial resolution of 20–50 nm but the focal length depends on the wavelength.

Classical x-ray spectroscopy because of its long history has been reviewed many times [17–20] and commercial instruments have been developed. The application of classical x-ray spectroscopy techniques to the new ultrabrilliant sources still faces some challenges but will lead to a new understanding of physical phenomena since for example ultrashort phenomena can be studied in real time.

19.2 Basic Equations

X-ray diffraction is due to constructive interference from radiation scattered from successive crystal planes. The crystal acts as a three-dimensional grating for x-rays where the grating element is the distance between the lattice planes. The basic equation for crystal optics is the Bragg law [1]:

$$m\lambda = 2d_{hkl} \sin \Theta_{hkl}, \tag{1}$$

where

m	is the order of diffraction,
λ	is the wavelength,
d	is the net plane spacing,
h, k, l	are the Miller indices of this plane, and
Θ_{hkl}	the coangle of incidence, i.e., between the net planes and the incident ray.

Because the crystal refractive index is very slightly less than unity, a refraction correction should be included in Eq. (1) for high precision measurements, as discussed by, for example, Compton and Allison [21].

A fundamental difference from a grating is that diffraction takes place from the volume of the crystal, not just on its surface, and this imposes the extra condition that the angles of incidence and reflection to the lattice planes are equal. Note that in the general case the crystal may be cut asymmetrically, that is, the lattice planes make an angle α to the crystal surface. In this case the incident and diffracted rays still make the Bragg angle Θ to the lattice planes, but angles $(\Theta + \alpha)$ or $(\Theta - \alpha)$ to the crystal surface.

The Bragg law shows immediately that the longest wavelength that can be diffracted is $\lambda_{max} = 2d$. Equation (1) should not be understood as a strict delta function. A sharply defined wavelength diffraction occurs over a finite narrow range of angles, or alternatively over a narrow range of wavelengths for a given angle.

The Bragg law can be stated in differential form:

$$\frac{\Delta \lambda}{\lambda} = \frac{\Delta \Theta}{\tan \Theta} = \frac{\Delta x / L}{\tan \Theta}. \tag{2}$$

The first two terms link wavelength and angle coordinates and the third term makes the link with position coordinates. If a beam is diffracted by a flat crystal and then strikes a detector at a distance L from the source, the distance Δx is the projected length on the detector corresponding to an incremental angle.

Real x-ray optical systems may use a number of crystals, which may be curved, cut asymmetrically, or both. The behavior of such systems can be complicated, and diagrammatic methods have been developed to visualize their properties and have proved useful in optimizing experimental design. The earliest method is the wavelength-angle diagram introduced by DuMond [22] where, for example, the radiation passed through a complete optical system can be represented as the intersection of bands representing a beam of radiation and the acceptance window of an optical component. Three-dimensional DuMond diagrams allowing for two perpendicular direction coordinates for the beams have been described by Xu and Li [23] and Xu et al. [24]. A related approach is the matrix optics formalism used elsewhere in optics. This generally uses two coordinates, the position and angle of a ray, but for application to x-ray optics it is helpful to include wavelength as a third coordinate because it is coupled to angle through the Bragg equation. These are the three coordinates used in Eq. (2). Matsushita and Kaminaga [25] and Matsushita and Hashizume [26] have described a matrix formalism where a beam is described in three-dimensional phase space defined by position, direction with respect to the beam axis, and wavelength. The transmission of this beam and the effect of optical components on it are described by transformation matrices acting on points in phase space. This can be combined with three-dimensional bands showing acceptance windows, and the intensity of the transmitted beam can thus be estimated from the volume occupied in phase space.

19.3 Flat Crystal Spectrometers

19.3.1 Fundamental Properties

Flat crystal spectrometers generally contain one or two crystals, depending on the application. The properties of the important double crystal spectrometer have been discussed in detail by Compton and Allison [21], who have given an expression (their Eq. 9.51) for the transmitted intensity. Their expression assumes that the center of the first crystal is set at the exact Bragg angle and the second crystal is offset by an angle β. In the antiparallel configuration the spectral profile $J(\lambda - \lambda_0)$ of a source can be determined by varying β and measuring the total diffracted intensity $I(\beta)$, which is

$$I(\beta) = \int_{-\phi_m}^{\phi_m} \int_{\lambda_{\min}}^{\lambda_{\max}} \int_{-\alpha_m}^{\alpha_m} G(\alpha, \phi) J(\lambda - \lambda_0) C_1 \left[\alpha - \frac{\phi^2}{2} \tan \Theta_0 - (\lambda - \lambda_0) \left(\frac{\partial \Theta}{\partial \lambda} \right)_{\lambda_0} \right]$$
$$\times C_2 \left[\beta - \alpha - \frac{\phi^2}{2} \tan \Theta_0 - (\lambda - \lambda_0) \left(\frac{\partial \Theta}{\partial \lambda} \right)_{\lambda_0} \right] d\alpha \, d\lambda \, d\phi, \qquad (3)$$

where $G(\alpha, \phi)$ describes the source intensity profile in terms of the divergence α in the dispersion plane and ϕ perpendicular to it, and $C_{1,2}[\cdots]$ are the reflection curves for crystals 1 and 2.

Equation (3) for the intensity can be written in terms of wavelength by using Eq. (2). The transmitted profile is wider than the incident profile $J(\lambda - \lambda_0)$ because the latter is convolved with the divergence function $G(\alpha, \phi)$ and the reflection curves $C_{1,2}$. To minimize the broadening introduced by the convolution the divergence and crystal reflection functions should be made as narrow as possible. Recently Härtwig et al. ([27], Eq. 1) have given a more general expression for the total diffracted intensity, which includes in addition the position dependence of source intensity and detector efficiency. These authors discuss each term of the expression in detail and how measurements of line profiles can be optimized. The corresponding expression for a single crystal was previously given by Härtwig and Grosswig ([28], Eq. 1).

Reflection curve C shows the variation of the relative reflected intensity with the angle incidence of an ideally monochromatic plane x-ray beam. The main parameters of reflection curves are the peak reflectivity $C_{\max}$, the FWHM $\Delta\Theta_c$ and the integrated reflectivity R_{int}, defined as the integral of the reflectivity C over all angles of incidence. There are two limiting theoretical approximations. One limit is the dynamical theory of x-ray interference, which assumes a perfect crystal, that is, where the structure is perfect over a range of millimeters such as in freshly cleaved natural calcite crystals or artificially grown silicon crystals. A detailed theoretical treatment was first given by Darwin [29] and independently by Ewald [30], where absorption in the crystal was ignored. The theory was extended to include absorption by Prins [31]. It has been discussed in the books of Zachariasen [32] and von Laue [33]; modern treatments have been given by James [34] and Batterman and Cole [35], and a comprehensive review given in the book by Pinsker [36]. Examples of reflection curves in dynamic theory are given in Figs. 2a and b. Figure 2a shows the theoretical reflection curves with and without absorption for the 111 reflection from silicon for a wavelength of 0.154 nm. This is well below the K absorption edge at 0.6742 nm. The curve for no absorption has a maximum of unity as expected theoretically. Figure 2b shows the corresponding curve including absorption for 0.6 nm, much nearer but still below the K edge. The absorption is now much stronger and the reflectivity reduced. The opposite limiting approximation is the kinematic theory of x-ray interference or "mosaic" model proposed by Darwin [37] for very imperfect crystals. Here the crystal is assumed to consist of a mosaic of independently diffracting domains. The dimensions of each domain are much smaller than the x-ray extinction length in the crystal, and the degree of misorientation between the domains is much greater than the reflection curve width. Both theories have been usefully summarized by Burek [38] who explains how the predicted reflection properties depend on the model used, gives theoretical expressions, and discusses the further assumptions needed to measure reflection curves experimentally. This article is a valuable source of information for the lattice spacing, reflectivities, and resolving powers of many crystals used in practical work.

In classical x-ray spectroscopy where the crystal is rocked through the reflection range the source spectral profile $J(\lambda - \lambda_0)$ can be obtained by deconvolution of Eq. (3). In most cases the resolution is degraded by source size effects, but these can be reduced if the dispersion of the spectrometer is maximized by a proper choice of crystal, or if a bent crystal is used. For x-ray spectroscopy of the new flash sources it is obviously not practical to rock the crystal in the nanosecond or even picosecond timescale of the source, and spectra are normally recorded by area detectors.

19.3.2 Choice of Crystal

A wide range of crystals is now available. For a given application the basic condition is that the double lattice spacing $2d$ must exceed the wavelength of interest. The wavelength should preferably be not much less than $2d$ because then the Bragg angle is large and the spectral dispersion $\Delta\Theta/\Delta\lambda$ is also large. For quantitative work the crystal should not contain elements with absorption edges in the wavelength range of interest and hence not suffer from abrupt changes in reflectivity there.

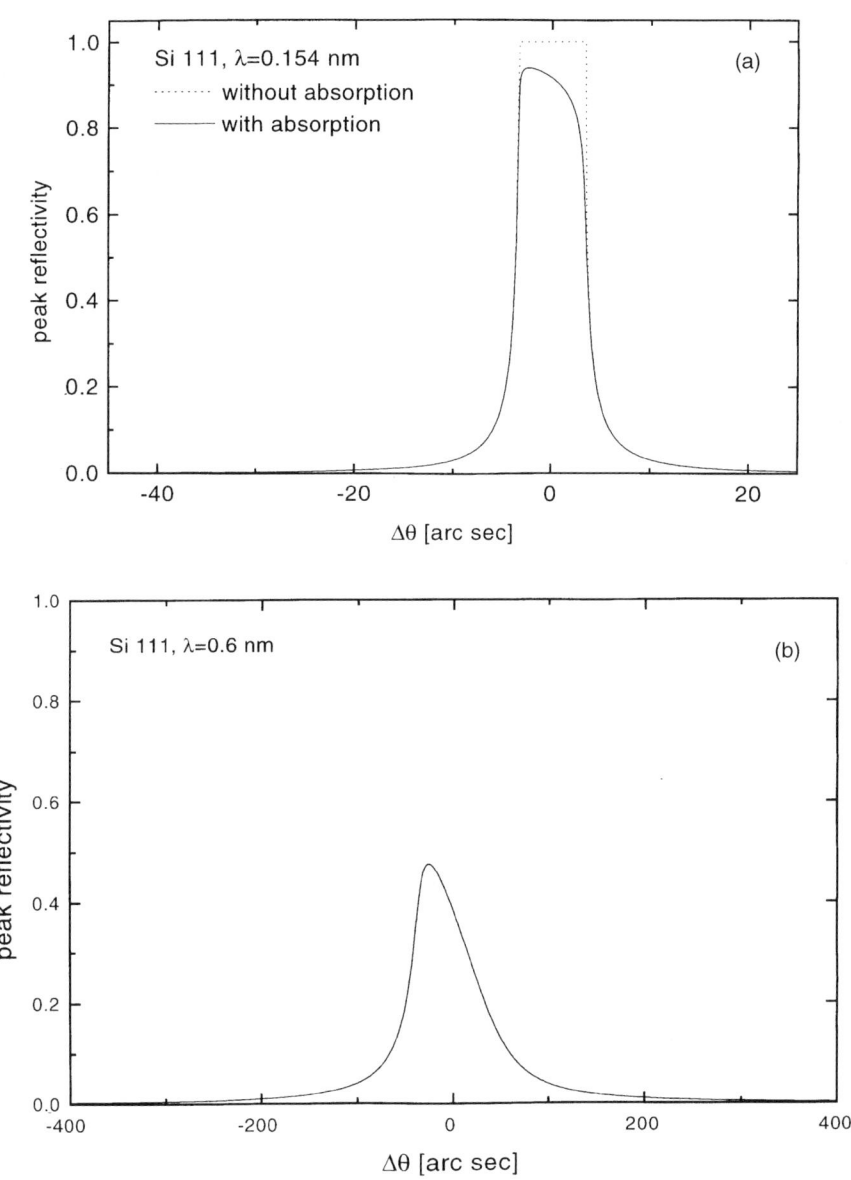

FIG. 2. Reflection curve for silicon (reflection 111) from flat crystal at wavelengths of (a) 0.154 nm and (b) 0.6 nm. $\Delta\theta$ is the difference from the Bragg angle, corrected for refraction. The dotted curve in part a shows the case where absorption is neglected. (Courtesy G. Hölzer.)

Acid phthalate crystals with $2d$ values of about 2.6 to 2.7 nm have been available for many years and x-ray crystal optics is now being extended to longer wavelengths. An important advance is the development of the organic crystal OHM (octadecyl hydrogen maleate) with $2d = 6.35$ nm. This has been used by Fan *et al.* [39] to record the spectrum of a laser-produced plasma in the range of 4.3–5.1 nm with a spectral resolution of 1100. Alternative approaches are Langmuir–Blodgett multilayer films [40], and sputtered or evaporated multilayer mirrors, which are physically more stable and provide high reflectivity [41, 42]. However, the comparative study carried out by Barnsley *et al.* [43] shows that the spectral resolution for multilayers is more than an order of magnitude lower than for an OHM crystal.

19.3.3 Special Spectrometers

Special configurations of flat crystals can be used for absolute wavelength measurements for example by Fraenkel [44] who exploited the phenomenon of double reflection first discovered by Renninger [45]. Monoliths where x-rays are reflected from two accurately parallel faces with known separation [46, 47] can be used over a much wider wavelength range. Two sets of spectra are recorded on a detector and if their separations can be measured, absolute wavelengths can be determined from the monolith dimensions and the $2d$ spacing of the material. This method can achieve an overall wavelength uncertainty as low as 10^{-5}. A further advantage is that the lattice spacing can be calibrated by a second crystal and the lattice spacing of the latter can be linked to basic length standards to an accuracy of 10^{-8} by a combination of optical and x-ray interferometry [48–50].

He *et al.* [51] have applied a double crystal spectrometer originally described by Hrdy [52] to the diagnostics of laser-produced plasmas. This provides excellent spatial resolution in the dispersion plane, limited only by the crystal rocking curve width, and excellent spectral resolution perpendicular to this plane. The use of two crystals reduces the image intensity but this need not be a problem with an intense source.

19.4 Bent Crystal Spectrometers

Crystals bent in one dimension are commonly used in advanced x-ray spectroscopy instruments. Although the focusing properties of concave crystals are analogous to those of concave gratings there is the extra condition that the angles of incidence and reflection are equal. Johann [8] described a spectrometer with a concave cylindrical crystal. Rays from an extended source, or a point source on the Rowland circle, are approximately refocused on this circle. The focusing may be made exact in the Johansson configuration [10], but this is difficult to prepare accurately because although the crystal is bent to a radius R

the surface is cut to a radius $R/2$, that is, it cuts into the lattice planes. The convex cylindrical crystal configuration is easy to build but is defocusing and hence cannot be used for imaging; moreover line intensities in the diffracted spectrum are weak. It does, however, have a wide spectral range and so has been used for line identification work with intense sources. Goetz et al. [53] have compared the flat, concave, and convex crystals in terms of the ratio $\Delta\alpha/\Delta\Theta$ where $\Delta\alpha$ is the change in the divergence angle of an incident ray for a change $\Delta\Theta$ in the Bragg angle. As discussed by Henke et al. [54] the intensity of a spectral line recorded by a fixed single-crystal analyzer is proportional to this ratio. DuMond [55] described a concave crystal spectrometer, which forms an extended image from a point source. In the von Hamos configuration [56] the crystal is cylindrically bent and the source and image planes lie on the cylinder axis. This produces a line focus along the axis. The main features of these configurations have been reviewed by [19] among others.

Two-dimensional focusing clearly requires bending the crystal in two dimensions. The simplest such crystal form is the spherical. For imaging purposes this has the major disadvantage that the crystal must be used away from normal incidence to satisfy the Bragg law and this introduces astigmatism, often severe. This may be seen from the basic lens equation, analogous to the Coddington equations already established in 1829 for refraction [57]:

$$1/u + 1/v = 2/(R \sin \Theta) \quad \text{in the meridional plane, and}$$
$$1/u + 1/v = (2 \sin \Theta)/R \quad \text{in the sagittal plane.}$$

Thus the focal length $R/2$ in the sagittal plane is longer than that in the meridional plane by a factor $1/\sin^2 \Theta$. This problem can be solved with a toroidally bent crystal where the sagittal radius of curvature R_s and the meridional radius R_m are related by $R_s = R_m \sin^2 \Theta$, so that the focal lengths in the two directions are equal. However, it is difficult to bend the crystal to the precise form required and two-dimensional focusing can only be achieved at the one wavelength satisfying this condition. Ellipsoidal crystals have the property that if a source is placed at one of the confocal points, radiation of all wavelengths is focused at the other but the focusing degrades away from the confocal points. An important advantage of ellipsoidal crystals is that the pathlength between the foci is constant for all rays so that ultrashort pulses are not temporally stretched by differing pathlengths outside the crystal. A detailed consideration of time-dependent x-ray diffraction shows that no further pulse stretching is introduced by the section of the optical path within the crystal [58].

Bending a crystal distorts the crystal and if the distortion is small the reflection properties can be calculated by applying the dynamical theory to slightly deformed crystals as described in [59, 60]. Figures 3a and b show theoretical reflection curves corresponding to the flat crystal curves in Figs. 2a and b, but here for a crystal bent to a radius of curvature of $R = 1.0$ m. In

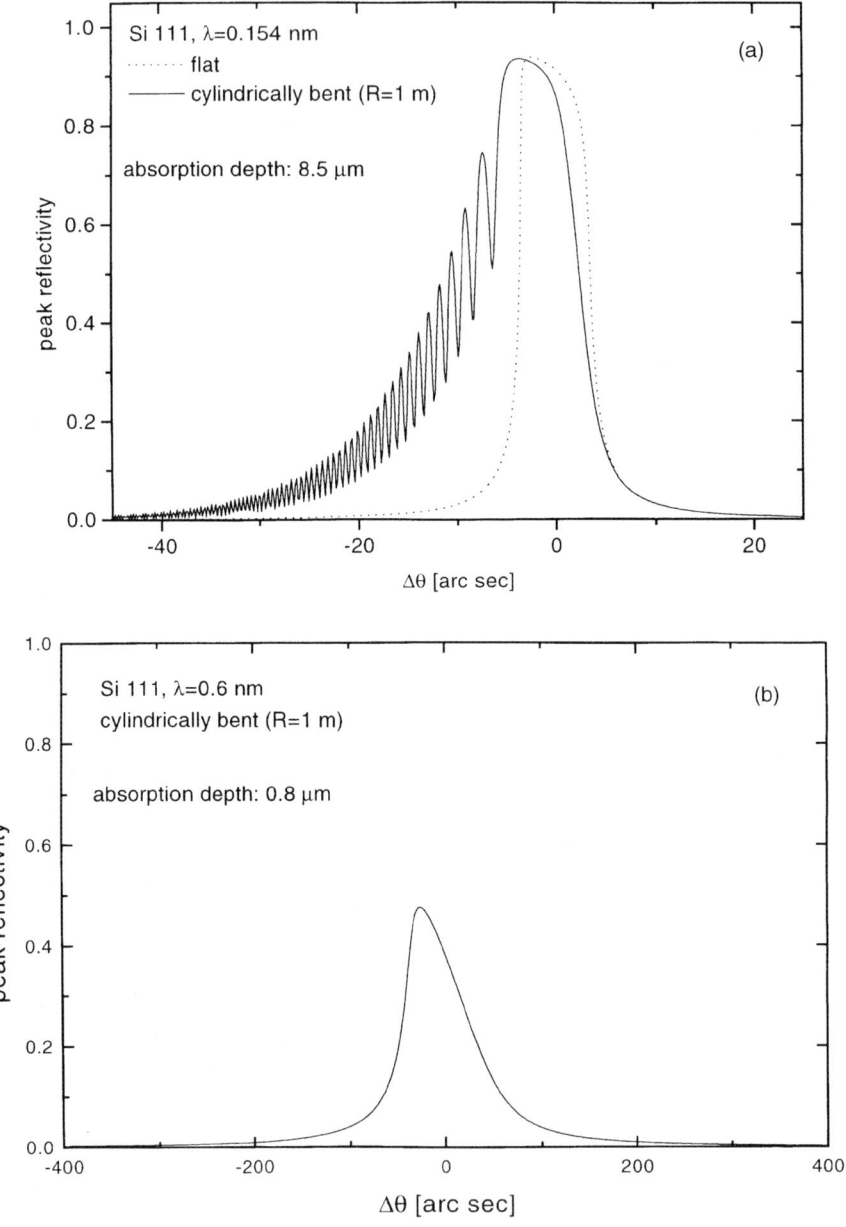

FIG. 3. (a) Reflection curve for a silicon (reflection 111) crystal bent to a radius of curvature of 1.0 m. Also includes for comparison a dotted curve showing the corresponding curve from Fig. 2a for a flat crystal. (b) This reflection curve is almost identical to that in Fig. 2b because the absorption depth is only 0.85 μm and absorption is much stronger than extinction. (Courtesy G. Hölzer.)

general, distortions broaden the reflection curve, make it asymmetric, and introduce a tail with oscillations [61, 62] as can be seen from Fig. 3a. The integrated reflectivity is also increased.

19.4.1 System Design

The diagrammatic phase volume method introduced by Matsushita and Hashizume may be useful to visualize the general features of more complex systems. In a detailed design study, parameters such as image intensity, spatial and spectral resolution, and aberrations should be considered. Chukhovskii et al. [63] describe a rigorous solution of the Maxwell equation with Fresnel diffraction but the amount of computation required is prohibitive. A quick impression can be obtained with adequate accuracy by ray-tracing methods [64, 65]. These consider sources as a point or grid of points, each emitting a fan of rays, and then trace each ray through an x-ray optics system to the detector plane.

It is still an art to bend a crystal to the exact form required and obtain the expected reflectivity and imaging properties. The most frequently applied techniques among others are four-bar bending of a rectangular crystalline plate and bending the tip of a triangular plate. Higher precision can be achieved if the thin crystal waver is glued on to a high-precision glass former [66]. The perfection of crystal bending should always be checked by x-ray topography and diffractometry methods.

High-resolution x-ray spectroscopy now has many important and diverse applications, which cannot all be reviewed here. This technique is essential to understand the space and time evolution of the high-temperature plasmas in research into inertial confinement fusion (e.g., [67]), x-ray laser, and the interaction of ultrashort, intense laser pulses with matter (e.g., [68]). An important applied analytical technique is EXAFS (x-ray absorption fine structure) [69]. Scanning bent crystal monochromators designed for EXAFS studies have been described by Marcus et al. [70], Cohen and Deslattes [71], and Khalid et al. [72]. X-ray imaging techniques can be applied to astronomy as discussed by Schnopper [73]; Culhane et al. [74] used a convexly bent crystal spectrometer on the Japanes YOHKOH satellite to study solar x-ray radiation. Crystal optics can be expected to be widely applied in the future, for example, in the New Ignition Facility, the proposed American large laser facilities for inertial fusion research.

References

1. W. L. Bragg, *Proc. Cambridge Philos. Soc.* **17**, 43 (1912).
2. G. Hildebrandt, *Cryst. Res. Technol.* **28**, 747 (1993).
3. P. Goby, *C. R. Acad. Sci. (Paris)* **156**, 686 (1913).
4. J. Rutherford and E. N. da C. Andrade, *Phil. Mag.* **27**, 854 (1914).

5. M. de Broglie and F. A. Lindemann, *C. R. Acad. Sci. (Paris)* **158**, 944 (1914).
6. H. Rohmann, *Z. Phys.* **15**, 510 (1914).
7. J. W. M. DuMond and H. A. Kirkpatrick, *Rev. Sci. Instrum.* **1**, 88 (1930).
8. H. H. Johann, *Z. Phys.* **69**, 185 (1931).
9. Y. Cauchois, *C. R. Acad. Sci. (Paris)* **194**, 362 (1932).
10. T. Johansson, *Z. Phys.* **82**, 507 (1933).
11. G. Hildebrand, *Forschr. d. Phys.* **4**, 1 (1956).
12. A. G. Michette, *Rep. Prog. Phys.* **51**, 1525 (1988).
13. T. Wilhein, R. Häßner, D. Altenbernd, U. Teubner, W. Theobald, E. Förster, and R. Sauerbrey, *J. Opt. Soc. Am. B* **15** (March 1998).
14. C. Wülker, W. Theobald, F. P. Schäfer, and J. S. Bakos, *Phys. Rev. E.* **50**, 4920 (1994).
15. U. Teubner, W. Theobald, and C. Wülker, *J. Phys. B* **29**, 4333 (1996).
16. G. M. Shimkaveg, K. M. Gäbel, S. E. Grantham, R. E. Hannon, Jr., and M. C. Richardson, *Proc. SPIE* **2523**, 299 (1995).
17. A. E. Sandström, "Handbuch der Physik," Vol. 30 (S. Flügge, ed.), Springer-Verlag, Berlin, pp. 78–245 (1957).
18. M. A. Blokhin, "Methods of X-Ray Spectroscopic Research," Chap. 4, Pergamon, New York (1965).
19. E. P. Bertin, "Principles and Practice of X-Ray Spectrometric Analysis," 2nd ed., Plenum, New York (1975).
20. B. K. Agarwal, "X-Ray Spectroscopy," 2nd ed., Springer-Verlag, Berlin, p. 114 (1989).
21. A. H. Compton and S. K. Allison, "X-Rays in Theory and Experiment," 2nd ed., Van Nostrand, Princeton, NJ (1935).
22. J. W. M. DuMond, *Phys. Rev.* **52**, 872 (1937).
23. S. Xu and R. Li, *J. Appl. Cryst.* **21**, 213 (1988).
24. S. Xu, J. Chen, and R. Li, *J. Appl. Cryst.* **21**, 218 (1988).
25. T. Matsushita and U. Kaminaga, *J. Appl. Cryst.* **13**, 465 and 472 (1980).
26. T. Matsushita and H. Hashizume, in "Handbook on Synchrotron Radiation," Vol. 1 (E. E. Koch, ed.), North-Holland, Amsterdam, pp. 261–314 (1983).
27. J. Härtwig, G. Hölzer, J. Wolf, and E. Förster, *J. Appl. Cryst.* **26**, 539 (1993).
28. J. Härtwig and S. Grosswig, *Phys. Status Solidi A* **115**, 369 (1989).
29. C. G. Darwin, *Phil. Mag.* **27**, 315 (1914); C. G. Darwin, *Phil. Mag.* **27**, 675 (1914).
30. P. P. Ewald, *Ann. Physik* **54**, 519 (1917).
31. J. A. Prins, *Z. Phys.* **47**, 479 (1928).
32. W. H. Zachariasen, "Theory of X-Ray Diffraction in Crystals," Wiley, New York, p. 168 (1945).
33. M. von Laue, "Röntgenstrahl-Interferenzen," 3rd ed., Akad. verl. Ges., Frankfurt/Main (1960).
34. R. W. James, *Solid State Phys.* **15**, 53 (1963).
35. B. W. Batterman and H. Cole, *Rev. Mod. Phys.* **36**, 681 (1964).
36. Z. G. Pinsker, "Dynamical Scattering of X-Rays in Crystals," Springer-Verlag, Berlin (1978).
37. C. G. Darwin, *Phil. Mag.* **43**, 800 (1922).
38. A. Burek, *Space Sci. Instrum.* **2**, 53 (1976).
39. P. Z. Fan, E. E. Fill, and G. Tietang, *Rev. Sci. Instrum.* **67**, 786 (1996).
40. B. L. Henke, *Adv. X-Ray Analysis* **7**, 460 (1964).
41. T. C. Huang, A. Fung, and R. L. White, *X-Ray Spectrom.* **18**, 53 (1989).

42. W. Moos, A. P. Zwicker, S. P. Regan, and M. Finkenthal, *Rev. Sci. Instrum.* **61**, 2733 (1990).
43. R. Barnsley, S. N. Lea, A. Patel, and N. J. Peacock, in "Tenth International Colloquium on UV and X-Ray Spectroscopy of Astrophysical and Laboratory Plasmas," Cambridge University Press (1992).
44. B. S. Fraenkel, *J. X-Ray Sci. Technol.* **5**, 341 (1995).
45. M. Renninger, *Z. Phys.* **106**, 141 (1937).
46. A. V. Rode, A. M. Maksimchuk, G. V. Sklizkov, A. Ridgeley, C. Danson, N. Rizvi, E. Förster, K. Goetz, and I. Uschmann, *J. X-Ray Sci. Techol.* **2**, 149 (1990).
47. D. Klöpfel, G. Hölzer, E. Förster, and P. Beiersdörfer, *Rev. Sci. Instrum.* **68**, 3669 (1997).
48. R. D. Deslattes, E. G. Kessler, W. C. Sauder, and A. Henins, *Ann. Phys.* **129**, 378 (1980).
49. R. D. Deslattes, *Nucl. Instrum. Methods Phys. Res.* **B31**, 51 (1988).
50. P. Becker, K. Dorenwendt, G. Ebeling, R. Lauer, W. Lucas, R. Probst, H. J. Rademacher, G. Reim, P. Seyfried, and H. Sigert, *Phys. Rev. Lett.* **46**, 1540 (1981).
51. H. He, J. S. Wark, E. Förster, I. Uschmann, O. Renner, M. Kopecky, and W. Blyth, *Rev. Sci. Instrum.* **64**, 26 (1993).
52. J. Hrdy, *Czech J. Phys.* **B18**, 532 (1968).
53. K. Goetz, Yu. A. Mikhailov, S. A. Pikuz, G. V. Sklizkov, A. Ya. Faenov, S. I. Fedotov, E. Förster, and P. Zaumseil, *Instr. Exp. Techniques* **21**, 771 (1978). (Translated from the original Russian by the Instrument Society of America. In the English version the names of the first and last two authors have been transliterated as Getts, Ferster, and Tsaumzail.)
54. B. L. Henke, H. T. Yamada, and T. J. Tanaka, *Rev. Sci. Instrum.* **54**, 1311 (1983).
55. J. W. M. DuMond, *Rev. Sci. Instrum.* **18**, 626 (1947).
56. L. von Hamos, *Am. Mineral* **23**, 215 (1938).
57. R. Kingslake, *Optics and Photonics News* 20 (Aug. 1994).
58. J. S. Wark and H. He, *Laser Part. Beams* **12**, 507 (1994).
59. S. Takagi, *Acta Crystallogr.* **15**, 1311 (1962).
60. D. Taupin, Ph.D. Thesis, University of Paris (1964).
61. J. E. White, *J. Appl. Phys.* **21**, 855 (1950).
62. D. W. Berreman, *Phys. Rev. B* **14**, 4313 (1976).
63. F. N. Chukhovskii, W. Z. Chang, and E. Förster, *J. Appl. Phys.* **77**, 1843 and 1849 (1995).
64. S. Morita, *Jpn. J. Appl. Phys.* **22**, 1030 (1984).
65. M. Dirksmöller, O. Rancu, I. Uschmann, P. Renaudin, C. Chenais-Popovics, J.-C. Gauthier, and E. Förster, *Opt. Commun.* **118**, 379 (1995).
66. E. Förster, K. Gäbel, and I. Uschmann, *Rev. Sci. Instrum.* **63**, 5012 (1992).
67. I. Uschmann, E. Förster, H. Nishimura, K. Fujita, Y. Kato, and S. Nakai, *Rev. Sci. Instrum.* **66**, 734 (1995).
68. P. Gibbon and E. Förster, *Plas. Phys. Control. Fusion* **38**, 769 (1996).
69. P. Eisenberger and B. M. Kincaid, *Science* **200**, 1441 (1978).
70. M. Marcus, L. S. Powers, A. R. Storm, B. M. Kincaid, and B. Chance, *Rev. Sci. Instrum.* **51**, 1023 (1980).
71. G. G. Cohen and R. D. Deslattes, *Nucl. Instrum. Methods* **193**, 33 (1982).
72. S. Khalid, R. Emrich, R. Dujari, J. Schultz, and J. R. Katzer, *Rev. Sci. Instrum.* **53**, 22 (1982).
73. H. W. Schnopper, *Appl. Opt.* **20**, 1089 (1981).
74. J. L. Culhane *et al.*, *Solar Phys.* **136**, 89 (1991).

INDEX

A

Abbe sine condition, 155–156
Aberration
  definition, 151
  nonsymmetrical systems, 163–166
  primary, 154
  reduction, 172–173
  representation, 172–173
  secondary, 158
  spherical, 170
  third order, 361–364
  zone plate, 296–298
Aberration correction, 30
  circular mirror, 170–172
  See also Coma correction; Astigmatism correction
Absolute flux, laser plasma, 85–86
Absorption coefficient, calculation, 258
Accelerator, characteristics, 2
Aluminum, reflectance spectra, 206–207
Aluminum oxide, reflectance spectra, 222
Amplitude gratings, 397
Ancillary components, beamline design, 35–36
Aperture stops, 150–151
Application
  laser plasma XUV source, 88–89
  wall stabilized arc, 46
  x-ray laser, 107–109
Astigmatism
  definition, 156
  Rowland circle, 367
  singlehanded mirror, 160–161
Astigmatism correction
  beamline design, 28
  glancing incidence spherical mirror, 162–163
Astigmatism curvature, 367
ASTRID storage ring
  beamline design, 30–32
  efficiency, 32–35
Asymmetrical optical system, 148–149
  aberration, 163–164

Atomic scattering factor, 260
Axisymmetrical optical system aberration, 152–153

B

Beam divergence, monochromator, 183
Beamline characteristics, 27
Beamline design
  ancillary components, 35–36
  ASTRID storage ring, 30–32
  efficiency, optimization, 35
  process, 27–30
  radiation protection, 36
Beamline development, 27
Bending magnet
  brightness, 11–12
  power, 10
  radiation, 3–7
Bent crystal spectrometers, 408–411
Bent mirror, spectroscopic application, 179–180
Berlin electron storage ring (BESSY), 71
  conditions for comparison standard, 130
  primary source standard, 125
Beryllium
  reflectance spectra, 208–209
  transition radiation, 94–95
Bessel functions, 5–6
Blackbody radiators, 120–121
Blaze grating, 370
  efficiency, 386–394
Boltzmann equation, 50
Boltzmann plot, 52
Boltzmann population ratio, XUV laser, 102
Boundary imperfections, 275
Bragg-Fresnel lens, 296
Bragg law
  crystal optics, 403–404
  multilayer gratings, 380
Brehmsstrahlung, 36
Brightness calculation

415

bending magnet, 11–12
undulators, 17
wigglers, 14

C

Calibration
  electron-beam excitation, 78–79
  extreme UV, 133
  laser plasma source, 86
  plasma source, 130
  secondary source standard, 127
  soft x-ray, 129–131
  spectrometers, 71–72
  ultraviolet, 138–143
  vacuum UV, 138–143
  See also Source standard
Capillary discharge, 59–60
Carbon, reflectance spectra, 223–224
Cathode erosion
  electron beam excitation, 76
  hollow cathode, 69–71
  Penning discharge, 76
Caustic curve, 166–167
Charged particle oscillators, 38
Chromium, reflectance spectra, 218
Circular polarization, 7–10
  piezo-optical effect, 252
  by reflection, 246–251
  by transmission, 252–253
Coddington equation, spherical reflector, 159
Cold cathode glow discharge, 39
Coma
  definition, 155, 168
  non-symmetrical systems, 168–170
Coma correction
  circular mirror, 171–172
  toroidal optics, 173–175
Complex index of refraction, 227
Constricted glow discharge, 41
Copper, reflectance spectra, 211–212
Coronal diagnostics spectrometer (CDS), 72
  calibration, 135, 138
Critical angle, 263
  interference filters, 338
  mirror (CAM), 338–340
Crystal optics, history, 401–402
Curvature of field, 156–158
Curvature of field correction, 172

Cutoff wavelength, 305
  temperature effect, 317–320

D

Debye-Waller factor, 276
Design
  extreme ultraviolet spectroscopy, 146–147
  hollow cathode, 65–70
  Penning discharge, 74
  wall stabilized arc, 46–48
Design, geometric
  dielectric barrier discharge, 44
Dichroic filter, 313
Dichroic polarizer, 232
Dielectric barrier discharge, 41–45
Dielectric constant, calculation, 258
Dielectric filters, 305–314
Diffraction grating, 357–358, 370–375, 380
  astigmatism, 367–368
  beamline design, 28–30
  dispersion, 364–365
  efficiency, 370–375
  equation, 357–358
  focal curves, 365–367
  history, 347–348
  measure of efficiency, 194
  ray tracing, 358–361
  resolving power, 368–370
Diffractive focusing with a zone plate, 293–295
Dipole pattern, 2–3
Dispersion, 257–259, 364–365
Distortion, 153, 170, 172
Drude model of dispersion, 257

E

Electrical circuit, dielectric barrier discharge, 42
Electron beam excitation, 76–80
Electron-beam trap (EBIT), 54–56
Electron bunching, 1, 25
  description, 124
Electron cyclotron resonance (ECR), 52–54
Electron impact ionization, 40
Ellipsometry, 234
Emission spectra

laser plasma source, 86–87
Enveloping curvature, 367–368
Extreme Ultraviolet (EUV)
  design, 146–147
  instrumentation, 133–135, 138
  mirror optics, 145

F

Fermat's principle, 355–357
Field stops, 150–151
Flat crystal spectrometers, 405–408
Flux, 6, 60
  absolute, in laser plasma, 85–86
  brightness ratio, 21
  transversely coherent, 21
Flux calculation, 6
  electron-beam excitation, 77
Focal curves, 365–367
Four mirror circular polarizer, 251
Four mirror polarizer, 239–242
  adjustable, 241–242
Fourth-generation sources, 24–25
Fresnel equation, 265
Fresnel formulas, 273–274
Fresnel lens, 289
Fresnel reflection coefficients, 280
Fresnel rhomb, 247–248

G

Gain coefficients, saturated laser, 104
Gases as transmitting media, 326–329
Geometrical optics, 150–158
Glow discharges, 37–41
Gold, reflectance spectra, 209–210
Goniometer. See reflectometer
Graded period multilayer filters, 335–338

H

High-order harmonic generation, 109–115
  disagreement with lowest order perturbation theory, 110–111
  features from rare gases, 109–110
  theory, 111–113
  vs. x-ray lasers, 115

VUV beam properties, 113–115
High reflectivity mirrors, design, 279–281
Hollow cathode, 65–72
  low current, 66
Holographic grating, aspheric wavefront, 352–353
Holographic grating, spherical wavefront, 350–352
Homocentric pencil of rays, 151–152
Hydrogen plasma, continuum emission, 121–122

I

Inductively coupled plasma (ICP), 57–59
Insertion devices, 12–14, 20–21
Instrumentation, radiometric characterization, 131–132
Interference polarizer, 242–243
Ionization states, laser produced plasmas, 84
  laser produced plasmas, 84
Iridium, reflectance spectra, 213–214

K

Kirkpatrick-Baez mirror system, 162–163, 179
Kramers-Kronig relation, 268

L

Laminar grating, 394–398
Laser/matter interaction, 84–85
Laser plasma, 83–90
Law of reflection, 147
Linear polarization, 228–229
Linear polarization analyzer, 245–246
Liquid droplet target sources, 88
Lithography, 353–354
Lowest order perturbation theory, 112
  disagreement in high-order harmonic generation, 110–111

M

Magnetic circular dichroism (MCD), 7–10, 253
Mirror fabrication, 175–179

418 INDEX

Mirror of circular cross section, aberration, 164
Mirror optics, 145
Molybdenum, reflectance spectra, 218
Monochromator layout, 30
Multilayer
   construction, 285–286
   performance, 285–286
Multilayer coatings, 271–272
Multilayer filters, 282–283, 330–338
Multilayer gratings, 379
   design, 380–385
Multilayer polarizer, 244–245, 284–285
Multilayer zone plate, 296
Multiparticle coherent emission, 2
Multiplate resonant reflector (MRR), 340–343

N

Nickel, reflectance spectra, 217–218

O

Obliquity of field
   circular mirror, 170
   non-symmetrical system, 164–165
Optical constants
   calculation, 260
   determination, 267–269
Organic/metal film filters, 324–326
Osmium, reflectance spectra, 214–215
Ozone production, 41

P

Paraxial optics, 148–150
Penning discharge, 72–76
Perturbation calculation, 112
Phase difference, 248–250
Phase retarder, 284–285
Phase-space area, 21, 23–24
Piezo optical effect, 252
Pile-of-plates polarizer, 229–230
Pinholes in thin metal films, 323–324

Planck radiation law, 49
Plane diffraction grating, 29–30
Plane mirror
   extreme ultraviolet beam deflection, 161
   x-ray beam deflection, 161
Platinum, reflectance spectra, 212–213
Polarization, 227
   characteristics of toroidal grating
     monochromator (TGM), 135
   circular, 7–10
   circular by reflection, 246–251
   circular by transmission, 252–253
   insertion devices, 20–21
   linear, 228–229
   maximum, 236
   mirror/spectrometer combination, 132
   secondary source standard, 128–129
   synchotron radiation, 123
   ultraviolet source, 141
   vacuum ultraviolet source, 141
Polarizer
   dichroic, 232
   four mirror, 239–242
   interference, 242–243
   multilayer, 244–245
   pile-of-plates, 229–230
   pseudo-Brewster angle, 244–245
   reflecting, 232–242
   Rochon prism, 229
   thin film, 230–231
   three mirror, 239–242
   transmitting, 229–232
   Wollaston prism, 229
Population inversion
   collisional scheme, 103
   recombination scheme, 102–103
   XUV lasers, 101–103
Power
   angle dependence, 20
   insertion device, 18
Pressure requirement, 65
   limitations, 65
   Penning discharge, 73
Primary source standards, 122–125
Principal ray, definition, 151
Principle surface, definition, 156
Pseudo-Brewster angle polarizers, 244–245
Pulsed lasers, 101
Pumping systems, 65
Pupils, 150–151

Q

Quartz, transmission, 310–312

R

Radiation damage to dielectric material, 314–317
Radiation pattern. See Dipole pattern
Radiation protection, beamline design, 36
Radiometric output, stabilization, 48–49
Radiometric properties, determining, 119
Rare gas
  high order harmonic, 109
  spectral distribution, 43
Ray aberrations, 151–152
Ray tracing, 32–35
  aberration reduction, 172–173
  through a grating, 358
Reflectance
  diffuse, 200–203
  specular, 184
Reflectance spectra, 205–206
Reflecting filters on substrates, 329–338
Reflecting interference filter, 340
Reflecting polarizer, 232–242
Reflection coefficient, 265
Reflection grating, efficiency, 370
Reflection law, 147
Reflectivity, multilayer coatings, 271–272, 276
Reflectometer
  definition, 183
  design, 187–200
  different angle of incidence, 192–193
  fixed angle of incidence, 184–185
  high vacuum, 193–195
  instrumentation, 191–192
  mechanical deviation error, 188–190
  mechanical displacement error, 187–188
  oblique angle of incidence, 185–187
  systematic errors, 187–191
  ultrahigh vacuum, 195–200
Reflectometry facilities, 203
Resolution, 282–283
Resolving power, 368–370
Rhodium, reflectance spectra, 215–216
Rms divergence, calculation, 6–7
Rochon prism polarizer, 229
Rowland circle, 28

definition, 366
monochromator, 131
non-symmetrical system, 164–166

S

Sagittal ray, definition, 159
Saha equation, 50
Saturated lasers, 104–106
Scattering, 259–260
  boundary roughness, 276–278
Schrodinger equation, high-order harmonic generation, 112
Secondary source standard, 125–127
  calibration, 127–129
Silicon, reflectance spectra, 207–208
Silicon carbide, reflectance spectra, 224
Silicon dioxide, reflectance spectra, 222
Silver
  reflectance spectra, 210–211
  transition radiation, 93–94
Snell's law, 273
Soft x-ray, source comparison, 129–133
Solar Ultraviolet Measurements of Emitted Radiation (SUMER), 72
  calibration, 135, 138
Source standard, primary vs. secondary, 126–127
Spectral line intensity, 52
Spectral sensitivity calibration, 71–72
SPECTROSIL, 309, 326
Specular reflectance, definition, 184
Spherical aberration, 154
  circular mirror, 170–171
  term derivation, 166–170
Spherical aberration correction, 173–175
Spherical grating monochromator (SGM), 28, 29
Spot diagram formulas, 363
Stops, 150–151
Storage ring
  development, 2
  as source standard, 122–125
Stray light, reflectance, 191
Supermirrors, 283–284
SUPRASIL, 309
Synchrotron radiation, 2–3
  components, 123

definition, 1
emission, 2–3
as primary source standard, 122–125
sources, 1

T

Tangential meridional plane, definition, 159
Tangent ray, definition, 159
Tantalum, reflectance spectra, 219–221
Target debris, laser plasma source, 87–88
Thin film polarizer, 230–231
Thin lens
    Gaussian formula, 149
    sign convention, 149
Thin metal film filters, 320–324
Thin organic film filters, 324–326
Third order aberration theory, 361–364
Three mirror circular polarizer, 251
Three mirror polarizer, 239–242
    adjustable, 240–241
Toroidal grating monochromator (TGM), 28–29
    polarization, 135
Toroidal mirror, astigmatism correction, 162–163
Toroidal optics, 173–175
Transition radiation, 93–99
    features, 97–98
    maximum, 96
Transmission coefficient, 265
Transmission grating, 353–355
    efficiency, 373–375
Transverse spatial coherence, 21–24
Tungsten, reflectance spectra, 216–217
Twiss parameters, 23

U

Ultraviolet source comparison with BESSY, 138–143
Undulators, 2, 15–18
    polarization, 20–21

V

Vacuum ultraviolet, 101–115
comparison with BESSY, 138–143
lasers, 101–115
mirror optics, 145
Varied-line-spacing (VLS) grating, 348–350
Vertical angular distribution, calculation, 6
Vignetting, 151

W

Wall stabilized arc, 45–52
    continuum emission, 121–122
Wavefront aberrations, 151–152
Wave propogation in a solid, 260–264
Wave reflection, 264–267
Wave transmission, 264–267
Wigglers, 2, 14
    polarization, 20–21
Windows, cleanliness, 307–308
Wollaston prism polarizer, 229
Wolter optics, 173–175

X

X-ray laser
    applications, 107–109
    efficiency, 105
    vs. high-order harmonic generation, 115
XUV emission from laser plasma
    applications, 84
XUV lasers, 101–109

Y

Yttrium x-ray laser, characteristics, 106

Z

Zinc sulfide, reflectance spectra, 221
Zinc x-ray laser, characteristics, 106
Zone plate
    aberrations, 296–298
    construction, 298–301
    multilayer, 296
Zone plate geometry, 289–292

QC 457 .V32 1998

Vacuum ultraviolet
spectroscopy I

DATE DUE